Evaluation and Utilization of Bioethanol Fuels. I.

This book aims to inform readers about the recent developments in the evaluation and utilization of bioethanol fuels. It covers the evaluation and utilization of bioethanol fuels in general, gasoline fuels, nanotechnology applications in bioethanol fuels, utilization of bioethanol fuels in transport engines, evaluation of bioethanol fuels, utilization of bioethanol fuels in general, and development and utilization of bioethanol fuel sensors.

This book is the fifth volume in the ***Handbook of Bioethanol Fuels*** (Six-Volume Set). It indicates that research on the evaluation and utilization of bioethanol fuels has intensified in recent years to become a major part of bioenergy and biofuels research together primarily with biodiesel, biohydrogen, and biogas research as a sustainable alternative to crude oil-based gasoline and petrodiesel fuels as well as natural gas and syngas.

This book is a valuable resource for stakeholders primarily in the research fields of energy and fuels, chemical engineering, environmental science and engineering, biotechnology, microbiology, chemistry, physics, mechanical engineering, agricultural sciences, food science and engineering, materials science, biochemistry, genetics, and molecular biology, plant sciences, water resources, economics, business and management, transportation science and technology, ecology, public, environmental and occupational health, social sciences, toxicology, multidisciplinary sciences, and humanities, among others.

Evaluation and Utilization of Bioethanol Fuels. I.

Evaluation of Bioethanol Fuels, Transport Engines, and Bioethanol Sensors

Edited by
Ozcan Konur

CRC Press
Taylor & Francis Group
Boca Raton London New York

CRC Press is an imprint of the
Taylor & Francis Group, an **informa** business

Designed cover image: © Shutterstock

First edition published 2024
by CRC Press
2385 NW Executive Center Drive, Suite 320, Boca Raton FL 33431

and by CRC Press
4 Park Square, Milton Park, Abingdon, Oxon, OX14 4RN

CRC Press is an imprint of Taylor & Francis Group, LLC

ISBN: 978-1-032-12756-9 (hbk)
ISBN: 978-1-032-12869-6 (pbk)
ISBN: 978-1-003-22656-7 (ebk)

DOI: 10.1201/9781003226567

Typeset in Times
by codeMantra

Contents

PART 27 Utilization of Bioethanol Fuels

PART 28 Bioethanol Fuel Sensors

Preface

The recent supply shocks caused first by the COVID-19 pandemic and later by the Ukrainian war have shown that biofuels such as bioethanol, biohydrogen, biogas, biosyngas, and biodiesel fuels could play a vital role in maintaining the energy security and indirectly food security at the global scale. These shocks have also resulted in the need for further setup of the incentive structures for the production and consumption of bioethanol fuels in blends with crude oil-based gasoline, petrodiesel, or liquefied natural gas (LNG) in gasoline and diesel engines and their direct utilization in direct ethanol fuel cells (DEFCs) and in the production of biohydrogen fuels for fuel cells and valuable biochemicals from bioethanol fuels.

Thus, it is essential to assess research on the production, evaluation, and utilization of bioethanol fuels from a wide range of biomass including first generation starch and sugar feedstocks, wood, and grass; second generation lignocellulosic biomass including waste biomass and agricultural residues such as starch feedstock residues and sugar feedstock residues; and third generation algal biomass.

Thus, this six-volume *Handbook of Bioethanol Fuels* assesses research on the production, evaluation, and utilization of bioethanol fuels and presents a representative sample of this interdisciplinary research population with a sample of 110 chapters (Table 1.1).

The first two volumes provide an overview of research on the fundamental processes for bioethanol fuel production with a sample of 39 chapters: Pretreatments of the biomass, hydrolysis of the pretreated biomass, microbial fermentation of the hydrolysates with yeasts, and separation and distillation of the bioethanol fuels from the fermentation broth. They also provide an overview of research on bioethanol fuels and production processes for bioethanol fuels (Tables 1.2 and 1.3).

The third and fourth volumes provide an overview of research on the production of bioethanol fuels from the non-waste and waste biomass, respectively, with a sample of 36 chapters. In this context, the third volume covers the production of bioethanol fuels from first generation starch feedstocks and sugar feedstocks, grass biomass, wood biomass, cellulose, biosyngas, and third generation algae (Table 1.4), while the fourth volume covers the production of second generation bioethanol fuels from residual sugar feedstocks, residual starch feedstocks, food waste, industrial waste, urban waste, forestry waste, and lignocellulosic biomass at large (Table 1.5). They also provide an overview of research on feedstock-based bioethanol fuels, non-waste feedstock-based bioethanol fuels, and second generation waste biomass-based bioethanol fuels (Tables 1.4 and 1.5).

Finally, the fifth and sixth volumes provide an overview of research on the evaluation and utilization of bioethanol fuels with a sample of 37 chapters. In this context, the fifth volume covers the evaluation and utilization of bioethanol fuels in general, gasoline fuels, nanotechnology applications in bioethanol fuels, utilization of bioethanol fuels in transport engines, evaluation of bioethanol fuels, utilization of bioethanol fuels, and development and utilization of bioethanol fuel sensors (Table 1.6). Furthermore, the sixth volume of this handbook provides an overview of research on the country-based experience of bioethanol fucls at large, Chinese, the US, and European experience of bioethanol fuels, production of bioethanol fuel-based biohydrogen fuels for fuel cells, bioethanol fuel cells, and bioethanol fuel-based biochemicals with a sample of 19 chapters (Table 1.7).

Thus, the fifth volume of this handbook provides an overview of research on the evaluation and utilization of bioethanol fuels in general, gasoline fuels, nanotechnology applications in bioethanol fuels, utilization of bioethanol fuels in the transport engines, evaluation of bioethanol fuels, utilization of bioethanol fuels, and development and utilization of bioethanol fuel sensors with a sample of 18 chapters (Table 1.6).

Hence, the fifth volume indicates that research on the evaluation and utilization of bioethanol fuels has intensified in recent years to become a major part of bioenergy and biofuels research together primarily with biodiesel, biohydrogen, and biogas research as a sustainable alternative to crude oil-based gasoline and petrodiesel fuels as well as natural gas and syngas.

The fifth volume also indicates that the production of bioethanol fuels and their derivatives and coproducts in a biorefinery context increase the utility of bioethanol fuels and reduce their production cost in relation to the crude oil, natural gas, syngas-, and coal-based fuels as well as other biofuels such as biodiesel fuels.

The fifth volume also indicates that bioethanol fuels are primarily used in transport engines such as gasoline and diesel engines blending gasoline and petrodiesel fuels as well as LNG, directly in DEFCs as an alternative to crude oil- and natural gas-based fuels, the production of biohydrogen fuels for fuel cells as an alternative to their direct use in the DEFCs, and the production of biochemicals as an alternative to crude oil-based chemicals. Furthermore, bioethanol sensors are developed and utilized to sense bioethanol fuels with the increasing utilization of bioethanol fuels.

The fifth volume also indicates that research on the evaluation of bioethanol fuels has progressed in the research fronts of the environmental impact of bioethanol fuels, technoeconomics of bioethanol fuels, and, to a lesser extent, energy assessments of bioethanol fuels, and policy and economic issues in bioethanol fuel production. Furthermore, the secondary research fronts for the first research front have been emissions during the production and utilization of bioethanol fuels and, to a lesser extent, life cycle assessment of bioethanol fuels and other environmental issues.

The fifth volume also indicates that hydrolysis of the biomass, microbial hydrolysate fermentation, and separation and distillation of bioethanol fuels from fermentation broth together with the biomass pretreatments are the fundamental production processes for bioethanol fuel production from the first, second, and third generation feedstocks, making bioethanol fuels more competitive in relation to crude oil- and natural gas-based fuels.

The fifth volume also indicates that a small number of documents, authors, institutions, publication years, source titles, countries, Scopus subject categories, Scopus keywords, and research fronts have shaped the research on the evaluation and utilization of bioethanol fuels.

The fifth volume also indicates that the level of funding for the research on evaluation and utilization of bioethanol fuels has not been sufficient with the resulting loss of momentum in the research output in recent years. Thus, there is a crucial need to improve the incentive structures for the major stakeholders such as researchers and their institutions as well as source titles and academic databases to improve the volume and quality of the research output in these fields. This is a crucial need to maintain the energy security and indirectly food security at a global scale in light of the recent supply shocks caused by the COVID-19 pandemic and the Ukrainian war.

The fifth volume also indicates that the contribution of social sciences and humanities to research in these fields have been minimal in part due to the restrictive editorial policies of the source titles in these fields towards social science- and humanities-based interdisciplinary studies. Thus, there is ample room to improve incentive structures for the inclusion of social sciences and humanities into these fields.

The fifth volume also indicates that China, Europe as a whole, and the USA have been major producers of research in these research fields, and there has been heavy competition among them in terms of both volume and citation impact of the research output. The USA and Europe as a whole have had a higher citation impact in relation to China, benefiting from their first-mover advantage starting their research in these fields in the 1970s. China as a late mover have had more intensive research funding initiatives in relation to the USA and Europe, improving its both research output and citation impact through the provision of the efficient incentive structures for its major stakeholders in the last two decades. In this way, China might also overtake both the USA and Europe in terms of citation impact of the research output in addition to the volume of the research output in future.

This handbook at large and fifth volume are a valuable resource for stakeholders primarily in the research fields of energy and fuels, chemical engineering, environmental science and engineering, biotechnology, microbiology, chemistry, physics, mechanical engineering, agricultural sciences, food science and engineering, materials science, biochemistry, genetics, and molecular biology, plant sciences, water resources, economics, business and management, transportation science and technology, ecology, public, environmental and occupational health, social sciences, toxicology, multidisciplinary sciences, and humanities, among others.

Ozcan Konur

Acknowledgments

This handbook has been a multi-stakeholder project from its conception to its publication. CRC Press and Taylor & Francis Group have been the major stakeholders in financing and executing it. Marc Gutierrez has been the executive director of the project. A large number of teams from the Publisher has contributed immensely to the production of the handbook. Only a limited number of authors have participated in this project due to the low level of incentives, compared to journals. A small number of highly cited scholars has shaped the research on bioethanol fuels. The contribution of all these and other stakeholders to this handbook has been greatly acknowledged.

Editor

Ozcan Konur has interdisciplinary research interests and has published primarily in the areas of bioenergy and biofuels, algal bioenergy and biofuels, nanoenergy and nanofuels, nanobiomedicine, algal biomedicine, disability studies, higher education, biodiesel fuels, algal biomass, lignocellulosic biomass, scientometrics, and bioethanol fuels. He has edited a book titled *Bioenergy and Biofuels* (CRC Press, 2018), a handbook titled *Handbook of Algal Science, Technology, and Medicine* (Elsevier, 2020), and a handbook titled *Handbook of Biodiesel and Petrodiesel Fuels: Science, Technology, Health, and Environment* (CRC Press, 2021) in three volumes.

Contributors

Luís A. B. Cortez
Interdisciplinary Center for Energy Planning (NIPE)
University of Campinas
Campinas, Brazil

José Goldemberg
Institute of Energy and Environment
University of Sao Paulo
Sao Paulo, Brazil

Ozcan Konur
(Formerly) Department of Materials Engineering
Ankara Yildirim Beyazit University
Ankara, Turkey

Jose R. Moreira
Institute of Energy and Environment
University of Sao Paulo
Sao Paulo, Brazil

Part 24

Introduction to the Evaluation and Utilization of Bioethanol Fuels

76 Bioethanol Fuel Evaluation and Utilization

Scientometric Study

Ozcan Konur
(Formerly) Ankara Yildirim Beyazit University

76.1 INTRODUCTION

The crude oil-based gasoline fuels (Ma et al., 2002; Newman and Kenworthy, 1989) have been widely used in the transportation sector since the 1920s. However, there have been great public concerns over the adverse environmental and human impact of these fuels (Hill et al., 2006, 2009). Hence, biomass-based bioethanol fuels (Alvira et al., 2010; Hill et al., 2006; Konur, 2012e, 2015, 2019, 2020a) have increasingly been used in blending gasoline fuels (Hsieh et al., 2002; Najafi et al., 2009) and in the fuel cells (Antolini, 2007, 2009). Additionally, bioethanol fuels have been used to produce valuable biochemicals (Liu and Hensen, 2013; Wang et al., 2004) in a biorefinery (Maity, 2015a,b) context.

The primary focus of the research in this area has been the utilization of bioethanol fuels in fuel cells (Antolini, 2007; Liang et al., 2009), evaluation of bioethanol fuels (Farrell et al., 2006; Hill et al. 2006), utilization of bioethanol fuels in the transport engines (Hansen et al., 2005; Kohse-Hoinghaus et al., 2010), production of biohydrogen fuels from bioethanol fuels (Haryanto et al., 2005; Ni et al., 2007), and to a lesser extent the experiences of countries (Macedo et al., 2008; Sheehan et al., 2003), development and utilization of bioethanol sensors (Liu et al., 2005; Wan et al., 2004), and production of biochemicals from bioethanol fuels (Liu and Hensen, 2013; Wang et al., 2004). Additionally, the research on the gasoline fuels (Khalili et al., 1995; Song, 2003) and applications of nanotechnology (Murdoch et al., 2011; Wan et al., 2004) in this field has also been related to this field.

However, it is essential to develop efficient incentive structures (North, 1991) for the primary stakeholders to enhance the research in this field (Konur, 2000, 2002a–c, 2006a,b, 2007a,b). The scientometric analysis has been used in this context to inform the primary stakeholders about the current state of the research in a selected research field (Garfield, 1955; Konur, 2011, 2012a–i, 2015, 2018a,b, 2019, 2020a).

As there have been no scientometric studies on the bioethanol evaluation and utilization, this book chapter presents a scientometric study of the research in bioethanol evaluation and utilization. It examines the scientometric characteristics of both the sample and population data presenting scientometric characteristics of these both datasets in the order of documents, authors, publication years, institutions, funding bodies, source titles, countries, Scopus subject categories, keywords, and research fronts.

76.2 MATERIALS AND METHODS

The search for this study was carried out using Scopus database (Burnham, 2006) in October 2021.

As a first step for the search of the relevant literature, the keywords were selected using the first most-cited 200 papers for each substituent research front. The selected keyword list was optimized to obtain a representative sample of papers for each research field and they are integrated to form the keyword list for this study. This keyword list was provided in the appendices of Konur (2023a–j) for future replication studies.

DOI: 10.1201/9781003226567-104

As a second step, two sets of data were used for this study. First, a population sample of over 15,000 papers was used to examine the scientometric characteristics of the population data. Secondly, a sample of 100 most-cited papers was used to examine the scientometric characteristics of these citation classics with over 277 citations each.

The scientometric characteristics of these both sample and population datasets were presented in the order of documents, authors, publication years, institutions, funding bodies, source titles, countries, Scopus subject categories, keywords, and research fronts.

Lastly, the key scientometric findings for both datasets were discussed to highlight the research landscape for the bioethanol evaluation and utilization. Additionally, a number of brief conclusions were drawn and a number of relevant recommendations were made to enhance the future research landscape.

76.3 RESULTS

76.3.1 The Most Prolific Documents in the Bioethanol Evaluation and Utilization

The information on the types of documents for both datasets is given in Table 76.1. The articles and review papers dominate both the sample papers, while articles and conference papers dominate population datasets. Further, review papers and conference papers gave a surplus and deficit, respectively.

It is also interesting to note that all of the papers in the sample dataset were published in the journals, while only 96% of the papers were published in the journals for the population dataset. Furthermore, 1.7% and 2.3% of the population papers were published in books and book series, respectively.

76.3.2 The Most Prolific Authors in the Bioethanol Evaluation and Utilization

The information about the most prolific 31 authors with at least two sample papers and four population papers each is given in Table 76.2. Additionally, 11 authors with at least 30 population papers are also listed in this table.

The most prolific authors are Claude Lamy and Christophe Coutanceau with four sample papers each. Xenophon E. Verykios, Changwei Xu, and El Mustapha Belgsir follow these top authors with three sample papers each.

TABLE 76.1
Documents in the Bioethanol Evaluation and Utilization

Documents	Sample Dataset (%)	Population Dataset (%)	Surplus (%)
Article	83	85.6	−2.6
Review	13	2.2	10.8
Conference paper	3	8.6	−5.6
Short Survey	1	0.2	0.8
Book chapter	0	2.0	−2.0
Note	0	0.8	−0.8
Letter	0	0.4	−0.4
Editorial	0	0.1	−0.1
Book	0	0.1	−0.1
Sample size	100	15,008	

TABLE 76.2
Most Prolific Authors in the Bioethanol Evaluation and Utilization

No.	Author Name	Author Code	Sample Papers	Population Papers (%)	Institution	Country	Res. Front
1	Lamy, Claude	7007017658	4	0.1	Univ. Poitiers	France	Fuel cells
2	Coutanceau, Christophe	8714035200	4	0.1	Univ. Poitiers	France	Fuel cells
3	Verykios, Xenophon E.	35551305100	3	0.1	Univ. Patras	Greece	Biohydrogen fuels
4	Xu, Changwei	9248835900	3	0.1	Jinan Univ.	China	Fuel cells
5	Belgsir, El Mustapha	6701740559	3	0.0	Univ. Poitiers	France	Fuel cells
6	Llorca, Jordi	26039349400	2	0.4	Tech. Univ. Catalonia	Spain	Biohydrogen fuels
7	Tsiakaras, Panagiotis	7003948427	2	0.2	Univ. Thessalia	Greece	Fuel cells
8	Zhao, Tianshou	13004121800	2	0.2	Hong Kong Univ. Sci. Technol.	China	Fuel cells
9	Sun, Shi-Gang	7404510197	2	0.1	Xiamen Univ.	China	Fuel cells
10	Zacchi, G	7006727748	2	0.1	Lund Univ.	Sweden	Evaluation
11	Galbe, M	7003788758	2	0.1	Lund Univ.	Sweden	Evaluation
12	Song, Shuqin	7403349881	2	0.1	Chinese Acad. Sci.	China	Fuel cells
13	Aden, Andy	35324090200	2	0.1	NREL	US	Evaluation
14	Rakopoulos, Constantine D.	6506189320	2	0.1	Natl. Tech. Univ. Athens	Greece	Engines
15	Wang, Hong	55963856800	2	0.1	Dongguan Univ. Technol.	China	Fuel cells
16	Zhou, Zhi-You	7406098551	2	0.1	Xiamen Univ.	China	Fuel cells
17	Rakopoulos, Dimitrios C.	6603012578	2	0.1	Natl. Tech. Univ. Athens	Greece	Engines
18	Behm, Rolf J.	36885065400	2	0.1	Univ. Ulm	Germany	Fuel cells
19	Xu, Jianbo	22636447700	2	0.1	Hong Kong Univ. Sci. Technol.	China	Fuel cells
20	Wymann, CE	7004396809	2	0.1	Univ. California Riverside	USA	Evaluation
21	Jusys, Zenonas	7003450828	2	0.1	Univ. Ulm	Germany	Fuel cells
22	Kondarides, Dimitris, I	6603982077	2	0.0	Univ. Patras	Greece	Biohydrogen fuels
23	Wang, Hongsen	24400286300	2	0.0	Univ. Ulm	Germany	Fuel cells
24	Westbrook, Charles K.	7102354977	2	0.0	Lawrence Livermore Natl. Lab.	USA	Engines
25	Kammen, Daniel M.	7004564550	2	0.0	Univ. California Berkeley	USA	Evaluation
26	Liu, Yingliang	57192569158	2	0.0	Jinan Univ.	China	Fuel cells
27	O'Hare, Michael	35778797400	2	0.0	Univ. California Berkeley	USA	Evaluation
28	Plevin, Richard J.	12040501900	2	0.0	Univ. California Berkeley	USA	Evaluation
29	Rousseau, Severine	8714035100	2	0.0	Univ. Poitiers	France	Fuel cells
30	Tian, Na	57203214786	2	0.0	Xiamen Univ.	China	Fuel cells
31	Zhen, Huang	57198607015	2	0.0	Shanghai Jiao Tong Univ.	China	Engines
32	Neto, Almir O.	7102618978	0	0.4	Nuclear Ener. Res. Inst.	Brazil	Fuel cells
33	Spinace, Estevam V.	6602499706	0	0.3	Nuclear Ener. Res. Inst.	Brazil	Fuel cells
34	Noronha, Fabio V.	7004514118	1	0.3	Natl. Technol. Inst.	Brazil	Biohydrogen fuels
35	Mattos, Lisiane V.	7005199691	1	0.2	Natl. Technol. Inst.	Brazil	Biohydrogen fuels

(Continued)

TABLE 76.2 (*Continued*)
Most Prolific Authors in the Bioethanol Evaluation and Utilization

No.	Author Name	Author Code	Sample Papers	Population Papers (%)	Institution	Country	Res. Front
36	Palma, Vincenzo	7006256714	0	0.2	Univ. Salerno	Italy	Biohydrogen fuels
37	Gonzalez, Ernesto R.	57199756260	1	0.2	Univ. Sao Paulos	Brazil	Fuel cells
38	Linardi, Marcelo	6603870993	0	0.2	Nuclear Ener. Res. Inst.	Brazil	Fuel cells
39	Tremiliosi-Filho, Germano	6701503483	0	0.2	Univ. Sao Paulo	Brazil	Fuel cells
40	Du, Yukou	7402894251	0	0.2	Soochow Univ.	China	Fuel cells
41	Rossetti, Ilenia	6701909842	0	0.2	Univ. Milan	Italy	Biohydrogen fuels
42	Sodre, Jose R.	6701797187	0	0.2	Pontificial Cath. Univ. Minas Gerais	Brazil	Engines

Author code: the unique code given by Scopus to the authors. Sample papers: the number of papers authored in the sample dataset. Population papers: the number of papers authored in the population dataset.

The most prolific institution for the sample dataset is the University of Poitiers with four authors. The University of California Berkeley, University of Ulm, and Xiamen University follow this top institution with three authors each. In total, 17 institutions house these prolific authors.

The most prolific countries for the sample dataset are China, the USA, and Greece with ten, six, and five authors, respectively. France, Germany, and Spain follow these top countries with at least two authors each. In total, seven countries house these authors.

The most prolific research front is the fuel cells with 17 authors. The other prolific research fronts are the evaluation of bioethanol fuels, utilization of bioethanol fuels in gasoline and diesel engines, and utilization of bioethanol fuels for biohydrogen fuel production with seven, four, and three papers each.

On the other hand, there is significant gender deficit (Beaudry and Lariviere, 2016; Lariviere et al., 2013; Macaluso et al., 2016) for the sample dataset as surprisingly nearly all of these top researchers are male.

76.3.3 The Most Prolific Research Output by Years in Bioethanol Evaluation and Utilization

Information about papers published between 1970 and 2021 is given in Figure 76.1. This figure clearly shows that the bulk of the research papers in the population dataset were published primarily in the 2010s with 61.1% of the population dataset. The publication rates for the 2020s, 2000s, 1990s, 1980s, 1970s, and pre-1970s were 13.4%, 17.1%, 3.9%, 2.5%, 0.9%, and 0.7%, respectively.

Similarly, the bulk of the research papers in the sample dataset were published in the 2000s and 2010s with 69% and 26% of the sample dataset, respectively. The publication rates for the 1990s, 1980s, and 1970s were 5%, 4%, and 2% of the sample papers, respectively.

The most prolific publication years for the population dataset were after 2012 with at least 6.0% each year of the dataset. Similarly, 68% of the sample papers were published between 2004 and 2010. Further, the number of population papers rose from 2003 to 2013 and thereafter it steadied around 6% of the population papers for each year, losing its momentum.

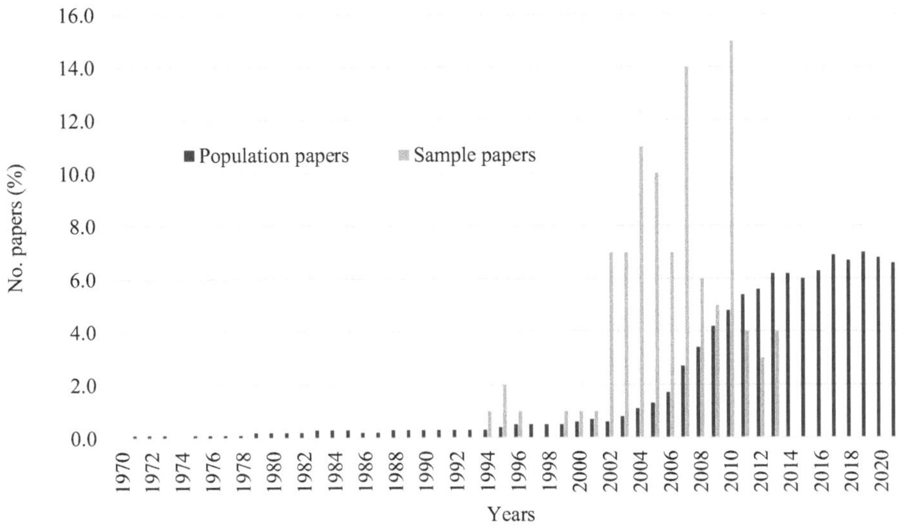

FIGURE 76.1 The research output by years regarding the bioethanol evaluation and utilization.

76.3.4 The Most Prolific Institutions in the Bioethanol Evaluation and Utilization

Information about the most prolific 22 institutions publishing papers on the bioethanol evaluation and utilization with at least two sample papers and 0.2% of the population papers each is given in Table 76.3. Additionally, nine institutions with at least 0.6% of the population dataset are included in this table.

The most prolific institutions are the University of Sao Paulo, Chinese Academy of Sciences, and University of Poitiers with five papers each. Sun Yat-Sen University and Shanghai Jiao Tong University follow these top institutions with four and three sample papers, respectively.

The top country for these most prolific institutions is the USA and China with seven and six institutions, respectively. Next, Brazil follows these countries with three institutions. In total, nine countries house these top institutions.

On the other hand, the institutions with the most citation impact are University of Poitiers and Sun Yat-Sen University with 4.6% and 4.7% surplus, respectively. Similarly, the institutions with the least impact are Jilin University, National Council of Scientific and Technical Research, Nuclear and Energy Research Institute, Paulista State University, and Scientific Research National Center with at least 0.6% deficit each.

76.3.5 The Most Prolific Funding Bodies in the Bioethanol Evaluation and Utilization

Information about the most prolific six funding bodies funding at least two sample papers and 1.5% of the population papers each is given in Table 76.4. Additionally, eight funding bodies with at least 0.9% of the population papers are listed in this table.

The most prolific funding bodies are the National Natural Science Foundation of China and the European Commission with six and five sample papers, respectively. The US Department of Energy and the National Science Foundation follow these top funding bodies with four and three sample papers, respectively. Further, the National Natural Science Foundation of China is the largest funder of the population papers with 11.3% funding rate.

It is notable that 36% and 43.8% of the sample and population papers are funded, respectively.

The most prolific countries for these top funding bodies are Brazil and China with two funding bodies each. In total, three countries and the European Union house these top funding bodies.

TABLE 76.3
The Most Prolific Institutions in Bioethanol Evaluation and Utilization

No.	Institutions	Country	Sample Papers (%)	Population Papers (%)	Surplus (%)
1	Univ. Sao Paulo	Brazil	5	2.8	2.2
2	Chinese Acad. Sci.	China	5	2.5	2.5
3	Univ. Poitiers	France	5	0.4	4.6
4	Sun Yat-Sen Univ.	China	4	0.3	3.7
5	Shanghai Jiao Tong Univ.	China	3	0.6	2.4
6	Tsinghua Univ.	China	2	1.0	1.0
7	Fed. Univ. Rio de Janeiro	Brazil	2	0.9	1.1
8	Univ. Illinois U. C.	USA	2	0.7	1.3
9	Lund Univ.	Sweden	2	0.6	1.4
10	Xiamen Univ.	China	2	0.6	1.4
11	Natl. Technol. Inst.	Brazil	2	0.5	1.5
12	Natl. Cheng Kung Univ.	Taiwan	2	0.4	1.6
13	Purdue Univ.	USA	2	0.4	1.6
14	Natl. Renew. Ener. Lab.	USA	2	0.4	1.6
15	Univ. Minnesota	USA	2	0.4	1.6
16	Univ. California Berkeley	USA	2	0.3	1.7
17	Nanyang Technol. Univ.	Singapore	2	0.3	1.7
18	Cornell Univ.	USA	2	0.3	1.7
19	Hong Kong Univ. Sci. Technol.	China	2	0.3	1.7
20	Univ. Thessaly	Greece	2	0.3	1.7
21	Univ. Utrecht	Netherlands	2	0.2	1.8
22	USDA ARS	USA	2	0.2	1.8
23	Tianjin Univ.	China	1	1.4	−0.4
24	State Univ. Campinas	Brazil	1	1.2	−0.2
25	Natl. Res. Counc.	Italy	1	0.8	0.2
26	Chulalongkorn Univ.	Thailand	1	0.6	0.4
27	Jilin Univ.	China	0	0.9	−0.9
28	Natl. Counc. Sci. Tech. Res.	Argentina	0	0.6	−0.6
29	Paulista State Univ.	Brazil	0	0.6	−0.6
30	Nuclear Ener. Res. Inst.	Brazil	0	0.6	−0.6
31	Sci. Res. Natl. Ctr.	France	0	0.6	−0.6

The funding bodies with the most citation impact are the European Commission and the US Department of Energy with 3.5% and 2.2% surplus, respectively. Similarly, the funding bodies with the least citation impact are the National Natural Science Foundation of China and the National Council for Science and Technology Development with at least 5.3% and 2.4% deficit, respectively.

76.3.6 The Most Prolific Source Titles in the Bioethanol Evaluation and Utilization

Information about the most prolific 16 source titles publishing at least two sample papers and 0.3% of the population papers each in bioethanol evaluation and utilization is given in Table 76.5. Additionally, eight journals with at least 0.7% of the population papers are also listed in this table.

The most prolific source titles are 'Fuel' and 'Bioresource Technology' with seven sample papers each. The 'Journal of Power Sources', 'International Journal of Hydrogen Energy', 'Applied Catalysis B Environmental', and 'Biomass and Bioenergy' follow these top titles with at least five sample papers each.

TABLE 76.4
The Most Prolific Funding Bodies in Bioethanol Evaluation and Utilization

No.	Funding Bodies	Country	Sample Paper No. (%)	Population Paper No. (%)	Surplus (%)
1	National Natural Science Foundation of China	China	6	11.3	–5.3
2	European Commission	Europe	5	1.5	3.5
3	US Dept. Energy	USA	4	1.8	2.2
4	National Science Foundation	USA	3	1.6	1.4
5	Coord. Improv. Higher Educ. Person.	Brazil	2	2.3	–0.3
6	Sao Paulo State Res. Found	Brazil	2	2.3	–0.3
7	Natl. Counc. Sci. Technol. Dev.	Brazil	1	3.4	–2.4
8	Minist. Sci. Technol. Innov.	Brazil	1	2.6	–1.6
9	Ministry Educ.	China	1	2.0	–1.0
10	Fund. Res. Fund. Central Univ.	China	1	1.8	–0.8
11	Minist. Sci. Technol.	China	0	1.5	–1.5
12	Natl. Res. Found. Korea	S. Korea	0	1.0	–1.0
13	Natl. Key Res. Devnt. Pr.	China	0	1.0	–1.0
14	Minist. Educ. Cult. Sci. Technol.	Japan	0	0.9	–0.9

TABLE 76.5
The Most Prolific Source Titles in Bioethanol Evaluation and Utilization

No.	Source Titles	Sample Papers (%)	Population Papers (%)	Surplus (%)
1	Fuel	7	3.0	4.0
2	Bioresource Technology	7	0.9	6.1
3	Journal of Power Sources	6	1.8	4.2
4	International Journal of Hydrogen Energy	5	4.2	0.8
5	Applied Catalysis B Environmental	5	1.2	3.8
6	Biomass and Bioenergy	5	0.8	4.2
7	Journal of the American Chemical Society	4	0.3	3.7
8	Sensors and Actuators B Chemical	3	2.5	0.5
9	Electrochimica Acta	2	1.4	0.6
10	Energy and Fuels	2	1.2	0.8
11	Applied Energy	2	1.0	1.0
12	Journal of Catalysis	2	0.8	1.2
13	Renewable Energy	2	0.7	1.3
14	Journal of Electroanalytical Chemistry	2	0.6	1.4
15	Chemical Engineering Journal	2	0.5	1.5
16	Energy Conversion and Management	2	0.5	1.5
17	Journal of Cleaner Production	1	1.1	–0.1
18	Catalysis Today	1	1.0	0.0
19	RSC Advances	1	0.9	0.1
20	SAE Technical Papers	0	4.7	–4.7
21	Energy	0	1.1	–1.1
22	Applied Catalysis A General	0	1.0	–1.0
23	Industrial and Engineering Chemistry Research	0	0.8	–0.8
24	Applied Surface Science	0	0.7	–0.7

On the other hand, the source title with the most citation impact is the 'Bioresource Technology' with at least 6.1% surplus. The 'Journal of Power Sources', Biomass and Bioenergy', and 'Fuel' follow this top title with at least 4.0% surplus each. Similarly, the source title with the least impact is the 'SAE Technical Papers' with 4.7% deficit. 'Energy', 'Applied Catalysis A General', and 'Industrial and Engineering Chemistry Research' follow this title with at least 1.1% deficit each.

76.3.7 The Most Prolific Countries in the Bioethanol Evaluation and Utilization

Information about the most prolific 19 countries publishing at least two sample papers and 0.6% of the population papers each in bioethanol evaluation and utilization is given in Table 76.6. Additionally, ten countries with at least 1.2% of the population papers are listed in this table.

The most prolific countries are the USA and China with 29 and 24 sample papers, respectively. Brazil, Greece, and France are the other prolific countries with at least seven sample papers each. China is also the largest producer of the population papers with 22.4% publication rate as a single country. Further, 11 European countries listed in Table 76.6 produce 38% and 25% of the sample and population papers, respectively, making them the largest producer of the sample and population papers as a whole.

TABLE 76.6
The Most Prolific Countries in the Bioethanol Evaluation and Utilization

No.	Countries	Sample Papers (%)	Population Papers (%)	Surplus (%)
1	USA	29	17.0	12.0
2	China	24	22.4	1.6
3	Brazil	8	9.7	–1.7
4	Greece	8	1.0	7.0
5	France	7	2.5	4.5
6	Spain	4	3.9	0.1
7	Italy	3	3.7	–0.7
8	Germany	3	2.8	0.2
9	Netherlands	3	1.3	1.7
10	India	2	7.6	–5.6
11	UK	2	3.6	–1.6
12	Taiwan	2	1.7	0.3
13	Sweden	2	1.6	0.4
14	Australia	2	1.5	0.5
15	Turkey	2	1.4	0.6
16	S. Africa	2	0.8	1.2
17	Portugal	2	0.7	1.3
18	Singapore	2	0.7	1.3
19	Switzerland	2	0.6	1.4
20	Japan	1	4.6	–3.6
21	Iran	1	2.5	–1.5
22	Thailand	1	2.0	–1.0
23	Malaysia	1	1.5	–0.5
24	Poland	1	1.3	–0.3
25	Argentina	1	1.2	–0.2
26	Mexico	1	1.2	–0.2
27	Canada	0	3.2	–3.2
28	S. Korea	0	3.1	–3.1
29	Russia	0	1.6	–1.6

TABLE 76.7
The Most Prolific Scopus Subject Categories in the Bioethanol Evaluation and Utilization

No.	Scopus Subject Categories	Sample Papers (%)	Population Papers (%)	Surplus (%)
1	Energy	45	30.4	14.6
2	Chemistry	44	36.6	7.4
3	Chemical Engineering	44	31.9	12.1
4	Environmental Science	29	21.7	7.3
5	Engineering	19	30.5	−11.5
6	Materials Science	15	18.9	−3.9
7	Physics and Astronomy	13	18.9	−5.9
8	Biochemistry, Genetics and Molecular Biology	8	7.3	0.7
9	Agricultural and Biological Sciences	8	6.3	1.7
10	Multidisciplinary	4	0.9	3.1
11	Earth and Planetary Sciences	2	1.6	0.4
12	Social Sciences	1	2.6	−1.6
13	Business, Management and Accounting	1	2.4	−1.4
14	Immunology and Microbiology	1	2.3	−1.3
15	Economics, Econometrics and Finance	0	2.5	−2.5
16	Mathematics	0	1.6	−1.6
17	Computer Science	0	1.5	−1.5

On the other hand, the countries with the most citation impact are the USA and Greece with 12% and 7% surplus, respectively. France and the Netherlands follow these countries with at least 1.7% surplus each. Similarly, the country with the least citation impact is India with 5.6% deficit. Japan, Canada, and South Korea follow India with at least 3.1% deficit.

76.3.8 The Most Prolific Scopus Subject Categories in the Bioethanol Evaluation and Utilization

Information about the most prolific 11 Scopus subject categories indexing at least 5% and 1.3% of the sample and population papers, respectively, is given in Table 76.7. Additionally, six categories with at least 1.5% of the population paper set is listed in this table.

The most prolific Scopus subject categories in the bioethanol evaluation and utilization are 'Energy', 'Chemistry', and 'Chemical Engineering' with 45, 44, and 44 sample papers, respectively. 'Environmental Science', 'Engineering', and 'Materials Science' follow these top categories with at least 15 sample papers each.

On the other hand, the Scopus subject categories with the most citation impact are the 'Energy' and 'Chemical Engineering' with 14.6% and 12.1% surplus, respectively. Similarly, the Scopus subject category with the least citation impact is 'Engineering' with 11.5% deficit. 'Physics and Astronomy' and 'Materials Science' are the other categories with large deficits.

76.3.9 The Most Prolific Keywords in the Bioethanol Evaluation and Utilization

Information about the keywords used in at least 5% and 3.9% of the sample or population papers, respectively, is given in Table 76.8. For this purpose, keywords related to the keyword set, given in the appendices of Konur (2023a–j), are selected from a list of the most prolific keyword set provided by Scopus database.

These keywords are grouped under the nine headings: bioethanol fuels, biomass, evaluation of bioethanol fuels, experience of countries, utilization of bioethanol fuels in fuel cells, transport engines, the production of biohydrogen fuels and biochemical, and the utilization of bioethanol sensors.

TABLE 76.8
The Most Prolific Keywords in Bioethanol Evaluation and Utilization

No.	Keywords	Sample Papers (%)	Population Papers (%)	Surplus (%)
1.	**Bioethanol Fuels**			
	Ethanol	89	77.5	11.5
	Direct ethanol fuel cells	16	4.5	11.5
	Ethanol fuels	12	11.3	0.7
	Ethanol oxidation	10	5.9	4.1
	Bioethanol	8	10.0	−2.0
	Ethanol production	6	3.0	3.0
	Ethanol electro-oxidation	5	4.5	0.5
	Ethanol steam reforming		4.1	−4.1
	Ethanol oxidation reaction		3.7	−3.7
2.	**Biomass**			
	Biomass	17	3.8	13.2
	Cellulose	11	2.2	8.8
	Zea	10	2.3	7.7
	Lignin	8		8.0
	Corn	6		6.0
	Sugarcane	5	1.7	3.3
	Lignocellulose	5		5.0
3.	**Evaluation**			
	Greenhouse gases	10	4.4	5.6
	Gas emissions	10	2.8	7.2
	Economics	9	1.6	7.4
	Costs	8	2.5	5.5
	Carbon	7	4.6	2.4
	Environmental impact	7	2.0	5.0
	Emissions	6	1.3	4.7
	Environment	6		6.0
	Hydrolysis	5	1.9	3.1
	Energy efficiency	5	1.7	3.3
	Air pollution	5		5.0
4.	**Experience**			
	Brazil	5	1.8	1.2
5.	**Fuel Cells**			
	Fuel cells	25	7.0	18.0
	Catalysts	22	10.8	11.2
	Oxidation	22	9.9	12.1
	Catalysis	20	4.3	15.7
	Platinum	15	6.2	8.8
	Electrocatalysts	12	5.6	6.4
	Electrooxidation	11	5.7	5.3
	Catalyst activity	10	8.2	1.8

(Continued)

TABLE 76.8 (*Continued*)
The Most Prolific Keywords in Bioethanol Evaluation and Utilization

No.	Keywords	Sample Papers (%)	Population Papers (%)	Surplus (%)
	Palladium	9	3.5	5.5
	Electrochemistry	9	1.5	7.5
	Nickel	7	3.9	3.1
	Zinc oxide	6	3.1	2.9
	Tin	5	1.8	3.2
	Nanoparticles		5.1	−5.1
6.	**Engines**			
	Diesel engines	17	7.7	9.3
	Gasoline	15	9.7	5.3
	Carbon dioxide	14	6.1	7.9
	Diesel fuels	14	2.3	11.7
	Carbon monoxide	12	3.6	8.4
	Exhaust emissions	11	1.4	9.6
	Combustion	9	6.6	2.4
	Fuel consumption	9		9.0
	Biodiesel	6	2.4	3.6
	Engine performance	6	1.5	4.5
	Engine cylinders	5	3.1	1.9
	Fuel additives	5		5.0
7.	**Biochemicals**			0.0
	Methanol	9	3.7	5.3
	Acetaldehyde	5	3.4	1.6
8.	**Biohydrogen Fuels**			0.0
	Hydrogen	13	5.1	7.9
	Steam reforming	12	6.3	5.7
	Hydrogen production	11	6.4	4.6
	Reforming reactions	9	1.4	7.6
9.	**Sensors**			0.0
	Chemical sensors	5	3.9	1.1
	Gas detectors		3.9	−3.9

There are eight prolific keywords used related to the bioethanol fuels: ethanol with 89 HCPs and direct ethanol fuel cells, ethanol fuels, and ethanol oxidation with at least 16, 12, and 10 HCPs each. It is notable that bioethanol keyword appears in the sample paper keyword list with around 8% and 10% of the sample papers for the keywords of bioethanol and bio-ethanol, respectively.

The prolific keywords related to the biomass are biomass, cellulose, and zea with 17, 11, and 10 HCPs, respectively, while the prolific keywords related to the evaluation of bioethanol fuels are greenhouse gases, gas emissions, and economics with at least ten HCPs each.

The prolific keywords related to the countries and continents is Brazil with four HCPs while the prolific keywords related to fuel cells are fuel cells, catalysts, oxidation, and catalysis with 20–25 HCPs each.

TABLE 76.9
The Most Prolific Research Fronts in Bioethanol Evaluation and Utilization

No.	Research Fronts	Reviewed Papers (%)
1	Utilization of bioethanol fuels in fuel cells	29
2	Evaluation of bioethanol fuels	26
3	Utilization of bioethanol fuels in transport engines	19
4	Production of biohydrogen fuels from bioethanol fuels	19
5	Experience of countries	10
6	Bioethanol sensors	8
7	Production of biochemical from bioethanol fuels	4

Sample papers: the sample of the most-cited 100 papers.

The prolific keywords related to the transport engines are diesel engines, gasoline, carbon dioxide, diesel fuels, carbon monoxide, and exhaust emissions with 11–17 HCPs each while the prolific keywords related to the production of biochemical from bioethanol fuels are methanol and acetaldehyde with 5–9 HCPs each.

The prolific keywords related to the production of biohydrogen fuels from bioethanol fuels are hydrogen, steam reforming, and hydrogen production with 11–13 HCPs each. Further, the prolific keyword related to bioethanol sensors is chemical sensors with five HCPs.

76.3.10 The Most Prolific Research Fronts in Bioethanol Evaluation and Utilization

Information about the most prolific research fronts for the sample papers in bioethanol fuel evaluation and utilization is given in Table 76.9.

As Table 76.9 shows, there are four primary research fronts for these HCPs: utilization of bioethanol fuels in fuel cells, evaluation of bioethanol fuels, utilization of bioethanol fuels transport engines, and production of biohydrogen fuels from bioethanol fuels with 29, 26, 19, and 19 HCPs, respectively. The other research fronts are experience of countries, bioethanol sensors, and production of biochemicals from bioethanol fuels with ten, eight, and four HCPs, respectively.

76.4 DISCUSSION

76.4.1 Introduction

The crude oil-based gasoline fuels have been widely used in the transportation sector since the 1920s. However, there have been great public concerns over the adverse environmental and human impact of these fuels. Hence, biomass-based bioethanol fuels have increasingly been used in blending gasoline fuels. Additionally, bioethanol fuels have been used to produce valuable biochemicals in a biorefinery context.

The primary focus of the research in this area has been the utilization of bioethanol fuels in fuel cells, evaluation of bioethanol fuels, utilization of bioethanol fuels transport engines, production of biohydrogen fuels from bioethanol fuels, and to a lesser extent the experiences of countries, development and utilization of bioethanol sensors, and production of biochemical from bioethanol fuels. Additionally, the research on the gasoline fuels and applications of nanotechnology in this field has also been related to this field.

However, it is essential to develop efficient incentive structures for the primary stakeholders to enhance the research in this field. The scientometric analysis has been used in this context to inform the primary stakeholders about the current state of the research in a selected research field.

As a first step for the search of the relevant literature, the keywords were selected using the first most-cited 200 papers for each substituent research front. The selected keyword list was optimized to obtain a representative sample of papers for each research field and they are integrated to form the keyword list for this study. This keyword list was provided in the appendices of Konur (2023a–j) for future replication studies.

As a second step, two sets of data were used for this study. First, a population sample of over 15,000 papers was used to examine the scientometric characteristics of the population data. Secondly, a sample of 100 most-cited papers was used to examine the scientometric characteristics of these citation classics with over 277 citations each.

The scientometric characteristics of these both sample and population datasets were presented in the order of documents, authors, publication years, institutions, funding bodies, source titles, countries, Scopus subject categories, keywords, and research fronts.

Lastly, the key scientometric findings for both datasets were discussed to highlight the research landscape for the bioethanol evaluation and utilization. Additionally, a number of brief conclusions were drawn and a number of relevant recommendations were made to enhance the future research landscape.

76.4.2 The Most Prolific Documents in the Bioethanol Evaluation and Utilization

The articles and review papers dominate both the sample and population datasets, while the review papers and book chapters have a surplus and deficit, respectively.

Scopus differs from the Web of Science database in differentiating and showing articles and conference papers published in the journals separately. Similarly, Scopus differs from Web of science database in introducing 'short surveys'. Hence, the total number of articles and review papers in the sample dataset are 86% and 14%, respectively.

It is observed during the search process that there has been inconsistency in the classification of the documents in Scopus as well as in other databases such as Web of Science. This is especially relevant for the classification of papers as reviews or articles as the papers not involving a literature review may be erroneously classified as a review paper. There is also a case of review papers being classified as articles. For example, although there are 14 review papers and short surveys as classified by the Scopus database, 18 of the sample papers are review papers based on the literature reviews.

In this context, it would be helpful to provide a classification note for the published papers in the books and journals at the first instance. It would also be helpful to use the document types listed in Table 76.1 for this purpose. Book chapters may also be classified as articles or reviews as an additional classification to differentiate review chapters from the experimental chapters as it is done by the Web of Science. It would be further helpful to additionally classify the conference papers as articles or review papers as well as it is done in the Web of Science database.

76.4.3 The Most Prolific Authors in the Bioethanol Evaluation and Utilization

There have been most prolific 31 authors with at least two sample papers each as given in Table 76.2. These authors have shaped the development of the research in this field.

The most prolific authors are Claude Lamy, Christophe Coutanceau, and to a lesser extent Xenophon E. Verykios, Changwei Xu, and El Mustapha Belgsir.

It is important to note the inconsistencies in indexing of the author names in Scopus and other databases. It is especially an issue for the names with more than two components such as 'Judge Alex de Camp Llorca'. The probable outcomes are 'Llorca, J.A.D.C.', 'de Camp Llorca, J.A.', or 'Camp Llorca, J.A.D.'. The first choice is the gold standard of the publishing sector as the last word in the name is taken as the last name. In most of the academic databases such as PUBMED, EBSCO databases, this version is used predominantly. The second choice is a strong alternative while the last choice is an undesired outcome as two last words are taken as the last name. It is good practice to combine the words of the last name by a hyphen: 'Camp-Lorca, J.A.D.'. It is notable that inconsistent indexing of the author names may cause substantial inefficiencies in the search process for the papers as well as allocating credit to the authors as there are different author entries for each outcome in the databases.

There is also a case of shortening Chinese names. For example, 'Yuoyang Yang is often shortened as 'Yang, Y.', 'Yang, Y.-Y.', and 'Yang Y.Y.' as it is done in the Web of Science database as well. However, the gold stand in this case is 'Yang Y' where the last word is taken as the last name and the first word is taken as a single forename. In most of the academic databases such as Pubmed and EBSCO, this first version is used predominantly. However, it makes sense to use the third option to differentiate Chinese names efficiently. 'Yang Y.Y.'. Therefore, there have been difficulties to locate papers for the Chinese authors. In such cases, the use of the unique author codes provided for each author by the Scopus database has been helpful.

There is also a difficulty in allowing credit for the authors especially for the authors with common names such as 'Wang, Y.', or 'Huang, Y.' or 'Zhu, Y.' in conducting scientometric studies. These difficulties strongly influence the efficiency of the scientometric studies as well as allocating credit to the authors as there are the same author entries for different authors with the same name, for example, 'Wang Y.' in the databases.

In this context, the coding of authors in Scopus database is a welcome innovation compared to the other databases such as Web of Science. In this process, Scopus allocates a unique number to each author in the database (Aman, 2018). However, there might still be substantial inefficiencies in this coding system especially for common names. For example, some of the papers for a certain author maybe allocated to another researcher with a different author code. It is possible that Scopus uses a number of software programs to differentiate the author names and the program may not be false-proof (Kim, 2018).

In this context, it does not help that author names are not given in full in some journals and books. This makes difficult to differentiate authors with common names and makes the scientometric studies further difficult in the author domain. Therefore, the author names should be given in all books and journals at the first instance. There is also a cultural issue where some authors do not use their full names in their papers. Instead, they use initials for their forenames: 'Aden, A.P.' or just 'Aden' instead of 'Aden, Alas Padras'.

There are also inconsistencies in naming of the authors with more than two components by the authors themselves in journal papers and book chapters. For example, 'Aden, A.P.C.' might be given as 'Aden, A', 'Aden, P', 'Aden, P.C.', or 'Aden C.' in the journals and books. This also makes the scientometric studies difficult in the author domain. Hence, contributing authors should use their name consistently in their publications.

The other critical issue regarding the author names is the spelling of the author names in the national spellings (e.g., Üzümcü, Çağla) rather than in the English spellings (e.g., Uzumcu, Cagla) in Scopus database. Scopus differs from the Web of science database and many other databases in this respect where the author names are given only in the English spellings. It is observed that national spellings of the author names do not help in conducting scientometric studies as well in allocating credits to the authors as sometimes there are the different author entries for the English and National spellings in the Scopus database.

The most prolific institutions for the sample dataset are the University of Poitiers, and to a lesser extent University of California Berkeley, University of Ulm, and Xiamen University. The most prolific countries for the sample dataset are China, the USA, Greece, and to a lesser extent France, Germany, and Spain.

It is also notable that there is significant gender deficit for the sample dataset as surprisingly nearly all of these top researchers are male. This finding is the most thought-provoking with strong public policy implications. Hence, institutions, funding bodies, and policy makers should take efficient measures to reduce the gender deficit in this field as well as other scientific fields with strong gender deficit. In this context, it is worth to note the level of representation of the researchers from the minority groups in science on the basis of race, sexuality, age, and disability, besides the gender (Blankenship, 1993; Dirth and Branscombe, 2017; Konur, 2000, 2002a–c, 2016a,b, 2017a,b).

76.4.4 The Most Prolific Research Output by Years in the Bioethanol Evaluation and Utilization

The research output observed between 1970 and 2021 is illustrated in Figure 76.1. This figure clearly shows that the bulk of the research papers in the population dataset were published primarily in the 2010s and to a lesser extent 2000s and the early 2020s. Similarly, the bulk of the research papers in the sample dataset were published in the 2000s and to a lesser extent 2010s.

These data suggest that the most-cited sample and population papers were primarily published in the 2000s, 2010s, and the early 2020s. These are the thought-provoking findings as there has been no significant research in this field prior to 2005, but there has been significant research boom after that. In this context, the increasing public concerns about climate change (Change, 2007), greenhouse gas emissions (Carlson et al., 2017), and global warming (Kerr, 2007) have been certainly behind the boom in the research in this field in the last two decades.

Based on these findings, the size of the population papers likely to more than double in the current decade, provided that the public concerns about climate change, greenhouse gas emissions, and global warming are translated efficiently to the research funding in this field. However, it should be noted that the research output lost for the population papers stagnated after 2013.

76.4.5 The Most Prolific Institutions in the Bioethanol Evaluation and Utilization

The most prolific 22 institutions publishing papers on the bioethanol evaluation and utilization with at least two sample papers and 0.2% of the population papers each given in Table 76.3 have shaped the development of the research in this field.

The most prolific institutions are the University of Sao Paulo, Chinese Academy of Sciences, University of Poitiers, and to a lesser extent Sun Yat-Sen University and Shanghai Jiao Tong University. Further, the top countries for these most prolific institutions are the USA and China, and to a lesser extent Brazil. In total, nine countries house these top institutions.

On the other hand, the institutions with the most citation impact are University of Poitiers and Sun Yat-Sen University, while the institutions with the least impact are Jilin University, National Council of Scientific and Technical Research, Nuclear and Energy Research Institute, Paulista State University, and Scientific Research National Center.

76.4.6 The Most Prolific Funding Bodies in the Bioethanol Evaluation and Utilization

The most prolific 12 funding bodies funding at least two sample papers and 1.5% of the population papers each are given in Table 76.4. It is notable that 36% and 44% of the sample and population papers are funded, respectively.

The most prolific funding bodies are the National Natural Science Foundation of China, the European Commission, and to a lesser extent the US Department of Energy and the National Science Foundation. The most prolific countries for these top funding bodies are Brazil and China. In total, three countries and the European Union house these top funding bodies. The heavy funding by the Brazilian and Chinese funding bodies are notable as these countries are major producers of the research in this field.

These findings on the funding of the research in this field suggest that the level of the funding mostly in the last two decades has been largely instrumental in enhancing the research in this field (Ebadi and Schiffauerova, 2016) in light of North's institutional framework (North, 1991). However, there is ample room to increase this funding rate.

76.4.7 The Most Prolific Source Titles in Bioethanol Evaluation and Utilization

The most prolific 16 source titles publishing at least two sample papers and 0.3% of the population papers each in bioethanol evaluation and utilization have shaped the development of the research in this field (Table 76.5).

The most prolific source titles are 'Fuel', 'Bioresource Technology', and to a lesser extent 'Journal of Power Sources', 'International Journal of Hydrogen Energy', 'Applied Catalysis B Environmental', and 'Biomass and Bioenergy'.

On the other hand, the source titles with the most citation impact are the 'Bioresource Technology', and to a lesser extent 'Journal of Power Sources', Biomass and Bioenergy', and 'Fuel'. Similarly, the source title with the least impact is the 'SAE Technical Papers', and to a lesser extent 'Energy', 'Applied Catalysis A General', and 'Industrial and Engineering Chemistry Research'.

It is notable that these top source titles are primarily related to the energy. This finding suggests that the journals in these fields have significantly shaped the development of the research in this field as they focus on the evaluative studies for the country-based experiences of bioethanol fuels.

76.4.8 The Most Prolific Countries in Bioethanol Evaluation and Utilization

The most prolific 19 countries publishing at least two papers and 0.9% of the population papers each have significantly shaped the development of the research in this field (Table 76.6).

The most prolific countries are the USA, China, and to a lesser extent Brazil, Greece, and France. On the other hand, the countries with the most citation impact are the USA, Greece, and to a lesser extent France and the Netherlands. Similarly, the countries with the least citation impact are India, and to a lesser extent Japan, Canada, and South Korea.

The close examination of these findings suggests that the Europe, USA, China, and to a lesser extent Brazil are the major producers of the research in this field. It is a fact that the USA has been a major player in science (Leydesdorff and Wagner, 2009; Leydesdorff et al., 2014). The USA has further developed a strong research infrastructure to support its corn- and grass-based bioethanol industry (Vadas et al., 2008).

However, China has been a rising star in scientific research in competition with the USA and Europe (Leydesdorff and Zhou, 2005). China is also a major player in this field as a major producer of bioethanol (Li and Chan-Halbrendt, 2009).

Next, Europe has been a persistent player in the scientific research in competition with both the USA and China (Leydesdorff, 2000). Europe has also been a persistent producer of bioethanol along with the USA and Brazil (Gnansounou, 2010).

Additionally, Brazil has also been a persistent player in scientific research at a moderate level (Glanzel et al., 2006). Brazil has also developed a strong research infrastructure to support its biomass-based bioethanol industry (Macedo et al., 2008).

76.4.9 The Most Prolific Scopus Subject Categories in Bioethanol Evaluation and Utilization

The most prolific 11 Scopus subject categories indexing at least 6% and 0.9% of the sample and population papers, respectively, given in Table 76.7 have shaped the development of the research in this field.

The most prolific Scopus subject categories in the bioethanol evaluation and utilization are 'Energy', 'Chemistry', 'Chemical Engineering', and to a lesser extent 'Environmental Science', 'Engineering', and 'Materials Science'.

On the other hand, the Scopus subject categories with the most citation impact are the 'Energy' and 'Chemical Engineering', while the Scopus subject categories with the least citation impact are 'Engineering' and to a lesser extent 'Physics and Astronomy' and 'Materials Science'.

These findings are thought provoking suggesting that the primary subject categories are 'Environmental Science', 'Chemistry', and 'Energy'. The other key finding is that social sciences are relatively well represented in both the sample and population papers, unlike most fields in bioethanol fuels. These findings are not surprising as the key research fronts in this field are the evaluative studies on the bioethanol fuels.

76.4.10 The Most Prolific Keywords in Bioethanol Evaluation and Utilization

A limited number of keywords have shaped the development of the research in this field as shown in Table 76.8 and the appendices of Konur (2023a–j).

These keywords are grouped under the nine headings: bioethanol fuels, biomass, evaluation of bioethanol fuels, experience of countries, utilization of bioethanol fuels in fuel cells, transport engines, the production of biohydrogen fuels and biochemical, and the utilization of bioethanol sensors. These prolific keywords highlight the key research fronts in this field and reflect well the keywords used in the sample and population papers.

76.4.11 The Most Prolific Research Fronts in Bioethanol Evaluation and Utilization

As Table 76.9 shows, there are four primary research fronts for these HCPs: utilization of bioethanol fuels in fuel cells, evaluation of bioethanol fuels, utilization of bioethanol fuels transport engines, and production of biohydrogen fuels from bioethanol fuels. The other research fronts are experience of countries, bioethanol sensors, and production of biochemicals from bioethanol fuels.

As the bioethanol fuels are one of the most-used biofuels in the gasoline and diesel engines, it is not surprising that the research in this field is substantial. However, it is further interesting to note the production of biohydrogen fuels for the fuel cells and the utilization of bioethanol fuels directly in the fuel cells. With the recent societal focus on the utilization of biofuels for the transportation, it is likely that the research in these fields would continue to be substantial in the future.

There is an increasing trend for the evaluation of bioethanol fuels in relation to the crude oil-based diesel and gasoline fuels using a number of social science-based analytical tools such as technoeconomic and life cycle studies of the bioethanol fuels. With the recent societal focus on the utilization of biofuels for the transportation, it is likely that the research in this field would continue to be substantial in the future. In this content, there is also an increasing trend to narrate the experiences of the countries relating to both production and utilization of bioethanol fuels. The experiences of the major producers and consumers of bioethanol fuels such as the USA, China, Brazil, Canada, and the European countries are often narrated with a focus on the development of national policies to provide an effective set of incentive structures for the major stakeholders.

The production of biochemicals from the bioethanol fuels in a biorefinery context is a useful strategy to reduce the production cost of the bioethanol fuels. The other research front is related to bioethanol sensors with a focus on the development of effective materials for high bioethanol selectivity.

In the end, these most-cited papers in this field hint that the efficiency of bioethanol fuels could be optimized using the structure, processing, and property relationships of the bioethanol fuels and their derivatives (Formela et al., 2016; Konur, 2018a, 2020, 2021a–d; Konur and Matthews, 1989).

76.5 CONCLUSION AND FUTURE RESEARCH

The research on the bioethanol evaluation and utilization has been mapped through a scientometric study of both sample and population datasets.

The critical issue in this study has been to obtain a representative sample of the research as in any other scientometric study. Therefore, the keyword set has been carefully devised and optimized after a number of runs in the Scopus database. It is a representative sample of the wider population studies which include large numbers of countries in this field.

The other issue has been the selection of a multidisciplinary database to carry out the scientometric study of the research in this field. For this purpose, Scopus database has been selected. The journal coverage of this database has been wider than that of Web of Science.

The key scientometric properties of the research in this field have been determined and discussed in this book chapter. It is evident that a limited number of documents, authors, institutions, publication periods, institutions, funding bodies, source titles, countries, Scopus subject categories, keywords, and research fronts have shaped the development of the research in this field.

There is ample scope to increase the efficiency of the scientometric studies in this field in the author and document domains by developing consistent policies and practices in both domains across all the academic databases. In this respect, authors, journals, and academic databases have a lot to do. Furthermore, the significant gender deficit as in most scientific fields emerges as a public policy issue. The potential deficits on the basis of age, race, disability, and sexuality need also to be explored in this field as in other scientific fields.

The research in this field has boomed in the 2000s, 2010s, and the early 2020s possibly promoted by the public concerns on global warming, greenhouse gas emissions, and climate change. The institutions from Brazil, China, and the USA have mostly shaped the research in this field.

The relatively low funding rate of 36% and 44% for sample and population papers, respectively, suggests that funding in this field significantly enhanced the research in this field primarily in the 2010s, possibly more than doubling in the current decade. However, there is ample room for more funding.

The most prolific journals have been primarily related to the energy. This finding suggests that the journals in this have significantly shaped the development of the research in this field as they focus on the evaluative studies for the utilization of bioethanol fuels.

The USA, China, Brazil, and European countries have been the major producers of the research in this field as the major producers and users of bioethanol fuels from different types of biomasses such as corn, sugarcane, and grass as well as other types of biomasses. These countries have well-developed research infrastructure in bioethanol fuels.

The primary subject categories have been Energy, Chemistry, Chemical Engineering, and to a lesser extent Environmental Science, Engineering, and Materials Science as the focus of the sample papers has been on the utilization of bioethanol fuels. Although a number of social science-based analytical tools are used in the evaluation of bioethanol fuels, it is notable that social sciences and humanities are not well represented as it would be expected.

Ethanol is more popular than bioethanol as a keyword with strong implications for the search strategy. In other words, the search strategy using only bioethanol keyword would not be much helpful. It is also recommended that following the increasing trend in this field, the term 'bioethanol fuels' is used rather than 'ethanol' or 'bio-ethanol', or 'bioethanol', or 'fuel ethanol' in the titles of the papers by their authors.

These keywords are grouped under the nine headings: bioethanol fuels, biomass, evaluation of bioethanol fuels, experience of countries, utilization of bioethanol fuels in fuel cells, transport engines, the production of biohydrogen fuels and biochemical, and the utilization of bioethanol sensors. It is important to note that these groups of keywords highlight the potential primary research fronts for these fields for both sample and population papers.

These findings are thought-provoking. The focus of these most-cited 100 papers is on the evaluation and utilization of bioethanol fuels highlighting strong structure–processing–property relationships for bioethanol fuels and their derivatives.

Thus, the scientometric analysis has a great potential to gain valuable insights into the evolution of the research in this field as in other scientific fields.

It is recommended that further scientometric studies are carried out about the other aspects of both evaluation and utilization of bioethanol fuels. It is further recommended that reviews of the most-cited papers are carried out for each research front to complement these scientometric studies. Next, the scientometric studies of the hot papers in these primary fields are carried out.

ACKNOWLEDGMENTS

The contribution of the highly cited researchers in the field of the bioethanol evaluation and utilization has been gratefully acknowledged.

REFERENCES

Alvira, P., E. Tomas-Pejo, M. Ballesteros and M. J. Negro. 2010. Pretreatment technologies for an efficient bioethanol production process based on enzymatic hydrolysis: A review. *Bioresource Technology* 101:4851–4861.

Aman, V. 2018. Does the Scopus author ID suffice to track scientific international mobility? A case study based on Leibniz laureates. *Scientometrics* 117:705–720.

Antolini, E. 2007. Catalysts for direct ethanol fuel cells. *Journal of Power Sources* 170:1–12.

Antolini, E. 2009. Palladium in fuel cell catalysis. *Energy and Environmental Science* 2:915–931.

Beaudry, C. and V. Lariviere. 2016. Which gender gap? Factors affecting researchers' scientific impact in science and medicine. *Research Policy* 45:1790–1817.

Blankenship, K. M. 1993. Bringing gender and race in: US employment discrimination policy. *Gender & Society* 7:204–226.

Burnham, J. F. 2006. Scopus database: A review. *Biomedical Digital Libraries* 3:1–8.

Carlson, K. M., J. S. Gerber and D. Mueller, et al. 2017. Greenhouse gas emissions intensity of global croplands. *Nature Climate Change* 7:63–68.

Change, C. 2007. Climate change impacts, adaptation and vulnerability. *Science of the Total Environment* 326:95–112.

Dirth, T. P. and N. R. Branscombe. 2017. Disability models affect disability policy support through awareness of structural discrimination. *Journal of Social Issues* 73:413–442.

Ebadi, A. and A. Schiffauerova. 2016. How to boost scientific production? A statistical analysis of research funding and other influencing factors. *Scientometrics* 106:1093–1116.

Farrell, A. E., R. J. Plevin and B. T. Turner, et al. 2006. Ethanol can contribute to energy and environmental goals. *Science* 311:506–508.

Formela, K., A. Hejna, L. Piszczyk, M. R. Saeb and X. Colom. 2016. Processing and structure-property relationships of natural rubber/wheat bran biocomposites. *Cellulose* 23:3157–3175.

Garfield, E. 1955. Citation indexes for science. *Science* 122:108–111.

Glanzel, W., J. Leta and B. Thijs. 2006. Science in Brazil. Part 1: A macro-level comparative study. *Scientometrics* 67:67–86.

Gnansounou, E. 2010. Production and use of lignocellulosic bioethanol in Europe: Current situation and perspectives. *Bioresource Technology* 101:4842–4850.

Hansen, A. C., Q. Zhang and P. W. L. Lyne. 2005. Ethanol-diesel fuel blends: A review. *Bioresource Technology* 96:277–285.

Haryanto, A., S. Fernando, N. Murali and S. Adhikari. 2005. Current status of hydrogen production techniques by steam reforming of ethanol: A review. *Energy and Fuels* 19:2098–2106.

Hill, J., E. Nelson, D. Tilman, S. Polasky and D. Tiffany. 2006. Environmental, economic, and energetic costs and benefits of biodiesel and ethanol biofuels. *Proceedings of the National Academy of Sciences of the United States of America* 103:11206–11210.

Hill, J., S. Polasky and E. Nelson, et al. 2009. Climate change and health costs of air emissions from biofuels and gasoline. *Proceedings of the National Academy of Sciences of the United States of America* 106:2077–2082.

Hsieh, W. D., R. H. Chen, T. L. Wu and T. H. Lin. 2002. Engine performance and pollutant emission of an SI engine using ethanol-gasoline blended fuels. *Atmospheric Environment* 36:403–410.

Kerr, R. A. 2007. Global warming is changing the world. *Science* 316:188–190.

Khalili, N. R., P. A. Scheff and T. M. Holsen. 1995. PAH source fingerprints for coke ovens, diesel and, gasoline engines, highway tunnels, and wood combustion emissions. *Atmospheric Environment* 29:533–542.
Kim, J. 2018. Evaluating author name disambiguation for digital libraries: A case of DBLP. *Scientometrics* 116:1867–1886.
Kohse-Hoinghaus, K., P. Osswald and T. A. Cool, et al. 2010. Biofuel combustion chemistry: From ethanol to biodiesel. *Angewandte Chemie: International Edition* 49:3572–3597.
Konur, O. 2000. Creating enforceable civil rights for disabled students in higher education: An institutional theory perspective. *Disability & Society* 15:1041–1063.
Konur, O. 2002a. Access to nursing education by disabled students: Rights and duties of nursing programs. *Nurse Education Today* 22:364–374.
Konur, O. 2002b. Assessment of disabled students in higher education: Current public policy issues. *Assessment and Evaluation in Higher Education* 27:131–152.
Konur, O. 2002c. Access to employment by disabled people in the UK: Is the Disability Discrimination Act working? *International Journal of Discrimination and the Law* 5:247–279.
Konur, O. 2006a. Participation of children with dyslexia in compulsory education: Current public policy issues. *Dyslexia* 12:51–67.
Konur, O. 2006b. Teaching disabled students in higher education. *Teaching in Higher Education* 11:351–363.
Konur, O. 2007a. A judicial outcome analysis of the Disability Discrimination Act: A windfall for the employers? *Disability & Society* 22:187–204.
Konur, O. 2007b. Computer-assisted teaching and assessment of disabled students in higher education: The interface between academic standards and disability rights. *Journal of Computer Assisted Learning* 23:207–219.
Konur, O. 2011. The scientometric evaluation of the research on the algae and bio-energy. *Applied Energy* 88:3532–3540.
Konur, O. 2012a. Prof. Dr. Ayhan Demirbas' scientometric biography. *Energy Education Science and Technology Part A: Energy Science and Research* 28:727–738.
Konur, O. 2012b. The evaluation of the biogas research: A scientometric approach. *Energy Education Science and Technology Part A: Energy Science and Research* 29:1277–1292.
Konur, O. 2012c. The evaluation of the global energy and fuels research: A scientometric approach. *Energy Education Science and Technology Part A: Energy Science and Research* 30:613–628.
Konur, O. 2012d. The evaluation of the research on the biodiesel: A scientometric approach. *Energy Education Science and Technology Part A: Energy Science and Research* 28:1003–1014.
Konur, O. 2012e. The evaluation of the research on the bioethanol: A scientometric approach. *Energy Education Science and Technology Part A: Energy Science and Research* 28:1051–1064.
Konur, O. 2012f. The evaluation of the research on the biofuels: A scientometric approach. *Energy Education Science and Technology Part A: Energy Science and Research* 28:903–916.
Konur, O. 2012g. The evaluation of the research on the biohydrogen: A scientometric approach. *Energy Education Science and Technology Part A: Energy Science and Research* 29:323–338.
Konur, O. 2012h. The evaluation of the research on the microbial fuel cells: A scientometric approach. *Energy Education Science and Technology Part A: Energy Science and Research* 29:309–322.
Konur, O. 2012i. The scientometric evaluation of the research on the production of bioenergy from biomass. *Biomass and Bioenergy* 47:504–515.
Konur, O. 2015. Current state of research on algal bioethanol. In *Marine Bioenergy: Trends and Developments*, Eds. S. K. Kim and C. G. Lee, pp. 217–244. Boca Raton, FL: CRC Press.
Konur, O., Ed. 2018a. *Bioenergy and Biofuels*. Boca Raton, FL: CRC Press.
Konur, O. 2018b. Bioenergy and biofuels science and technology: Scientometric overview and citation classics. In *Bioenergy and Biofuels*, Ed. O. Konur, pp. 3–63. Boca Raton, FL: CRC Press.
Konur, O. 2019. Cyanobacterial bioenergy and biofuels science and technology: A scientometric overview. In *Cyanobacteria: From Basic Science to Applications*, Eds. A. K. Mishra, D. N. Tiwari and A. N. Rai, pp. 419–442. Amsterdam: Elsevier.
Konur, O. 2020a. The scientometric analysis of the research on the bioethanol production from green macroalgae. In *Handbook of Algal Science, Technology and Medicine,* Ed. O. Konur, pp. 385–401. London: Academic Press.
Konur, O., Ed. 2020b. *Handbook of Algal Science, Technology and Medicine.* London: Academic Press.
Konur, O., Ed. 2021a. *Handbook of Biodiesel and Petrodiesel Fuels: Science, Technology, Health, and Environment*. Boca Raton, FL: CRC Press.
Konur, O., Ed. 2021b. *Handbook of Biodiesel and Petrodiesel Fuels: Science, Technology, Health, and Environment. Volume 1. Biodiesel Fuels: Science, Technology, Health, and Environment.* Boca Raton, FL: CRC Press.

Konur, O., Ed. 2021c. *Handbook of Biodiesel and Petrodiesel Fuels: Science, Technology, Health, and Environment. Volume 2. Biodiesel Fuels Based on the Edible and Nonedible Feedstocks, Wastes, and Algae: Science, Technology, Health, and Environment.* Boca Raton, FL: CRC Press.

Konur, O., Ed. 2021d. *Handbook of Biodiesel and Petrodiesel Fuels: Science, Technology, Health, and Environment. Volume 3. Petrodiesel Fuels: Science, Technology, Health, and Environment.* Boca Raton, FL: CRC Press.

Konur, O. 2023a. Gasoline fuels: Scientometric study. In *Evaluation and Utilization of Bioethanol Fuels. I.: Evaluation of Bioethanol Fuels, Transport Engines, and Bioethanol Sensors. Handbook of Bioethanol Fuels Volume 5*, Ed. O. Konur, pp. 87–106. Boca Raton, FL: CRC Press.

Konur, O. 2023b. Nanotechnology applications in bioethanol fuels: Scientometric study. In *Evaluation and Utilization of Bioethanol Fuels. I.: Evaluation of Bioethanol Fuels, Transport Engines, and Bioethanol Sensors. Handbook of Bioethanol Fuels Volume 5*, Ed. O. Konur, pp. 120–139. Boca Raton, FL: CRC Press.

Konur, O. 2023c. Utilization of bioethanol fuels in the transport engines: Scientometric study. In *Evaluation and Utilization of Bioethanol Fuels. I.: Evaluation of Bioethanol Fuels, Transport Engines, and Bioethanol Sensors. Handbook of Bioethanol Fuels Volume 5*, Ed. O. Konur, pp. 157–175. Boca Raton, FL: CRC Press.

Konur, O. 2023d. Evaluation of bioethanol fuels: Scientometric study. In *Evaluation and Utilization of Bioethanol Fuels. I.: Evaluation of Bioethanol Fuels, Transport Engines, and Bioethanol Sensors. Handbook of Bioethanol Fuels Volume 5*, Ed. O. Konur, pp. 195–213. Boca Raton, FL: CRC Press.

Konur, O. 2023e. Utilization of bioethanol fuels: Scientometric study. In *Evaluation and Utilization of Bioethanol Fuels. I.: Evaluation of Bioethanol Fuels, Transport Engines, and Bioethanol Sensors. Handbook of Bioethanol Fuels Volume 5*, Ed. O. Konur, pp. 271–295. Boca Raton, FL: CRC Press.

Konur, O. 2023f. Bioethanol fuel sensors: Scientometric study. In *Evaluation and Utilization of Bioethanol Fuels. I.: Evaluation of Bioethanol Fuels, Transport Engines, and Bioethanol Sensors. Handbook of Bioethanol Fuels Volume 5*, Ed. O. Konur, pp. 317–334. Boca Raton, FL: CRC Press.

Konur, O. 2023g. Country-based experience of bioethanol fuels: Review. In *Evaluation and Utilization of Bioethanol Fuels. II.: Biohydrogen Fuels, Fuel Cells, Biochemicals, and Country Experiences. Handbook of Bioethanol Fuels Volume 6*, Ed. O. Konur, pp. 26–41. Boca Raton, FL: CRC Press.

Konur, O. 2023h. Bioethanol fuel-based biohydrogen fuels: Scientometric study. In *Evaluation and Utilization of Bioethanol Fuels. II.: Biohydrogen Fuels, Fuel Cells, Biochemicals, and Country Experiences. Handbook of Bioethanol Fuels Volume 6*, Ed. O. Konur, pp. 215–236. Boca Raton, FL: CRC Press.

Konur, O. 2023i. Bioethanol fuel cells: Scientometric study. In *Evaluation and Utilization of Bioethanol Fuels. II.: Biohydrogen Fuels, Fuel Cells, Biochemicals, and Country Experiences. Handbook of Bioethanol Fuels Volume 6*, Ed. O. Konur, pp. 277–297. Boca Raton, FL: CRC Press.

Konur, O. 2023j. Bioethanol fuel-based biochemical production: Scientometric study. In *Evaluation and Utilization of Bioethanol Fuels. II.: Biohydrogen Fuels, Fuel Cells, Biochemicals, and Country Experiences. Handbook of Bioethanol Fuels Volume 6*, Ed. O. Konur, pp. 317–337. Boca Raton, FL: CRC Press.

Konur, O. and F. L. Matthews. 1989. Effect of the properties of the constituents on the fatigue performance of composites: A review. *Composites* 20:317–328.

Leydesdorff, L. 2000. Is the European Union becoming a single publication system? *Scientometrics* 47:265–280.

Leydesdorff, L. and C. Wagner. 2009. Is the United States losing ground in science? A global perspective on the world science system. *Scientometrics* 78:23–36.

Leydesdorff, L. and P. Zhou. 2005. Are the contributions of China and Korea upsetting the world system of science? *Scientometrics* 63:617–630.

Leydesdorff, L., C. S. Wagner and L. Bornmann. 2014. The European Union, China, and the United States in the top-1% and top-10% layers of most-frequently cited publications: Competition and collaborations. *Journal of Informetrics* 8:606–617.

Li, S. Z. and C. Chan-Halbrendt. 2009. Ethanol production in (the) People's Republic of China: Potential and technologies. *Applied Energy* 86:S162–S169.

Liang, Z. X., T. S. Zhao, J. B. Xu and L. D. Zhu. 2009. Mechanism study of the ethanol oxidation reaction on palladium in alkaline media. *Electrochimica Acta* 54:2203–2208.

Liu, J., X. Wang, Q. Peng and Y. Li. 2005. Vanadium pentoxide nanobelts: Highly selective and stable ethanol sensor materials. *Advanced Materials* 17:764–767.

Liu, P. and E. J. M. Hensen. 2013. Highly efficient and robust Au/$MgCuCr2O_4$ catalyst for gas-phase oxidation of ethanol to acetaldehyde. *Journal of the American Chemical Society* 135:14032–14035.

Ma, X., L. Sun and C. Song. 2002. A new approach to deep desulfurization of gasoline, diesel fuel and jet fuel by selective adsorption for ultra-clean fuels and for fuel cell applications. *Catalysis Today* 77:107–116.

Macedo, I. C., J. E. A. Seabra and J. E. A. R. Silva. 2008. Green house gases emissions in the production and use of ethanol from sugarcane in Brazil: The 2005/2006 averages and a prediction for 2020. *Biomass and Bioenergy* 32:582–595.

Maity, S. K. 2015a. Opportunities, recent trends and challenges of integrated biorefinery: Part I. *Renewable and Sustainable Energy Reviews* 43:1427–1445.
Maity, S. K. 2015b. Opportunities, recent trends and challenges of integrated biorefinery: Part II. *Renewable and Sustainable Energy Reviews* 43:1446–1466.
Murdoch, M., G. I. N. Waterhouse and M. A. Nadeem, et al. 2011. The effect of gold loading and particle size on photocatalytic hydrogen production from ethanol over Au/TiO_2 nanoparticles. *Nature Chemistry* 3:489–492.
Najafi, G., B. Ghobadian and T. Tavakoli, et al. 2009. Performance and exhaust emissions of a gasoline engine with ethanol blended gasoline fuels using artificial neural network. *Applied Energy* 86:630–639.
Newman, P. W. G. and J. R. Kenworthy. 1989. Gasoline consumption and cities: A comparison of U.S. cities with a global survey. *Journal of the American Planning Association* 55:24–37.
Ni, M., D. Y. C. Leung and M. K. H. Leung. 2007. A review on reforming bio-ethanol for hydrogen production. *International Journal of Hydrogen Energy* 32:3238–3247.
North, D. C. 1991. Institutions. *Journal of Economic Perspectives* 5:97–112.
Sheehan, J., A. Aden and K. Paustian, et al. 2003. Energy and environmental aspects of using corn stover for fuel ethanol. *Journal of Industrial Ecology* 7:117–146.
Song, C. 2003. An overview of new approaches to deep desulfurization for ultra-clean gasoline, diesel fuel and jet fuel. *Catalysis Today* 86:211–263.
Vadas, P. A., K. H. Barnett and D. J. Undersander 2008. Economics and energy of ethanol production from alfalfa, corn, and switchgrass in the Upper Midwest, USA. *Bioenergy Research* 1:44–55.
Wan, Q., Q. H. Li and Y. J. Chen, et al. 2004. Fabrication and ethanol sensing characteristics of ZnO nanowire gas sensors. *Applied Physics Letters* 84:3654–3656.
Wang, H., Z. Jusys and R. J. Behm. 2004. Ethanol electrooxidation on a carbon-supported Pt catalyst: Reaction kinetics and product yields. *Journal of Physical Chemistry B* 108:19413–19424.

77 Bioethanol Fuel Evaluation and Utilization

Review

Ozcan Konur
(Formerly) Ankara Yildirim Beyazit University

77.1 INTRODUCTION

The crude oil-based gasoline fuels (Ma et al., 2002; Newman and Kenworthy, 1989) have been widely used in the transportation sector since the 1920s. However, there have been great public concerns over the adverse environmental and human impact of these fuels (Hill et al., 2006, 2009). Hence, biomass-based bioethanol fuels (Hill et al., 2006; Konur, 2012, 2015, 2019, 2020a) have increasingly been used in blending gasoline fuels (Hsieh et al., 2002; Najafi et al., 2009), in the fuel cells (Antolini, 2007, 2009), and in the biochemical production (Angelici et al., 2013; Morschbacker, 2009) in a biorefinery context (Fernando et al., 2006).

The primary focus of the research in this area has been the utilization of bioethanol fuels in fuel cells (Antolini, 2007; Liang et al., 2009), evaluation of bioethanol fuels (Farrell et al., 2006; Hill et al. 2006), utilization of bioethanol fuels transport engines (Hansen et al., 2005; Kohse-Hoinghaus et al., 2010), production of biohydrogen fuels from bioethanol fuels for the fuel cells (Haryanto et al., 2005; Ni et al., 2007), and to a lesser extent the experiences of countries (Macedo et al., 2008; Sheehan et al., 2003), development and utilization of bioethanol sensors (Liu et al., 2005; Wan et al., 2004), and production of biochemicals from bioethanol fuels (Liu and Hensen, 2013; Wang et al., 2004). Additionally, the research on the gasoline fuels (Khalili et al., 1995; Song, 2003) and applications of nanotechnology (Murdoch et al., 2011; Wan et al., 2004) in this field has also been related to this field.

However, it is essential to develop efficient incentive structures (North, 1991) for the primary stakeholders to enhance the research in this field (Konur, 2000, 2002a–c, 2006a,b, 2007a,b). Although there has been a number of review papers on the evaluation and utilization of bioethanol fuels (Antolini, 2007; Hansen et al., 2005; Haryanto et al., 2005; Kohse-Hoinghaus et al., 2010), there has been no review of the most-cited 25 articles in this field.

Thus, this book chapter presents a review of the most-cited 25 articles in these fields. Then, it discusses the key findings of these highly influential papers and comments on the future research priorities in this field.

77.2 MATERIALS AND METHODS

The search for this study was carried out using Scopus database (Burnham, 2006) in October 2021.

As a first step for the search of the relevant literature, the keywords were selected using the first most-cited 200 papers for each research front. The selected keyword list was optimized to obtain a representative sample of papers for the searched research field (Konur, 2023a–j). This keyword list for each research front was collected to form a combined keyword set for all research fronts and this combined set was provided in the appendix of the constituent studies for future replication studies (Konur, 2023a–j).

DOI: 10.1201/9781003226567-105

As a second step, a sample dataset was used for this study. The first 25 articles in the sample of 100 most-cited papers with at least 278 citations each were selected for the review study. Key findings from each paper were taken from the abstracts of these papers and were discussed. Additionally, a number of brief conclusions were drawn and a number of relevant recommendations were made to enhance the future research landscape.

77.3 RESULTS

The brief information about 25 most-cited papers with at least 451 citations each on the evaluation and utilization of bioethanol fuels is given below.

77.3.1 Utilization of Bioethanol Fuels in Fuel Cells

The brief information about ten prolific studies with at least 451 citations each on the utilization of bioethanol fuels in fuel cells is given in Table 77.1. Furthermore, the brief notes on the contents of these studies are also given below.

Marinov (1999) developed a chemical kinetic model for high temperature bioethanol oxidation in a paper with 669 citations. They used laminar flame speed data, ignition delay data, and bioethanol oxidation product profiles from a jet-stirred and turbulent flow reactor. They found good agreement in modeling of the data sets obtained from the five different experimental systems. They observed that the high temperature bioethanol oxidation exhibited strong sensitivity to the falloff kinetics of bioethanol decomposition, branching ratio selection for $C_2H_5OH+OH \leftrightarrow$ products, and reactions involving the hydroperoxyl radical.

TABLE 77.1
Utilization of Bioethanol Fuels in Fuel Cells

No.	Papers	Catalysts	Parameters	Country	Cits
1	Marinov (1999)	Na	High temperature bioethanol oxidation, kinetic models	USA	669
2	Liang et al. (2009)	Pd	Bioethanol oxidation reaction (BOR) mechanisms, electrocatalytic activity	China	658
3	Kowal et al. (2009)	C/Pt-Rh-SnO_2	Bioethanol oxidation, electrocatalytic activity, catalyst types	USA	618
4	Zhou et al. (2003)	C/PtSn, C/PtRu, C/PtW, C/PtPd	Bioethanol electrooxidation, additives, DEFC performance, temperature effect, catalyst types	China	583
5	Lamy et al. (2004)	Pt-Sn	Bioethanol electrooxidation, electrocatalytic activity	France	550
6	Klosek and Raftery (2002)	V-TiO_2	Bioethanol photooxidation	USA	537
7	Dong et al. (2010)	Gr/Pt NP, Gr/Pt-Ru NP, C/Pt NP, C/Pt-Ru NP	Bioethanol electrooxidation, electrocatalytic activity, catalyst support types, nanocatalysts, catalyst types	USA	530
8	Vigier et al. (2004)	Pt, PtSn	Bioethanol electrooxidation mechanisms, catalyst types, electrocatalytic activity	France	467
9	Xu et al. (2007a)	C/Pt, C/Pd	Bioethanol electrooxidation, catalyst types, C black, C microsphere supports	China	463
10	Xu et al. (2007b)	Pd nanowire arrays	Bioethanol electrooxidation, electrocatalytic activity, nanocatalysts	China	451

Na: Non available.

Liang et al. (2009) evaluated bioethanol oxidation reaction (BOR) mechanisms on palladium (Pd) electrode in alkaline media using the cyclic voltammetry method in a paper with 658 citations. They found that the dissociative adsorption of bioethanol proceeded rather quickly and the rate-determining step was the removal of the adsorbed ethoxy by the adsorbed hydroxyl on the Pd electrode. The adsorption of OH^- ions followed the Temkin-type isotherm on the Pd electrode. Finally, at higher potentials, the reaction kinetics was affected by the adsorption of the OH^- ions and by the formation of the inactive oxide layer on the Pd electrode.

Kowal et al. (2009) evaluated the C/platinum (Pt)–rhodium (Rh)–tin dioxide (SnO_2) electrocatalysts for the bioethanol oxidation in a paper with 618 citations. They found that this electrocatalyst effectively split the C–C bond in bioethanol at room temperature in acid solutions, facilitating its oxidation at low potentials to CO_2. They reasoned that this superb electrocatalytic activity was due to the specific property of each of its constituents, induced by their interactions.

Zhou et al. (2003) evaluated the C/PtSn, C/PtRu, C/PtW, C/PtPd catalyst systems for bioethanol electrooxidation in a paper with 583 citations. They found that the presence of Sn, ruthenium (Ru), and tungsten (W) enhanced the activity of Pt towards bioethanol electrooxidation: $Pt_1Sn_1/C>Pt_1Ru_1/C>Pt_1W_1/C>Pt_1Pd_1/C>Pt/C$. Moreover, PtRu/C further modified by W and molybdenum (Mo) showed improved bioethanol electrooxidation activity, but its direct ethanol fuel cell (DEFC) performance was lower than that measured for PtSn/C. Furthermore, the single DEFC with Pt_1Sn_1/C or Pt_3Sn_2/C or Pt_2Sn_1/C showed better performances than those with Pt_3Sn_1/C or Pt_4Sn_1/C. The latter two DEFCs exhibited higher performances than the single DEFC using Pt_1Ru_1/C. They reasoned that this distinct difference in DEFC performance between these catalysts was due to the bifunctional mechanism and to the electronic interaction between Pt and additives. Finally, at lower temperatures or at low current density regions, the Pt_3Sn_2/C was more suitable for the DEFC. At 75C, the single DEFC with Pt_3Sn_2/C showed a comparable performance to that with Pt_2Sn_1/C, but at higher temperature of 90C, the latter was much better performance.

Lamy et al. (2004) evaluated the Pt–Sn electrocatalysts for the bioethanol electrooxidation in a paper with 550 citations. They found that the overall electrocatalytic activity was greatly enhanced at low potentials. The optimum composition in Sn was in the range 10–20 at.%. With this composition, they observed that poisoning by adsorbed CO was greatly reduced leading to a significant enhancement of the electrode activity. However, the oxidation of ethanol was not complete leading to the formation of C_2 products.

Klosek and Raftery (2002) evaluated the visible light-driven vanadium (V)-doped TiO_2 photocatalyst for the photooxidation of bioethanol fuels in a paper with 537 citations. They found that this catalyst photooxidized bioethanol fuels to produce mostly CO_2 with small amounts of acetaldehyde, formic acid, and CO under visible irradiation. Under UV irradiation, this catalyst had comparable activity and product distribution as a similarly prepared TiO_2 thin-film monolayer catalyst.

Dong et al. (2010) evaluated the graphene (Gr)-supported Pt and Pt-Ru nanoparticles (NPs) with high electrocatalytic activity of bioethanol electrooxidation in a paper with 530 citations. They found that these nanocatalysts enhanced efficiency for ethanol electrooxidation with regard to diffusion efficiency, oxidation potential, forward oxidation peak current density, and the ratio of the forward peak current density to the reverse peak current density, compared to the C black catalyst supports. The forward peak current density of bioethanol electrooxidation for Gr- and C black-supported Pt nanoparticles was 16.2 and 13.8 mA/cm^2, respectively; and the ratios were 3.66 and 0.90, respectively.

Vigier et al. (2004) evaluated the mechanisms of bioethanol electrooxidation on Pt and PtSn catalysts in a paper with 467 citations. First, they established the beneficial effect of Sn for ethanol electrooxidation where the PtSn catalyst activity was almost double that on Pt. They identified adsorbed CO, adsorbed CH_3CO, CH_3CHO, CH_3COOH, and CO_2 as reaction products and intermediates. They established that two effects were involved in ethanol electrooxidation on PtSn: the bifunctional mechanism and the ligand effect. The presence of Sn allowed bioethanol to adsorb dissociatively, then to break the C–C bond, at lower potentials and with a higher selectivity than on pure Pt. It then allowed the formation of acetic acid at lower potentials than on Pt alone.

Xu et al. (2007a) evaluated bioethanol electrooxidation on Pt and Pd electrocatalysts with C microsphere and carbon black supports in alkaline media in a paper with 463 citations. They found that these electrocatalyst systems gave better performance than that those with C black supports. Pd showed excellently higher activity and better steady-state electrolysis than Pt for bioethanol electrooxidation in alkaline media. Finally, there was a synergistic effect by the interaction between Pd and C microspheres while the Pd supported on C microspheres had excellent electrocatalytic properties.

Xu et al. (2007b) evaluated Pd nanowire arrays (NWAs) as electrocatalysts for bioethanol electrooxidation in DEFCs in a paper with 451 citations. They observed that Pd nanowires (NWs) were highly ordered with uniform diameter and length. Further, the NWs were uniform, well isolated, parallel to one another, and standing vertically to the electrode substrate surface while the Pd NWAs exhibited a face-centered cubic (FCC) lattice structure. Finally, they observed that the high electrocatalytic activity of the Pd NWA electrode resulted in its superior performance for the electrooxidation reaction of bioethanol fuels in DEFC.

77.3.2 Evaluation of Bioethanol Fuels

The brief information about six prolific studies with at least 520 citations each on the evaluation of bioethanol fuels is given in Table 77.2. Furthermore, the brief notes on the contents of these studies are also given below.

Farrell et al. (2006) evaluated the net energy and greenhouse gas (GHG) emission balance of bioethanol fuels in a paper with 2074 citations. They found that current first generation corn grain-based ethanol technologies were much less crude oil-intensive than crude oil-based gasoline fuels but had GHG emissions similar to those of gasoline fuels. They developed new metrics that measured specific resource inputs and recommended further research into environmental metrics. They also recommended the use of lignocellulosic biomass-based bioethanol production from such as corn stovers.

Hill et al. (2006) evaluated the net energy, GHG emission balance, and economics of the first generation bioethanol biofuels from corn grains through a life cycle assessment (LCA) compared to biodiesel fuels from soybeans in a paper with 1989 citations. They laid down the rules for the bioethanol fuels: They should have a net energy balance, have environmental benefits, have economic competitiveness, have large-scale production, and have no competition to the food production. They

TABLE 77.2
Evaluation of Bioethanol Fuels

No.	Papers	Biomass	Parameters	Country	Cits.
1	Farrell et al. (2006)	Corn grains, corn stovers	Net energy balance, GHG emission balance	USA	2074
2	Hill et al. (2006)	Corn grains, corn stovers	Net energy balance, GHG emission balance, economics, LCA	USA	1989
3	Hamelinck et al. (2005)	Lignocellulosic biomass	Techno-economics, technological advances, production costs	Netherlands	1173
4	Schmer et al. (2008)	Switchgrass	Net energy balance, production costs, GHG emissions, biomass yield	USA	800
5	Macedo et al. (2008)	Sugarcane, sugarcane bagasse	Net energy balance, GHG emissions, CO_2 emissions, E25, E100	Brazil	609
6	Wingren et al. (2003)	Softwood	Techno-economics of softwood-based bioethanol fuels, production costs, SSF, SHF, energy consumption, stillage recycling	Sweden	520

Cits.: The number of the citations received by each paper.

found that bioethanol fuels had 25% more net energy balance compared to biodiesel fuels with 93% net energy balance. The production and combustion of bioethanol and biodiesel fuels resulted in a fall in 12% and 41% GHG emissions, respectively, compared to gasoline and petrodiesel fuels. Biodiesel fuels also released less air pollutants per net energy gain than bioethanol fuels due to the lower agricultural inputs and more efficient conversion of feedstocks to biodiesel fuels. However, both biofuels could not replace much crude oil-based fuels without endangering food supplies. Even dedicating all U.S. corn and soybean production to biofuels could meet only 12% of gasoline demand and 6% of petrodiesel demand. High production costs made biofuels unprofitable without subsidies, while only biodiesel fuels provided sufficient environmental advantages for subsidies. They recommended the use lignocellulosic biomass-based bioethanol fuels from waste biomass such as corn stovers instead of corn grains.

Hamelinck et al. (2005) evaluated the techno-economic performance of bioethanol fuels from lignocellulosic biomass in a paper with 1,173 citations. They found that the currently available technology, based on dilute acid hydrolysis, had about 35% efficiency (higher heating value, HHV) from biomass to bioethanol fuels. The overall efficiency, with the coproduction of power from lignin, was about 60%. However, improvements in pretreatments and biotechnology could bring the bioethanol efficiency to 48% and the overall process efficiency to 68%. They estimated current investment costs at 2.1 k€/kWHHV at 400 MWHHV input, 2,000 ton dry/day input). Hence, an advanced technology in a five times larger plant with 2 GWHHV could have investments of 900 k€/kWHHV. They assessed that the combined effect of higher hydrolysis–fermentation efficiency, lower specific capital investments, increase of scale, and cheaper biomass feedstock costs from 3 to 2 €/GJHHV could bring the bioethanol production costs from 22 €/GJHHV in the next 5 years, to 13 €/GJ over the 10–15 year timescale, and down to 8.7 €/GJ in 20 or more years.

Schmer et al. (2008) evaluated the net energy balance, production costs, and GHG emissions of switchgrass (*Panicum virgatum* L.)-based bioethanol fuels in a paper with 800 citations. They used 3–9 ha on marginal cropland. They found that the annual biomass yields of were 5.2–11.1 Mg/ha with a net energy yield (NEY) of 60 GJ/ha·year. This process produced 540% more renewable than nonrenewable energy consumed. Switchgrass monocultures produced 93% more biomass yield and a NEY than previous estimates. Further, the estimated average GHG emissions from these bioethanol fuels were 94% lower than estimated GHG from gasoline fuels. They reasoned that the improved genetics and agronomics might further enhance energy sustainability and bioethanol yield of switchgrass.

Macedo et al. (2008) evaluated the net energy and GHG emission balance in the production and use of first- and second generation bioethanol fuels from sugarcane and sugarcane bagasse in Brazil in a paper with 609 citations. They used two scenarios: 2005/2006 for a sample of mills processing up to 100 million tons of sugarcane per year and 2020. They found that the net energy balance was 9.3 for 2005/2006 and might reach 11.6 in 2020. The total GHG emissions were 436 kg CO_2 eq/m^3 bioethanol for 2005/2006, decreasing to 345 kg CO_2 eq/m^3 in the 2020 scenario. However, the avoided GHG emissions depended on the final use. For E100 use in Brazil, they were 2,181 kg CO_2 eq/m^3 bioethanol in 2005/2006, and for E25, they were 2,323 kg CO_2 eq/m^3 bioethanol. Both values would increase about 26% for 2020 mostly due to the large increase in sales of power surpluses. Finally, there was high impact of sugarcane productivity and bioethanol yield variation on these balances and of sugarcane bagasse and power surpluses on GHG emissions avoidance.

Wingren et al. (2003) evaluated the techno-economics of softwood-based bioethanol fuels in a paper with 520 citations. They used the simultaneous saccharification and fermentation (SSF) and the separate hydrolysis and fermentation (SHF) as scenarios for bioethanol production. They found that the bioethanol production costs for the SSF and SHF cases were 0.57 and 0.63 USD/L, respectively, due to the lower capital cost and higher the overall ethanol yield for the SSF case. They recommended lowering the energy consumption in the process through running the enzymatic hydrolysis or the SSF step at a higher substrate concentration and by recycling the process streams. For example, running SSF with use of 8% rather than 5% nonsoluble solid material would result in a 19% decrease in production cost. Furthermore, if after distillation 60% of the stillage stream was recycled back to the SSF step, the production cost would be further reduced by 14%. The cumulative effect of these various improvements was in a production cost of 0.42 USD/L for the SSF process.

77.3.3 Utilization of Bioethanol Fuels for Biohydrogen Production for Fuel Cells

The brief information about six prolific studies with at least 491 citations each on the utilization of bioethanol fuels for biohydrogen production for fuel cells is given in Table 77.3. Furthermore, the brief notes on the contents of these studies are also given below.

Murdoch et al. (2011) evaluated the effect of gold (Au) loading and NP size on photocatalytic hydrogen production from bioethanol fuels over Au/TiO_2 NPs in a paper with 937 citations. They showed that Au NPs in the size range 3–30 nm on TiO_2 were very active in biohydrogen production. Au NPs of similar size on anatase NPs resulted in a rate two orders of photoreaction magnitude higher than that that for Au on rutile NPs. Surprisingly, Au NP size did not affect the photoreaction rate over the 3–12 nm range. They obtained the high biohydrogen yield with these nanocatalysts.

Deluga et al. (2004) evaluated biohydrogen production from bioethanol fuels by autothermal reforming in a paper with 889 citations. They converted the bioethanol and bioethanol–water mixtures into biohydrogen with nearly 100% selectivity and at least 95% conversion by catalytic partial oxidation on rhodium (Rh)–ceria (CeO_2) catalysts. They carried out rapid vaporization and mixing with air with an automotive fuel injector at temperatures sufficiently low and times sufficiently fast that homogeneous reactions producing biochemicals such as bioethylene could be minimized.

TABLE 77.3
Utilization of Bioethanol Fuels for Biohydrogen Production for Fuel Cells

No.	Papers	Catalysts	Parameters	Country	Cits
1	Murdoch et al. (2011)	Au/TiO_2 NPs	Photocatalytic hydrogen production, catalyst loading, NP size, nanocatalysts	UK	937
2	Deluga et al. (2004)	Rh/CeO_2	Biohydrogen production by autothermal reforming	Greece	889
3	Liguras et al. (2003)	Rh, Ru, Pt, Pd/Al_2O_3, MgO, TiO_2	Biohydrogen production by steam reforming, catalyst loading, bioethanol conversions and biohydrogen selectivities, catalyst types	Greece	597
4	Fatsikostas and Verykios (2004)	Ni/γ-Al_2O_3, Ni/La_2O_3, Ni/La_2O_3-γ-Al_2O	Steam reforming of bioethanol fuels, catalyst support types	Greece	537
5	Bamwenda et al. (1995)	Au/TiO_2, Pt/TiO_2	Photocatalytic biohydrogen fuel production, photocatalytic activity, catalyst types, preparation method	Japan	533
6	Llorca et al. (2002)	Co/MgO, Co/γ-Co/Al_2O_3, Co/SiO_2, Co/TiO_2, Co/V_2O_5, Co/ZnO, Co/La_2O_3, Co/CeO_2	Steam reforming of bioethanol fuels, catalyst support types, catalytic performance, bioethanol selectivity	Spain	491

Liguras et al. (2003) evaluated biohydrogen production by steam reforming of bioethanol over noble metal catalysts such as Rh, Ru, Pt, and Pd with catalyst supports such as Al_2O_3, MgO, and TiO_2 in a paper with 597 citations. They used the temperature range of 600°C–850°C and the metal loading of 0–5 wt.%. They found that for low-loaded catalysts, Rh was significantly more active and selective toward biohydrogen formation compared to Ru, Pt, and Pd. The catalytic performance of Rh and, particularly, Ru was significantly improved with increasing metal loading, leading to higher bioethanol conversion and biohydrogen selectivities at given reaction temperatures. The catalytic activity and selectivity of high-loaded Ru catalysts was comparable to that of Rh. They observed that, under certain reaction conditions, the 5% Ru/Al_2O_3 catalyst system completely converted bioethanol with selectivities toward biohydrogen above 95% with methane coproduct.

Fatsikostas and Verykios (2004) evaluated the reaction network of steam reforming of bioethanol fuels over Ni-based catalysts in a paper with 535 citations. They used alumina (γ-Al_2O_3), lanthana (La_2O_3), and La_2O_3/γ-Al_2O_3 as catalyst supports. They found that bioethanol fuels interacted strongly with alumina and less strongly with lanthana. In the presence of Ni, catalytic activity was shifted toward lower temperatures. In addition to the above reactions, reforming, water–gas shift, methanation, and carbon deposition contributed significantly to product distribution. They then found that the rate of carbon deposition was a strong function of the carrier, the steam-to-bioethanol ratio, and reaction temperature as the presence of lanthana on the catalyst, high steam-to-bioethanol ratio, and high temperature offer enhanced resistance toward carbon deposition.

Bamwenda et al. (1995) evaluated the photocatalytic biohydrogen fuel production from a water–ethanol solution using Au/TiO_2 and Pt/TiO_2 catalyst systems in a paper with 533 citations. They deposited Au and Pt using TiO_2 powders in aqueous suspensions by deposition–precipitation (DP), impregnation (IMP), photodeposition (FD) and, in the case of Au, by mixing TiO_2 with colloidal Au suspensions (MIX). The main reaction products were hydrogen, methane, carbon dioxide, and acetaldehyde. They found that the overall activity of Au samples was generally about 30% lower than that of Pt samples. The activity of Au samples strongly depended on the method of preparation and decreased in the order Au/TiO_2-FD > Au/TiO_2-DP > Au/TiO_2-IMP > Au/TiO_2-MIX. The activities of the Pt samples were less sensitive to the preparation method and decreased in the order Pt/TiO_2-FD > Pt/TiO_2-DP ≈ Pt/TiO_2-IMP. Au and Pt precursors calcined in air at 573 K showed the highest activity towards H_2 generation, followed by a decline in activity with increasing calcination temperature. The H_2 yield was dependent on the metal content on TiO_2 and showed a maximum in the ranges 0.3–1 wt.% Pt and 1–2 wt.% Au. Finally, the rate of H_2 production was strongly dependent on the initial pH of the suspension as pH values in the range four to seven gave better yields, whereas highly acidic and basic suspensions resulted in a considerable decrease in the H_2 yield.

Llorca et al. (2002) evaluated the steam reforming of bioethanol fuels over supported cobalt (Co) catalysts to produce biohydrogen fuels in a paper with 491 citations. They used temperature range of 573–723 K at atmospheric pressure. They prepared catalysts (1%) by impregnation of $Co_2(CO)_8$ on MgO, γ-Al_2O_3, SiO_2, TiO_2, V_2O_5, ZnO, La_2O_3, CeO_2, and Sm_2O_3. They found that bioethanol steam reforming took place largely over ZnO-, La_2O_3-, Sm_2O_3-, and CeO_2-supported catalysts where CO-free biohydrogen was produced. Depending on the catalyst support, they identified different Co-based phases: metallic Co particles, Co_2C, CoO, and La_2CoO_4. Furthermore, the extent and nature of carbon deposition depended on the sample and on the reaction temperature. Finally, ZnO-supported samples showed the best catalytic performances, while under 100% bioethanol conversion, they obtained selectivity up to 73.8% to H_2 and 24.2% to CO_2.

77.3.4 Other Issues for the Evaluation and Utilization of Bioethanol Fuels

The brief information about three prolific studies with at least 491 citations each on the ethanol sensors and utilization of bioethanol fuels in the transport engines is given in Table 77.4. Furthermore, the brief notes on the contents of these studies are also given below. However, there were no papers in the fields of utilization of bioethanol fuels for biochemical production and country-based experiences of bioethanol fuels.

TABLE 77.4
Other Issues for the Evaluation and Utilization of Bioethanol Fuels

No.	Papers	Materials	Parameters	Country	Cits,
1	Wan et al. (2004)	ZnO nanowires	Nanosensors	China	1847
2	Liu et al. (2005)	V_2O_5 nanobelts	Nanosensors	China	508
3	Hsieh et al. (2002)	Bioethanol-gasoline fuel blends	Properties, performance, gas emissions, heating values, octane numbers, Reid vapor pressure, torque output, fuel consumption, CO, CO_2, HC, NO_x emissions, bioethanol content, E0, E5, E10, E20, E30 blends	Taiwan	491

Cits.: The number of the citations received by each paper.

77.3.4.1 Bioethanol Sensors

Wan et al. (2004) evaluated zinc oxide (ZnO) nanowire-based bioethanol sensors in a paper with 1,847 citations. They used the microelectromechanical system (MEMS) technology to fabricate these sensors. They found that these highly sensitive nanosensors exhibited a very high sensitivity to bioethanol gas with fast response time at 300°C.

Liu et al. (2005) evaluated the single-crystalline divanadium pentoxide (V_2O_5) nanobelts as bioethanol sensor materials in a paper with 508 citations. They produced these nanobelts by a simple mild hydrothermal method with high yield. They found that the sensors fabricated using these nanobelts showed great potential for the detection of bioethanol molecules at a relatively low temperature. Finally, there were no problems of interference with bioethanol due to relative humidity and other gases.

77.3.4.2 Utilization of Bioethanol Fuels in Transportation

Hsieh et al. (2002) evaluated the properties, engine performance, and gas emissions of an spark ignition (SI) engine using bioethanol–gasoline fuel blends as a function of bioethanol content for E0, E5, E10, E20, and E30 in a paper with 491 citations. They found that with increasing the bioethanol content, the heating value of the blended fuels was decreased, while the octane number of the blended fuels increased. While with increasing the bioethanol content, the Reid vapor pressure of the blended fuels initially increased to a maximum at 10% bioethanol addition, and then decreased. Furthermore, using these blends, torque output and fuel consumption of the engine slightly increased while CO and hydrocarbon (HC) emissions decreased dramatically as a result of the ethanol addition; and CO_2 emission increased due to the improved combustion. Finally, nitrogen oxide (NO_x) emission depended on the engine operating conditions rather than the bioethanol content.

77.4 DISCUSSION

77.4.1 Introduction

The crude oil-based gasoline fuels have been widely used in the transportation sector since the 1920s. However, there have been great public concerns over the adverse environmental and human impact of these fuels. Hence, biomass-based bioethanol fuels have increasingly been used in blending gasoline and petrodiesel fuels, in the fuel cells, and in the biochemical production in a biorefinery context.

The research in the field of utilization and evaluation of bioethanol fuels has also intensified in recent years. The primary focus of the research in this area has been the utilization of bioethanol fuels in fuel cells, evaluation of bioethanol fuels, utilization of bioethanol fuels transport engines, production of biohydrogen fuels from bioethanol fuels, and to a lesser extent the experiences of countries, development and utilization of bioethanol sensors, and production of biochemicals from bioethanol fuels. Additionally, the research on the gasoline fuels and applications of nanotechnology in this field has also been related to this field.

However, it is essential to develop efficient incentive structures for the primary stakeholders to enhance the research in this field. Although there has been a number of review papers on the evaluation and utilization of bioethanol fuels, there has been no review of the most-cited 25 articles in this field.

Thus, this book chapter presents a review of the most-cited 25 articles in these fields. Then, it discusses the key findings of these highly influential papers and comments on the future research priorities in this field.

As a first step for the search of the relevant literature, the keywords were selected using the first most-cited 200 papers for each research front. The selected keyword list was optimized to obtain a representative sample of papers for the searched research field (Konur, 2023a–j). This keyword list for each research front was collected to form a combined keyword set for all research fronts and this combined set was provided in the appendix of the constituent studies for future replication studies (Konur, 2023a–j).

As a second step, a sample dataset was used for this study. The first 25 articles in the sample of 100 most-cited papers with at least 278 citations each were selected for the review study. Key findings from each paper were taken from the abstracts of these papers and were discussed. Additionally, a number of brief conclusions were drawn and a number of relevant recommendations were made to enhance the future research landscape.

As Table 77.5 shows, there are four primary research fronts for these most-cited 25 papers: utilization of bioethanol fuels in fuel cells, evaluation of bioethanol fuels, utilization of bioethanol fuels for biohydrogen production for fuel cells, and other issues for the evaluation and utilization of bioethanol fuels.

TABLE 77.5
The Evaluation and Utilization of Bioethanol Fuels: Research Fronts

No.	Research Fronts	I100 Sample (%)	Sample Papers (%)	Reviewed Papers (%)
1	Utilization of bioethanol fuels in fuel cells	27.6	30	40
2	Evaluation of bioethanol fuels	20.9	26	24
3	Utilization of bioethanol fuels for biohydrogen production for fuel cells	19.1	19	24
4	Other issues for the evaluation and utilization of bioethanol fuels	47.4	40	12
	Bioethanol sensors	12.0	7	8
	Utilization of bioethanol fuels in transportation such as gasoline engines	19.4	19	4
	Utilization of bioethanol fuels for biochemical production	6.3	4	0
	Country-based experiences of bioethanol fuels	9.7	10	0
	Sample size	774	100	25

Reviewed papers: 25 papers reviewed in this study. Sample papers: the sample of the most-cited 100 papers. I100 sample: the sample of papers with at least 100 citations each.

77.4.2 Utilization of Bioethanol Fuels in Fuel Cells

Over 27%, 30%, and 40% of the I100 sample, 100 most-cited paper sample, and reviewed papers were related to the utilization of bioethanol fuels in fuel cells (Table 77.5). These studies were carried out in the USA, China, and France with four, four, and two papers, respectively (Table 77.1).

A number of catalysts were used in these studies: Pt (Dong et al., 2010; Kowal et al., 2009; Lamy et al., 2004; Vigier et al., 2004; Xu et al., 2007a; Zhou et al., 2003); Pd (Liang et al., 2009; Xu et al., 2007a,b; Zhou et al., 2003), Ru (Dong et al., 2010; Zhou et al., 2003), Sn (Vigier et al., 2004; Zhou et al., 2003), Rh (Kowal et al., 2009), W (Zhou et al., 2003), and V (Klosek and Raftery, 2002). Only Dong et al. (2010) and Xu et al. (2007b) considered the applications of nanotechnology in these catalysts.

Four studies focused on the C as catalyst supports (Dong et al., 2010; Kowal et al., 2009; Xu et al., 2007a; Zhou et al., 2003). Additionally, Dong et al. (2010) used graphene as a catalyst support compared to C catalyst supports.

These studies focused mostly on the electrooxidation of bioethanol fuels (Dong et al., 2010; Kowal et al., 2009; Lamy et al., 2004; Liang et al., 2009; Vigier et al., 2004; Xu et al., 2007a,b; Zhou et al., 2003). Additionally, Marinov (1999) and Klosek and Raftery (2002) studied the oxidation and photooxidation of bioethanol fuels, respectively. The activity and mechanisms of electrooxidation of bioethanol fuels were extensively evaluated in these studies.

These findings are thought-provoking in seeking ways to optimize the utilization of bioethanol fuels in fuel cells. The catalysts, catalyst supports, mechanisms and activity of bioethanol electrooxidation, and applications of nanotechnology in catalysts emerge as the key issues for this field.

77.4.3 Evaluation of Bioethanol Fuels

Around 21%, 26%, and 24% of the I100 sample, 100 most-cited paper sample, and reviewed papers were related to the evaluation of bioethanol fuels (Table 77.5). These studies were carried out in the USA, Brazil, Netherlands, and Sweden with three, one, one, and one papers, respectively (Table 77.2).

A number of feedstocks were used in these studies: corn grains and corn stovers (Farrell et al., 2006; Hill et al., 2006), switchgrass (Schmer et al., 2008), softwood (Wingren et al., 2003), and sugarcane and sugarcane bagasse (Macedo et al., 2008).

Four studies focused on the net energy balance and GHG emissions (Farrell et al., 2006; Hill et al., 2006; Schmer et al., 2008; Macedo et al., 2008), while Hamelinck et al. (2005) and Wingren et al. (2003) evaluated techno-economics of bioethanol fuels and production costs.

These findings are thought-provoking in seeking ways to evaluate bioethanol fuels. The feedstocks, net energy balance, GHG emissions, techno-economics, and production costs emerge as the key issues for this field.

77.4.4 Utilization of Bioethanol Fuels for Biohydrogen Production for Fuel Cells

Around 19%, 19%, and 24% of the I100 sample, 100 most-cited paper sample, and reviewed papers were related to the utilization of bioethanol fuels for biohydrogen production for the fuel cells (Table 77.5). These studies were carried out in the Greece, Japan, Spain, and the UK with three, one, one, and one papers, respectively (Table 77.3).

A number of catalysts were used in these studies: Au (Bamwenda et al., 1995; Murdoch et al., 2011); Rh (Deluga et al., 2004; Liguras et al., 2003); Ru (Liguras et al., 2003), Pt (Bamwenda et al., 1995; Liguras et al., 2003); Pd (Liguras et al., 2003); Ni (Fatsikostas and Verykios, 2004); and Co (Llorca et al., 2002).

Similarly, a number of catalyst supports were used in these studies: TiO_2 (Bamwenda et al., 1995; Liguras et al., 2003; Llorca et al., 2002; Murdoch et al., 2011); CeO_2 (Deluga et al., 2004; Llorca et al. (2002); Al_2O_3 (Liguras et al., 2003; Llorca et al., 2002); MgO (Liguras et al., 2003; Llorca et al., 2002);

γ-Al_2O_3 (Fatsikostas and Verykios, 2004); La_2O_3 (Fatsikostas and Verykios, 2004; Llorca et al., 2002); La_2O_3-γ-Al_2O (Fatsikostas and Verykios, 2004); SiO_2 (Llorca et al., 2002); V_2O_5 (Llorca et al., 2002); and ZnO (Llorca et al., 2002). Only Murdoch et al. (2011) considered the nanotechnology applications.

These studies focused mostly on the steam reforming of bioethanol fuels (Fatsikostas and Verykios, 2004; Liguras et al., 2003; Llorca et al., 2002), photocatalytic hydrogen production (Bamwenda et al., 1995; Murdoch et al., 2011), and autothermal reforming of bioethanol fuels (Deluga et al. 2004).

The key issues considered were the catalyst loading (Liguras et al., 2003; Murdoch et al., 2011); catalyst types (Bamwenda et al., 1995; Liguras et al., 2003); catalyst support types (Fatsikostas and Verykios, 2004; Liguras et al., 2003; Llorca et al., 2002); nanostructured catalyst systems (Murdoch et al., 2011), preparation methods (Bamwenda et al., 1995); bioethanol conversion and biohydrogen selectivities (Liguras et al., 2003; Llorca et al., 2002), and catalytic activity (Bamwenda et al., 1005; Llorca et al., 2002).

These findings are thought-provoking in seeking ways to optimize the utilization of bioethanol fuels for biohydrogen production for fuel cells. The catalyst types, catalyst support types, nanostructured catalyst systems, steam and autothermal reforming of bioethanol fuels, photocatalytic biohydrogen production, catalyst loading, preparation methods, bioethanol conversion, biohydrogen selectivities, and catalytic activity emerge as the key issues for this field.

77.4.5 Other Issues for the Evaluation and Utilization of Bioethanol Fuels

Around 47%, 40%, and 12% of the I100 sample, 100 most-cited paper sample, and reviewed papers were related to the other issues for the evaluation and utilization of bioethanol fuels (Table 77.5). These studies were carried out in China and Taiwan (Table 77.3).

There were two and one papers on the bioethanol sensors and the utilization of bioethanol fuels in the transport engines, respectively (Table 77.5). The latter research front was underrepresented in the reviewed papers by 15%. Furthermore, there were no papers on the utilization of bioethanol fuel for the biochemical production and country-based experiences of bioethanol fuels.

77.4.5.1 Bioethanol Sensors

Around 12%, 7%, and 8% of the I100 sample, 100 most-cited paper sample, and reviewed papers were related to the bioethanol sensors (Table 77.5). Both studies were carried out in China (Table 77.4).

Two catalysts were used in these studies: ZnO nanowires (Wan et al., 2004) and V_2O_5 nanobelts (Liu et al., 2005). These findings are thought-provoking in seeking ways to optimize the development of bioethanol sensors using nanomaterials.

77.4.5.2 Utilization of Bioethanol Fuels in Transportation

Around, 19%, 19%, and 4% of the I100 sample, 100 most-cited paper sample, and reviewed papers were related to the utilization of bioethanol fuels in the transportation such as gasoline or diesel engines (Table 77.5). This study was carried out in the Taiwan (Table 77.4).

A wide range of issues were covered in this paper: properties, performance, gas emissions, heating values, octane numbers, Reid vapor pressure, torque output, fuel consumption, CO, CO_2, HC, NO_x emissions, bioethanol content, E0, E5, E10, E20, and E30 blends.

77.4.6 The Overall Remarks

Although there are seven research fronts in the field of evaluation and utilization of bioethanol fuels, only three of them have importance presented among the reviewed papers: utilization of bioethanol fuels in fuel cells, evaluation of bioethanol fuels, and utilization of bioethanol fuels for biohydrogen production for fuel cells with 40%, 24%, and 24%, respectively (Table 77.5). The other research fronts were bioethanol sensors, utilization of bioethanol fuels in transport engines such as gasoline or diesel engines. Considering that biohydrogen fuels are used ultimately in the fuel cells for electric vehicles, the share of the papers related to fuel cells rises to 64% of the reviewed paper sample, compared to 47% and 59% of the I100 sample and 100 most cited paper sample, respectively. In light of North's institutional framework (North, 1991), it could be said that these three major research fronts have greater societal importance compared to the two other minor research fields.

The findings related to the utilization of bioethanol fuels in the fuel cells are thought-provoking in seeking ways to optimize the utilization of bioethanol fuels in fuel cells. The catalyst types (de Boer et al., 1965), catalyst support types (Arai and Machida, 1996), mechanisms and activity of bioethanol electrooxidation (Bagotzky and Vassilyev, 1967), and applications of nanotechnology in catalysts (Fihri et al., 2011) emerge as the key issues for this field.

The findings related to the evaluation of bioethanol fuels are thought-provoking in seeking ways to evaluate bioethanol fuels. The biomass feedstocks (Bergmann et al., 2013), net energy balance (Huettner, 1976), GHG emissions (Burnham et al., 2012), techno-economics (Callon, 1990), life cycle analysis (LCA) (Ayres, 1995), economic analysis (Landes, 1971), and production costs (Blumenfeld et al., 1985) emerge as the key issues for this field.

The findings related to the utilization of bioethanol fuels for biohydrogen fuel production for fuel cells are thought-provoking in seeking ways to optimize the utilization of bioethanol fuels for biohydrogen production for fuel cells. The catalyst types (de Boer et al., 1965), catalyst support types (Arai and Machida, 1996), nanostructured catalyst systems (Fihri et al., 2011), steam reforming (Palo et al., 2007) and autothermal reforming (Ayabe et al., 2003) of bioethanol fuels, photocatalytic biohydrogen production (Tong et al., 2012), catalyst loading (Jusys et al., 2003), bioethanol conversion (Maurya et al., 2015), and catalytic activity (Thiele et al., 1939) emerge as the key issues for this field.

A wide range of issues were covered in the paper related to the utilization of bioethanol fuels in gasoline and petrodiesel engines: fuel properties (Dooley et al., 2010), fuel performance (Crookes, 2006), heating values (Nhuchhen and Salam, 2012), octane numbers (Albahri, 2003), Reid vapor pressure (Cooper et al., 1995), torque output (Ahmadi et al., 2015), fuel consumption (Ramos-Paja et al., 2008), CO (Muller and Stavrakou, 2005), CO_2 (Davis et al., 2011), unburned hydrocarbons (UHC) (Stanglmaier et al., 1999), and NO_x emissions (Muller and Stavrakou, 2005), bioethanol content (Yunoki and Saito, 2009), E0, E5, E10, E20, and E30 blends.

Similarly, the focus for these hot papers related to bioethanol sensors is on the development of nanomaterials for bioethanol sensors. The research on the chemical sensors and biosensors has been very intense (Chaubey and Malhotra, 2002). In this context, the research on the ZnO-based sensors has also been very intense (Kumar et al., 2006). Nanomaterials have been widely used in the development of sensors (Howes et al., 2014).

Although there were no reviewed papers on both the utilization of bioethanol fuel for the biochemical production (Angelici et al., 2013; Liu and Hensen, 2013; Wang et al., 2004) and country-based experiences of bioethanol fuels (Macedo et al., 2008; Sheehan et al., 2003), these fields have great importance for the historical development of bioethanol industry and research. It is very import to produce biochemicals within the biorefinery context to reduce the production costs of bioethanol fuels as well as to develop human- and environment-friendly biochemicals as an alternative to crude oil-based chemicals. Similarly, the studies on the latter research front emphasize the importance of proper incentive structures for the efficient development of bioethanol industry and research in light of North's institutional framework (North, 1991).

Since the studies in each of the seven research fields take gasoline fuels as a base case, the review of the gasoline fuels as a whole was provided in Konur (2023a). The key research fronts for gasoline fuels are the desulfurization of gasoline fuels, properties, combustion, performance, and emissions of gasoline fuels and their blends, gasoline economics, gasoline fuel production, and upgrading.

It is important to note that Hill et al. (2006) laid down the rules for the environment- and human-friendly production of bioethanol fuels: They should have a net energy balance, have environmental benefits, have economic competitiveness, have large-scale production, and have no competition to the food production. The driving force for the development of the research and industry in this field has been the public concerns about climate change (Change, 2007), GHG emissions (Carlson et al., 2017), and global warming (Kerr, 2007).

The key findings from these studies also hint that bioethanol fuels have comparable net energy gain, better emissions, and comparable economic competitiveness following the list of the issues given by Hill et al. (2006). As expected, the emissions, net energy, and economic competitiveness of second generation bioethanol fuels such as food waste-based bioethanol fuels are better than those of the first generation bioethanol fuels such as corn- and sugarcane-based bioethanol fuels and much better than those of gasoline or petrodiesel fuels in general.

It is expected the nanotechnology (Geim, 2009; Geim and Novoselov, 2010) would play more vital role in the development of catalyst systems for the electrooxidation of bioethanol fuels, production of biohydrogen fuels from bioethanol fuels, and development of bioethanol sensors.

In the end, these most-cited papers in this field hint that the efficiency of the bioethanol, biohydrogen, and biochemical production could be optimized using the structure, processing, and property relationships of the bioethanol fuels and their derivatives (Formela et al., 2016; Konur, 2018, 2020b, 2021a–d; Konur and Matthews, 1989).

These reviewed studies also show the importance of the incentive structures for the development of bioethanol industry and research at a global scale in light of North's institutional framework (North, 1991). In this context, the major producers and users of bioethanol fuels such as Brazil, the USA, China, and Europe had developed strong incentive structures for the effective development of bioethanol industry and research.

77.5 CONCLUSION AND FUTURE RESEARCH

The brief information about the key research fronts covered by the 25 most-cited papers with at least 451 citations each is given under four headings: utilization of bioethanol fuels in fuel cells, evaluation of bioethanol fuels, utilization of bioethanol fuels for biohydrogen production for fuel cells, and other issues for the evaluation and utilization of bioethanol fuels (Table 77.5). The last group of papers include bioethanol sensors, utilization of bioethanol fuels in transport engines such as gasoline engines, utilization of bioethanol fuels for biochemical production, and country-based experiences of bioethanol fuels. In light of North's institutional framework (North, 1991), it could be said that these three major research fronts have greater societal importance compared to the four other minor research fields.

The key findings on these research fronts should be read in light of the increasing public concerns about climate change, GHG emissions, and global warming as these concerns have been certainly behind the boom in the research in this field in the last two decades.

These findings confirm that bioethanol fuels are a viable alternative to crude oil-based gasoline and petrodiesel fuels, have a net energy gain, have environmental benefits impacting favorably global warming, GHG emissions, and climate change, are economically competitive, and are producible in large quantities without reducing food supplies especially using the second generation bioethanol feedstocks such as lignocellulosic feedstocks using the criteria introduced by Hill et al. (2006).

There is a persistent trend for the bioethanol research to focus more on the second and third generation feedstocks such as sugarcane bagasse and algae and to move away from the first generation bioethanol feedstocks such as corn and sugarcane. These feedstocks would certainly impact the properties, engine performance, emissions as well as the net energy gain, economic competitiveness, and emissions of the resulting bioethanol fuels and their blends. This trend should be followed closely by the major stakeholders in the future.

The findings related to the utilization of bioethanol fuels in the fuel cells are thought-provoking in seeking ways to optimize the utilization of bioethanol fuels in fuel cells. The catalysts, catalyst supports, mechanisms and activity of bioethanol electrooxidation, and applications of nanotechnology in catalysts emerge as the key issues for this field.

The findings related to the evaluation of bioethanol fuels are thought-provoking in seeking ways to evaluate bioethanol fuels. The biomass feedstocks, net energy balance, GHG emissions, techno-economics, life cycle analysis (LCA), economic analysis, and production costs emerge as the key issues for this field.

The findings related to the utilization of bioethanol fuels for biohydrogen fuel production for fuel cells are thought-provoking in seeking ways to optimize the utilization of bioethanol fuels for biohydrogen production for fuel cells. The catalyst types, catalyst support types, nanostructured catalyst systems, steam reforming and autothermal reforming of bioethanol fuels, photocatalytic biohydrogen production, catalyst loading, preparation methods, bioethanol conversion, biohydrogen selectivities, and catalytic activity emerge as the key issues for this field.

A wide range of issues were covered in the paper related to the utilization of bioethanol fuels in gasoline and petrodiesel engines: properties, performance, heating values, octane numbers, Reid vapor pressure, torque output, fuel consumption, CO, CO_2, HC, and NO_x emissions, bioethanol content, E0, E5, E10, E20, and E30 blends. Similarly, the focus is on the development of nanomaterials for bioethanol sensors.

Although there were no reviewed papers on both the utilization of bioethanol fuel for the biochemical production and country-based experiences of bioethanol fuels, these fields have great importance for the historical development of bioethanol industry and research. It is very import to produce biochemicals within the biorefinery context to reduce the production costs of bioethanol fuels as well as to develop human- and environment-friendly biochemicals as an alternative to crude oil-based chemicals. Similarly, the studies on the latter research front emphasize the importance of proper incentive structures for the efficient development of bioethanol industry and research in light of North's institutional framework (North, 1991).

It is important to note that Hill et al. (2006) laid down the rules for the environment and human friendly production of bioethanol fuels: They should have a net energy balance, have environmental benefits, have economic competitiveness, have large-scale production, and have no competition to the food production. The driving force for the development of the research and industry in this field has been the public concerns about climate change, GHG emissions, and global warming.

The key findings from these studies also hint that bioethanol fuels have comparable net energy gain, better emissions, and comparable economic competitiveness following the list of the issues given by Hill et al. (2006). As expected, the emissions, net energy, and economic competitiveness of second generation bioethanol fuels are better than those of the first generation bioethanol fuels and much better than those of gasoline or petrodiesel fuels in general.

It is expected the nanotechnology would play more vital role in the development of catalyst systems for the electrooxidation of bioethanol fuels, production of biohydrogen fuels from bioethanol fuels, and development of bioethanol sensors.

In the end, these most-cited papers in this field hint that the efficiency of the bioethanol, biohydrogen, and biochemical production could be optimized using the structure, processing, and property relationships of the bioethanol fuels.

These reviewed studies also show the importance of the incentive structures for the development of bioethanol industry and research at a global scale in light of North's institutional framework (North, 1991). In this context, the major producers and users of bioethanol fuels such as Brazil, the USA, China, and Europe had developed strong incentive structures for the effective development of bioethanol industry and research.

It is recommended that such review studies should be performed for the other research fronts on both the production and utilization of bioethanol fuels complementing the constituent scientometric studies for the evaluation and utilization of bioethanol fuels (Konur, 2023a–j).

ACKNOWLEDGMENTS

The contribution of the highly cited researchers in the field of the evaluation and utilization of bioethanol fuels has been gratefully acknowledged.

REFERENCES

Ahmadi, M. H., M. A. Ahmadi, S. A. Sadatsakkak and M. Feidt. 2015. Connectionist intelligent model estimates output power and torque of stirling engine. *Renewable and Sustainable Energy Reviews* 50:871–883.

Albahri, T. A. 2003. Structural group contribution method for predicting the octane number of pure hydrocarbon liquids. *Industrial & Engineering Chemistry Research* 42:657–662.

Angelici, C., B. M. Weckhuysen and P. C. A. Bruijnincx. 2013. Chemocatalytic conversion of ethanol into butadiene and other bulk chemicals. *ChemSusChem* 6:1595–1614.

Antolini, E. 2007. Catalysts for direct ethanol fuel cells. *Journal of Power Sources* 170:1–12.

Antolini, E. 2009. Palladium in fuel cell catalysis. *Energy and Environmental Science* 2:915–931.

Arai, H. and M. Machida. 1996. Thermal stabilization of catalyst supports and their application to high-temperature catalytic combustion. *Applied Catalysis A: General* 138:161–176.

Ayabe, S., H. Omoto and T. Utaka, et al. 2003. Catalytic autothermal reforming of methane and propane over supported metal catalysts. *Applied Catalysis A: General* 241:261–269.

Ayres, R. U. 1995. Life cycle analysis: A critique. *Resources, Conservation and Recycling* 14:199–223.

Bagotzky, V. S. and Y. B. Vassilyev. 1967. Mechanism of electro-oxidation of methanol on the platinum electrode. *Electrochimica Acta* 12:1323–1343.

Bamwenda, G. R., S. Tsubota, T. Nakamura and M. Haruta. 1995. Photoassisted hydrogen production from a water-ethanol solution: A comparison of activities of Au/TiO_2 and Pt/TiO_2. *Journal of Photochemistry and Photobiology, A: Chemistry* 89:177–189.

Bergmann, J. C., D. D. Tupinamba and O. Y. A. Costa, et al. 2013. Biodiesel production in Brazil and alternative biomass feedstocks. *Renewable and Sustainable Energy Reviews* 21:411–420.

Blumenfeld, D. E., L. D. Burns, J. D. Diltz and C. F. Daganzo. 1985. Analyzing trade-offs between transportation, inventory and production costs on freight networks. *Transportation Research Part B: Methodological* 19:361–380.

Burnham, A., J. Han, and C. E. Clark, et al. 2012. Life-cycle greenhouse gas emissions of shale gas, natural gas, coal, and petroleum. *Environmental Science & Technology* 46:619–627.

Burnham, J. F. 2006. Scopus database: A review. *Biomedical Digital Libraries* 3:1–8.

Callon, M. 1990. Techno-economic networks and irreversibility. *Sociological Review* 38:132–161.

Carlson, K. M., J. S. Gerber and N. D. Mueller, et al. 2017. Greenhouse gas emissions intensity of global croplands. *Nature Climate Change* 7:63–68.

Change, C. 2007. Climate change impacts, adaptation and vulnerability. *Science of the Total Environment* 326:95–112.

Chaubey, A. and B. Malhotra. 2002. Mediated biosensors. *Biosensors and Bioelectronics* 17: 441–456.

Cooper, J. B., K. L. Wise, J. Groves and W. T. Welch. 1995. Determination of octane numbers and Reid vapor pressure of commercial petroleum fuels using FT-Raman spectroscopy and partial least-squares regression analysis. *Analytical Chemistry* 67:4096–4100.

Crookes, R. J. 2006. Comparative bio-fuel performance in internal combustion engines. *Biomass and Bioenergy* 30:461–468.

Davis, S. J., G. P. Peters and K. Caldeira. 2011. The supply chain of CO_2 emissions. *Proceedings of the National Academy of Sciences* 108:18554–18559.

de Boer, J. H., B. G. Linsen and T. J. Osinga. 1965. Studies on pore systems in catalysts: VI. The universal t curve. *Journal of Catalysis* 4:643–648.

Deluga, G. A., J. R. Salge, L. D. Schmidt and X. E. Verykios. 2004. Renewable hydrogen from ethanol by autothermal reforming. *Science* 303:993–997.

Dong, L., R. R. S. Gari, Z. Li, M. M. Craig and S. Hou. 2010. Graphene-supported platinum and platinum-ruthenium nanoparticles with high electrocatalytic activity for methanol and ethanol oxidation. *Carbon* 48:781–787.

Dooley, S., S. H. Won and M. Chaos, et al. 2010. A jet fuel surrogate formulated by real fuel properties. *Combustion and Flame* 157:2333–2339.

Farrell, A. E., R. J. Plevin and B. T. Turner, et al. 2006. Ethanol can contribute to energy and environmental goals. *Science* 311:506–508.

Fatsikostas, A. N. and X. E. Verykios. 2004. Reaction network of steam reforming of ethanol over Ni-based catalysts. *Journal of Catalysis* 225:439–452.

Fernando, S., S. Adhikari, C. Chandrapal and M. Murali. 2006. Biorefineries: Current status, challenges, and future direction. *Energy & Fuels* 20:1727–1737.

Fihri, A., M. Bouhrara, B. Nekoueishahraki, J. M. Basset and V. Polshettiwar. 2011. Nanocatalysts for Suzuki cross-coupling reactions. *Chemical Society Reviews* 40:5181–5203.

Formela, K., A. Hejna, L. Piszczyk, M. R. Saeb and X. Colom. 2016. Processing and structure-property relationships of natural rubber/wheat bran biocomposites. *Cellulose* 23:3157–3175.

Geim, A. K. 2009. Graphene: Status and prospects. *Science* 324:1530–1534.
Geim, A. K. and K. S. Novoselov. 2007. The rise of graphene. *Nature Materials* 6:183–191.
Hamelinck, C. N., G. van Hooijdonk and A. P. C. Faaij. 2005. Ethanol from lignocellulosic biomass: Techno-economic performance in short-, middle- and long-term. *Biomass and Bioenergy* 28:384–410.
Hansen, A. C., Q. Zhang and P. W. L. Lyne. 2005. Ethanol-diesel fuel blends: A review. *Bioresource Technology* 96:277–285.
Haryanto, A., S. Fernando, N. Murali and S. Adhikari. 2005. Current status of hydrogen production techniques by steam reforming of ethanol: A review. *Energy and Fuels* 19:2098–2106.
Hill, J., E. Nelson, D. Tilman, S. Polasky and D. Tiffany. 2006. Environmental, economic, and energetic costs and benefits of biodiesel and ethanol biofuels. *Proceedings of the National Academy of Sciences of the United States of America* 103:11206–11210.
Hill, J., S. Polasky and E. Nelson, et al. 2009. Climate change and health costs of air emissions from biofuels and gasoline. *Proceedings of the National Academy of Sciences of the United States of America* 106:2077–2082.
Howes, P. D., R. Chandrawati and M. M. Stevens. 2014. Colloidal nanoparticles as advanced biological sensors. *Science* 346:1247390.
Hsieh, W. D., R. H. Chen, T. L. Wu and T. H. Lin. 2002. Engine performance and pollutant emission of an SI engine using ethanol-gasoline blended fuels. *Atmospheric Environment* 36:403–410.
Huettner, D. A. 1976. Net energy analysis: An economic assessment. *Science* 192:101–104.
Jusys, Z., J. Kaiser, J. and R. J. Behm. 2003. Methanol electrooxidation over Pt/C fuel cell catalysts: Dependence of product yields on catalyst loading. *Langmuir* 19:6759–6769.
Kerr, R. A. 2007. Global warming is changing the world. *Science* 316:188–190.
Khalili, N. R., P. A. Scheff and T. M. Holsen. 1995. PAH source fingerprints for coke ovens, diesel and, gasoline engines, highway tunnels, and wood combustion emissions. *Atmospheric Environment* 29:533–542.
Klosek, S. and D. Raftery. 2002. Visible light driven V-doped TiO_2 photocatalyst and its photooxidation of ethanol. *Journal of Physical Chemistry B* 105:2815–2819.
Kohse-Hoinghaus, K., P. Osswald and T. A. Cool, et al. 2010. Biofuel combustion chemistry: From ethanol to biodiesel. *Angewandte Chemie: International Edition* 49:3572–3597.
Konur, O. 2000. Creating enforceable civil rights for disabled students in higher education: An institutional theory perspective. *Disability & Society* 15:1041–1063.
Konur, O. 2002a. Access to nursing education by disabled students: Rights and duties of nursing programs. *Nurse Education Today* 22:364–374.
Konur, O. 2002b. Assessment of disabled students in higher education: Current public policy issues. *Assessment and Evaluation in Higher Education* 27:131–152.
Konur, O. 2002c. Access to employment by disabled people in the UK: Is the Disability Discrimination Act working? *International Journal of Discrimination and the Law* 5:247–279.
Konur, O. 2006a. Participation of children with dyslexia in compulsory education: Current public policy issues. *Dyslexia* 12:51–67.
Konur, O. 2006b. Teaching disabled students in higher education. *Teaching in Higher Education* 11:351–363.
Konur, O. 2007a. A judicial outcome analysis of the Disability Discrimination Act: A windfall for the employers? *Disability & Society* 22:187–204.
Konur, O. 2007b. Computer-assisted teaching and assessment of disabled students in higher education: The interface between academic standards and disability rights. *Journal of Computer Assisted Learning* 23:207–219.
Konur, O. 2012. The evaluation of the research on the bioethanol: A scientometric approach. *Energy Education Science and Technology Part A: Energy Science and Research* 28:1051–1064.
Konur, O. 2015. Current state of research on algal bioethanol. In *Marine Bioenergy: Trends and Developments*, Ed. S. K. Kim and C. G. Lee, pp. 217–244. Boca Raton, FL: CRC Press.
Konur, O., Ed. 2018. *Bioenergy and Biofuels*. Boca Raton, FL: CRC Press.
Konur, O. 2019. Cyanobacterial bioenergy and biofuels science and technology: A scientometric overview. In *Cyanobacteria: From Basic Science to Applications*, Ed. A. K. Mishra, D. N. Tiwari and A. N. Rai, pp. 419–442. Amsterdam: Elsevier.
Konur, O. 2020a. The scientometric analysis of the research on the bioethanol production from green macroalgae. In *Handbook of Algal Science, Technology and Medicine,* Ed. O. Konur, pp. 385–401. London: Academic Press.
Konur, O., Ed. 2020b. *Handbook of Algal Science, Technology and Medicine.* London: Academic Press.
Konur, O., Ed. 2021a. *Handbook of Biodiesel and Petrodiesel Fuels: Science, Technology, Health, and Environment*. Boca Raton, FL: CRC Press.

Konur, O., Ed. 2021b. *Handbook of Biodiesel and Petrodiesel Fuels: Science, Technology, Health, and Environment. Volume 1. Biodiesel Fuels: Science, Technology, Health, and Environment.* Boca Raton, FL: CRC Press.

Konur, O., Ed. 2021c. *Handbook of Biodiesel and Petrodiesel Fuels: Science, Technology, Health, and Environment. Volume 2. Biodiesel Fuels Based on the Edible and Nonedible Feedstocks, Wastes, and Algae: Science, Technology, Health, and Environment.* Boca Raton, FL: CRC Press.

Konur, O., Ed. 2021d. *Handbook of Biodiesel and Petrodiesel Fuels: Science, Technology, Health, and Environment. Volume 3. Petrodiesel Fuels: Science, Technology, Health, and Environment.* Boca Raton, FL: CRC Press.

Konur, O. 2023a. Gasoline fuels: Scientometric study. In *Evaluation and Utilization of Bioethanol Fuels. I.: Evaluation of Bioethanol Fuels, Transport Engines, and Bioethanol Sensors. Handbook of Bioethanol Fuels Volume 5*, Ed. O. Konur, pp. 87–106. Boca Raton, FL: CRC Press.

Konur, O. 2023b. Nanotechnology applications in bioethanol fuels: Scientometric study. In *Evaluation and Utilization of Bioethanol Fuels. I.: Evaluation of Bioethanol Fuels, Transport Engines, and Bioethanol Sensors. Handbook of Bioethanol Fuels Volume 5*, Ed. O. Konur, pp. 120–139. Boca Raton, FL: CRC Press.

Konur, O. 2023c. Utilization of bioethanol fuels in the transport engines: Scientometric study. In *Evaluation and Utilization of Bioethanol Fuels. I.: Evaluation of Bioethanol Fuels, Transport Engines, and Bioethanol Sensors. Handbook of Bioethanol Fuels Volume 5*, Ed. O. Konur, pp. 157–175. Boca Raton, FL: CRC Press.

Konur, O. 2023d. Evaluation of bioethanol fuels: Scientometric study. In *Evaluation and Utilization of Bioethanol Fuels. I.: Evaluation of Bioethanol Fuels, Transport Engines, and Bioethanol Sensors. Handbook of Bioethanol Fuels Volume 5*, Ed. O. Konur, pp. 195–213. Boca Raton, FL: CRC Press.

Konur, O. 2023e. Utilization of bioethanol fuels: Scientometric study. In *Evaluation and Utilization of Bioethanol Fuels. I.: Evaluation of Bioethanol Fuels, Transport Engines, and Bioethanol Sensors. Handbook of Bioethanol Fuels Volume 5*, Ed. O. Konur, pp. 271–295. Boca Raton, FL: CRC Press.

Konur, O. 2023f. Bioethanol fuel sensors: Scientometric study. In *Evaluation and Utilization of Bioethanol Fuels. I.: Evaluation of Bioethanol Fuels, Transport Engines, and Bioethanol Sensors. Handbook of Bioethanol Fuels Volume 5*, Ed. O. Konur, pp. 317–334. Boca Raton, FL: CRC Press.

Konur, O. 2023g. Country-based experience of bioethanol fuels: Review. In *Evaluation and Utilization of Bioethanol Fuels. II.: Biohydrogen Fuels, Fuel Cells, Biochemicals, and Country Experiences. Handbook of Bioethanol Fuels Volume 6*, Ed. O. Konur, pp. 26–41. Boca Raton, FL: CRC Press.

Konur, O. 2023h. Bioethanol fuel-based biohydrogen fuels: Scientometric study. In *Evaluation and Utilization of Bioethanol Fuels. II.: Biohydrogen Fuels, Fuel Cells, Biochemicals, and Country Experiences. Handbook of Bioethanol Fuels Volume 6*, Ed. O. Konur, pp. 215–236. Boca Raton, FL: CRC Press.

Konur, O. 2023i. Bioethanol fuel cells: Scientometric study. In *Evaluation and Utilization of Bioethanol Fuels. II.: Biohydrogen Fuels, Fuel Cells, Biochemicals, and Country Experiences. Handbook of Bioethanol Fuels Volume 6*, Ed. O. Konur, pp. 277–297. Boca Raton, FL: CRC Press.

Konur, O. 2023j. Bioethanol fuel-based biochemical production: Scientometric study. In *Evaluation and Utilization of Bioethanol Fuels. II.: Biohydrogen Fuels, Fuel Cells, Biochemicals, and Country Experiences. Handbook of Bioethanol Fuels Volume 6*, Ed. O. Konur, pp. 317–337. Boca Raton, FL: CRC Press.

Konur, O. and F. L. Matthews. 1989. Effect of the properties of the constituents on the fatigue performance of composites: A review. *Composites* 20:317–328.

Kowal, A., M. Li and M. Shao, et al. 2009. Ternary Pt/Rh/SnO_2 electrocatalysts for oxidizing ethanol to CO_2. *Nature Materials* 8:325–330.

Kumar, N., A. Dorfman and J. I. Hahm. 2006. Ultrasensitive DNA sequence detection using nanoscale ZnO sensor arrays. *Nanotechnology* 17:2875.

Lamy, C., S. Rousseau, E. M. Belgsir, C. Coutanceau and J. M. Leger. 2004. Recent progress in the direct ethanol fuel cell: Development of new platinum-tin electrocatalysts. *Electrochimica Acta* 49:3901–3908.

Landes, W. M. 1971. An economic analysis of the courts. *Journal of Law and Economics* 14:61–107.

Liang, Z. X., T. S. Zhao, J. B. Xu and L. D. Zhu. 2009. Mechanism study of the ethanol oxidation reaction on palladium in alkaline media. *Electrochimica Acta* 54:2203–2208.

Liguras, D. K., D. L. Kondarides and X. E. Verykios. 2003. Production of hydrogen for fuel cells by steam reforming of ethanol over supported noble metal catalysts. *Applied Catalysis B: Environmental* 43:345–354.

Liu, J., X. Wang, Q. Peng and Y. Li. 2005. Vanadium pentoxide nanobelts: Highly selective and stable ethanol sensor materials. *Advanced Materials* 17:764–767.

Liu, P. and E. J. M. Hensen. 2013. Highly efficient and robust Au/$MgCuCr2O_4$ catalyst for gas-phase oxidation of ethanol to acetaldehyde. *Journal of the American Chemical Society* 135:14032–14035.

Llorca, J., N. Homs, J. Sales and P. R. de la Piscina. 2002. Efficient production of hydrogen over supported cobalt catalysts from ethanol steam reforming. *Journal of Catalysis* 209:306–317.

Ma, X., L. Sun and C. Song. 2002. A new approach to deep desulfurization of gasoline, diesel fuel and jet fuel by selective adsorption for ultra-clean fuels and for fuel cell applications. *Catalysis Today* 77:107–116.

Macedo, I. C., J. E. A. Seabra and J. E. A. R. Silva. 2008. Green house gases emissions in the production and use of ethanol from sugarcane in Brazil: The 2005/2006 averages and a prediction for 2020. *Biomass and Bioenergy* 32:582–595.

Marinov, N. M. 1999. A detailed chemical kinetic model for high temperature ethanol oxidation. *International Journal of Chemical Kinetics* 31:183–220.

Maurya, D. P., A. Singla and S. Negi. 2015. An overview of key pretreatment processes for biological conversion of lignocellulosic biomass to bioethanol. *3 Biotech* 5:597–609.

Morschbacker, A. 2009. Bio-ethanol based ethylene. *Polymer Reviews* 49:79–84.

Muller, J. F. and T. Stavrakou. 2005. Inversion of CO and NO_x emissions using the adjoint of the IMAGES model. *Atmospheric Chemistry and Physics* 5:1157–1186.

Murdoch, M., G. I. N. Waterhouse and M. A. Nadeem, et al. 2011. The effect of gold loading and particle size on photocatalytic hydrogen production from ethanol over Au/TiO_2 nanoparticles. *Nature Chemistry* 3:489–492.

Najafi, G., B. Ghobadian and T. Tavakoli, et al. 2009. Performance and exhaust emissions of a gasoline engine with ethanol blended gasoline fuels using artificial neural network. *Applied Energy* 86:630–639.

Newman, P. W. G. and J. R. Kenworthy. 1989. Gasoline consumption and cities: A comparison of U.S. cities with a global survey. *Journal of the American Planning Association* 55:24–37.

Nhuchhen, D. R. and P. A. Salam. 2012. Estimation of higher heating value of biomass from proximate analysis: A new approach. *Fuel* 99:55–63.

Ni, M., D. Y. C. Leung and M. K. H. Leung. 2007. A review on reforming bio-ethanol for hydrogen production. *International Journal of Hydrogen Energy* 32:3238–3247.

North, D. C. 1991. Institutions. *Journal of Economic Perspectives* 5:97–112.

Palo, D. R., R. A. Dagle and J. D. Holladay. 2007. Methanol steam reforming for hydrogen production. *Chemical Reviews* 107:3992–4021.

Ramos-Paja, C. A., C. Bordons, C., A. Romero, R. Giral and L. Martínez-Salamero. 2008. Minimum fuel consumption strategy for PEM fuel cells. *IEEE Transactions on Industrial Electronics* 56:685–696.

Schmer, M. R., K. P. Vogel, R. B. Mitchell and R. K. Perrin. 2008. Net energy of cellulosic ethanol from switchgrass. *Proceedings of the National Academy of Sciences of the United States of America* 105:464–469.

Sheehan, J., A. Aden and K. Paustian, et al. 2003. Energy and environmental aspects of using corn stover for fuel ethanol. *Journal of Industrial Ecology* 7:117–146.

Song, C. 2003. An overview of new approaches to deep desulfurization for ultra-clean gasoline, diesel fuel and jet fuel. *Catalysis Today* 86:211–263.

Stanglmaier, R. H., J. Li and R. D. Matthews. 1999. The effect of in-cylinder wall wetting location on the HC emissions from SI engines. SAE Transactions 1999:01-0502.

Thiele, E. W. 1939. Relation between catalytic activity and size of particle. *Industrial & Engineering Chemistry* 31:916–920.

Tong, H., S. Ouyang and Y. Bi, et al. 2012. Nano-photocatalytic materials: Possibilities and challenges. *Advanced Materials* 24:229–251.

Vigier, F., C. Coutanceau, F. Hahn, E. M. Belgsir and C. Lamy. 2004. On the mechanism of ethanol electro-oxidation on Pt and PtSn catalysts: Electrochemical and *in situ* IR reflectance spectroscopy studies. *Journal of Electroanalytical Chemistry* 563:81–89.

Wan, Q., Q. H. Li and Y. J. Chen, et al. 2004. Fabrication and ethanol sensing characteristics of ZnO nanowire gas sensors. *Applied Physics Letters* 84:3654–3656.

Wang, H., Z. Jusys and R. J. Behm. 2004. Ethanol electrooxidation on a carbon-supported Pt catalyst: Reaction kinetics and product yields. *Journal of Physical Chemistry B* 108:19413–19424.

Wingren, A., M. Galbe and G. Zacchi. 2003. Techno-economic evaluation of producing ethanol from softwood: Comparison of SSF and SHF and identification of bottlenecks. *Biotechnology Progress* 19:1109–1117.

Xu, C., H. Wang, P. K. Shen and S. P. Jiang. 2007b. Highly ordered Pd nanowire arrays as effective electrocatalysts for ethanol oxidation in direct alcohol fuel cells. *Advanced Materials* 19:4256–4259.

Xu, C., L. Cheng, P. Shen and Y. Liu. 2007a. Methanol and ethanol electrooxidation on Pt and Pd supported on carbon microspheres in alkaline media. *Electrochemistry Communications* 9:997–1001.

Yunoki, S. and M. Saito. 2009. A simple method to determine bioethanol content in gasoline using two-step extraction and liquid scintillation counting. *Bioresource Technology* 100:6125–6128.

Zhou, W., Z. Zhou and S. Song, et al. 2003. Pt based anode catalysts for direct ethanol fuel cells. *Applied Catalysis B: Environmental* 46:273–285.

78 Bioethanol Fuel Evaluation and Utilization

Hot Papers, Scientometric Study

Ozcan Konur
(Formerly) Ankara Yildirim Beyazit University

78.1 INTRODUCTION

Crude oil-based gasoline fuels (Ma et al., 2002; Newman and Kenworthy, 1989) have been widely used in the transportation sector since the 1920s. However, there have been great public concerns over the adverse environmental and human impact of these fuels (Hill et al., 2006, 2009). Hence, biomass-based bioethanol fuels (Hill et al., 2006; Konur, 2012e, 2015, 2019, 2020a) have increasingly been used in blending gasoline fuels (Hsieh et al., 2002; Najafi et al., 2009), in fuel cells (Antolini, 2007, 2009), and in biochemical production (Angelici et al., 2013; Morschbacker, 2009) in a biorefinery context (Fernando et al., 2006; Huang et al., 2008).

Research in the field of utilization and evaluation of bioethanol fuels has also intensified in recent years. The primary focus of research in this area has been the utilization of bioethanol fuels in fuel cells (Antolini, 2007; Liang et al., 2009), evaluation of bioethanol fuels (Farrell et al., 2006; Hill et al. 2006), utilization of bioethanol fuels in transport engines (Hansen et al., 2005; Kohse-Hoinghaus et al., 2010), production of biohydrogen fuels from bioethanol fuels (Haryanto et al., 2005; Ni et al., 2007), and to a lesser extent the experiences of countries (Macedo et al., 2008; Sheehan et al., 2003), development and utilization of bioethanol sensors (Liu et al., 2005; Wan et al., 2004), and production of biochemicals from bioethanol fuels (Liu and Hensen, 2013; Wang et al., 2004). Additionally, the research on gasoline fuels (Khalili et al., 1995; Song, 2003) and applications of nanotechnology (Murdoch et al., 2011; Wan et al., 2004) has also been related to this field.

However, it is essential to develop efficient incentive structures (North, 1991) for the primary stakeholders to enhance the research in this field (Konur, 2000, 2002a–c, 2006a,b, 2007a,b). The scientometric analysis has been used in this context to inform the primary stakeholders about the current state of research in a selected research field (Garfield, 1955; Konur, 2011, 2012a–i, 2015, 2018a,b, 2019, 2020a).

As there have been no scientometric studies on bioethanol evaluation and utilization, this book chapter presents a scientometric study of the research in bioethanol evaluation and utilization, published between 2016 and 2021. It examines the scientometric characteristics of both the sample and population data, presenting scientometric characteristics of both datasets in the order of documents, authors, publication years, institutions, funding bodies, source titles, countries, Scopus subject categories, keywords, and research fronts.

78.2 MATERIALS AND METHODS

The search for this study was carried out using the Scopus database for the period between 2016 and 2021 (Burnham, 2006) in October 2021.

As the first step for the search of the relevant literature, the keywords were selected using the first 200 most cited papers for each research front. The selected keyword list was optimized to obtain a representative sample of papers for the searched research field (Konur, 2023a–j). This keyword list for each research front was collected to form a combined keyword set for all seven research fronts, and this combined set was provided in the appendices of (Konur, 2023a–j) for future replication studies.

As the second step, two sets of data were used for this study. First, a population sample of over 6,000 papers was used to examine the scientometric characteristics of the population data. Secondly, a sample of 100 most cited papers were used to examine the scientometric characteristics of these citation classics with over 69 citations each.

The scientometric characteristics of both sample and population datasets were presented in the order of documents, authors, publication years, institutions, funding bodies, source titles, countries, Scopus subject categories, keywords, and research fronts.

Lastly, the key scientometric findings for both datasets were discussed to highlight the research landscape for bioethanol evaluation and utilization. Additionally, a number of brief conclusions were drawn and a number of relevant recommendations were made to enhance the future research landscape.

78.3 RESULTS

78.3.1 The Most Prolific Documents in Bioethanol Evaluation and Utilization: Hot Papers

The information on the types of documents for both datasets is given in Table 78.1. Articles and review papers dominate the sample papers while articles dominate population datasets. Further, review papers have a surplus while articles, conference papers, and book chapters have a deficit each.

It is also interesting to note that all of the papers in the sample dataset were published in journals, while only 97.2% of the papers were published in journals for the population dataset. Furthermore, 1.5% and 1.4% of the population papers were published in books and book series, respectively.

78.3.2 The Most Prolific Authors in Bioethanol Evaluation and Utilization: Hot Papers

The information about the 28 most prolific authors with at least two sample papers and five population papers each is given in Table 78.2a.

The most prolific author is Harish with four sample papers. Haifeng Liu, Yu Chen, Siti Kartom Kamarudin, Elena Pastor, Bo Jiang, Yanqiong Li, and Abdi H. Sebayang are the other prolific authors with two sample papers and with at least 11 population papers each.

TABLE 78.1
Documents in the Bioethanol Evaluation and Utilization: Hot Papers

Documents	Sample Dataset (%)	Population Dataset (%)	Surplus (%)
Article	88	91.9	−3.9
Review	12	2.4	9.6
Conference paper	0	3.3	−3.3
Book chapter	0	2.2	−2.2
Editorial	0	0.1	−0.1
Letter	0	0.1	−0.1
Note	0	0.1	−0.1
Short survey	0	0.1	−0.1
Book	0	0.0	0.0
Sample size	100	100	

TABLE 78.2A
Most Prolific Authors in the Bioethanol Evaluation and Utilization: Sample Dataset: Hot Papers

No.	Author Name	Author Code	Sample Papers No.	Population Papers No.	Institution	Country	Research Front
1	Venu, Harish	57189525542	4	7	Anna Univ.	India	Engines
2	Liu, Haifeng	56126891200	2	16	Tianjin Univ.	China	Engines
3	Chen, Yu	55831306300	2	14	Shaanxi Normal Univ.	China	Fuel cells
4	Kamarudin, Siti Kartom	6506009910	2	13	Univ. Kebangsaan	Malaysia	Fuel cells
5	Pastor, Elena	14047153200	2	13	Univ. Laguna	Spain	Fuel cells
6	Jiang, Bo	56870692600	2	11	Dalian Univ. Technol.	China	Biohydrogen fuels
7	Li, Yanqiong	55660924100	2	11	Chongqing Univ. Arts Sci.	China	Sensors
8	Sebayang, Abdi H.	39262519300	2	11	Univ. Malaysia	Malaysia	Engines
9	Lu, Geyu	7403460117	2	10	Jilin Univ.	China	Sensors
10	Min, Yulin	24554570700	2	10	Shanghai Univ. Elect. Power	China	Fuel cells
11	Yao, Mingfa	7403317253	2	10	Tianjin Univ.	China	Engines
12	Zheng, Zunqing	7403007391	2	10	Tianjin Univ.	China	Engines
13	Dou, Binlin	6701717734	2	9	Univ. Shanghai Sci. Technol.	China	Biohydrogen fuels
14	Xu, Qunjie	7403744157	2	9	Shanghai Univ. Elect. Power	China	Fuel cells
15	Chen, Haisheng	57192536693	2	8	Chinese Acad. Sci.	China	Biohydrogen fuels
16	Fan, Jinchen	54683952700	2	8	Shanghai Univ. Elect. Power	China	Fuel cells
17	Mahlia, Teuku M. I.	56997615100	2	8	Univ. Technol. Sydney	Australia	Engines
18	Mamat, Rizalman	37057681900	2	8	Univ. Malaysia Pahang	Malaysia	Engines
19	Park, Sunghoon	57206638688	2	8	Inha Univ.	S. Korea	Sensors
20	Rizo, Ruben	55840187600	2	7	Univ. Laguna	Spain	Fuel cells
21	Huang, Yu	13408109800	2	6	Univ. Calif. Berkeley	USA	Fuel cells
22	Sun, Peng	35254303000	2	6	Jilin Univ.	China	Sensors
23	Wang, Kaiqiang	56600956300	2	6	Dalian Univ. Technol.	China	Biohydrogen fuels
24	Iodice, Paolo	55348632700	2	5	Univ. Napoli	Italy	Engines
25	Li, Qiaoxia	55821435100	2	5	Shanghai Univ. Elect. Power	China	Fuel cells
26	Li, Yadong	57192004602	2	5	Tsinghua Univ.	China	Fuel cells
27	Wu, Tong	57188647061	2	5	Shanghai Univ. Elect. Power	China	Fuel cells
28	Xu, Yujie	8451810000	2	5	Chinese Acad. Sci.	China	Biohydrogen fuels

Author code: the unique code given by Scopus to the authors. Sample papers: the number of papers authored in the sample dataset. Population papers: the number of papers authored in the population dataset.

The most prolific institutions for the sample dataset are the Shanghai University of Electric Power and Tianjin University with five and three authors, respectively. The Chinese Academy of Sciences, Dalian University of Technology, Jilin University, and University of Laguna are the other prolific institutions with two authors each. In total 18 institutions house these prolific authors.

The most prolific country for the sample dataset is China with 18 authors. Malaysia and Spain are the other prolific countries with three and two authors, respectively. In total, eight countries house these authors.

The most prolific research fronts are fuel cells and engines with 11 and eight authors, respectively. The other prolific research fronts are biohydrogen fuels and sensors with five and four authors, respectively.

On the other hand, there is significant gender deficit (Beaudry and Lariviere, 2016) for the sample dataset as surprisingly nearly all of these top researchers are male.

Additionally, data for the prolific authors with less than two sample papers and with at least 13 population papers each are listed in Table 78.2b.

TABLE 78.2B
Most Prolific Authors in the Bioethanol Evaluation and Utilization-Population Dataset: Hot Papers

No.	Author Name	Author Code	Sample Papers	Population Papers	Institution	Country	Research Front
1	Du, Yukou	7402894251	0	26	Soochow Univ.	China	Fuel cells
2	Soloviev, Sergiy O.	15048785900	0	22	Natl. Acad. Sci.	Ukraine	Biochemicals
3	Kyriienko, Pavlo I.	55314336200	0	22	Natl. Acad. Sci.	Ukraine	Biochemicals
4	Ruocco, Concetta	56574025000	0	21	Univ. Salerno	Italy	Biohydrogen fuels
5	Rossetti, Ilenia	6701909842	0	21	Univ. Milan	Italy	Biohydrogen Fuels
6	Palma, Vincenzo	7006256714	0	21	Univ. Salerno	Italy	Biohydrogen Fuels
7	Larina, Olga V.	56911170000	0	19	Natl. Acad. Sci.	Ukraine	Biochemicals
8	Jongsomjit, Bunjerd	6603065177	0	19	Chulalongkorn Univ.	Thailand	Biochemicals
9	Llorca, Jordi	26039349400	0	18	Tech. Univ. Catalonia	Spain	Biohydrogen fuels
10	Lawler, Benjamin	42661890400	0	17	Stony Brook Univ.	USA	Engines
11	Bonomi, Antonio	7004767629	0	17	Univ. Campinas	Brazil	Evaluation
12	Zhao, Hua	55541772200	0	16	Brunel Univ.	UK	Engines
13	Wu, Xianyuan	57192088078	0	16	Zhejiang Univ. Technol.	China	Biochemicals
14	Ricca, Antonio	39161920500	0	16	Univ. Salerno	Italy	Biohydrogen fuels
15	Noronha, Fabio V.	7004514118	0	16	Natl. Technol. Inst.	Brazil	Biohydrogen fuels
16	Martins, Mario E. S.	55582321300	0	16	Fed. Univ. Santa Maria	Brazil	Engines
17	Zeng, Wen	35436737200	1	15	Chongqing Univ. Arts Sci.	China	Sensors
18	Tripodi, Antonio	57192387878	0	15	Univ. Milan	Italy	Biohydrogen fuels
19	Li, Yuqiang	56095804300	1	15	Central S University	China	Engines
20	Xie, Xianmei	14046469100	0	14	Taiyuan Univ. Technol.	China	Biohydrogen fuels
21	Ramis, Gianguido	7003665935	0	14	Univ. Genoa	Italy	Biohydrogen fuels
22	Neto, Almir O.	35458306500	0	14	Nuclear Ener. Res. Inst.	Brazil	Fuel cells
23	Lee, Chia-Fon	55680638300	1	14	Univ. Illinois	USA	Engines
24	An, Xia	15049196100	0	14	Taiyuan Univ. Technol.	China	Biohydrogen fuels
25	Xu, Hui	57191485966	0	13	Soochow Univ.	China	Fuel cells

(Continued)

TABLE 78.2B (*Continued*)
Most Prolific Authors in the Bioethanol Evaluation and Utilization-Population Dataset: Hot Papers

No.	Author Name	Author Code	Sample Papers	Population Papers	Institution	Country	Research Front
26	Salau, Nina PG	13611985700	0	13	Fed. Univ. Santa Maria	Brazil	engines
27	Pickup, Peter G.	7006750642	0	13	Memorial Univ.	Canada	Fuel cells
28	Parreira, Luanna S.	7102618978	0	13	Univ. Sao Paulo	Brazil	Fuel cells
29	Noroozifar, Meissam	57204665058	0	13	Univ. Sistan Baluchestan	Iran	Fuel cells
30	Greluk, Magdalena	29767519200	0	13	Univ. Maria Curie-Skłodowska	Poland	Biohydrogen fuels
31	Gan, Yunhua	7102646620	0	13	S. China Univ. Technol.	China	Engines
32	Cao, Jianliang	23033231900	0	13	Henan Polytech Univ.	China	Sensors

Author code: the unique code given by Scopus to the authors. Sample papers: the number of papers authored in the sample dataset. Population papers: the number of papers authored in the population dataset.

The most prolific authors are Yukou Du, Sergiy O. Soloviev, Pavlo I. Kyriienko, Concetta Ruocco, Ilenia Rossetti, and Vincenzo Palma, with at least 21 population papers each.

The most prolific institutions for the population dataset are the National Academy of Sciences of Ukraine and University of Salerno with three authors each. The Federal University of Santa Maria, Soochow University, Taiyuan University of Technology, and University of Milan are the other prolific institutions with two authors each. In total 24 institutions house these prolific authors.

The most prolific countries for the population dataset are China, Brazil, and Italy with eight, six, and six authors, respectively. Ukraine and the USA are the other prolific countries with two authors each. In total, 11 countries house these authors.

The most prolific research front is biohydrogen fuels with 11 authors. The other prolific research fronts are engines, fuel cells, biochemicals, and sensors with seven, six, five, and two authors, respectively.

On the other hand, there is significant gender deficit for the sample dataset as surprisingly nearly all of these top researchers are male.

78.3.3 The Most Prolific Research Output by Years in Bioethanol Evaluation and Utilization: Hot Papers

Information about papers published between 2016 and 2021 is given in Figure 78.1. This figure clearly shows that the bulk of the research papers in the population dataset were published primarily in the 2010s with 65.8% of the population dataset. The publication rate for the 2020s, was 33.5%. Similarly, all of the research papers in the sample dataset were published in the 2010s.

The yearly publication rates for the population dataset changed between 15.5% and 17.1% while the yearly publication rates for the sample papers were 49%, 33%, and 13% for 2016, 2017, and 2018, respectively.

78.3.4 The Most Prolific Institutions in Bioethanol Evaluation and Utilization: Hot Papers

Information about the 21 most prolific institutions publishing papers on the bioethanol evaluation and utilization with at least two sample papers and 0.3% of the population papers each is given in Table 78.3. Additionally, 15 institutions with at least 0.7% of the population dataset are included in this table.

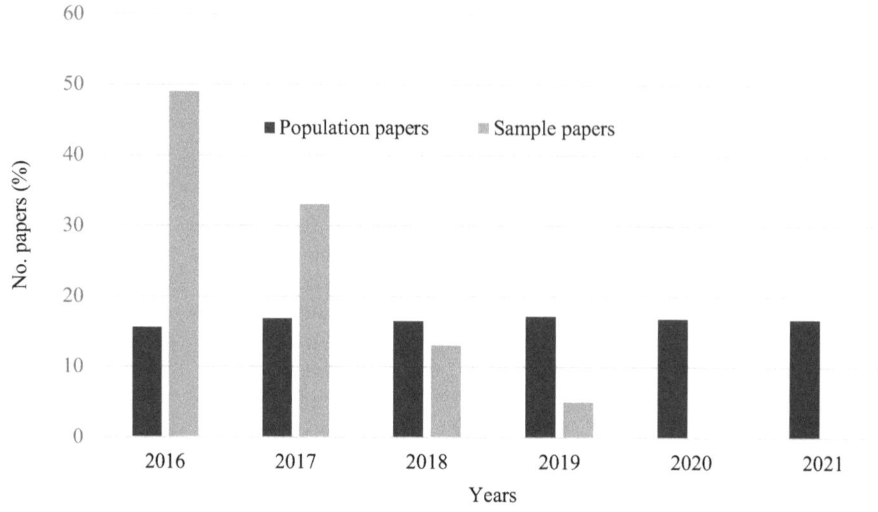

FIGURE 78.1 The research output by years regarding the bioethanol evaluation and utilization: Hot papers.

TABLE 78.3
The Most Prolific Institutions in Bioethanol Evaluation and Utilization: Hot Papers

No.	Institutions	Country	Sample Papers (%)	Population Papers (%)	Surplus (%)
1	Chinese Acad. Sci.	China	5	2.5	2.5
2	Tianjin Univ.	China	4	1.8	2.2
3	Tsinghua Univ.	China	4	1.1	2.9
4	CSIC	Spain	4	1.0	3.0
5	Soochow Univ.	China	3	0.9	2.1
6	Dalian Univ. Technol.	China	3	0.6	2.4
7	Beijing Univ. Chem. Technol.	China	3	0.5	2.5
8	Shaanxi Normal Univ.	China	3	0.4	2.6
9	Anna Univ.	India	3	0.3	2.7
10	Shanghai Univ. Electr. Power	China	3	0.3	2.7
11	Jilin Univ.	China	2	1.0	1.0
12	Paulista State Univ.	Brazil	2	0.9	1.1
13	Univ. Xiamen	China	2	0.9	1.1
14	Univ. Illinois U. C.	USA	2	0.8	1.2
15	Univ. Malaya	Malaysia	2	0.5	1.5
16	Wuhan Univ. Technol.	China	2	0.5	1.5
17	Chongqing Univ.	China	2	0.4	1.6
18	Iowa State Univ.	USA	2	0.3	1.7
19	Malaysia Technol. Univ.	Malaysia	2	0.3	1.7
20	Univ. Kebangsaan Malaysia	Malaysia	2	0.3	1.7
21	Univ. Malaysia Pohang	Malaysia	2	0.3	1.7
22	Univ. Sao Paulo	Brazil	1	2.6	−1.6
23	State Univ. Campinas	Brazil	1	1.3	−0.3

(*Continued*)

TABLE 78.3 (*Continued*)
The Most Prolific Institutions in Bioethanol Evaluation and Utilization: Hot Papers

No.	Institutions	Country	Sample Papers (%)	Population Papers (%)	Surplus (%)
24	Fed. Univ. Rio de Janeiro	Brazil	0	1.0	−1.0
25	CNR Natl Res. Counc.	Italy	0	0.9	−0.9
26	Beijing Inst. Technol.	China	1	0.8	0.2
27	Chulalongkorn Univ.	Thailand	0	0.8	−0.8
28	Shanghai Jiao Tong Univ.	China	0	0.8	−0.8
29	Taiyuan Univ. Technol.	China	1	0.8	0.2
30	CNRS	France	0	0.7	−0.7
31	Fed. Univ. Rio Grande do Sul	Brazil	0	0.7	−0.7
32	Huazhong Univ. Sci. Technol	China	1	0.7	0.3
33	Jiangsu Univ.	China	0	0.7	−0.7
34	Natl. Acad. Sci.	Ukraine	0	0.7	−0.7
35	Univ. Sci. Technol. China	China	0	0.7	−0.7
36	Xi'an Jiaotong Univ.	China	0	0.7	−0.7

The most prolific institution is the Chinese Academy of Sciences with five papers. Tianjin University, Tsinghua University, and the CSIC follow this top institution with four sample papers each.

The top country for these most prolific institutions is China with 12 institutions. Next, Malaysia and the USA follow China with four and two institutions, respectively. In total, six countries house these top institutions.

On the other hand, the institutions with the most citation impact are CSIC and Tsinghua University with 3.0% and 2.9% surplus, respectively. Similarly, the institutions with the least impact are Jilin University, Paulista State University, and University of Xiamen with at least 1.0% deficit each.

78.3.5 The Most Prolific Funding Bodies in Bioethanol Evaluation and Utilization: Hot Papers

Information about the 16 most prolific funding bodies funding at least two sample papers and 1.0% of the population papers each is given in Table 78.4. Additionally, five funding bodies with at least 1.0% of the population papers are listed in this table.

The most prolific funding body is the National Natural Science Foundation of China with 25 sample papers. The Ministry of Science and Technology and Ministry of Education of China are the other prolific funding bodies with five and four sample papers, respectively.

It is notable that 75% and 64.2% of the sample and population papers are funded, respectively.

The most prolific country for these top funding bodies is China with seven funding bodies. The other prolific countries are Brazil, the USA, and European Union with at least two sample papers. In total, five countries and the European Union house these top funding bodies.

The funding body with the most citation impact is the National Natural Science Foundation of China with 5.6% surplus. The other prolific funding bodies are the Ministry of Science and Technology of China and the Ministry of Economics and Competitiveness of Spain with 4.8% and 3.0% surplus, respectively. Similarly, the funding bodies with the least citation impact are the Sao Paulo State Research Foundation and the Ministry of Science Technology, and Innovation of Brazil with at least 0.7% deficit each.

TABLE 78.4
The Most Prolific Funding Bodies in Bioethanol Evaluation and Utilization: Hot Papers

No.	Funding Bodies	Country	Sample Paper No. (%)	Population Paper No. (%)	Surplus (%)
1	Natl. Natr. Sci. Found. China	China	25	19.4	5.6
2	Minist. Sci. Technol.	China	7	2.2	4.8
3	Ministry Educ.	China	5	2.7	2.3
4	Natl. Counc. Sci. Technol. Dev.	Brazil	4	4.2	−0.2
5	Fund. Res. Fund. Central Univ.	China	4	3.7	0.3
6	US Dept. Energy	USA	4	2.3	1.7
7	Natl. Sci. Found.	USA	4	2.2	1.8
8	Europ. Commis.	Europe	4	1.8	2.2
9	Priority Acad. Prg. Devnt Jiangsu High. Educ. Inst.	China	4	1.1	2.9
10	Minist. Econ. Compet.	Spain	4	1.0	3.0
11	Sao Paulo State Res. Found	Brazil	2	2.8	−0.8
12	Minist. Sci. Technol. Innov.	Brazil	2	2.7	−0.7
13	Natl. Res. Found.	S. Korea	2	1.6	0.4
14	China Postdoc. Sci. Found.	China	2	1.5	0.5
15	Europ. Reg. Devnt. Fund	EU	2	1.4	0.6
16	China Scholr. Counc.	China	2	1.2	0.8
17	Coord. Impov. Higher Educ. Person.	Brazil	1	3.6	−2.6
18	Natl. Key Res. Devnt. Pr.	China	1	1.9	−0.9
19	Natl. Res. Found. Korea	S. Korea	0	1.6	−1.6
20	Japan Soc. Prom. Sci.	Japan	0	1.0	−1.0
21	Minist. Educ. Cult. Sci. Technol.	Japan	0	1.0	−1.0

78.3.6 The Most Prolific Source Titles in Bioethanol Evaluation and Utilization: Hot Papers

Information about the 15 most prolific source titles publishing at least two sample papers and 0.5% of the population papers each in bioethanol evaluation and utilization is given in Table 78.5. Additionally, 14 journals with at least 0.7% of the population papers are also listed in this table.

The most prolific source titles are 'Sensors and Actuators B Chemical', 'Applied Catalysis B Environmental', and 'ACS Applied Materials and Interfaces' with at least seven sample papers each. 'Renewable and Sustainable Energy Reviews', ACS Catalysis', 'Fuel', and 'Energy Conversion and Management' follow these top titles with at least five sample papers each.

On the other hand, the source titles with the most citation impact are the 'ACS Applied Materials and Interfaces' and 'Applied Catalysis B Environmental' with at least 6.1% surplus each. The 'Sensors and Actuators B Chemical', 'ACS Catalysis', and 'Renewable and Sustainable Energy Reviews' follow this top title with at least 5.3% surplus each. Similarly, the source title with the least impact is the 'International Journal of Hydrogen Energy' with 2.2% deficit. 'Fuel' and 'Bioresource Technology' follow this title with at last 1.2% surplus each.

TABLE 78.5
The Most Prolific Source Titles in Bioethanol Evaluation and Utilization: Hot Papers

No.	Source Titles	Sample Papers (%)	Population Papers (%)	Surplus (%)
1	Sensors and Actuators B Chemical	8	2.4	5.6
2	Applied Catalysis B Environmental	7	0.9	6.1
3	ACS Applied Materials and Interfaces	7	0.8	6.2
4	Renewable and Sustainable Energy Reviews	6	0.7	5.3
5	ACS Catalysis	6	0.6	5.4
6	Fuel	5	4.5	0.5
7	Energy Conversion and Management	5	0.6	4.4
8	Energy	4	1.5	2.5
9	Applied Energy	3	1.1	1.9
10	Renewable Energy	3	1.1	1.9
11	Applied Thermal Engineering	3	0.7	2.3
12	International Journal of Hydrogen Energy	2	4.2	−2.2
13	Bioresource Technology	2	0.8	1.2
14	Journal of Power Sources	2	0.6	1.4
15	Fuel Processing Technology	2	0.5	1.5
16	SAE Technical Papers	0	2.3	−2.3
17	Journal of Cleaner Production	0	2.0	−2.0
18	Electrochimica Acta	1	1.2	−0.2
19	RSC Advances	0	1.2	−1.2
20	Applied Surface Science	0	1.1	−1.1
21	Energy and Fuels	0	1.1	−1.1
22	Energies	0	1.0	−1.0
23	Journal of Alloys and Compounds	1	0.9	0.1
24	Journal of Materials Science Materials in Electronics	0	0.9	−0.9
25	Applied Catalysis A General	1	0.8	0.2
26	Catalysts	0	0.8	−0.8
27	Combustion and Flame	1	0.8	0.2
28	Catalysis Today	1	0.7	0.3
29	Materials Letters	1	0.7	0.3

78.3.7 The Most Prolific Countries in Bioethanol Evaluation and Utilization: Hot Papers

Information about the 18 most prolific countries publishing at least two sample papers and 1.1% of the population papers each in bioethanol evaluation and utilization is given in Table 78.6. Additionally, ten countries with at least 1.0% of the population papers are listed in this table.

The most prolific country is China with 38 sample papers. The USA, India, Malaysia, and Spain are the other prolific countries with at least six sample papers each. Further, eight European countries listed in Table 78.6 produce 29% and 18% of the sample and population papers, respectively. China is also the largest producer of the population papers with a publication rate of 29.7%.

On the other hand, the countries with the most citation impact are China and Malaysia with 8.3% and 5.6% surplus, respectively. The USA, Spain, and Australia follow these countries with at least 3.5% surplus each. Similarly, the country with the least citation impact is Brazil with 5.5% deficit. Iran, India, and France follow Brazil with at least 1.3% deficit each.

TABLE 78.6
The Most Prolific Countries in the Bioethanol Evaluation and Utilization: Hot Papers

No.	Countries	Sample Papers (%)	Population Papers (%)	Surplus (%)
1	China	38	29.7	8.3
2	USA	16	11.9	4.1
3	India	10	10.5	−0.5
4	Malaysia	8	2.4	5.6
5	Spain	6	2.4	3.6
6	Brazil	5	10.5	−5.5
7	UK	5	3.9	1.1
8	Italy	5	3.4	1.6
9	S. Korea	5	3.4	1.6
10	Australia	5	1.5	3.5
11	Germany	4	2.4	1.6
12	Iran	3	4.3	−1.3
13	Turkey	3	1.5	1.5
14	France	2	2.0	0.0
15	Indonesia	2	1.5	0.5
16	Poland	2	1.5	0.5
17	Taiwan	2	1.4	0.6
18	Sweden	2	1.1	0.9
19	Japan	0	3.1	−3.1
20	Canada	1	2.8	−1.8
21	Thailand	0	2.4	−2.4
22	S. Arabia	1	1.9	−0.9
23	Mexico	0	1.6	−1.6
24	Russia	0	1.6	−1.6
25	Egypt	0	1.2	−1.2
26	Colombia	1	1.1	−0.1
27	Netherlands	0	1.1	−1.1
28	Vietnam	1	1.0	0.0

78.3.8 The Most Prolific Scopus Subject Categories in Bioethanol Evaluation and Utilization: Hot Papers

Information about the eight most prolific Scopus subject categories indexing at least 10% and 5.1% of the sample and population papers, respectively, is given in Table 78.7. Additionally, eight subject categories with at least 1.2% of the population paper set are listed in this table.

The most prolific Scopus subject categories in the bioethanol evaluation and utilization are 'Energy', 'Chemical Engineering', and 'Engineering' with 39, 37, and 31 sample papers, respectively. 'Chemistry', Materials Science', Environmental Science', and 'Physics and Astronomy' follow these top categories with at least 17 sample papers each.

On the other hand, the Scopus subject categories with the most citation impact are the 'Energy', 'Chemical Engineering', and 'Materials Science' with at least 2.8% surplus each. Similarly, the Scopus subject categories with the least citation impact are 'Chemistry' and 'Physics and Astronomy' with at least 4.9% deficit each.

TABLE 78.7
The Most Prolific Scopus Subject Categories in the Bioethanol Evaluation and Utilization: Hot Papers

No.	Scopus Subject Categories	Sample Papers (%)	Population Papers (%)	Surplus (%)
1	Energy	39	34.6	4.4
2	Chemical engineering	37	33.4	3.6
3	Engineering	31	31.8	−0.8
4	Chemistry	27	34.6	−7.6
5	Materials science	26	23.2	2.8
6	Environmental science	20	19.5	0.5
7	Physics and astronomy	17	21.9	−4.9
8	Biochemistry, genetics, and molecular biology	10	5.1	4.9
9	Agricultural and biological sciences	0	4.9	−4.9
10	Business, management, and accounting	0	3.3	−3.3
11	Mathematics	0	2.7	−2.7
12	Social sciences	0	2.6	−2.6
13	Computer science	1	2.2	−1.2
14	Economics, econometrics, and finance	0	2.0	−2.0
15	Earth and planetary sciences	0	1.6	−1.6
16	Immunology and microbiology	1	1.2	−0.2

78.3.9 The Most Prolific Keywords in Bioethanol Evaluation and Utilization: Hot Papers

Information about the keywords used in at least 5% of the sample or population papers each is given in Table 78.8. For this purpose, keywords related to the keyword set given in the appendices of Konur (2023a–j) are selected from a list of the most prolific keyword set provided by the Scopus database.

These keywords are grouped under nine headings: Bioethanol fuels, biomass, evaluation, experiences, fuel cells, engines, biochemicals, biohydrogen fuels, and sensors.

There are 15 keywords used related to bioethanol fuels: The prolific keywords are ethanol, ethanol fuels, direct ethanol fuel cells, ethanol oxidation reaction, bioethanol, and ethanol electro-oxidation with at least ten sample papers each. It is notable that bioethanol keyword appears in the sample paper keyword list with around 13% of the sample papers.

The prolific keywords related to biomass are biomass, cellulose, and lignocellulose with at least four papers each while those related to evaluation are costs, emissions, and energy efficiency with at least six sample papers each.

The prolific keyword related to experiences is Brazil, while those related to the fuel cells are catalysts, electrocatalysts, nanoparticles, oxidation, fuel cells, catalyst activity, electro-oxidation, and zinc oxide with at least ten sample papers each. Those related to engines are diesel engines, engines, combustion, carbon dioxide, exhaust emissions, biodiesel, nitrogen oxides, and carbon monoxide with at least 12 citations each while those related to biochemicals are methanol, diethyl ethers, acetone, butanol, and butenes with at least five citations each.

Finally, those related to biohydrogen fuels are hydrogen production, steam reforming, and hydrogen with at least four citations each while those related to sensors are gas detectors, chemical sensors, gas sensors, gas sensing properties, and sensing performance with at least five citations each.

TABLE 78.8
The Most Prolific Keywords in Bioethanol Evaluation and Utilization: Hot Papers

No.	Keywords	Sample Papers (%)	Population Papers (%)	Surplus (%)
1.	**Bioethanol Fuels**			
	Ethanol	99	82.7	16.3
	Ethanol fuels	25	12.8	12.2
	Direct ethanol fuel cells	15	5.4	9.6
	Ethanol oxidation reaction	14	6.6	7.4
	Bioethanol	13	12.4	0.6
	Ethanol electro-oxidation	10	5.5	4.5
	Ethanol oxidation	8	6.0	2.0
	Ethanol concentrations	6	3.0	3.0
	Ethanol steam reforming	5	3.8	1.2
2.	**Biomass**			
2.1.	Biomass	7	3.4	3.6
2.2.	Cellulose	4	1.9	2.1
2.3.	Lignocellulose	4	0.0	4.0
2.4.	Agricultural wastes	3	0.0	3.0
3.	**Evaluation**			
	Costs	9	2.8	6.2
	Emissions	8	3.9	4.1
	Energy efficiency	6	1.7	4.3
4.	**Experience**			
	Brazil	5	0.0	3.0
5.	**Fuel Cells**			
	Catalysts	26	11.5	14.5
	Electrocatalysts	19	7.1	11.9
	Nanoparticles	16	6.6	9.4
	Oxidation	15	10.3	4.7
	Fuel cells	15	5.4	9.6
	Catalyst activity	11	9.5	1.5
	Electro-oxidation	11	7.5	3.5
	Zinc oxide	10	3.7	6.3
	Nickel	8	3.9	4.1
	Platinum	7	4.7	2.3
	Palladium	7	4.1	2.9
	Catalyst selectivity	7	3.7	3.3
	Graphene	7	2.9	4.1
	Platinum alloys	7	2.1	4.9
	Alkaline fuel cells	6	0.0	6.0
	Catalytic oxidation	5	3.8	1.2
	Electrocatalytic activity	5	3.6	1.4
	Electrodes	5	3.2	1.8
	Catalysis	5	3.0	2.0
	Catalyst supports	5	0.9	4.1
	Hydrothermal synthesis	5	0.0	5.0
	Electrocatalysis	4	3.0	1.0

(Continued)

TABLE 78.8 (*Continued*)
The Most Prolific Keywords in Bioethanol Evaluation and Utilization: Hot Papers

No.	Keywords	Sample Papers (%)	Population Papers (%)	Surplus (%)
	Nanorods	4	1.6	2.4
	Alumina	4	1.5	2.5
	Nanowires	4	1.5	2.5
	Catalytic performance	3	2.9	0.1
	Electrocatalytic performance	3	1.6	1.4
	Catalyst deactivation	3	0.0	3.0
	Copper	3	0.0	3.0
	Nanocatalysts		2.9	–2.9
6.	**Engines**			
	Diesel engines	26	9.4	16.6
	Engines	17	4.9	12.1
	Combustion	14	8.1	5.9
	Carbon dioxide	14	6.1	7.9
	Exhaust emissions	14	0.0	14.0
	Biodiesel	13	3.2	9.8
	Nitrogen oxides	12	3.5	8.5
	Carbon monoxide	12	3.1	8.9
	Ignition	9	4.9	4.1
	Engine cylinders	9	2.8	6.2
	Gasoline	8	9.0	–1.0
	Direct injection	8	3.9	4.1
	Engine performance	8	1.7	6.3
	Fuel consumption	8	0.0	8.0
	Blending	6	3.5	2.5
	Diesel fuels	6	1.7	4.3
	Heat release rate	6	0.0	6.0
	Brake specific fuel consumption	5	0.0	5.0
	Brakes	5	0.0	5.0
	Single cylinder diesel engine	5	0.0	5.0
7.	**Biochemicals**			
	Methanol	7	3.6	3.4
	Diethyl ethers	7	0.0	7.0
	Acetone	6	3.3	2.7
	Butanol	6	0.0	6.0
	Butenes	5	0.0	5.0
	Ethylene		3.7	–3.7
8.	**Biohydrogen Fuels**			
	Hydrogen production	6	6.1	–0.1
	Steam reforming	6	6.1	–0.1
9.	**Sensors**			
	Gas detectors	11	6.5	4.5
	Chemical sensors	8	5.0	3.0
	Gas sensors	5	2.8	2.2
	Gas sensing properties	5	2.0	3.0
	Sensing performance	5	1.7	3.3

TABLE 78.9
The Evaluation and Utilization of Bioethanol Fuels: Research Fronts

No.	Research Fronts	I100 Sample (%)	Sample Papers (%)
1	Utilization of bioethanol fuels in transportation such as gasoline engines	14.3	28
2	Utilization of bioethanol fuels in fuel cells	31.4	22
3	Bioethanol Sensors	22.9	21
4	Other issues for the evaluation and utilization of bioethanol fuels	37.2	33
	Evaluation of bioethanol fuels	8.6	11
	Utilization of bioethanol fuels for biohydrogen production for fuel cells	14.3	9
	Utilization of bioethanol fuels for biochemical production	8.6	8
	Country-based experiences of bioethanol fuels.	5.7	5
	Sample size	735	100

Sample papers: the sample of the 100 most cited papers. I100 sample: the sample of papers with at least 100 citations each.

On the other hand, the keywords with the most citation impact are diesel engines, ethanol, catalysts, exhaust emissions, ethanol fuels, engines, and electrocatalysts with at least 11.9% surplus each. Similarly, the keywords with the least citation impact are ethylene, bio-ethanol production, greenhouse gases, gas emissions, nanocatalysts, and ethanol production.

78.3.10 The Most Prolific Research Fronts in Bioethanol Evaluation and Utilization

Information about the most prolific research fronts for the sample papers in bioethanol evaluation and utilization is given in Table 78.9. As Table 78.9 shows, there are three primary research fronts for these 100 most cited papers: Utilization of bioethanol fuels in transportation such as gasoline engines, utilization of bioethanol fuels in fuel cells, and bioethanol sensors with 28, 22, and 21 highly cited papers (HCPs) each. The other research fronts are evaluation of bioethanol fuels, utilization of bioethanol fuels for biohydrogen production for fuel cells, utilization of bioethanol fuels for biochemical production, and country-based experiences of bioethanol fuels with 11, 9, 8, and 5 HCPs, respectively. It is also notable that 43 HCPs were involved with the applications of nanotechnology in this field.

78.4 DISCUSSION

78.4.1 Introduction

Crude oil-based gasoline fuels have been widely used in the transportation sector since the 1920s. However, there have been great public concerns over the adverse environmental and human impact of these fuels. Hence, biomass-based bioethanol fuels have increasingly been used in blending gasoline fuels in fuel cells and in biochemical production in a biorefinery context.

Research in the field of utilization and evaluation of bioethanol fuels has also intensified in recent years. The primary focus of the research in this area has been the utilization of bioethanol fuels in fuel cells, evaluation of bioethanol fuels, utilization of bioethanol fuels in transport engines, production of biohydrogen fuels from bioethanol fuels, and to a lesser extent the experiences of countries, development and utilization of bioethanol sensors, and production of biochemicals from bioethanol fuels. Additionally, research on gasoline fuels and applications of nanotechnology has also been related to this field.

However, it is essential to develop efficient incentive structures for the primary stakeholders to enhance the research in this field. The scientometric analysis has been used in this context to inform the primary stakeholders about the current state of research in a selected research field.

As there have been no scientometric studies on bioethanol evaluation and utilization, this book chapter presents a scientometric study of the research in bioethanol evaluation and utilization. It examines the scientometric characteristics of both the sample and population data presenting scientometric characteristics of both datasets in the order of documents, authors, publication years, institutions, funding bodies, source titles, countries, Scopus subject categories, keywords, and research fronts.

As the first step for the search of the relevant literature, the keywords were selected using the first 200 most cited papers for each research front. The selected keyword list was optimized to obtain a representative sample of papers for the searched research fields (Konur, 2023a–j). This keyword list for each research front was collected to form a combined keyword set for all research fronts, and this combined set was provided in the appendices of Konur (2023a–j) for future replication studies.

As the second step, two sets of data were used for this study. First, a population sample of over 15,000 papers were used to examine the scientometric characteristics of the population data. Secondly, a sample of 100 most cited papers were used to examine the scientometric characteristics of these citation classics with over 69 citations each.

The scientometric characteristics of both sample and population datasets were presented in the order of documents, authors, publication years, institutions, funding bodies, source titles, countries, Scopus subject categories, keywords, and research fronts.

Lastly, key scientometric findings for both datasets were discussed to highlight the research landscape for bioethanol evaluation and utilization. Additionally, a number of brief conclusions were drawn and a number of relevant recommendations were made to enhance the future research landscape.

78.4.2 The Most Prolific Documents in Bioethanol Evaluation and Utilization

Articles and review papers dominate both the sample papers while articles and conference papers dominate population datasets (Table 78.1). Further, review papers and conference papers have a surplus and deficit, respectively.

Scopus differs from the Web of Science database in differentiating and showing articles and conference papers published in the journals separately. Similarly, Scopus differs from the Web of Science database in introducing short surveys. Hence, the total number of articles and review papers in the sample dataset are 88% and 12%, respectively.

It is observed during the search process that there has been inconsistency in the classification of documents in Scopus as well as in other databases such as Web of Science. This is especially relevant for the classification of papers as reviews or articles as the papers not involving a literature review may be erroneously classified as a review paper. There is also a case of review papers being classified as articles. For example, although there are 12 review papers and short surveys as classified by the Scopus database, nine of the sample papers are review papers based on the literature reviews. The primary reason of the over-indexing of the papers as reviews by the database is that some of the evaluative studies are classed as reviews.

In this context, it would be helpful to provide a classification note for the published papers in the books and journals at the first instance. It would also be helpful to use the document types listed in Table 78.1 for this purpose. Book chapters may also be classified as articles or reviews as an additional classification to differentiate review chapters from the experimental chapters as it is done by the Web of Science. It would be further helpful to additionally classify the conference papers as articles or review papers as well as it is done in the Web of Science database.

78.4.3 The Most Prolific Authors in Bioethanol Evaluation and Utilization

There have been 28 most prolific authors with at least two sample papers and four population papers each as given in Table 78.2a. These authors have shaped the development of the research in this field. Additionally, 32 authors with at least 12 population papers are also listed in Table 78.32b.

The most prolific authors are Harish and to a lesser extent Haifeng Liu, Yu Chen, Siti Kartom Kamarudin, Elena Pastor, Bo Jiang, Yanqiong Li, and Abdi H. Sebayang.

The most prolific institutions for the sample dataset are the Shanghai University of Electric Power, Tianjin University, and to a lesser extent Chinese Academy of Sciences, Dalian University of Technology, Jilin University, and University of Laguna.

The most prolific country for the sample dataset are China and to a lesser extent Malaysia and Spain. The most prolific research fronts are the utilization of bioethanol fuels in fuel cells and transport engines, and to a lesser extent the biohydrogen fuels and sensors.

It is important to note the inconsistencies in indexing of the author names in Scopus and other databases. It is especially an issue for the names with more than two components such as 'Judge Alex de Camp Pastor'. The probable outcomes are 'Pastor, J.A.D.C.', 'de Camp Pastor, J.A.', or 'Camp Pastor, J.A.D.'. The first choice is the gold standard of the publishing sector as the last word in the name is taken as the last name. In most of the academic databases such as Pubmed, EBSCO databases, this version is used predominantly. The second choice is a strong alternative while the last choice is an undesired outcome as two last words are taken as the last name. It is good practice to combine the words of the last name by a hyphen: 'Camp-Pastor, J.A.D.'. It is notable that inconsistent indexing of the author names may cause substantial inefficiencies in the search process for the papers as well as allocating credit to the authors as there are different author entries for each outcome in the databases.

There are also inconsistencies in the shortening Chinese names. For example, 'Yuoyang Liu' is often shortened as 'Liu, Y.', 'Liu, Y.-Y.', and 'Liu Y.Y.' as it is done in the Web of Science database as well. However, the gold stand in this case is 'Liu Y' where the last word is taken as the last name and the first word is taken as a single forename. In most of the academic databases such as Pubmed and EBSCO, this first version is used predominantly. However, it makes sense to use the third option to differentiate Chinese names efficiently: Liu Y.Y.'. Therefore, there have been difficulties to locate papers for the Chinese authors. In such cases, the use of the unique author codes provided for each author by the Scopus database has been helpful.

There is also a difficulty in allowing credit for the authors, especially for the authors with common names such as 'Wang, Y', or 'Huang, Y' or 'Zhu, Y.' in conducting scientometric studies. These difficulties strongly influence the efficiency of the scientometric studies as well as allocating credit to the authors as there are the same author entries for different authors with the same name, for example 'Wang Y', in the databases.

In this context, the coding of authors in the Scopus database is a welcome innovation compared to the other databases such as Web of Science. In this process, Scopus allocates a unique number to each author in the database (Aman, 2018). However, there might still be substantial inefficiencies in this coding system, especially for common names. For example, some of the papers for a certain author maybe allocated to another researcher with a different author code. It is possible that Scopus uses a number of software programs to differentiate the author names and the program may not be false-proof (Kim, 2018).

In this context, it does not help that author names are not given in full in some journals and books. This makes it difficult to differentiate authors with common names and makes the scientometric studies further difficult in the author domain. Therefore, the author names should be given in all books and journals at the first instance. There is also a cultural issue where some authors do not use their full names in their papers. Instead, they use initials for their forenames: 'Pastor, A.P.' or just 'Pastor' instead of 'Pastor, Alas Padras'.

There are also inconsistencies in naming of the authors with more than two components by the authors themselves in journal papers and book chapters. For example, 'Alaspanda, A.P.C.', 'Sakoura, C.E.', and 'Mentaslo, S.J.' might be given as 'Alaspanda, A', 'Sakoura, C', or 'Mentaslo, S.' in the journals and books. This also makes the scientometric studies difficult in the author domain. Hence, contributing authors should use their name consistently in their publications.

The other critical issue regarding the author names is the inconsistencies in the spelling of the author names in the national spellings (e.g., Çakırsöğüt, Übüz) rather than in the English spellings (e.g., Cakirsogut, Ubuz) in the Scopus database. Scopus differs from the Web of science database and many other databases in this respect where the author names are given only in the English spellings. It is observed that national spellings of the author names do not help in conducting scientometric studies as well in allocating credits to the authors as sometimes there are the different author entries for the English and National spellings in the Scopus database.

It is also notable that there is significant gender deficit for the sample dataset as surprisingly nearly all of these top researchers are male. This finding is the most thought-provoking with strong public policy implications. Hence, institutions, funding bodies, and policymakers should take efficient measures to reduce the gender deficit in this field as well as other scientific fields with strong gender deficit. In this context, it is worth noting the level of representation of the researchers from the minority groups in science on the basis of race, sexuality, age, and disability, besides the gender (Blankenship, 1993; Dirth and Branscombe, 2017; Konur, 2000, 2002a–c, 2016a,b, 2017a,b).

78.4.4 The Most Prolific Research Output by Years in Bioethanol Evaluation and Utilization

The research output observed between 2016 and 2021 is illustrated in Figure 78.1. This figure clearly shows that the bulk of the research papers in the population dataset were published primarily in the 2010s and the early 2020s with 66% and 34% of the population dataset, respectively. Similarly, the bulk of the research papers in the sample dataset were published in the 2010s.

In this context, the increasing public concerns about climate change (Change, 2007), greenhouse gas emissions (Carlson et al., 2017), and global warming (Kerr, 2007) have been certainly behind the boom in the research in this field in the last two decades.

Based on these findings, the size of the population papers likely to more than double in the current decade, provided that the public concerns about climate change, greenhouse gas emissions, and global warming are translated efficiently to the research funding in this field.

78.4.5 The Most Prolific Institutions in Bioethanol Evaluation and Utilization

The 21 most prolific institutions publishing papers on the bioethanol evaluation and utilization with at least two sample papers and 0.3% of the population papers each given in Table 78.3 have shaped the development of the research in this field. Additionally, 15 institutions with at least 0.7% of the population dataset are included in this table.

The most prolific institutions are the Chinese Academy of Sciences, and to a lesser extent Tianjin University, Tsinghua University, and the CSIC. The top countries for these most prolific institutions are China and to a lesser extent Malaysia and the USA.

On the other hand, the institutions with the most citation impact are CSIC and Tsinghua University. Similarly, the institutions with the least impact are Jilin University, Paulista State University, and University of Xiamen. These data suggest that the most productive institutions are from China, the USA, European countries, and Malaysia.

78.4.6 The Most Prolific Funding Bodies in Bioethanol Evaluation and Utilization

The 16 most prolific funding bodies funding at least two sample papers and 1.0% of the population papers each are given in Table 78.4. Additionally, five funding bodies with at least 1.0% of the population papers are listed in this table.

The most prolific funding bodies are the National Natural Science Foundation of China, and to a lesser extent Ministry of science and Technology and Ministry of Education of China. It is notable that 75% and 64% of the sample and population papers are funded, respectively.

The most prolific countries for these top funding bodies are China, and to a lesser extent Brazil, the USA and European Union. The funding bodies with the most citation impact are the National

Natural Science Foundation of China, and to a lesser extent Ministry of Science and Technology of China and the Ministry of Economics and Competitiveness of Spain. Similarly, the funding bodies with the least citation impact are the Sao Paulo State Research Foundation and the Ministry of Science Technology, and Innovation of Brazil.

These findings on the funding of research in this field suggest that the level of funding is highly intensive and it has been largely instrumental in enhancing the research in this field (Ebadi and Schiffauerova, 2016) in light of North's institutional framework (North, 1991). However, considering the relatively high levels of funding at the macroscale, there is still ample room to enhance funding in this field.

78.4.7 The Most Prolific Source Titles in Bioethanol Evaluation and Utilization

The 15 most prolific source titles publishing at least two sample papers and 0.5% of the population papers each in bioethanol evaluation and utilization have shaped the development of the research in this field (Table 78.5). Additionally, 11 journals with at least 0.5% of the population papers are also listed in this table.

The most prolific source titles are 'Sensors and Actuators B Chemical', 'Applied Catalysis B Environmental', 'ACS Applied Materials and Interfaces', and to a lesser extent 'Renewable and Sustainable Energy Reviews', ACS Catalysis', 'Fuel', and 'Energy Conversion and Management'.

On the other hand, the source titles with the most citation impact are the 'ACS Applied Materials and Interfaces', 'Applied Catalysis B Environmental', and to a lesser extent 'Sensors and Actuators B Chemical', 'ACS Catalysis', and 'Renewable and Sustainable Energy Reviews'. Similarly, the source titles with the least impact are the 'International Journal of Hydrogen Energy' and to a lesser extent 'Fuel' and 'Bioresource Technology'.

It is notable that these top source titles are related to the energy, sensors, and materials. This finding suggests that the journals in this field have significantly shaped the development of the research in this field as they focus on the energy- and materials-related aspects of bioethanol fuels.

78.4.8 The Most Prolific Countries in Bioethanol Evaluation and Utilization

The 18 most prolific countries publishing at least two papers and 1.1% of the population papers each have significantly shaped the development of the research in this field (Table 78.6). Additionally, ten countries with at least 1.1% of the population papers are listed in this table.

The most prolific countries are the China and to a lesser extent USA, India, Malaysia, and Spain. On the other hand, the countries with the most citation impact are the China, Malaysia, and to a lesser extent the USA, Spain, and Australia. Similarly, the countries with the least citation impact are Brazil and to a lesser extent Iran, India, and France.

Close examination of these findings suggests that China, the USA, Malaysia, and European countries are the major producers of the research in this field. It is a fact that the USA has been a major player in science (Leydesdorff and Wagner, 2009; Leydesdorff et al., 2014). The USA has further developed a strong research infrastructure to support its corn- and grass-based bioethanol industry (Vadas et al., 2008).

However, China has been a rising star in scientific research in competition with the USA and Europe (Leydesdorff and Zhou, 2005). China is also a major player in this field as a major producer of bioethanol (Li and Chan-Halbrendt, 2009).

Next, Europe has been a persistent player in the scientific research in competition with both the USA and China (Leydesdorff, 2000). Europe has also been a persistent producer of bioethanol along with the USA and Brazil (Gnansounou, 2010).

Additionally, Brazil has also been a persistent player in scientific research at a moderate level (Glanzel et al., 2006). Brazil has also developed a strong research infrastructure to support its biomass-based bioethanol industry (Macedo et al., 2008).

78.4.9 The Most Prolific Scopus Subject Categories in Bioethanol Evaluation and Utilization

The ten most prolific Scopus subject categories indexing at least 10% and 5.1% of the sample and population papers, respectively, given in Table 78.7 have shaped the development of the research in this field. Additionally, eight categories with at least 1.1% of the population paper set are listed in this table.

The most prolific Scopus subject categories in the bioethanol evaluation and utilization are 'Energy', 'Chemical Engineering', 'Engineering', and to a lesser extent 'Chemistry', Materials Science', Environmental Science', and 'Physics and Astronomy'. On the other hand, the Scopus subject categories with the most citation impact are the 'Energy', 'Chemical Engineering', and 'Materials Science'. Similarly, the Scopus subject categories with the least citation impact are 'Physics and Astronomy' and 'Chemistry'.

These findings are thought-provoking, suggesting that the primary subject categories are energy, chemical engineering, and engineering in general. The other key finding is that social sciences are relatively well represented in the population papers, unlike most fields in bioethanol fuels. These findings are not surprising as one of the key research fronts in this field is the evaluative studies on the bioethanol fuels which used social-science-oriented analytical tools such as technoeconomic and life cycle tools.

78.4.10 The Most Prolific Keywords in Bioethanol Evaluation and Utilization

A limited number of keywords have shaped the development of the research in this field as shown in Table 78.8 and the appendices of Konur (2023a–j).

These keywords are grouped under the nine headings: Bioethanol fuels, biomass, evaluation, experiences, fuel cells, engines, biochemicals, biohydrogen fuels, and sensors.

The findings suggest that it is necessary to determine the keyword set carefully to locate the relevant research in each of these research fronts. Additionally, the size of the samples for each keywords highlights the intensity of the research in the relevant research areas for both sample and population papers.

78.4.11 The Most Prolific Research Fronts in Bioethanol Evaluation and Utilization

As Table 78.9 shows, there are three primary research fronts for these 100 most cited papers: Utilization of bioethanol fuels in transportation such as gasoline engines, utilization of bioethanol fuels in fuel cells, and bioethanol sensors. The other research fronts are evaluation of bioethanol fuels, utilization of bioethanol fuels for biohydrogen production for fuel cells, utilization of bioethanol fuels for biochemical production, and country-based experiences of bioethanol fuels.

In recent years, bioethanol fuels have emerged as primary fuels for fuel cells (Antolini, 2007). The research in this field has focused on the development of catalysts and catalyst supports for the oxidation of bioethanol fuels for fuel cells. Due to the strong demand for the environment- and human-friendly transport vehicles, the demand for the extensive research in this field will likely continue in the future.

As bioethanol fuels have been a strong alternative for the crude-oil based gasoline fuels, the studies evaluating bioethanol fuels at large have been another strong research field (Farrell et al., 2006; Hamelinck et al., 2005; Hill et al., 2006). The research in this field has focused on the evaluation of the technoeconomic and environmental performance of bioethanol fuels and their derivatives. Due to the strong demand for the environment- and human-friendly transport vehicles, the demand for the extensive research in this field will likely continue in the future as well.

The use of biohydrogen fuels produced from bioethanol fuels in fuel cells as an alternative to direct bioethanol fuels has increased in recent years (Haryanto et al., 2005). The research in this field has focused on the development of catalysts and catalyst supports for the biohydrogen fuel production for the fuel cells. Due to the strong demand for the environment- and human-friendly transport vehicles, the demand for the extensive research in this field will likely continue in the future as well.

The main use of bioethanol fuels has been in the transport engines such as gasoline or diesel engines (Hansen et al., 2005; Kohse-Hoinghaus et al., 2010). The research in this field has focused on the combustion, performance, and emissions of bioethanol fuels blended with crude oil-based gasoline or petrodiesel fuels. Due to the strong demand for the environment- and human-friendly transport vehicles, the demand for the extensive research in this field will likely continue in the future as well.

Another intense research field has been the development and use of bioethanol fuel sensors (Liu et al., 2005; Wan et al., 2005). The research in this field has focused on the development of nanomaterials and conventional materials for the development of sensors. Due to the strong demand for the environment- and human-friendly transport vehicles, the demand for the extensive research in this field will likely continue in the future as well.

Since the production of biochemicals in the biorefinery context has been a way to reduce the total production cost of bioethanol fuels, the research in this area has also intensified in recent years (Angelici et al., 2013; Liu and Hensen, 2013). A wide range of biochemicals have been produced from bioethanol fuels using a number of processes. Due to the strong demand for the environment- and human-friendly biochemicals, the demand for the extensive research in this field will likely continue in the future as well.

Finally, a strong stream of the research in this field has focused on the country-based experiences of bioethanol fuels (Macedo et al., 2008; Sheehan et al., 2003). The research in this field has been primarily related to the major producers and users of bioethanol fuels such as Brazil, Europe, China, and the USA. These studies have highlighted the development of the efficient incentive structures in these countries to improve the research on bioethanol fuels as well as bioethanol production. Due to the strong demand for the environment- and human-friendly bioethanol fuels, the demand for the extensive research in this field will likely continue in the future as well.

It is also notable that 43 HCPs were involved with the applications of nanotechnology in this field. The related fields are the development and utilization of ethanol nanosensors, utilization of bioethanol fuels in transportation such as gasoline engines and fuel cells, bioethanol sensors, and the production of biochemicals and biohydrogen from bioethanol fuels.

The related findings are thought-provoking in seeking ways to optimize the bioethanol utilization and evaluation at the global scale. It is clear that all of these research fronts have public importance and merit substantial funding and other incentives.

In the end, these most cited papers in this field hint that the efficiency of bioethanol fuels and their derivatives could be optimized using the structure, processing, and property relationships of bioethanol fuels and their derivatives (Formela et al., 2016; Konur, 2018a, 2020b, 2021a–d; Konur and Matthews, 1989).

78.5 CONCLUSION AND FUTURE RESEARCH

The research on the bioethanol evaluation and utilization has been mapped through a scientometric study of both sample and population datasets.

The critical issue in this study has been to obtain a representative sample of the research as in any other scientometric study. Therefore, the keyword set has been carefully devised and optimized after a number of runs in the Scopus database. It is a representative sample of the wider population studies.

The other issue has been the selection of a multidisciplinary database to carry out the scientometric study of the research in this field. For this purpose, the Scopus database has been selected. The journal coverage of this database has been wider than that of Web of Science.

The key scientometric properties of the research in this field have been determined and discussed in this book chapter. It is evident that a limited number of documents, authors, institutions, publication periods, institutions, funding bodies, source titles, countries, Scopus subject categories, keywords, and research fronts have shaped the development of the research in this field.

There is ample scope to increase the efficiency of the scientometric studies in this field in the author and document domains by developing consistent policies and practices in both domains across all the academic databases. In this respect, it seems that authors, journals, and academic databases have a lot to do. Furthermore, the significant gender deficit as in most scientific fields emerges as a public policy issue. The potential deficits on the basis of age, race, disability, and sexuality need also to be explored in this field as in other scientific fields.

The institutions from China, Malaysia, the USA, and European countries have mostly shaped the research in this field. The relatively high funding rate of 75% and 64% for sample and population papers, respectively, suggests that funding in this field significantly enhanced the research in this field, possibly more than doubling in the current decade. However, it is evident that there is ample room for more funding and other incentives to further enhance the research in this field. It is also notable that the funding rate is relatively high to that for the bioethanol production, which usually is below 50%. This suggests the strategic importance attributed to the studies on the utilization and evaluation of bioethanol fuels compared to the production of bioethanol fuels.

The most prolific journals have been mostly related to energy, sensors, and materials as the focus of the sample papers has been on the development of catalysts and catalyst supports for the development and utilization of bioethanol derivatives.

China, Europe, and to a lesser extent USA, India, and Malaysia have been the major producers of the research in this field as the major producers and users of bioethanol fuels from different types of biomass such as corn, sugarcane, and grass as well as other types of biomass. It is evident that these countries have well-developed research infrastructure in bioethanol fuels and their derivatives. However, the dominance of this field by China merits further attention as China has heavily funded the research particularly on the derivatives of bioethanol fuels.

The primary subject categories have been energy, chemical engineering, and engineering in general. The other key finding is that social sciences are relatively well represented in the population papers, unlike most fields in bioethanol fuels. These findings are not surprising as one of the key research fronts in this field is the evaluative studies on bioethanol fuels which used social-science-oriented analytical tools such as technoeconomic and life cycle tools. However, it is notable that some social-science-based studies such as user, economic, and scientometric studies are not well presented in this field.

It emerges that ethanol is more popular than bioethanol as a keyword with strong implications for the search strategy. In other words, the search strategy using only bioethanol keyword would not be much helpful. It is also recommended that the term ‘ bioethanol fuels’ rather than ‘ethanol’, ‘bio-ethanol’, ‘bioethanol’, and ‘ethanol fuels’ is used in the titles of the papers by their authors to improve the search efficiency and hence the healthy development of the research in this field.

The keywords indexed by the Scopus database are grouped under the nine headings: Bioethanol fuels, biomass, evaluation, experiences, fuel cells, engines, biochemicals, biohydrogen fuels, and sensors. These groups of keywords highlight the potential primary research fronts for these fields for both sample and population papers.

These findings are thought-provoking. The focus of these 100 most cited papers is utilization of bioethanol fuels in transportation such as gasoline engines, utilization of bioethanol fuels in fuel cells, bioethanol sensors, and to a lesser extent evaluation of bioethanol fuels, utilization of bioethanol fuels for biohydrogen production for fuel cells, utilization of bioethanol fuels for biochemical production, and country-based experiences of bioethanol fuels. These studies highlight strong structure–processing–property relationships for bioethanol fuels and their derivatives.

Thus, the scientometric analysis has a great potential to gain valuable insights into the evolution of the research in this field as in other scientific fields.

It is recommended that further scientometric studies are carried out about the other aspects of both production and utilization of bioethanol fuels. It is further recommended that reviews of the most cited papers are carried out for each research front to complement these scientometric studies. Next, the scientometric studies of the hot papers in these primary fields are carried out.

ACKNOWLEDGMENTS

The contribution of the highly cited researchers in the field of bioethanol evaluation and utilization has been gratefully acknowledged.

REFERENCES

Aman, V. 2018. Does the Scopus author ID suffice to track scientific international mobility? A case study based on Leibniz laureates. *Scientometrics* 117:705–720.
Angelici, C., B. M. Weckhuysen and P. C. A. Bruijnincx. 2013. Chemocatalytic conversion of ethanol into butadiene and other bulk chemicals. *ChemSusChem* 6:1595–1614.
Antolini, E. 2007. Catalysts for direct ethanol fuel cells. *Journal of Power Sources* 170:1–12.
Antolini, E. 2009. Palladium in fuel cell catalysis. *Energy and Environmental Science* 2:915–931.
Beaudry, C. and V. Lariviere. 2016. Which gender gap? Factors affecting researchers' scientific impact in science and medicine. *Research Policy* 45:1790–1817.
Blankenship, K. M. 1993. Bringing gender and race in: US employment discrimination policy. *Gender & Society* 7:204–226.
Burnham, J. F. 2006. Scopus database: A review. *Biomedical Digital Libraries* 3:1–8.
Carlson, K. M., J. S. Gerber and D. Mueller, et al. 2017. Greenhouse gas emissions intensity of global croplands. *Nature Climate Change* 7:63–68.
Change, C. 2007. Climate change impacts, adaptation and vulnerability. *Science of the Total Environment* 326:95–112.
Dirth, T. P. and N. R. Branscombe. 2017. Disability models affect disability policy support through awareness of structural discrimination. *Journal of Social Issues* 73:413–442.
Ebadi, A. and A. Schiffauerova. 2016. How to boost scientific production? A statistical analysis of research funding and other influencing factors. *Scientometrics* 106:1093–1116.
Farrell, A. E., R. J. Plevin and B. T. Turner, et al. 2006. Ethanol can contribute to energy and environmental goals. *Science* 311:506–508.
Fernando, S., S. Adhikari, C. Chandrapal and M. Murali. 2006. Biorefineries: Current status, challenges, and future direction. *Energy & Fuels* 20:1727–1737.
Formela, K., A. Hejna, L. Piszczyk, M. R. Saeb and X. Colom. 2016. Processing and structure-property relationships of natural rubber/wheat bran biocomposites. *Cellulose* 23:3157–3175.
Garfield, E. 1955. Citation indexes for science. *Science* 122:108–111.
Glanzel, W., J. Leta and B. Thijs. 2006. Science in Brazil. Part 1: A macro-level comparative study. *Scientometrics* 67:67–86.
Gnansounou, E. 2010. Production and use of lignocellulosic bioethanol in Europe: Current situation and perspectives. *Bioresource Technology* 101:4842–4850.
Hamelinck, C. N., G. van Hooijdonk and A. P. C. Faaij. 2005. Ethanol from lignocellulosic biomass: Techno-economic performance in short-, middle- and long-term. *Biomass and Bioenergy* 28:384–410.
Hansen, A. C., Q. Zhang and P. W. L. Lyne. 2005. Ethanol-diesel fuel blends: A review. *Bioresource Technology* 96:277–285.
Haryanto, A., S. Fernando, N. Murali and S. Adhikari. 2005. Current status of hydrogen production techniques by steam reforming of ethanol: A review. *Energy and Fuels* 19:2098–2106.
Hill, J., E. Nelson, D. Tilman, S. Polasky and D. Tiffany. 2006. Environmental, economic, and energetic costs and benefits of biodiesel and ethanol biofuels. *Proceedings of the National Academy of Sciences of the United States of America* 103:11206–11210.
Hill, J., S. Polasky and E. Nelson, et al. 2009. Climate change and health costs of air emissions from biofuels and gasoline. *Proceedings of the National Academy of Sciences of the United States of America* 106:2077–2082.
Hsieh, W. D., R. H. Chen, T. L. Wu and T. H. Lin. 2002. Engine performance and pollutant emission of an SI engine using ethanol-gasoline blended fuels. *Atmospheric Environment* 36:403–410.

Huang, H. J., S. Ramaswamy, U. W. Tschirner and B. V. Ramarao. 2008. A review of separation technologies in current and future biorefineries. *Separation and Purification Technology* 62:1–21.

Kerr, R. A. 2007. Global warming is changing the world. *Science* 316:188–190.

Khalili, N. R., P. A. Scheff and T. M. Holsen. 1995. PAH source fingerprints for coke ovens, diesel and, gasoline engines, highway tunnels, and wood combustion emissions. *Atmospheric Environment* 29:533–542.

Kim, J. 2018. Evaluating author name disambiguation for digital libraries: A case of DBLP. *Scientometrics* 116:1867–1886.

Kohse-Hoinghaus, K., P. Osswald and T. A. Cool, et al. 2010. Biofuel combustion chemistry: From ethanol to biodiesel. *Angewandte Chemie: International Edition* 49:3572–3597.

Konur, O. 2000. Creating enforceable civil rights for disabled students in higher education: An institutional theory perspective. *Disability & Society* 15:1041–1063.

Konur, O. 2002a. Access to nursing education by disabled students: Rights and duties of nursing programs. *Nurse Education Today* 22:364–374.

Konur, O. 2002b. Assessment of disabled students in higher education: Current public policy issues. *Assessment and Evaluation in Higher Education* 27:131–152.

Konur, O. 2002c. Access to employment by disabled people in the UK: Is the Disability Discrimination Act working? *International Journal of Discrimination and the Law* 5:247–279.

Konur, O. 2006a. Participation of children with dyslexia in compulsory education: Current public policy issues. *Dyslexia* 12:51–67.

Konur, O. 2006b. Teaching disabled students in higher education. *Teaching in Higher Education* 11:351–363.

Konur, O. 2007a. A judicial outcome analysis of the Disability Discrimination Act: A windfall for the employers? *Disability & Society* 22:187–204.

Konur, O. 2007b. Computer-assisted teaching and assessment of disabled students in higher education: The interface between academic standards and disability rights. *Journal of Computer Assisted Learning* 23:207–219.

Konur, O. 2011. The scientometric evaluation of the research on the algae and bio-energy. *Applied Energy* 88:3532–3540.

Konur, O. 2012a. Prof. Dr. Ayhan Demirbas' scientometric biography. *Energy Education Science and Technology Part A: Energy Science and Research* 28:727–738.

Konur, O. 2012b. The evaluation of the biogas research: A scientometric approach. *Energy Education Science and Technology Part A: Energy Science and Research* 29:1277–1292.

Konur, O. 2012c. The evaluation of the global energy and fuels research: A scientometric approach. *Energy Education Science and Technology Part A: Energy Science and Research* 30:613–628.

Konur, O. 2012d. The evaluation of the research on the biodiesel: A scientometric approach. *Energy Education Science and Technology Part A: Energy Science and Research* 28:1003–1014.

Konur, O. 2012e. The evaluation of the research on the bioethanol: A scientometric approach. *Energy Education Science and Technology Part A: Energy Science and Research* 28:1051–1064.

Konur, O. 2012f. The evaluation of the research on the biofuels: A scientometric approach. *Energy Education Science and Technology Part A: Energy Science and Research* 28:903–916.

Konur, O. 2012g. The evaluation of the research on the biohydrogen: A scientometric approach. *Energy Education Science and Technology Part A: Energy Science and Research* 29:323–338.

Konur, O. 2012h. The evaluation of the research on the microbial fuel cells: A scientometric approach. *Energy Education Science and Technology Part A: Energy Science and Research* 29:309–322.

Konur, O. 2012i. The scientometric evaluation of the research on the production of bioenergy from biomass. *Biomass and Bioenergy* 47:504–515.

Konur, O. 2015. Current state of research on algal bioethanol. In *Marine Bioenergy: Trends and Developments*, Eds. S. K. Kim and C. G. Lee, pp. 217–244. Boca Raton, FL: CRC Press.

Konur, O., Ed. 2018a. *Bioenergy and Biofuels*. Boca Raton, FL: CRC Press, Boca Raton, FL.

Konur, O. 2018b. Bioenergy and biofuels science and technology: Scientometric overview and citation classics. In *Bioenergy and Biofuels*, Ed. O. Konur, pp. 3–63. Boca Raton, FL: CRC Press.

Konur, O. 2019. Cyanobacterial bioenergy and biofuels science and technology: A scientometric overview. In *Cyanobacteria: From Basic Science to Applications*, Eds. A. K. Mishra, D. N. Tiwari and A. N. Rai, pp. 419–442. Amsterdam: Elsevier.

Konur, O. 2020a. The scientometric analysis of the research on the bioethanol production from green macroalgae. In *Handbook of Algal Science, Technology and Medicine,* Ed. O. Konur, pp. 385–401. London: Academic Press.

Konur, O., Ed. 2020b. *Handbook of Algal Science, Technology and Medicine.* London: Academic Press.

Konur, O., Ed. 2021a. *Handbook of Biodiesel and Petrodiesel Fuels: Science, Technology, Health, and Environment*. Boca Raton, FL: CRC Press.

Konur, O., Ed. 2021b. *Handbook of Biodiesel and Petrodiesel Fuels: Science, Technology, Health, and Environment. Volume 1. Biodiesel Fuels: Science, Technology, Health, and Environment.* Boca Raton, FL: CRC Press.

Konur, O., Ed. 2021c. *Handbook of Biodiesel and Petrodiesel Fuels: Science, Technology, Health, and Environment. Volume 2. Biodiesel Fuels Based on the Edible and Nonedible Feedstocks, Wastes, and Algae: Science, Technology, Health, and Environment.* Boca Raton, FL: CRC Press.

Konur, O., Ed. 2021d. *Handbook of Biodiesel and Petrodiesel Fuels: Science, Technology, Health, and Environment. Volume 3. Petrodiesel Fuels: Science, Technology, Health, and Environment.* Boca Raton, FL: CRC Press.

Konur, O. 2023a. Gasoline fuels: Scientometric study. In *Evaluation and Utilization of Bioethanol Fuels. I.: Evaluation of Bioethanol Fuels, Transport Engines, and Bioethanol Sensors. Handbook of Bioethanol Fuels Volume 5*, Ed. O. Konur, pp. 87–106. Boca Raton, FL: CRC Press.

Konur, O. 2023b. Nanotechnology applications in bioethanol fuels: Scientometric study. In *Evaluation and Utilization of Bioethanol Fuels. I.: Evaluation of Bioethanol Fuels, Transport Engines, and Bioethanol Sensors. Handbook of Bioethanol Fuels Volume 5*, Ed. O. Konur, pp. 120–139. Boca Raton, FL: CRC Press.

Konur, O. 2023c. Utilization of bioethanol fuels in the transport engines: Scientometric study. In *Evaluation and Utilization of Bioethanol Fuels. I.: Evaluation of Bioethanol Fuels, Transport Engines, and Bioethanol Sensors. Handbook of Bioethanol Fuels Volume 5*, Ed. O. Konur, pp. 157–175. Boca Raton, FL: CRC Press.

Konur, O. 2023d. Evaluation of bioethanol fuels: Scientometric study. In *Evaluation and Utilization of Bioethanol Fuels. I.: Evaluation of Bioethanol Fuels, Transport Engines, and Bioethanol Sensors. Handbook of Bioethanol Fuels Volume 5*, Ed. O. Konur, pp. 195–213. Boca Raton, FL: CRC Press.

Konur, O. 2023e. Utilization of bioethanol fuels: Scientometric study. In *Evaluation and Utilization of Bioethanol Fuels. I.: Evaluation of Bioethanol Fuels, Transport Engines, and Bioethanol Sensors. Handbook of Bioethanol Fuels Volume 5*, Ed. O. Konur, pp. 271–295. Boca Raton, FL: CRC Press.

Konur, O. 2023f. Bioethanol fuel sensors: Scientometric study. In *Evaluation and Utilization of Bioethanol Fuels. I.: Evaluation of Bioethanol Fuels, Transport Engines, and Bioethanol Sensors. Handbook of Bioethanol Fuels Volume 5*, Ed. O. Konur, pp. 317–334. Boca Raton, FL: CRC Press.

Konur, O. 2023g. Country-based experience of bioethanol fuels: Review. In *Evaluation and Utilization of Bioethanol Fuels. II.: Biohydrogen Fuels, Fuel Cells, Biochemicals, and Country Experiences. Handbook of Bioethanol Fuels Volume 6*, Ed. O. Konur, pp. 26–41. Boca Raton, FL: CRC Press.

Konur, O. 2023h. Bioethanol fuel-based biohydrogen fuels: Scientometric study. In *Evaluation and Utilization of Bioethanol Fuels. II.: Biohydrogen Fuels, Fuel Cells, Biochemicals, and Country Experiences. Handbook of Bioethanol Fuels Volume 6*, Ed. O. Konur, pp. 215–236. Boca Raton, FL: CRC Press.

Konur, O. 2023i. Bioethanol fuel cells: Scientometric study. In *Evaluation and Utilization of Bioethanol Fuels. II.: Biohydrogen Fuels, Fuel Cells, Biochemicals, and Country Experiences. Handbook of Bioethanol Fuels Volume 6*, Ed. O. Konur, pp. 277–297. Boca Raton, FL: CRC Press.

Konur, O. 2023j. Bioethanol fuel-based biochemical production: Scientometric study. In *Evaluation and Utilization of Bioethanol Fuels. II.: Biohydrogen Fuels, Fuel Cells, Biochemicals, and Country Experiences. Handbook of Bioethanol Fuels Volume 6*, Ed. O. Konur, pp. 317–337. Boca Raton, FL: CRC Press.

Konur, O. and F. L. Matthews. 1989. Effect of the properties of the constituents on the fatigue performance of composites: A review. *Composites* 20:317–328.

Leydesdorff, L. 2000. Is the European Union becoming a single publication system? *Scientometrics* 47:265–280.

Leydesdorff, L. and C. Wagner. 2009. Is the United States losing ground in science? A global perspective on the world science system. *Scientometrics* 78:23–36.

Leydesdorff, L. and P. Zhou. 2005. Are the contributions of China and Korea upsetting the world system of science? *Scientometrics* 63:617–630.

Leydesdorff, L., C. S. Wagner and L. Bornmann. 2014. The European Union, China, and the United States in the top-1% and top-10% layers of most-frequently cited publications: Competition and collaborations. *Journal of Informetrics* 8:606–617.

Li, S. Z. and C. Chan-Halbrendt. 2009. Ethanol production in (the) People's Republic of China: Potential and technologies. *Applied Energy* 86:S162–S169.

Liang, Z. X., T. S. Zhao, J. B. Xu and L. D. Zhu. 2009. Mechanism study of the ethanol oxidation reaction on palladium in alkaline media. *Electrochimica Acta* 54:2203–2208.

Liu, J., X. Wang, Q. Peng and Y. Li. 2005. Vanadium pentoxide nanobelts: Highly selective and stable ethanol sensor materials. *Advanced Materials* 17:764–767.

Liu, P. and E. J. M. Hensen. 2013. Highly efficient and robust Au/$MgCuCr_2O_4$ catalyst for gas-phase oxidation of ethanol to acetaldehyde. *Journal of the American Chemical Society* 135:14032–14035.
Ma, X., L. Sun and C. Song. 2002. A new approach to deep desulfurization of gasoline, diesel fuel and jet fuel by selective adsorption for ultra-clean fuels and for fuel cell applications. *Catalysis Today* 77:107–116.
Macedo, I. C., J. E. A. Seabra and J. E. A. R. Silva. 2008. Green house gases emissions in the production and use of ethanol from sugarcane in Brazil: The 2005/2006 averages and a prediction for 2020. *Biomass and Bioenergy* 32:582–595.
Morschbacker, A. 2009. Bio-ethanol based ethylene. *Polymer Reviews* 49:79–84.
Murdoch, M., G. I. N. Waterhouse and M. A. Nadeem, et al. 2011. The effect of gold loading and particle size on photocatalytic hydrogen production from ethanol over Au/TiO_2 nanoparticles. *Nature Chemistry* 3:489–492.
Najafi, G., B. Ghobadian and T. Tavakoli, et al. 2009. Performance and exhaust emissions of a gasoline engine with ethanol blended gasoline fuels using artificial neural network. *Applied Energy* 86:630–639.
Newman, P. W. G. and J. R. Kenworthy. 1989. Gasoline consumption and cities: A comparison of U.S. cities with a global survey. *Journal of the American Planning Association* 55:24–37.
Ni, M., D. Y. C. Leung and M. K. H. Leung. 2007. A review on reforming bio-ethanol for hydrogen production. *International Journal of Hydrogen Energy* 32:3238–3247.
North, D. C. 1991. Institutions. *Journal of Economic Perspectives* 5:97–112.
Sheehan, J., A. Aden and K. Paustian, et al. 2003. Energy and environmental aspects of using corn stover for fuel ethanol. *Journal of Industrial Ecology* 7:117–146.
Song, C. 2003. An overview of new approaches to deep desulfurization for ultra-clean gasoline, diesel fuel and jet fuel. *Catalysis Today* 86:211–263.
Vadas, P. A., K. H. Barnett and D. J. Undersander 2008. Economics and energy of ethanol production from alfalfa, corn, and switchgrass in the Upper Midwest, USA. *Bioenergy Research* 1:44–55.
Wan, Q., Q. H. Li and Y. J. Chen, et al. 2004. Fabrication and ethanol sensing characteristics of ZnO nanowire gas sensors. *Applied Physics Letters* 84:3654–3656.
Wang, H., Z. Jusys and R. J. Behm. 2004. Ethanol electrooxidation on a carbon-supported Pt catalyst: Reaction kinetics and product yields. *Journal of Physical Chemistry B* 108:19413–19424.

79 Bioethanol Fuel Evaluation and Utilization
Hot Papers, Review

Ozcan Konur
(Formerly) Ankara Yildirim Beyazit University

79.1 INTRODUCTION

Crude oil-based gasoline fuels (Ma et al., 2002; Newman and Kenworthy, 1989) have been widely used in the transportation sector since the 1920s. However, there have been great public concerns over the adverse environmental and human impact of these fuels (Hill et al., 2006, 2009). Hence, biomass-based bioethanol fuels (Hill et al., 2006; Konur, 2012, 2015, 2019, 2020a) have increasingly been used in blending gasoline fuels (Hsieh et al., 2002; Najafi et al., 2009), in fuel cells (Antolini, 2007, 2009), in biochemical production (Angelici et al., 2013; Morschbacker, 2009), and in biohydrogen fuel production (Haryanto et al., 2005; Murdoch et al., 2011) in a biorefinery context (Fernando et al., 2006; Huang et al., 2008).

Research in the fields of experiences of countries of bioethanol fuels (Macedo et al., 2008; Sheehan et al., 2003), evaluation of bioethanol fuels (Farrell et al., 2006; Hamelinck et al., 2005), production of biohydrogen fuels (Haryanto et al., 2005; Murdoch et al., 2011) and biochemicals (Angelici et al., 2013; Liu and Hensen, 2013) from bioethanol fuels, utilization of bioethanol fuels in transportation such as gasoline and diesel engines (Hansen et al., 2005; Hsieh et al., 2002; Kohse-Hoinghaus et al., 2010) and fuel cells (Antolini, 2007; Liang et al., 2009), and bioethanol sensors (Liu et al., 2005; Wan et al., 2004; Wang et al., 2012) has also intensified in recent years.

However, it is essential to develop efficient incentive structures (North, 1991) for the primary stakeholders to enhance the research in this field (Konur, 2000, 2002a–c, 2006a,b, 2007a,b). Although there have been a number of review papers on the evaluation and utilization of bioethanol fuels (Antolini, 2007; Hansen et al., 2005; Haryanto et al., 2005; Kohse-Hoinghaus et al., 2010), there has been no review of the 25 most cited articles in this field published between 2016 and 2021.

Thus, this book chapter presents a review of these 25 most cited articles in these fields published between 2016 and 2021 to highlight current research trends in these fields. Then, it discusses the key findings of these highly influential papers and comments on the future research priorities in this field.

79.2 MATERIALS AND METHODS

The search for this study was carried out using the Scopus database (Burnham, 2006) in October 2021.

As the first step for the search of the relevant literature, the keywords were selected using the first 200 most cited papers for each research front. The selected keyword list was optimized to obtain a representative sample of papers for each searched research field (Konur, 2023a–j). This keyword list for each research front was collected to form a combined keyword set for all research fronts, and these keyword sets were provided in the appendices of Konur (2023a–j) for future replication studies.

As the second step, a sample dataset was used for this study. The first 25 articles in the sample of 100 most cited papers with at least 101 citations each were selected for the review study.

 DOI: 10.1201/9781003226567-107

Key findings from each paper were taken from the abstracts of these papers and were discussed. Additionally, a number of brief conclusions were drawn and a number of relevant recommendations were made to enhance the future research landscape.

79.3 RESULTS

The brief information about 25 most cited hot papers with at least 101 citations each on the evaluation and utilization of bioethanol fuels is given below.

79.3.1 Utilization of Bioethanol Fuels in Fuel Cells: Hot Papers

The brief information about ten hot papers with at least 101 citations each on the utilization of bioethanol fuels in fuel cells is given in Table 79.1. Furthermore, the brief notes on the contents of these studies are also given below.

Chen et al. (2017) evaluated the bioethanol electrooxidation performance by 5-nm-sized Pd–Ni–P ternary nanoparticles (NPs) in a paper with 242 citations. They shortened the distance between Pd and Ni active sites for the Ni/Pd atomic ratio of 1:1. They found that the electrocatalytic activity was enhanced significantly compared to the Pd/C catalysts. They asserted that this improved electrocatalytic activity and stability were due to the promoted production of free OH radicals on Ni active sites, which facilitated the oxidative removal of carbonaceous poison and combination with CH_3CO radicals on adjacent Pd active sites.

Jiang et al. (2016) evaluated the bioethanol electrooxidation performance by ordered PdCu-based NPs in a paper with 206 citations. They developed PdCuM NPs with M=Co and Ni. They found that the ordered PdCuCo NPs exhibited better electrocatalytic activity and much enhanced stability

TABLE 79.1
Utilization of Bioethanol Fuels in Fuel Cells

No.	Papers	Catalysts	Parameters	Country	Cits.
1	Chen et al. (2017)	Pd–Ni–P NPs	Nanocatalysts, shortening of distance between Pd and Ni active sites	China	242
2	Jiang et al. (2016)	PdCuM NPs	Nanocatalysts, ordering effect, Cu and Ni effect	China	206
3	Mao et al. (2017)	Pt–Mo–Ni NWs	Ultrathin nanocatalyst effects	China	150
4	Huang et al. (2017)	Pd NCs, $Ni(OH)_2$ nanoflakes, graphene	Nanocatalysts, $Ni(OH)_2$ effect	China	135
5	Han et al. (2018)	PtRhCu CNBs	Nanocatalysts, composition effect, BOR	China	133
6	Huang et al. (2018)	2D PdAg alloy nanodendrites	Nanocatalysts, electrocatalytic activity, operation stability	China	124
7	Wu et al. (2018)	Pd NPs/TiO_2–BP nanoflakes, Pd/C	Nanocatalysts, catalyst support effect, electroactivity, durability, stability	China	119
8	Rizo et al. (2018)	Pt–Sn NPs	Nanocatalysts, morphology effect (core, subsurface, and skin)	Germany	102
9	Jana et al. (2016)	Pd_3Pb NCs, C/Pd	Nanocatalysts, surfactant effect, ordering effect	India	102
10	Qi et al. (2016)	PdAg/CNT	Nanocatalysts, mechanisms	USA	101

Cits.: The number of the citations received by each paper.

compared to disordered PdCuM NPs as well as the commercial Pt/C and Pd/C catalysts. They asserted that this improved electrocatalytic activity was due to the catalytically active hollow sites arising from the ligand effect and the compressive strain on the Pd surface as Cu, Co, and Ni had a smaller atomic size.

Mao et al. (2017) evaluated the bioethanol electrooxidation performance using ultrathin Pt–Mo–Ni nanowire (NW) catalysts in a paper with 150 citations. They developed these NWs with a diameter of 2.5 nm and lengths of up to several micrometers via an H_2-assisted solution route (HASR). They noted that these NWs with high numbers of surface atoms could increase the atomic efficiency of Pt and thus decrease the catalyst cost. Further, the incorporation of Ni could isolate Pt atoms on the surface and produce surface defects, leading to high catalytic activity. Finally, the incorporation of Mo could stabilize both Ni and Pt atoms, leading to high catalytic stability. They recommended that this HASR strategy could be used to synthesize a series of Pt–Mo–M (M=Fe, Co, Mn, Ru, etc.) NWs.

Huang et al. (2017) evaluated the promoting effect of $Ni(OH)_2$ on palladium (Pd) nanocrystals (NCs) on the bioethanol electrooxidation performance in alkaline solution to greatly improve operation durability in a paper with 135 citations. They obtained a hybrid electrocatalyst consisting of Pd NCs and defective $Ni(OH)_2$ nanoflakes with a graphene support by a two-step solution method. They found that this electrocatalyst had a high mass-specific peak current and excellent operational durability. They asserted that this great catalyst durability was due to the presence of $Ni(OH)_2$ alleviating the poisoning of Pd NCs by carbonaceous intermediates. The incorporation of $Ni(OH)_2$ also markedly shifted the reaction selectivity from the originally predominant C_2 pathway toward the more desirable C_1 pathway.

Han et al. (2018) evaluated the bioethanol electrooxidation performance using porous trimetallic PtRhCu cubic nanoboxes (CNBs) with a tunable Pt/Rh atomic ratio in a paper with 133 citations. They noted that these CNBs showed morphology- and composition-dependent electrocatalytic activity. They found that the composition-optimized $Pt_{54}Rh4Cu_{42}$ CNBs had excellent specific and mass activity and stability for the bioethanol oxidation reaction (BOR) due to its unique geometric structure and synergistic effects. They asserted that the hollow porous structure of the CNBs could effectively enhance the atomic utilization and mass transfer. The introduction of Cu improved the antipoisoning capability for CO while the introduction of Rh elevated the self-stability of these CNBs. Further, the introduction of Rh significantly promoted the cleavage of C–C bonds, leading to the transformation of the main catalytic pathway for BOR from the C_2 to C_1 pathway. Finally, the real concentration detection for C_2 products showed that these CNBs had a nearly 11.5-fold C_1 pathway enhancement compared to Pt NPs, showing an obvious selectivity enhancement for the C_1 pathway.

Huang et al. (2018) evaluated the bioethanol electrooxidation performance using two-dimensional PdAg alloy nanodendrites in a paper with 124 citations. They carried out BOR via the coreduction of Pd and Ag precursors in aqueous solution with a structural directing agent. They found that these electrocatalysts had enhanced electrocatalytic activity and excellent operation stability due to the combined electronic and structural effects.

Wu et al. (2018) evaluated the bioethanol electrooxidation performance using Pd NPs supported on anatase titanium dioxide (TiO_2) (ATN)–black phosphorus (BP) hybrids with BP nanoflakes in a paper with 119 citations. They noted that the structure of ATN–BP was beneficial for improving the electrolyte penetration and electron transportation and had a strong influence on the stripping of reactive intermediates through the synergistic interaction between Pd NPs and ATN–BP hybrids. They found that these electrocatalysts with heterointerfaces of Pd NPs, BP, and ATN had ultrahigh electroactivity and durability. Further, in the BOR, the Pd/ATN–BP catalyst achieved an electrochemically active surface area and a mass peak current density, which were significantly greater, respectively, than those of commercial Pd/C. These electrocatalysts also had remarkable stability with a high retention rate of the peak current density after a durability test.

Rizo et al. (2018) evaluated the bioethanol electrooxidation performance using Pt–Sn NPs in a paper with 102 citations. They found that the electrochemical activity of these NPs was about three times higher than that obtained with unshaped Pt–Sn NPs and six times higher than that of Pt nanocubes. In addition, this electrocatalyst preserved its morphology and remained well-dispersed on the carbon support, while a cubic (pure) Pt catalyst showed severe agglomeration of the NPs. Sn dissolved from the outer part of the shell after potential cycling, forming an ultrathin 0.5-nm Pt skin. They asserted that this particular atomic composition profile with a Pt-rich core, a Sn-rich subsurface layer, and a Pt-skin surface structure was responsible for the high activity and stability of this electrocatalyst.

Jana et al. (2016) evaluated the bioethanol electrooxidation performance using flower-like ordered intermetallic Pd_3Pb NCs in a paper with 102 citations. They obtained these NCs in different morphologies at relatively low temperature by polyol and hydrothermal methods both in presence and absence of a surfactant. They found that the as-synthesized ordered NCs had far superior electrocatalytic activity and durability toward bioethanol oxidation over Pd/C black. They asserted that the morphological variation of NCs played a crucial role in the electrocatalytic oxidation of bioethanol. Among the catalysts, the flower-like Pd_3Pb showed enhanced activity and stability in electrocatalytic bioethanol oxidation. The current density and mass activity of these flower-like Pd_3Pb catalysts were higher than that of Pd/C for bioethanol oxidation.

Qi et al. (2016) evaluated the bioethanol electrooxidation performance using PdAg/ultrafine 2.7-nm carbon nanotubes (CNT) in direct alcohol fuel cells in a paper with 101 citations. They found that in a half-cell system with three electrodes, the peak mass activity of PdAg/CNT was higher than the mass activity of Pd/CNT at the same applied potential. As Ag has excellent activity toward aldehyde oxidation, they asserted that the enhancement in bioethanol oxidation on PdAg/CNT was due to Ag's promotion of intermediate aldehyde oxidation.

79.3.2 Bioethanol Sensors: Hot Papers

The brief information about seven hot papers with at least 108 citations each on the bioethanol sensors in fuel cells is given in Table 79.2. Furthermore, the brief notes on the contents of these studies are also given below.

Wang et al. (2016a) evaluated the sensing properties of a bioethanol sensor based on 2.15 at% Al^{+3}-doped NiO nanorod-flowers by a facile solvothermal reaction in a paper with 229 citations. They found that this sensor had improved gas sensing properties compared to pure NiO nanorod-flowers. They asserted that the incorporation of Al ions with NiO NCs adjusted the carrier concentration and induced the change of the oxygen deficiency and chemisorbed oxygen of NiO nanorod-flowers.

TABLE 79.2
Bioethanol Sensors: Hot Papers

No.	Papers	Materials	Parameters	Country	Cits.
1	Wang et al. (2016a)	NiO nanorod-flowers	Al doping effect	China	229
2	Zhu et al. (2018)	ZnO NPs, nanoplates, and nanoflowers	Nanostructures, surfactant effect	China	213
3	Wang et al. (2016b)	3DOM ZnO	Nanostructure effect, doping effect	China	138
4	Lupan et al. (2016)	Copper oxide (Cu_xO_y) NCs	Nanostructures, TA effect, mixed phase effect, Zn doping effect	Germany	118
5	Liu et al. (2017)	ZnO NP/SnO_2 sphere nanocomposites	Nanocomposite sensing performance, detection limit	China	112
6	Salim and Lim (2016)	CSRR-loaded microfluidic bioethanol chemical sensor	Bioethanol concentration determination	S. Korea	111
7	Cinti et al. (2017)	Paper, CB–Prussian Blue NPs (PBNPs) nanocomposites	Nanocomposites, paper-based screen-printed electrode, sensitivity, detection limit	Italy	108

Cits.: The number of the citations received by each paper.

Zhu et al. (2018) evaluated the sensing properties of a bioethanol sensor based on the ZnO nanostructures in a paper with 213 citations. They obtained ZnO NPs, nanoplates, and nanoflowers by a facile hydrothermal route. They found that the nanoplates-assembled nanoflowers had significantly higher gas sensing performances compared to the other nanostructures. They asserted that this sensing performance was due to their hierarchical architectures with large specific area and abundant spaces for gas diffusion. Furthermore, they found that the concentration of surfactant cetrimonium bromide (CTAB) used had an essential effect on the ultimate morphology of the hierarchical nanoflowers.

Wang et al. (2016b) evaluated the sensing properties of a bioethanol sensor based on an indium (In)-doped three-dimensionally ordered macroporous (3DOM) ZnO in a paper with 138 citations. They obtained this ZnO by using a colloidal crystal templating method. They found that this ZnO with 5 at.% of In doping exhibited the highest bioethanol sensitivity, which was approximately three times higher than that of pure 3DOM ZnO. They asserted that this superb bioethanol sensitivity was due to the increase in the surface area and the electron carrier concentration as the In doping introduced more electrons into the matrix resulting in high bioethanol sensitivity.

Lupan et al. (2016) evaluated the sensing properties of a bioethanol sensor based on copper oxide (Cu_xO_y) NCs with mixed phases of CuO and Cu_2O in a paper with 118 citations. They obtained these NC thin films via a simple synthesis from chemical solutions (SCS) followed by two types of thermal annealing: rapid thermal annealing (RTA) and conventional thermal annealing (TA). They found the enhanced bioethanol sensing performances of the device structures based on synthesized Cu_xO_y NCs with one and two distinctly different phases: Cu_2O, CuO, and mixed phases CuO/Cu_2O. They observed a gradient in phase change of these NCs for annealed samples starting from CuO on the top to Cu_2O in their central region. They identified RTA effects on the gas response of the Cu_xO_y NCs as unprecedented selectivity and sensitivity to bioethanol vapors at different temperatures. They enhanced the response and recovery times for pure Cu_xO_y-based sensors significantly by Zn doping.

Liu et al. (2017) evaluated the sensing properties of a bioethanol sensor based on hollow ZnO NP/ SnO_2 sphere nanocomposite material in a paper with 112 citations. They obtained this nanocomposite material by the solution method. They observed ZnO NPs on the surface of SnO_2 hollow spheres and found that the surface oxygen chemisorbed ability of these nanocomposites was much higher than that of single-component SnO_2. Further, they found that these nanocomposites had an excellent bioethanol sensing performance compared to pristine SnO_2 at its optimum temperature. Finally, this sensor had a low detection limit at the parts per billion (ppb) level. They asserted that this enhanced sensing properties of this sensor was due to the formation of heterojunction and synergistic effect between SnO_2 sphere and ZnO NP.

Salim and Lim (2016) evaluated the sensing properties of a complementary split-ring resonator (CSRR)-loaded microfluidic bioethanol chemical sensor primarily to detect bioethanol concentration in a paper with 111 citations. First, they realized two tightly coupled concentric CSRRs loaded on a patch on a Rogers RT/Duroid 5870 substrate, and then integrated a microfluidic channel engraved on polydimethylsiloxane (PDMS) for bioethanol chemical sensor applications. They tested various bioethanol concentrations. They found that this sensor showed repeatability and successfully detected 10% bioethanol. They asserted that this was a miniaturized, non-contact, low-cost, reliable, reusable, and easily fabricated sensor using extremely small liquid volumes.

Cinti et al. (2017) evaluated the sensing properties of a bioethanol biosensor based on a paper-based screen-printed electrode (SPE) for beers in a paper with 108 citations. They used common office paper to fabricate the analytical device and compared it to polyester-based SPE. They found that paper had similar properties compared with polyester, highlighting suitability toward its utilization in sensor development, with the advantages of low cost and simple disposal by incineration. They then utilized a nanocomposite formed by carbon black (CB) and Prussian Blue NPs (PBNPs), CB/PBNPs, as an electrocatalyst to detect the hydrogen peroxide generated by the enzymatic reaction between alcohol oxidase (AO_x) and bioethanol. After optimizing the analytical parameters, such as pH, enzyme, concentration, and working potential, the developed biosensor allowed a facile quantification of bioethanol up to 0.058%vol. with a significant sensitivity and a detection limit.

79.3.3 Evaluation of Bioethanol Fuels: Hot Papers

The brief information about two hot papers with at least 114 citations each on the evaluation of bioethanol fuels in fuel cells is given in Table 79.3. Furthermore, the brief notes on the contents of these studies are also given below.

Liu et al. (2016) evaluated the cellulase enzyme costs for cellulose-based bioethanol production using Aspen Plus modeling in a paper with 132 citations. They used the minimum bioethanol selling price (MBSP) to underline the effects of varying enzyme supply modes, enzyme prices, process parameters, and enzyme loading on the enzyme cost. They found that the enzyme cost drove the bioethanol price below the minimum profit point when the enzyme was purchased from the available industrial enzyme market. They recommended that an innovative production of cellulase enzyme such as onsite enzyme production should be explored and tested at the industrial scale to ensure an economic enzyme supply for the future bioethanol production.

Mohsenzadeh et al. (2017) evaluated the technoeconomics of bioethylene production from bioethanol fuels in a paper with 114 citations. They used Aspen® Plus and Aspen Process Economic Analyzer with different qualities of bioethanol. They found that impurities in the bioethanol feed had no significant effect on the quality of the produced bioethylene. Further, the capacity of the bioethylene storage tank significantly affected the capital costs of the process.

79.3.4 Utilization of Bioethanol Fuels for Biohydrogen Production for Fuel Cells: Hot Papers

The brief information about two hot papers with at least 109 citations each on the utilization of bioethanol fuels for biohydrogen production for fuel cells is given in Table 79.4. Furthermore, the brief notes on the contents of these studies are also given below.

Ma et al. (2016) evaluated the catalytic performance of La-modified ordered mesoporous Ni-based (meso-xLaNiAl) catalysts for the biohydrogen fuel production from bioethanol steam reforming in a paper with 186 citations. For comparison, they used the conventional 0LaNiAl catalyst. They found that these nanocatalysts had excellent high specific surface areas, large pore volumes, and uniform

TABLE 79.3
Evaluation of Bioethanol Fuels: Hot Papers

No.	Papers	Biomass/Product	Parameters	Country	Cits.
1	Liu et al. (2016)	Cellulose	Cellulase enzyme cost effect	China	132
2	Mohsenzadeh et al. (2017)	Ethylene	Technoeconomics, impurities, storage tank capacity	Sweden	114

Cits.: The number of the citations received by each paper.

TABLE 79.4
Utilization of Bioethanol Fuels for Biohydrogen Production for Fuel Cells: Hot Papers

No.	Papers	Catalysts	Parameters	Country	Cits.
1	Ma et al. (2016)	Meso-xLaNiAl	Nanocatalysts, ordering effect, modifier effect, steam reforming	China	186
2	Zhao et al. (2017)	3D PdCu alloy NSs	Nanocatalysts, acetate and biohydrogen production, electrochemical reforming	China	109

Cits.: The number of the citations received by each paper.

pore sizes as this ordered mesostructure was beneficial to obtain and maintain the 4- to 6-nm Ni NPs. Consequently, these nanocatalysts exhibited superior initial catalytic activity compared to the 0LaNiAl catalyst, especially at higher temperatures. Particularly, this nanocatalyst had the highest amount of easily reduced Ni species and then the highest active surface areas. They asserted that the highest initial catalytic activity was mainly due to the highly dispersed Ni NPs and abundant active surface areas. This nanocatalyst also exhibited excellent long-term stability. Further, the presence of La modifiers enhanced the basicity of the nanocatalyst, strengthened the metal–support interaction, and cleaned the deposited carbon.

Zhao et al. (2017) evaluated the catalytic performance of self-supported 3D PdCu alloy nanosheets (NSs) as a bifunctional catalysts for the biohydrogen fuel and acetate production from bioethanol electrochemical reforming in a paper with 109 citations. They obtained this surfactant-free NS by a simple CO-assisted method. They found that this was an excellent bifunctional electrocatalyst for partial electrochemical reforming of bioethanol into biohydrogen and acetate. They asserted that Cu diffusion into the lattice of Pd induced the lattice distortion and thus made Pd NSs wrinkled, which was crucial for the successful preparation of 3D PdCu alloy NSs.

79.3.5 Utilization of Bioethanol Fuels in Transportation: Hot Papers

The brief information about three hot papers with at least 128 citations each on the utilization of bioethanol fuels in transportation such as gasoline or diesel engines is given in Table 79.5. Furthermore, the brief notes on the contents of these studies are also given below.

Venu and Madhavan (2017) evaluated the effect of 0%, 5%, and 10% diethyl ether (0/5/10 DEE) addition in bioethanol–biodiesel–petrodiesel (BBP – B20–B40–P40) compared to the methanol–biodiesel–petrodiesel (MBP – M20–B40–P40) ternary blends as an ignition enhancer in a diesel engine on the combustion, performance, and emissions of fuels in a paper with 147 citations. They found that the addition of DEE in BBP increased the combustion duration (CD), cylinder pressure, and brake-specific fuel consumption (BSFC) with reduced nitrogen oxides (NO_x), particulate matter (PM), and smoke emissions due to reduced ignition delay (ID) and higher latent heat evaporation. In comparison with BBP–10DEE, BBP–5DEE resulted in higher cylinder pressure, heat release rate (HRR), exhaust gas temperatures (EGTs) and NO_x with lowered CD, total hydrocarbon (THC) emissions, carbon dioxide (CO_2), and PM. This was due to improved fuel atomization and enhanced fuel spray characteristics. The peak HRR of BBP was highest compared to MBD–5DEE. Overall, BBP–5DEE had better engine performance, combustion, and emission characteristics than the pure BBP blend (0DEE).

Silitonga et al. (2018) evaluated the engine performance and exhaust emissions of bioethanol–biodiesel–petrodiesel blends using kernel-based extreme learning machine (K-ELM) in a paper with 128 citations. They found that the BSFC was lower while the brake thermal efficiency (BTE)

TABLE 79.5
The Utilization of Bioethanol Fuels in Gasoline and Diesel Engines: Hot Papers

No.	Papers	Bioethanol Blends	Parameters	Country	Cits.
1	Venu and Madhavan (2017)	Bioethanol–biodiesel–petrodiesel (BBP – B20–B40–P40) blend with 0/5/10 DEE	Bioethanol effect, DEE content effect	India	147
2	Silitonga et al. (2018)	Bioethanol–biodiesel–petrodiesel (BBP) blends	BSFC, BTE, CO, NO_x, smoke opacity, K-ELM method	Indonesia	130
3	Venu and Madhavan (2016)	Bioethanol–biodiesel–petrodiesel (BBP – B10–B20–P70) blends	Bioethanol effect, alumina NP effect, injection timing effect	India	128

Cits.: The number of the citations received by each paper.

was higher for these blends while the carbon monoxide (CO) emissions and smoke opacity were also lower for these blends. They asserted that the K-ELM method was a reliable method to estimate the engine performance and exhaust emission parameters of a single cylinder compression ignition engine fueled with these blends to reduce fuel consumption and exhaust emissions.

Venu and Madhavan (2016) evaluated the effect of alumina (Al_2O_3) NPs and injection timings (ITs) on the combustion, diesel engine performance, and exhaust emissions of BBP (B10–B20–P70) blends in a paper with 128 citations. They used three different ITs: original timing (ORG IT) of 23 deg before top dead center (BTDC), advanced timing (ADV IT) of 27 deg BTDC, and retarded timing (RET IT) of 19 deg BTDC. They blended 25 parts per million (ppm) alumina with this blend. They found that the alumina addition at ADV IT resulted in higher peak pressure and HRR occurring nearer to TDC, higher THC, higher CO, lower CO_2, higher NO_x, higher exhaust gas oxygen (EGO), higher CD, and lower ID. On the other hand, alumina addition in RET IT caused lower cylinder pressure and heat release away from TDC, followed by simultaneous reductions of THC, CO, NO_x, and smoke opacity. In addition, they observed higher levels of EGO and ID along with lowered BSFC and CD with alumina addition in RET IT. Overall, they concluded that the effect of 25 ppm alumina in RET IT of 19 deg BTDC resulted in better combustion, engine performance, and exhaust emissions.

79.3.6 Biochemical Production from Bioethanol Fuels: Hot Papers

The brief information about three hot papers with at least 109 citations each on the biochemical production from bioethanol fuels is given in Table 79.6. Furthermore, the brief notes on the contents of these studies are also given below.

Fu et al. (2017) evaluated the 1-butanol production from bioethanol fuels by Guerbet-type condensation reaction with manganese (Mn) catalysts in a paper with 145 citations. They found that this process proceeded selectively in the presence of a well-defined Mn pincer complex at the ppm level. The developed reaction resulted in a sustainable synthesis of 1-butanol with excellent turnover number and turnover frequency. They identified the essential role of the N–H moiety of the Mn catalysts and the major reaction intermediates related to the catalytic cycle for this reaction.

79.3.7 Country-Based Experiences of Bioethanol Fuels: Hot Papers

There was no hot paper published on the country-based experiences of bioethanol fuels as in the case of population studies published during the last half a century.

TABLE 79.6
Biochemical Production from Bioethanol Fuels: Hot Papers

No.	Papers	Biochemicals	Catalysts	Res. Front	Parameters	Country	Cits.
1	Fu et al. (2017)	1-Butanol	Mn	Biochemicals	Mn catalyst effect	China	145
2	Mohsenzadeh et al. (2017)	Ethylene	Na	Biochemicals, evaluation	Technoeconomics, impurities, storage tank capacity	Sweden	114
3	Zhao et al. (2017)	Acetate, H_2	3D PdCu alloy NSs	Biochemicals, biohydrogen fuels	Nanocatalysts, acetate and biohydrogen production, electrochemical reforming	China	109

Cits.: The number of the citations received by each paper.
Na, Not available.

79.4 DISCUSSION

79.4.1 Introduction

Crude oil-based gasoline fuels have been widely used in the transportation sector since the 1920s. However, there have been great public concerns over the adverse environmental and human impact of these fuels. Hence, biomass-based bioethanol fuels have increasingly been used in blending gasoline and petrodiesel fuels, in fuel cells, and in biochemical and biohydrogen fuel production in a biorefinery context.

Research in the fields of experiences of countries of bioethanol fuels, evaluation of bioethanol fuels, production of biohydrogen fuels and biochemicals from bioethanol fuels, utilization of bioethanol fuels in transportation such as gasoline and diesel engines, fuel cells, and bioethanol sensors has also intensified in recent years.

However, it is essential to develop efficient incentive structures for the primary stakeholders to enhance the research in this field. Although there have been a number of review papers on the evaluation and utilization of bioethanol fuels, there has been no review of the 25 most cited articles in this field published since 2015.

Thus, this book chapter presents a review of these 25 most cited articles in these fields to highlight the recent research trends in these fields. Then, it discusses the key findings of these highly influential papers and comments on the future research priorities in this field.

As the first step for the search of the relevant literature, the keywords were selected using the first 200 most cited papers for each research front. The selected keyword list was optimized to obtain a representative sample of papers for the searched research field (Konur, 2023a–j). This keyword list for each research front was collected to form a combined keyword set for all research fronts, and these individual keyword sets were provided in the appendices of Konur (2023a–j) for future replication studies.

As the second step, a sample dataset was used for this study. The first 25 articles in the sample of 100 most cited papers with at least 101 citations each were selected for the review study. Key findings from each paper were taken from the abstracts of these papers and were discussed. Additionally, a number of brief conclusions were drawn and a number of relevant recommendations were made to enhance the future research landscape.

As Table 79.7 shows, there are seven primary research fronts for these 25 most cited papers: Utilization of bioethanol fuels in fuel cells, bioethanol sensors, utilization of bioethanol fuels in transportation such as gasoline engines, utilization of bioethanol fuels for biochemical production, evaluation of bioethanol fuels, and utilization of bioethanol fuels for biohydrogen production for fuel cells. The data on the nanotechnology applications in these field are also given in this table due to its importance.

79.4.2 Utilization of Bioethanol Fuels in Fuel Cells: Hot Papers

Over 32.4%, 25%, and 40% of the I100 sample, 100 most cited paper sample, and reviewed papers were related to the utilization of bioethanol fuels in fuel cells, respectively (Table 79.7). Seven out of ten hot papers in this field were carried out in China (Table 79.1).

All these hot papers used a wide range of a nanocatalysts in contrast to only two papers related to nanocatalysts in the study covering the last half a century of studies in this field (Konur, 2023k,l): Pd–Ni–P NPs (Chen et al., 2017), PdCuM NPs (Jiang et al., 2016), Pt–Mo–Ni NWs (Mao et al., 2017), Pd NCs, $Ni(OH)_2$ nanoflakes, and graphene (Huang et al., 2017), PtRhCu CNBs (Han et al., 2018), 2D PdAg alloy nanodendrites (Huang et al., 2018), Pd NPs/TiO_2–BP nanoflakes, and Pd/C (Wu et al., 2018), Pt-Sn NPs (Rizo et al., 2018), Pd_3Pb NCs, and C/Pd (Jana et al., 2016), and PdAg/CNT (Qi et al., 2016).

TABLE 79.7
The Hot Papers in the Evaluation and Utilization of Bioethanol Fuels: Research Fronts

No.	Research Fronts	I100 Sample (%)	HP I100 Sample (%)	Sample Papers (%)	HP Sample Papers (%)	Reviewed Papers (%)	HP Reviewed Papers (%)
1	Utilization of bioethanol fuels in fuel cells	27.6	32.4	30	25	40	40
2	Bioethanol Sensors	12.0	23.5	7	21	8	28
3	Utilization of bioethanol fuels in transportation such as gasoline engines	19.4	14.7	19	28	4	12
4	Utilization of bioethanol fuels for biochemical production	6.3	8.8	4	8	0	12
5	Evaluation of bioethanol fuels	20.9	5.9	26	10	24	8
6	Utilization of bioethanol fuels for biohydrogen production for fuel cells	19.1	14.7	19	9	24	8
7	Country-based experiences of bioethanol fuels.	9.7	5.9	10	5	0	0
	Nanotechnology applications	18.1	67.6	14	20	20	80
	Sample size	775	34	100	100	25	25

Reviewed papers: 25 papers reviewed in this study. Sample papers: The sample of the 100 most cited papers. I100 sample: the sample of papers with at least 100 citations each.
HP, Hot papers.

It is notable that seven of these hot papers used Pd-based nanocatalysts (Chen et al., 2017; Huang et al., 2017, 2018; Jana et al., 2016; Jiang et al., 2016; Qi et al., 2016; Wu et al., 2018). On the other hand, only three studies used Pt-based nanocatalysts (Han et al., 2018; Mao et al., 2017; Rizo et al., 2018). Thus, there was a focus on these two catalyst systems unlike the study covering half a century of studies in this field where other catalysts such as Sn, Rh, and W were also considered. Additionally, these studies considered a range of metal additives: Ni (Chen et al., 2017; Huang et al., 2017; Mao et al., 2017), Cu (Han et al., 2018; Jiang et al., 2016), Rh (Han et al., 2018), Ag (Huang et al., 2018; Qi et al., 2016), Sn (Rizo et al., 2018), and Pb (Jana et al., 2016). The focus in these hot papers was on the catalysts themselves rather than catalyst supports unlike the study covering the last half a century of studies in this field (Konur, 2023k,l).

All these hot papers focused solely on the electrooxidation of bioethanol fuels unlike the study covering this last half a century of studies in this field. The activity and mechanisms of electrooxidation of bioethanol fuels were extensively evaluated in these studies. The studied parameters are listed in Table 79.1.

These findings are thought-provoking in seeking ways to optimize the utilization of bioethanol fuels in fuel cells. The catalysts, mechanisms, and activity of bioethanol electrooxidation, and applications of nanotechnology in catalysts emerge as the key issues for this field. It is notable that Pt- and Pd-based nanomaterials were used as nanocatalysts with a variety of metal additives in these hot papers published after 2015. In parallel with the developments in the nanomaterials and nanotechnology research, it is expected that a wider range of nanomaterials including the two-dimensional nanomaterials would be further used in this field in the future.

79.4.3 Bioethanol Sensors: Hot Papers

Around 23.5%, 21%, and 28% of the I100 sample, 100 most cited paper sample, and reviewed papers were related to the bioethanol sensors, respectively (Table 79.7). Four out of seven studies were carried out in China (Table 79.2). This field was overrepresented in the hot papers compared to the population studies carried out during the last half a century.

Six out of seven studies used nanocatalysts: (Cinti et al., 2017; Liu et al., 2017; Lupan et al., 2016; Wang et al. 2016a,b; Zhu et al., 2018). Among them, NPs (Cinti et al., 2017; Liu et al., 2017; Zhu et al., 2018), NCs (Lupan et al., 2016), nanoflowers (Wang et al., 2016a; Zhu et al., 2018), and nanocomposites (Cinti et al., 2017; Liu et al., 2017) were used.

The most used material was ZnO (Liu et al., 2017; Wang et al., 2016b; Zhu et al., 2018). The other materials were NiO (Wang et al., 2016a), Cu_xO_y (Lupan et al., 2016), SnO_2 (Liu et al., 2017), and paper (Cinti et al., 2017).

A wide range of parameters were studied in these hot papers, and they are listed in Table 79.2. These findings are thought-provoking in seeking ways to optimize the utilization of nanomaterials and conventional as sensor materials. The catalysts, doping and surfactant effects, and applications of nanotechnology in catalysts emerge as the key issues for this field. It is notable that primarily ZnO-based nanomaterials were used as nanocatalysts in these hot papers published after 2015. In parallel with the developments in the nanomaterials and nanotechnology research, it is expected that a wider range of nanomaterials including the two-dimensional nanomaterials would be further used in this field in the future.

79.4.4 Evaluation of Bioethanol Fuels: Hot Papers

Around 5.9%, 10%, and 8% of the I100 sample, 100 most cited paper sample, and reviewed papers were related to the evaluation of bioethanol fuels, respectively (Table 79.7). These studies were carried out in China and Sweden (Table 79.3). This field was underrepresented in the hot papers compared to the population studies carried out during the last half a century.

These studies were concerned with the technoeconomics of bioethanol production optimizing enzyme costs (Liu et al., 2016) and technoeconomics of bioethylene production from bioethanol fuels (Mohsenzadeh et al., 2017). These findings are thought-provoking in seeking ways to evaluate bioethanol fuels.

79.4.5 Utilization of Bioethanol Fuels for Biohydrogen Production for Fuel Cells: Hot Papers

Around 14.7%, 9%, and 8% of the I100 sample, 100 most cited paper sample, and reviewed papers were related to the utilization of bioethanol fuels for biohydrogen production for the fuel cells, respectively (Table 79.7). These studies were carried out in China (Table 79.4). This field was underrepresented in the hot papers compared to the population studies carried out during the last half a century.

A number of catalysts were used in these studies: Meso-xLaNiAl (Ma et al., 2016) and 3D PdCu alloy NSs (Zhao et al., 2017). This was in contrast to the population studies carried out during the last half a century where a wide range of catalysts were used. Similarly, unlike the population studies, the focus in these hot papers was on the nanocatalysts themselves rather than catalyst supports.

Both steam reforming (Ma et al., 2016) and electrochemical reforming (Zhao et al., 2017) of bioethanol fuels were used in these hot papers. These findings are thought-provoking in seeking ways to optimize the utilization of bioethanol fuels for biohydrogen production for fuel cells. A number of parameters listed in Table 79.4 emerge as the key issues for this field.

79.4.6 Utilization of Bioethanol Fuels in Transportation: Hot Papers

Around, 14.7%, 28%, and 12% of the I100 sample, 100 most cited paper sample, and reviewed papers were related to the utilization of bioethanol fuels in the transportation such as gasoline or diesel engines, respectively (Table 79.7). These studies were carried out in China and Sweden (Table 79.6). This field was overrepresented in the hot papers compared to the population studies carried out during the last half a century.

A wide range of issues related to the combustion, engine performance, and emissions of bioethanol blends with petrodiesel and biodiesel fuels were covered in these papers, and they are listed in Table 79.5 (Silitonga et al., 2018; Venu and Madhavan, 2016, 2017).

79.4.7 Biochemical Production from Bioethanol Fuels: Hot Papers

Around, 6.8%, 8%, and 12% of the I100 sample, 100 most cited paper sample, and reviewed papers were related to the biochemical production from bioethanol fuels, respectively (Table 79.7). These studies were carried out in Indonesia and India (Table 79.6). This field was overrepresented in the hot papers compared to the population studies carried out during the last half a century.

A wide range of issues related to the production of biochemicals from bioethanol fuels were covered in these hot papers, and they are listed in Table 79.6 (Fu et al., 2017; Mohsenzadeh et al., 2017; Zhao et al., 2017). The studied biochemicals were 1-butanol (Fu et al., 2017), ethylene (Mohsenzadeh et al., 2017), and acetate and H_2 (Zhao et al., 2017).

79.4.8 Country-based Experiences of Bioethanol Fuels

There was no hot paper published on the country-based experiences of bioethanol fuels as in the case of population studies published during the last half a century.

79.4.9 The Overall Remarks

Although there are seven research fronts in the field of evaluation and utilization of bioethanol fuels, only four of them have importance presented among the reviewed papers: Utilization of bioethanol fuels in fuel cells, bioethanol sensors, utilization of bioethanol fuels in transportation such as gasoline engines, and utilization of bioethanol fuels for biochemical production with 40%, 28%, 12%, and 12% of the reviewed hot papers, respectively (Table 79.7). The other research fronts were evaluation of bioethanol fuels, utilization of bioethanol fuels for biohydrogen production for fuel cells, and country-based experiences of bioethanol fuels.

Considering that biohydrogen fuels are used ultimately in the fuel cells for electric vehicles, the share of the papers related to fuel cells rises to 48% of the reviewed paper sample, compared to 47.1% and 34% of the I100 sample and 100 most cited paper sample, respectively. In light of North's institutional framework (North, 1991), it could be said that these four major research fronts have greater societal importance compared to the three other minor research fields.

The findings related to the direct utilization of bioethanol fuels in fuel cells are thought-provoking in seeking ways to optimize the utilization of bioethanol fuels in fuel cells. The catalyst types (de Boer et al., 1965), catalyst support types (Arai and Machida, 1996), mechanisms and activity of bioethanol electrooxidation (Bagotzky and Vassilyev, 1967), and applications of nanotechnology in catalysts (Fihri et al., 2011) emerge as the key issues for this field.

The findings related to the evaluation of bioethanol fuels are thought-provoking in seeking ways to evaluate bioethanol fuels. The biomass feedstocks (Bergmann et al., 2013), net energy balance (Cleveland, 2005), GHG emissions (Burnham et al., 2012), technoeconomics (Callon, 1990), life cycle analysis (LCA) (Ayres, 1995), economic analysis (Landes, 1971), and production costs (Blumenfeld et al., 1985) emerge as the key issues for this field.

The findings related to the utilization of bioethanol fuels for biohydrogen fuel production for fuel cells are thought-provoking in seeking ways to optimize the utilization of bioethanol fuels for biohydrogen production for fuel cells. The catalyst types (de Boer et al., 1965), catalyst support types (Arai and Machida, 1996), nanostructured catalyst systems (Fihri et al., 2011), steam reforming (Palo et al., 2007) and autothermal reforming (Ayabe et al., 2003) of bioethanol fuels, photocatalytic biohydrogen production (Tong et al., 2012), catalyst loading (Jusys et al., 2003), bioethanol conversion (Harun et al., 2014), and catalytic activity (Thiele et al., 1939) emerge as the key issues for this field.

A wide range of issues were covered in the papers related to the utilization of bioethanol fuels in gasoline and petrodiesel engines: Fuel properties (Dooley et al., 2010), fuel performance (Crookes, 2006), heating values (Nhuchhen and Salam, 2012), octane numbers (Albahri, 2003), Reid vapor pressure (Cooper et al., 1995), torque output (Ahmadi et al., 2015), fuel consumption (Ramos-Paja et al., 2008), CO (Muller and Stavrakou, 2005), CO_2 (Davis et al., 2011), HC (Stanglmaier, et al., 1999), NO_x emissions (Muller and Stavrakou, 2005), bioethanol content (Yunoki and Saito, 2009), and E0, E5, E10, E20, and E30 blends.

Similarly, the focus for the hot papers related to bioethanol sensors is on the development of nanomaterials for bioethanol sensors. Research on chemical sensors and biosensors has been very intense (Chaubey and Malhotra, 2002). In this context, research on the ZnO-based sensors has also been very intense (Kumar et al., 2006). Nanomaterials have been widely used in the development of sensors (Howes et al., 2014).

Both the utilization of bioethanol fuels for the biochemical production (Angelici et al., 2013; Liu and Hensen, 2013; Wang et al., 2004) and country-based experiences of bioethanol fuels (Macedo et al., 2008; Sheehan et al., 2003) have great importance for understanding the historical development of bioethanol industry and research.

It is very important to produce biochemicals within the biorefinery context to reduce the production costs of bioethanol fuels as well as to develop human- and environment-friendly biochemicals as an alternative to crude oil-based chemicals. Similarly, the studies on the latter research front emphasize the importance of proper incentive structures for the efficient development of bioethanol industry and research in light of North's institutional framework (North, 1991).

Since the studies in each of the seven research fields take gasoline fuels as a base case, the review of gasoline fuels as a whole was provided in Konur (2023a). The key research fronts for gasoline fuels are desulfurization of gasoline fuels, properties, combustion, performance, and emissions of gasoline fuels and their blends, gasoline economics, gasoline fuel production, and upgrading.

It is important to note that Hill et al. (2006) laid down the rules for the environment- and human-friendly production of bioethanol fuels: They should have a net energy balance, have environmental benefits, have economic competitiveness, have large scale production, and have no competition to the food production. The driving force for the development of the research and industry in this field has been the public concerns about climate change (Change, 2007), GHG emissions (Carlson et al., 2017), and global warming (Kerr, 2007).

The key findings from these studies also hint that bioethanol fuels have comparable net energy gain, better emissions, and comparable economic competitiveness following the list of the issues given by Hill et al. (2006). As expected, emissions, net energy, and economic competitiveness of second generation bioethanol fuels are better than those of the first generation bioethanol fuels and much better than those of gasoline or petrodiesel fuels in general.

It is expected that nanotechnology (Geim, 2009; Geim and Novoselov, 2010) would play a more vital role in the development of catalyst systems for the oxidation of bioethanol fuels for fuel cells, production of biohydrogen fuels and biochemicals from bioethanol fuels, and development of bioethanol sensors. As the data in Table 79.7 show that the research on the applications of nanomaterials has increased significantly since 2015 as the hot papers with nanotechnology applications were largely overrepresented. It is expected that this increasing trend would continue in the incoming decades.

In the end, these most cited papers in this field hint that the efficiency of the bioethanol, biohydrogen, and biochemical production as well as the development of bioethanol sensors and bioethanol blends for gasoline or diesel engines could be optimized using the structure, processing, and property relationships of the bioethanol fuels and their derivatives (Formela et al., 2016; Konur, 2018, 2020b, 2021a–d; Konur and Matthews, 1989).

These reviewed studies also show the importance of the incentive structures for the development of bioethanol industry and research at a global scale in light of North's institutional framework (North, 1991). In this context, it appears that the major producers and users of bioethanol fuels such as Brazil, the USA, China, and Europe have developed strong incentive structures for the effective development of bioethanol industry and research.

79.5 CONCLUSION AND FUTURE RESEARCH

The brief information about the key research fronts covered by the 25 most cited papers published between 2016 and 2021 with at least 101 citations each is given under seven headings for each underlying research front: Utilization of bioethanol fuels in fuel cells, evaluation of bioethanol fuels, utilization of bioethanol fuels for biohydrogen production for fuel cells, bioethanol sensors, utilization of bioethanol fuels in transportation such as gasoline engines, utilization of bioethanol fuels for biochemical production, and country-based experiences of bioethanol fuels (Table 79.7). In light of North's institutional framework (North, 1991), it could be said that some of these research fronts have greater societal importance compared to the other minor research fields.

The key findings on these research fronts should be read in light of the increasing public concerns about climate change, GHG emissions, and global warming as these concerns have been certainly behind the boom in the research in this field in the last two decades.

These findings confirm that bioethanol fuels are a viable alternative to crude oil-based gasoline and petrodiesel fuels, have a net energy gain, have environmental benefits impacting favorably global warming, GHG emissions, and climate change, are economically competitive, and are producible in large quantities without reducing food supplies, especially using the second generation bioethanol feedstocks such as lignocellulosic feedstocks using the criteria introduced by Hill et al. (2006).

There is a persistent trend for the bioethanol research to focus more on the second and third generation feedstocks such as sugarcane bagasse and algae and to move away from the first generation bioethanol feedstocks such as corn grains and sugarcane. These feedstocks would certainly impact the properties, engine performance, emissions as well as the net energy gain, economic competitiveness, and emissions of the resulting bioethanol fuels and their blends. This trend should be followed closely by the major stakeholders in the future.

The findings related to the utilization of bioethanol fuels in fuel cells are thought-provoking in seeking ways to optimize the utilization of bioethanol fuels in fuel cells. The catalysts, catalyst supports, mechanisms and activity of bioethanol electrooxidation, and applications of nanotechnology in catalysts emerge as the key issues for this field.

The findings related to the evaluation of bioethanol fuels are thought-provoking in seeking ways to evaluate bioethanol fuels. The biomass feedstocks, net energy balance, GHG emissions, techno-economics, life cycle analysis, economic analysis, and production costs emerge as the key issues for this field.

The findings related to the utilization of bioethanol fuels for biohydrogen fuel production for fuel cells are thought-provoking in seeking ways to optimize the utilization of bioethanol fuels for biohydrogen production for fuel cells. The catalyst types, catalyst support types, nanostructured catalyst systems, steam and autothermal reforming of bioethanol fuels, photocatalytic biohydrogen production, catalyst loading, preparation methods, bioethanol conversion, biohydrogen selectivities, and catalytic activity emerge as the key issues for this field.

A wide range of issues were covered in the paper related to the utilization of bioethanol fuels in gasoline and petrodiesel engines: Properties, performance, heating values, octane numbers, Reid vapor pressure, torque output, fuel consumption, CO, CO_2, HC, and NO_x emissions, and bioethanol content. Similarly, the focus is on the development of nanomaterials for bioethanol sensors.

Both the utilization of bioethanol fuel for the biochemical production and country-based experiences of bioethanol fuels have great importance for understanding the determinants of the historical development of bioethanol industry and research. It is very important to produce biochemicals within the biorefinery context to reduce the production costs of bioethanol fuels as well as to develop human- and environment-friendly biochemicals as an alternative to crude oil-based chemicals. Similarly, the studies on the latter research front emphasize the importance of proper incentive structures for the efficient development of bioethanol industry and research in light of North's institutional framework (North, 1991).

It is expected that the nanotechnology would play a more vital role in the development of catalyst systems for the oxidation of bioethanol fuels, production of biohydrogen fuels from bioethanol fuels, and development of bioethanol sensors and utilization of bioethanol blends in gasoline and diesel engines.

In the end, these most cited papers in this field hint that the efficiency of the bioethanol, biohydrogen, biochemical production, bioethanol sensor development, and the utilization of bioethanol blends in gasoline and diesel engines could be optimized using the structure, processing, and property relationships of the bioethanol fuels and their derivatives.

These reviewed studies also show the importance of the incentive structures for the development of bioethanol industry and research at a global scale in light of North's institutional framework (North, 1991). In this context, the major producers and users of bioethanol fuels such as Brazil, the USA, China, and Europe have developed strong incentive structures for the effective development of bioethanol industry and research.

It is recommended that such review studies should be performed for the other research fronts on both the production and utilization of bioethanol fuels complementing the corresponding scientometric studies for the evaluation and utilization of bioethanol fuels.

ACKNOWLEDGMENTS

The contribution of the highly cited researchers in the field of the evaluation and utilization of bioethanol fuels has been gratefully acknowledged.

REFERENCES

Ahmadi, M. H., M. A. Ahmadi, S. A. Sadatsakkak and M. Feidt. 2015. Connectionist intelligent model estimates output power and torque of Stirling engine. *Renewable and Sustainable Energy Reviews* 50:871–883.

Albahri, T. A. 2003. Structural group contribution method for predicting the octane number of pure hydrocarbon liquids. *Industrial & Engineering Chemistry Research* 42:657–662.

Angelici, C., B. M. Weckhuysen and P. C. A. Bruijnincx. 2013. Chemocatalytic conversion of ethanol into butadiene and other bulk chemicals. *ChemSusChem* 6:1595–1614.

Antolini, E. 2007. Catalysts for direct ethanol fuel cells. *Journal of Power Sources* 170:1–12.

Antolini, E. 2009. Palladium in fuel cell catalysis. *Energy and Environmental Science* 2:915–931.

Arai, H. and M. Machida. 1996. Thermal stabilization of catalyst supports and their application to high-temperature catalytic combustion. *Applied Catalysis A: General* 138:161–176.

Ayabe, S., H. Omoto and T. Utaka, et al. 2003. Catalytic autothermal reforming of methane and propane over supported metal catalysts. *Applied Catalysis A: General* 241:261–269.

Ayres, R. U. 1995. Life cycle analysis: A critique. *Resources, Conservation and Recycling* 14:199–223.

Bagotzky, V. S. and Y. B. Vassilyev. 1967. Mechanism of electro-oxidation of methanol on the platinum electrode. *Electrochimica Acta* 12:1323–1343.

Bergmann, J. C., D. D. Tupinamba and O. Y. A. Costa, et al. 2013. Biodiesel production in Brazil and alternative biomass feedstocks. *Renewable and Sustainable Energy Reviews* 21:411–420.

Blumenfeld, D. E., L. D. Burns, J. D. Diltz and C. F. Daganzo. 1985. Analyzing trade-offs between transportation, inventory and production costs on freight networks. *Transportation Research Part B: Methodological* 19:361–380.

Burnham, A., J. Han, and C. E. Clark, et al. 2012. Life-cycle greenhouse gas emissions of shale gas, natural gas, coal, and petroleum. *Environmental Science & Technology* 46:619–627.

Burnham, J. F. 2006. Scopus database: A review. *Biomedical Digital Libraries* 3:1–8.

Callon, M. 1990. Techno-economic networks and irreversibility. *Sociological Review* 38:132–161.

Carlson, K. M., J. S. Gerber and N. D. Mueller, et al. 2017. Greenhouse gas emissions intensity of global croplands. *Nature Climate Change* 7:63–68.

Change, C. 2007. Climate change impacts, adaptation and vulnerability. *Science of the Total Environment* 326:95–112.

Chaubey, A. and B. Malhotra. 2002. Mediated biosensors. *Biosensors and Bioelectronics* 17:441–456.

Chen, L., L. Lu and H. Zhu, et al. 2017. Improved ethanol electrooxidation performance by shortening Pd-Ni active site distance in Pd-Ni-P nanocatalysts. *Nature Communications* 8:14136.

Cinti, S., M. Basso, D. Moscone and F. Arduini. 2017. A paper-based nanomodified electrochemical biosensor for ethanol detection in beers. *Analytica Chimica Acta* 960:123–130.

Cleveland, C. J. 2005. Net energy from the extraction of oil and gas in the United States. *Energy* 30:769–782.

Cooper, J. B., K. L. Wise, J. Groves and W. T. Welch. 1995. Determination of octane numbers and Reid vapor pressure of commercial petroleum fuels using FT-Raman spectroscopy and partial least-squares regression analysis. *Analytical Chemistry* 67:4096–4100.

Crookes, R. J. 2006. Comparative bio-fuel performance in internal combustion engines. *Biomass and Bioenergy* 30:461–468.

Davis, S. J., G. P. Peters and K. Caldeira. 2011. The supply chain of CO_2 emissions. *Proceedings of the National Academy of Sciences* 108:18554–18559.

de Boer, J. H., B. G. Linsen and T. J. Osinga. 1965. Studies on pore systems in catalysts: VI. The universal t curve. *Journal of Catalysis* 4:643–648.

Dooley, S., S. H. Won and M. Chaos, et al. 2010. A jet fuel surrogate formulated by real fuel properties. *Combustion and Flame* 157:2333–2339.

Farrell, A. E., R. J. Plevin and B. T. Turner, et al. 2006. Ethanol can contribute to energy and environmental goals. *Science* 311:506–508.

Fernando, S., S. Adhikari, C. Chandrapal and M. Murali. 2006. Biorefineries: Current status, challenges, and future direction. *Energy & Fuels* 20:1727–1737.

Fihri, A., M. Bouhrara, B. Nekoueishahraki, J. M. Basset and V. Polshettiwar. 2011. Nanocatalysts for Suzuki cross-coupling reactions. *Chemical Society Reviews* 40:5181–5203.

Formela, K., A. Hejna, L. Piszczyk, M. R. Saeb and X. Colom. 2016. Processing and structure-property relationships of natural rubber/wheat bran biocomposites. *Cellulose* 23:3157–3175.

Fu, S., Z. Shao, Y. Wang and Q. Liu. 2017. Manganese-catalyzed upgrading of ethanol into 1-butanol. *Journal of the American Chemical Society* 139:11941–11948.

Geim, A. K. 2009. Graphene: Status and prospects. *Science* 324:1530–1534.

Geim, A. K. and K. S. Novoselov. 2007. The rise of graphene. *Nature Materials* 6:183–191.

Hamelinck, C. N., G. van Hooijdonk and A. P. C. Faaij. 2005. Ethanol from lignocellulosic biomass: Techno-economic performance in short-, middle- and long-term. *Biomass and Bioenergy* 28:384–410.

Han, S. H., H. M. Liu, P. Chen, J. X. Jiang and Y. Chen. 2018. Porous trimetallic PtRhCu cubic nanoboxes for ethanol electrooxidation. *Advanced Energy Materials* 8:1801326.

Hansen, A. C., Q. Zhang and P. W. L. Lyne. 2005. Ethanol-diesel fuel blends: A review. *Bioresource Technology* 96:277–285.

Harun, R., J. W. Yip and S. Thiruvenkadam, et al. 2014. Algal biomass conversion to bioethanol-a step-by-step assessment. *Biotechnology Journal* 9:73–86.

Haryanto, A., S. Fernando, N. Murali and S. Adhikari. 2005. Current status of hydrogen production techniques by steam reforming of ethanol: A review. *Energy and Fuels* 19:2098–2106.

Hill, J., E. Nelson, D. Tilman, S. Polasky and D. Tiffany. 2006. Environmental, economic, and energetic costs and benefits of biodiesel and ethanol biofuels. *Proceedings of the National Academy of Sciences of the United States of America* 103:11206–11210.

Hill, J., S. Polasky and E. Nelson, et al. 2009. Climate change and health costs of air emissions from biofuels and gasoline. *Proceedings of the National Academy of Sciences of the United States of America* 106:2077–2082.

Howes, P. D., R. Chandrawati and M. M. Stevens. 2014. Colloidal nanoparticles as advanced biological sensors. *Science* 346:1247390.

Hsieh, W. D., R. H. Chen, T. L. Wu and T. H. Lin. 2002. Engine performance and pollutant emission of an SI engine using ethanol-gasoline blended fuels. *Atmospheric Environment* 36:403–410.

Huang, H. J., S. Ramaswamy, U. W. Tschirner and B. V. Ramarao. 2008. A review of separation technologies in current and future biorefineries. *Separation and Purification Technology* 62:1–21.

Huang, W., X. Kang and C. Xu, et al. 2018. 2D PdAg alloy nanodendrites for enhanced ethanol electroxidation. *Advanced Materials* 30:1706962,

Huang, W., X. Y. Ma and H. Wang, et al. 2017. Promoting effect of $Ni(OH)_2$ on palladium nanocrystals leads to greatly improved operation durability for electrocatalytic ethanol oxidation in alkaline solution. *Advanced Materials* 29:1703057.

Jana, R., U. Subbarao and S. C. Peter. 2016. Ultrafast synthesis of flower-like ordered Pd_3Pb nanocrystals with superior electrocatalytic activities towards oxidation of formic acid and ethanol. *Journal of Power Sources* 301:160–169.

Jiang, K., P. Wang and S. Guo, et al. 2016. Ordered PdCu-based nanoparticles as bifunctional oxygen-reduction and ethanol-oxidation electrocatalysts. *Angewandte Chemie: International Edition* 55:9030–9035.

Jusys, Z., J. Kaiser and R. J. Behm. 2003. Methanol electrooxidation over Pt/C fuel cell catalysts: Dependence of product yields on catalyst loading. *Langmuir* 19:6759–6769.

Kerr, R. A. 2007. Global warming is changing the world. *Science* 316:188–190.

Kohse-Hoinghaus, K., P. Osswald and T. A. Cool, et al. 2010. Biofuel combustion chemistry: From ethanol to biodiesel. *Angewandte Chemie: International Edition* 49:3572–3597.

Konur, O. 2000. Creating enforceable civil rights for disabled students in higher education: An institutional theory perspective. *Disability & Society* 15:1041–1063.

Konur, O. 2002a. Access to nursing education by disabled students: Rights and duties of nursing programs. *Nurse Education Today* 22:364–374.

Konur, O. 2002b. Assessment of disabled students in higher education: Current public policy issues. *Assessment and Evaluation in Higher Education* 27:131–152.

Konur, O. 2002c. Access to employment by disabled people in the UK: Is the Disability Discrimination Act working? *International Journal of Discrimination and the Law* 5:247–279.

Konur, O. 2006a. Participation of children with dyslexia in compulsory education: Current public policy issues. *Dyslexia* 12:51–67.

Konur, O. 2006b. Teaching disabled students in higher education. *Teaching in Higher Education* 11:351–363.

Konur, O. 2007a. A judicial outcome analysis of the Disability Discrimination Act: A windfall for the employers? *Disability & Society* 22:187–204.

Konur, O. 2007b. Computer-assisted teaching and assessment of disabled students in higher education: The interface between academic standards and disability rights. *Journal of Computer Assisted Learning* 23:207–219.

Konur, O. 2012. The evaluation of the research on the bioethanol: A scientometric approach. *Energy Education Science and Technology Part A: Energy Science and Research* 28:1051–1064.

Konur, O. 2015. Current state of research on algal bioethanol. In *Marine Bioenergy: Trends and Developments*, Ed. S. K. Kim and C. G. Lee, pp. 217–244. Boca Raton, FL: CRC Press.

Konur, O., Ed. 2018. *Bioenergy and Biofuels*. Boca Raton, FL: CRC Press.

Konur, O. 2019. Cyanobacterial bioenergy and biofuels science and technology: A scientometric overview. In *Cyanobacteria: From Basic Science to Applications*, Eds. A. K. Mishra, D. N. Tiwari and A. N. Rai, pp. 419–442. Amsterdam: Elsevier.

Konur, O. 2020a. The scientometric analysis of the research on the bioethanol production from green macroalgae. In *Handbook of Algal Science, Technology and Medicine,* Ed. O. Konur, pp. 385–401. London: Academic Press.

Konur, O., Ed. 2020b. *Handbook of Algal Science, Technology and Medicine.* London: Academic Press.

Konur, O., Ed. 2021a. *Handbook of Biodiesel and Petrodiesel Fuels: Science, Technology, Health, and Environment.* Boca Raton, FL: CRC Press.

Konur, O., Ed. 2021b. *Handbook of Biodiesel and Petrodiesel Fuels: Science, Technology, Health, and Environment. Volume 1. Biodiesel Fuels: Science, Technology, Health, and Environment.* Boca Raton, FL: CRC Press.

Konur, O., Ed. 2021c. *Handbook of Biodiesel and Petrodiesel Fuels: Science, Technology, Health, and Environment. Volume 2. Biodiesel Fuels Based on the Edible and Nonedible Feedstocks, Wastes, and Algae: Science, Technology, Health, and Environment.* Boca Raton, FL: CRC Press.

Konur, O., Ed. 2021d. *Handbook of Biodiesel and Petrodiesel Fuels: Science, Technology, Health, and Environment. Volume 3. Petrodiesel Fuels: Science, Technology, Health, and Environment.* Boca Raton, FL: CRC Press.

Konur, O. 2023a. Gasoline fuels: Scientometric study. In *Evaluation and Utilization of Bioethanol Fuels. I.: Evaluation of Bioethanol Fuels, Transport Engines, and Bioethanol Sensors. Handbook of Bioethanol Fuels Volume 5*, Ed. O. Konur, pp. 87–106. Boca Raton, FL: CRC Press.

Konur, O. 2023b. Nanotechnology applications in bioethanol fuels: Scientometric study. In *Evaluation and Utilization of Bioethanol Fuels. I.: Evaluation of Bioethanol Fuels, Transport Engines, and Bioethanol Sensors. Handbook of Bioethanol Fuels Volume 5*, Ed. O. Konur, pp. 120–139. Boca Raton, FL: CRC Press.

Konur, O. 2023c. Utilization of bioethanol fuels in the transport engines: Scientometric study. In Evaluation and Utilization of Bioethanol Fuels. I.: Evaluation of Bioethanol Fuels, Transport Engines, and Bioethanol Sensors. Handbook of Bioethanol Fuels Volume 5, Ed. O. Konur, pp. 157–175. Boca Raton, FL: CRC Press.

Konur, O. 2023d. Evaluation of bioethanol fuels: Scientometric study. In *Evaluation and Utilization of Bioethanol Fuels. I.: Evaluation of Bioethanol Fuels, Transport Engines, and Bioethanol Sensors. Handbook of Bioethanol Fuels Volume 5*, Ed. O. Konur, pp. 195–213. Boca Raton, FL: CRC Press.

Konur, O. 2023e. Utilization of bioethanol fuels: Scientometric study. In *Evaluation and Utilization of Bioethanol Fuels. I.: Evaluation of Bioethanol Fuels, Transport Engines, and Bioethanol Sensors. Handbook of Bioethanol Fuels Volume 5*, Ed. O. Konur, pp. 271–295. Boca Raton, FL: CRC Press.

Konur, O. 2023f. Bioethanol fuel sensors: Scientometric study. In *Evaluation and Utilization of Bioethanol Fuels. I.: Evaluation of Bioethanol Fuels, Transport Engines, and Bioethanol Sensors. Handbook of Bioethanol Fuels Volume 5*, Ed. O. Konur, pp. 317–334. Boca Raton, FL: CRC Press.

Konur, O. 2023g. Country-based experience of bioethanol fuels: Review. In *Evaluation and Utilization of Bioethanol Fuels. II.: Biohydrogen Fuels, Fuel Cells, Biochemicals, and Country Experiences. Handbook of Bioethanol Fuels Volume 6*, Ed. O. Konur, pp. 26–41. Boca Raton, FL: CRC Press.

Konur, O. 2023h. Bioethanol fuel-based biohydrogen fuels: Scientometric study. In *Evaluation and Utilization of Bioethanol Fuels. II.: Biohydrogen Fuels, Fuel Cells, Biochemicals, and Country Experiences. Handbook of Bioethanol Fuels Volume 6*, Ed. O. Konur, pp. 215–236. Boca Raton, FL: CRC Press.

Konur, O. 2023i. Bioethanol fuel cells: Scientometric study. In *Evaluation and Utilization of Bioethanol Fuels. II.: Biohydrogen Fuels, Fuel Cells, Biochemicals, and Country Experiences. Handbook of Bioethanol Fuels Volume 6*, Ed. O. Konur, pp. 277–297. Boca Raton, FL: CRC Press.

Konur, O. 2023j. Bioethanol fuel-based biochemical production: Scientometric study. In *Evaluation and Utilization of Bioethanol Fuels. II.: Biohydrogen Fuels, Fuel Cells, Biochemicals, and Country Experiences. Handbook of Bioethanol Fuels Volume 6*, Ed. O. Konur, pp. 317–337. Boca Raton, FL: CRC Press.

Konur, O. 2023k. Bioethanol fuel evaluation and utilization: Scientometric study. In *Evaluation and Utilization of Bioethanol Fuels. I.: Evaluation of Bioethanol Fuels, Transport Engines, and Bioethanol Sensors. Handbook of Bioethanol Fuels Volume 5*, Ed. O. Konur, pp. 3–24. Boca Raton, FL: CRC Press.

Konur, O. 2023l. Bioethanol fuel evaluation and utilization: Review. In *Evaluation and Utilization of Bioethanol Fuels. I.: Evaluation of Bioethanol Fuels, Transport Engines, and Bioethanol Sensors. Handbook of Bioethanol Fuels Volume 5*, Ed. O. Konur, pp. 25–42. Boca Raton, FL: CRC Press.

Konur, O. and F. L. Matthews. 1989. Effect of the properties of the constituents on the fatigue performance of composites: A review. *Composites* 20:317–328.

Kumar, N., A. Dorfman and J. I. Hahm. 2006. Ultrasensitive DNA sequence detection using nanoscale ZnO sensor arrays. *Nanotechnology* 17:2875.

Landes, W. M. 1971. An economic analysis of the courts. *Journal of Law and Economics* 14:61–107.

Liang, Z. X., T. S. Zhao, J. B. Xu and L. D. Zhu. 2009. Mechanism study of the ethanol oxidation reaction on palladium in alkaline media. *Electrochimica Acta* 54:2203–2208.

Liu, G., J. Zhang and J. Bao. 2016. Cost evaluation of cellulase enzyme for industrial-scale cellulosic ethanol production based on rigorous Aspen Plus modeling. *Bioprocess and Biosystems Engineering* 39:133–140.

Liu, J., T. Wang and B. Wang, et al. 2017. Highly sensitive and low detection limit of ethanol gas sensor based on hollow ZnO/SnO_2 spheres composite material. *Sensors and Actuators, B: Chemical* 245:551–559.

Liu, J., X. Wang, Q. Peng and Y. Li. 2005. Vanadium pentoxide nanobelts: Highly selective and stable ethanol sensor materials. *Advanced Materials* 17:764–767.

Liu, P. and E. J. M. Hensen. 2013. Highly efficient and robust $Au/MgCuCr2O_4$ catalyst for gas-phase oxidation of ethanol to acetaldehyde. *Journal of the American Chemical Society* 135:14032–14035.

Lupan, O., V. Cretu and V. Postica, et al. 2016. Enhanced ethanol vapour sensing performances of copper oxide nanocrystals with mixed phases. *Sensors and Actuators, B: Chemical* 224:434–448.

Ma, H., L. Zeng and H. Tian, et al. 2016. Efficient hydrogen production from ethanol steam reforming over La-modified ordered mesoporous Ni-based catalysts. *Applied Catalysis B: Environmental* 181:321–331.

Ma, X., L. Sun and C. Song. 2002. A new approach to deep desulfurization of gasoline, diesel fuel and jet fuel by selective adsorption for ultra-clean fuels and for fuel cell applications. *Catalysis Today* 77:107–116.

Macedo, I. C., J. E. A. Seabra and J. E. A. R. Silva. 2008. Green house gases emissions in the production and use of ethanol from sugarcane in Brazil: The 2005/2006 averages and a prediction for 2020. *Biomass and Bioenergy* 32:582–595.

Mao, J., W. Chen and D. He, et al. 2017. Design of ultrathin Pt-Mo-Ni nanowire catalysts for ethanol electro-oxidation. *Science Advances* 3:1603068.

Mohsenzadeh, A., A. Zamani and M. J. Taherzadeh. 2017. Bioethylene production from ethanol: A review and techno-economical evaluation. *ChemBioEng Reviews* 4:75–91.

Morschbacker, A. 2009. Bio-ethanol based ethylene. *Polymer Reviews* 49:79–84.

Muller, J. F. and T. Stavrakou. 2005. Inversion of CO and NO_x emissions using the adjoint of the IMAGES model. *Atmospheric Chemistry and Physics* 5:1157–1186.

Murdoch, M., G. I. N. Waterhouse and M. A. Nadeem, et al. 2011. The effect of gold loading and particle size on photocatalytic hydrogen production from ethanol over Au/TiO_2 nanoparticles. *Nature Chemistry* 3:489–492.

Najafi, G., B. Ghobadian and T. Tavakoli, et al. 2009. Performance and exhaust emissions of a gasoline engine with ethanol blended gasoline fuels using artificial neural network. *Applied Energy* 86:630–639.

Newman, P. W. G. and J. R. Kenworthy. 1989. Gasoline consumption and cities: A comparison of U.S. cities with a global survey. *Journal of the American Planning Association* 55:24–37.

Nhuchhen, D. R. and P. A. Salam. 2012. Estimation of higher heating value of biomass from proximate analysis: A new approach. *Fuel* 99:55–63.

North, D. C. 1991. Institutions. *Journal of Economic Perspectives* 5:97–112.

Palo, D. R., R. A. Dagle and J. D. Holladay. 2007. Methanol steam reforming for hydrogen production. *Chemical Reviews* 107:3992–4021.

Qi, J., N. Benipal, C. Liang and W. Li. 2016. PdAg/CNT catalyzed alcohol oxidation reaction for high-performance anion exchange membrane direct alcohol fuel cell (alcohol=methanol, ethanol, ethylene glycol and glycerol). *Applied Catalysis B: Environmental* 199:494–503.

Ramos-Paja, C. A., C. Bordons, C., A. Romero, R. Giral and L. Martínez-Salamero. 2008. Minimum fuel consumption strategy for PEM fuel cells. *IEEE Transactions on Industrial Electronics* 56:685–696.

Rizo, R., R. M. Aran-Ais and E. Padgett, et al. 2018. Pt-rich$_{core}$/Sn-rich$_{subsurface}$/Pt$_{skin}$ nanocubes as highly active and stable electrocatalysts for the ethanol oxidation reaction. *Journal of the American Chemical Society* 140:3791–3797.

Salim, A. and S. Lim. 2016. Complementary split-ring resonator-loaded microfluidic ethanol chemical sensor. *Sensors* 16:1802.

Sheehan, J., A. Aden and K. Paustian, et al. 2003. Energy and environmental aspects of using corn stover for fuel ethanol. *Journal of Industrial Ecology* 7:117–146.

Silitonga, A. S., H. H. Masjuki and H. C. Ong, et al. 2018. Evaluation of the engine performance and exhaust emissions of biodiesel-bioethanol-diesel blends using kernel-based extreme learning machine. *Energy* 159:1075–1087.

Stanglmaier, R. H., J. Li and R. D. Matthews. 1999. The effect of in-cylinder wall wetting location on the HC emissions from SI engines. SAE Transactions 1999:01-0502.

Thiele, E. W. 1939. Relation between catalytic activity and size of particle. *Industrial & Engineering Chemistry* 31:916–920.

Tong, H., S. Ouyang and Y. Bi, et al. 2012. Nano-photocatalytic materials: Possibilities and challenges. *Advanced Materials* 24:229–251.

Venu, H. and V. Madhavan. 2016. Effect of Al_2O_3 nanoparticles in biodiesel-diesel-ethanol blends at various injection strategies: Performance, combustion and emission characteristics. *Fuel* 186:176–189.

Venu, H. and V. Madhavan. 2017. Influence of diethyl ether (DEE) addition in ethanol-biodiesel-diesel (EBD) and methanol-biodiesel-diesel (MBD) blends in a diesel engine. *Fuel* 189:377–390.

Wan, Q., Q. H. Li and Y. J. Chen, et al. 2004. Fabrication and ethanol sensing characteristics of ZnO nanowire gas sensors. *Applied Physics Letters* 84:3654–3656.

Wang, C., X. Cui and J. Liu, et al. 2016a. Design of superior ethanol gas sensor based on Al-doped NiO nanorod-flowers. *ACS Sensors* 1:131–136.

Wang, H., Z. Jusys and R. J. Behm. 2004. Ethanol electrooxidation on a carbon-supported Pt catalyst: Reaction kinetics and product yields. *Journal of Physical Chemistry B* 108:19413–19424.

Wang, Z., Z. Tian, D. Han and F. Gu. 2016b. Highly sensitive and selective ethanol sensor fabricated with In-doped 3DOM ZnO. *ACS Applied Materials and Interfaces* 8:5466–5474.

Wu, T., J. Fan, and Q. Li, et al. 2018. Palladium nanoparticles anchored on anatase titanium dioxide-black phosphorus hybrids with heterointerfaces: Highly electroactive and durable catalysts for ethanol electro-oxidation. *Advanced Energy Materials* 8:1701799.

Yunoki, S. and M. Saito. 2009. A simple method to determine bioethanol content in gasoline using two-step extraction and liquid scintillation counting. *Bioresource Technology* 100:6125–6128.

Zhao, X., L. Dai and Q. Qin, et al. 2017. Self-supported 3D PdCu alloy nanosheets as a bifunctional catalyst for electrochemical reforming of ethanol. *Small* 13:1602970.

Zhu, L., Y. Li and W. Zeng. 2018. Hydrothermal synthesis of hierarchical flower-like ZnO nanostructure and its enhanced ethanol gas-sensing properties. *Applied Surface Science* 427:281–287.

80 Gasoline Fuels
Scientometric Study

Ozcan Konur
(Formerly) Ankara Yildirim Beyazit University

80.1 INTRODUCTION

Crude oil-based gasoline fuels (Ma et al., 2002; Newman and Kenworthy, 1989) have been widely used in the transportation sector since the 1920s (Botsford, 1920; Brooks, 1915). However, there have been great public concerns over the adverse impact of these fuels on both the environment and human health (Hill et al., 2009; Miguel et al., 1998).

Hence, biomass-based bioethanol fuels (Hsieh et al., 2002; Najafi et al., 2009) and other bioalcohol fuels (Aleiferis and van Romunde, 2013; Christensen et al., 2011) have increasingly been used in blending gasoline fuels. In this context, there has been intense research in the research fronts of emissions (Miguel et al., 1998; Schauer et al., 2002), blends with bioethanol fuels and other bioalcohols, combustion (Kalghatgi et al., 2007; Mehl et al., 2011), economics (Borenstein et al., 1997; Newman and Kenworthy, 1989), production (Bjorgen et al., 2008; Meisel et al., 1976), and, to a lesser extent, the environmental impact (Hill et al., 2009; Moldovan et al., 2002), properties (Chica and Corma, 1999; Kelly et al., 1989), desulfurization (Ma et al., 2002; Song, 2003), and health impact (Hill et al., 2009; Jacobson, 2007) of gasoline fuels.

However, it is essential to develop efficient incentive structures (North, 1991) for the primary stakeholders to enhance the research in this field (Konur, 2000, 2002a–c, 2006a,b, 2007a,b). This is especially important as the funding rate in this field has been relative low.

The scientometric analysis has been used in this context to inform the primary stakeholders about the current state of the research in a selected research field (Garfield, 1955; Konur, 2011, 2012a–i, 2015, 2018a,b, 2019, 2020a,b, 2021a–d). However, there has been no scientometric study of the research in the field of gasoline fuels despite its great public importance.

This book chapter presents a scientometric study of the research in the field of the gasoline fuels. It examines the scientometric characteristics of both the sample and population data presenting the scientometric characteristics of these two datasets in the order of documents, authors, publication years, institutions, funding bodies, source titles, countries, Scopus subject categories, keywords, and research fronts.

80.2 MATERIALS AND METHODS

The search for this study was carried out using Scopus database (Burnham, 2006) in August 2021.

As a first step for the search of the relevant literature, the keywords were selected using the first most cited 500 papers. The selected keyword list was optimized to obtain a representative sample of papers for the searched research field. This keyword list was provided in the appendix for future replication studies. Additionally, the information about the most used keywords was given in the Section 80.3.9 to highlight the key research fronts in the Section 80.3.10.

As a second step, two sets of data were used for this study. First, a population sample of over 13,100 papers was used to examine the scientometric characteristics of the population data. Secondly, a sample of 100 most cited papers was used to examine the scientometric characteristics of these citation classics with over 180 citations each.

DOI: 10.1201/9781003226567-108

The scientometric characteristics of these both sample and population datasets were presented in the order of documents, authors, publication years, institutions, funding bodies, source titles, countries, Scopus subject categories, keywords, and research fronts.

Lastly, the key scientometric findings for both datasets were discussed to highlight the research landscape for the gasoline fuels. Additionally, a number of brief conclusions were drawn and a number of relevant recommendations were made to enhance the future research landscape.

80.3 RESULTS

80.3.1 The Most Prolific Documents in the Field of the Gasoline Fuels

The information on the types of documents for both datasets is given in Table 80.1. The articles, review papers, and conference papers published in the journals dominate the sample dataset while articles and conference papers dominate the population dataset.

Review papers and articles are overrepresented in the sample dataset while conference papers are overrepresented in the population dataset as shown in Table 80.1. Reviews and articles have a 5.3% and 17.4% surplus while conference papers have 17% deficit.

It is also interesting to note that all of the papers in the sample dataset were published in the journals, while only 95.8% of the papers were published in the journals for the population dataset. Furthermore, 3.1% and 1.0% of the population papers were published in books and book series, respectively.

80.3.2 The Most Prolific Authors in the Field of the Gasoline Fuels

The information about the most prolific 23 authors with at least two sample papers and with at least four population papers each is given in Table 80.2.

The most prolific authors are Robert A. Harley and Chunshan Song with four and three papers, respectively. All the other prolific authors have two papers each.

The most prolific institutions for the sample dataset are the Lawrence Livermore National Laboratory and University of California Berkeley with 3 authors each. The other prolific institutions with two papers each are the Chinese Academy of Sciences, the Desert Research Institute, Gazi University, Gubkin Russian State University of Oil and Gas, and Pennsylvania State University. In total, ten institutions house these authors.

The most prolific country for the sample dataset is the USA with 12 authors. This is followed by China, Turkey, Malaysia, and Russia with at least two authors each. In total, seven countries house these authors.

TABLE 80.1
Documents in the Field of the Gasoline Fuels

Documents	Sample Dataset (%)	Population Dataset (%)	Surplus (%)
Article	84	69.5	14.5
Conference paper	7	24.0	–17
Review	7	1.7	5.3
Note	1	2.3	–1.3
Short Survey	1	0.3	0.7
Book chapter	0	1.3	–1.3
Letter	0	0.8	–0.8
Editorial	0	0.2	–0.2
Book	0	0.0	0.0
Sample size	100	13,168	

TABLE 80.2
Most Prolific Authors in the Field of the Gasoline Fuels

No.	Author Name	Author Code	Sample Papers	Population Papers	Institution	Country	Fuels	Research Front
1	Harley, Robert A.	7102990151	4	18	Univ. Calif. Berkeley	USA	Gasoline	Emissions
2	Song, Chunshan	35376556000	3	12	Pennsylvania State Univ.	USA	Gasoline	Desulfurization
3	Xu, Hongming	16199134500	2	48	Birmingham Univ.	UK	Gasoline	Combustion
4	Mehl, Marco	7006214220	2	20	Lawrence Livermore Natl. lab.	USA	Gasoline	Properties, combustion
5	Pitz, William J.	7004802367	2	19	Lawrence Livermore Natl. lab.	USA	Gasoline	Properties, combustion
6	Masjuki Haji Hassan	57175108000	2	18	Univ. Malaya	Malaysia	Gasoline-bioalcohol blend	Emissions, performance
7	Canakci, Mustafa	6602870678	2	15	Kocaeli Univ.	Turkey	Gasoline-bioalcohol blend	Combustion, performance, emissions
8	Pitsch, Heinz	7003265812	2	14	Stanford Univ.	USA	Gasoline	Combustion
9	Kalam, Md Abul	55103352400	2	13	Univ. Malaya	Malaysia	Gasoline-bioalcohol blend	Emissions, performance
10	Yang, Yi	56784490100	2	12	Sandia Natl. Lab.	USA	Gasoline	Combustion
11	Kirchstetter, Thomas W	6604023055	2	9	Univ. Calif. Berkeley	USA	Gasoline	Emissions
12	Zielinska, Barbara*	7005780969	2	11	Desert Res. Inst.	USA	Gasoline	Emissions
13	Balabin, Roman M.	15049329900	2	8	Swiss Fed. Technol. Inst. Zurich	Switzerland	Gasoline	Properties
14	Goldstein, Allen H	7403063262	2	8	Univ. Calif. Berkeley	USA	Gasoline	Emissions
15	Sagebiel, John	7003501380	2	8	Desert Res. Inst.	USA	Gasoline-bioalcohol blend	Emissions
16	Westbrook, Charles K.	7102354977	2	7	Lawrence Livermore Natl. lab.	USA	Gasoline	Properties, combustion
17	Yucesu, Huseyin S.	6505773287	2	7	Gazi Univ.	Turkey	Gasoline-bioalcohol blend	Performance, emissions
18	Safieva, Ravilya Z.*	6602805500	2	5	Gubkin Russ. State Univ. Oil Gas	Russia	Gasoline	Properties
19	Topgul, Tolga	8367392400	2	5	Gazi Univ.	Turkey	Gasoline-bioalcohol blend	Performance, emissions
20	Li, Yan	55972807000	2	4	Chinese Acad. Sci.	China	Gasoline	Emissions
21	Lomakina, Ekaterina I.*	36721561800	2	4	Gubkin Russ. State Univ. Oil Gas	Russia	Gasoline	Properties
22	Ma, Xiaoliang	25226845100	2	4	Pennsylvania State Univ.	USA	Gasoline	Desulfurization
23	Tan, Mingguang	55633575300	2	4	Chinese Acad. Sci.	China	Gasoline	Emissions

Author code: The unique code given by Scopus to the authors. Sample papers: The number of papers authored in the sample dataset. Population papers: The number of papers authored in the population dataset. *: Female.

The most prolific research front for the sample dataset is the emissions of gasoline fuels with 11 authors. This is followed by the combustion, properties, performance, and desulfurization of gasoline fuels with seven, six, four, and two authors, respectively. Additionally, 13 and 6 of these prolific papers are concerned with gasoline fuels and gasoline-bioalcohol fuel blends, respectively.

On the other hand, there is a significant gender deficit (Beaudry and Lariviere, 2016) for the sample dataset as surprisingly nearly all of these top researchers are male.

80.3.3 The Most Prolific Research Output by Years in the Field of the Gasoline Fuels

Information about papers published between 1970 and 2021 is given in Figure 80.1. This figure clearly shows that the bulk of the research papers in the population dataset were published in the 2000s and 2010s with 22.3% and 42.7% of the population dataset, respectively. The publication rates were 5.6%, 3.3%, 6.8%, and 10.4% for the pre-1970s, 1970s, 1980s, and 1990s, respectively.

Similarly, the bulk of the research papers in the sample dataset were published in the 1990s, 2000s, and 2010s, with 21%, 49%, and 23% of the sample dataset, respectively. Furthermore, 1%, 2%, and 4% of the sample papers were published in the pre-1970s, 1970s, and 1980s, respectively.

The most prolific publication years for the population dataset were after 2012 with at least 3.8% each of the dataset. Similarly, 7% the sample papers each were published in 2004 and 2007. Additionally, there were 6 papers each published in 2005, 2006, 2008, and 2009.

80.3.4 The Most Prolific Institutions in the Field of the Gasoline Fuels

Information about the most prolific 21 institutions publishing papers on the gasoline fuels with at least two sample papers and at least 0.3% of the population papers each is given in Table 80.3.

The most prolific institutions are University of California Berkeley and Pennsylvania State University with five papers each. The Chinese Academy of Sciences and Stanford University follow these top institutions with four papers each.

The top countries for these most prolific institutions are the USA and China with eight and five institutions, respectively. Next, France follows these top countries with two institutions. In total, nine countries house these top institutions.

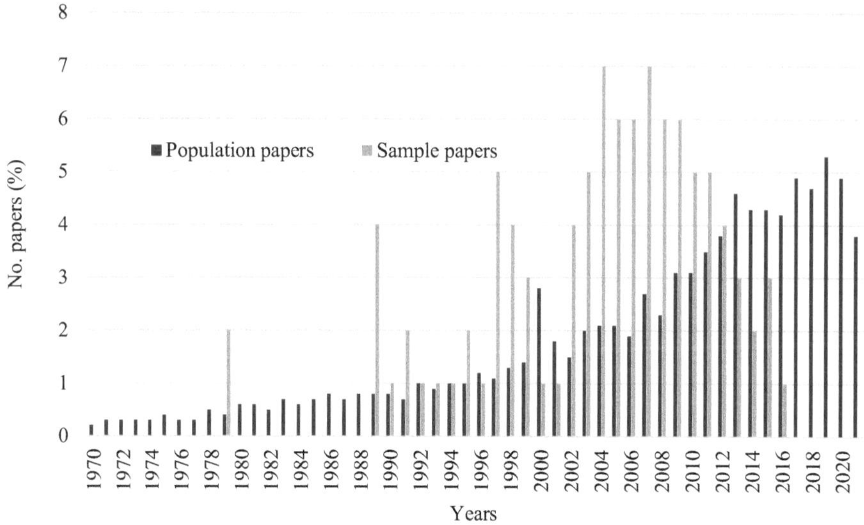

FIGURE 80.1 The research output by years regarding the gasoline fuels.

TABLE 80.3
The Most Prolific Institutions in the Field of the Gasoline Fuels

No.	Institutions	Country	Sample Papers (%)	Population Papers (%)	Surplus (%)
1	Univ. Calif. Berkeley	USA	5	0.6	4.4
2	Pennsylvania State Univ.	USA	5	0.3	4.7
3	Chinese Acad. Sci.	China	4	1.0	3.0
4	Stanford Univ.	USA	4	0.2	3.8
5	IFP New Energ.	France	3	0.7	2.3
6	Royal Dutch Shell	Netherlands	3	0.7	2.3
7	Lawrence Livermore Natl. Lab.	USA	3	0.3	2.7
8	Tianjin Univ.	China	2	1.8	0.2
9	Tsinghua Univ.	China	2	1.4	0.6
10	General Motors	USA	2	1.0	1.0
11	Exxon Mobil	USA	2	1.0	1.0
12	Toyota Motor	Japan	2	0.8	1.2
13	RWTH Aachen Univ.	Germany	2	0.6	1.4
14	Univ. Birmingham	UK	2	0.5	1.5
15	Xian Jiao Tong University	China	2	0.5	1.5
16	Beijing Univ. Technol.	China	2	0.5	1.5
17	Gubkin Natl. Univ. Oil Gas	Russia	2	0.4	1.6
18	CNRS	France	2	0.3	1.7
19	Natl. Renew. Energ. Lab.	USA	2	0.3	1.7
20	Univ. California, Riverside	USA	2	0.3	1.7
21	Univ. Malaya	Malaysia	2	0.2	1.8

On the other hand, the institutions with the most citation impact are the University of California Berkeley and Pennsylvania State University with over 4.4% surplus (Last column data in Table 80.3). The Chinese Academy of Sciences, Lawrence Livermore National Laboratory, IFP New Energies, and Royal Dutch Shell follow these top cited institutions with over 2.3% surplus each. Similarly, the institutions with the least impact are the Tianjin University, Tsinghua University, General Motors Co., and Exxon Mobil Co. with over 0.2% surplus each.

80.3.5 The Most Prolific Funding Bodies in the Field of the Gasoline Fuels

Information about the most prolific ten funding bodies funding at least two sample papers each is given in Table 80.4.

The most prolific funding bodies are the National Natural Science Foundation of China, the U.S. Department of Energy, and the U.S. Environmental Protection Agency funding over four sample papers each.

It is notable that only 33% and 22.9% of the sample and population papers are funded, respectively. The most prolific country for these top funding bodies is the USA with two funding bodies. In total, four countries and European Union house these top funding bodies.

80.3.6 The Most Prolific Source Titles in the Field of the Gasoline Fuels

Information about the most prolific 16 source titles publishing at least two papers and 0.3% of the population papers each in the field of the gasoline fuels is given in Table 80.5.

The most prolific source title is 'Fuel' with 11 sample papers. The 'Environmental Science and Technology' and 'Atmospheric Environment' follow this top journal with eight papers each.

TABLE 80.4
The Most Prolific Funding Bodies in the Field of the Gasoline Fuels

No.	Funding Bodies	Country	Sample Paper No. (%)	Population Paper No. (%)	Surplus (%)
1	National Natural Science Foundation of China	China	5	6.1	−1.1
2	U.S. Department of Energy	USA	4	1.8	2.2
3	U.S. Environmental Protection Agency	USA	4	0.4	3.6
4	National Science Foundation	USA	3	0.8	2.2
5	Engineering and Physical Sciences Research Council	UK	3	0.6	2.4
6	UK Research and Innovation	UK	3	0.5	2.5
7	Chinese Academy of Sciences	China	3	0.3	2.7
8	National Energy Technology Laboratory	USA	3	0.1	2.9
9	European Commission	EU	2	0.7	1.3
10	Univ. Malaya	Malaysia	2	0.1	1.9

TABLE 80.5
The Most Prolific Source Titles in the Field of the Gasoline Fuels

No.	Source Titles	Sample Papers (%)	Population Papers (%)	Surplus (%)
1	Fuel	11	4.5	6.5
2	Environmental Science and Technology	8	1.1	6.9
3	Atmospheric Environment	8	0.8	7.2
4	SAE Technical Papers	4	19.7	−15.7
5	Energy and Fuels	4	1.8	2.2
6	Applied Thermal Engineering	4	0.7	3.3
7	Energy Economics	4	0.7	3.3
8	Combustion and Flame	3	0.5	2.5
9	Renewable Energy	3	0.3	2.7
10	Science	3	0.2	2.8
11	Applied Energy	2	0.9	1.1
12	Energy Conversion and Management	2	0.8	1.2
13	Applied Catalysis A General	2	0.5	1.5
14	Catalysis Today	2	0.4	1.6
15	Proceedings of the Combustion Institute	2	0.4	1.6
16	Applied Catalysis B Environmental	2	0.3	1.7

On the other hand, the source title with the most citation impact is 'Atmospheric Environment' with 7.2% surplus. The 'Environmental Science and Technology' and 'Fuel' follow this top journal with at least 6.5% surplus each. Similarly, the source title with the least impact is the 'SAE Technical Papers' with 15.7% deficit.

It is notable that the bulk of these top source titles are indexed within the subject category of energy.

80.3.7 The Most Prolific Countries in the Field of the Gasoline Fuels

Information about the most prolific 17 countries publishing at least two papers each in the field of the gasoline fuels is given in Table 80.6.

The most prolific country is the USA with 46 sample papers. China, the UK, and Turkey follow this top country with 11, eight, and eight papers, respectively. Further, 10 European countries listed in Table 80.6 produce 35% and 16% of the sample and population papers, respectively.

TABLE 80.6
The Most Prolific Countries in the Field of the Gasoline Fuels

No.	Countries	Sample Papers (%)	Population Papers (%)	Surplus (%)
1	USA	46	27.2	18.8
2	China	11	16.0	–5.0
3	UK	8	5.9	2.1
4	Turkey	8	1.4	6.6
5	Germany	6	3.9	2.1
6	France	4	2.4	1.6
7	Sweden	4	1.5	2.5
8	Australia	4	1.2	2.8
9	Japan	3	5.9	–2.9
10	Canada	3	2.8	0.2
11	Spain	3	1.9	1.1
12	Russia	2	2.3	–0.3
13	Malaysia	2	1.0	1.0
14	Greece	2	0.7	1.3
15	Switzerland	2	0.7	1.3
16	Netherlands	2	0.6	1.4
17	Norway	2	0.2	1.8

TABLE 80.7
The Most Prolific Scopus Subject Categories in the Field of the Gasoline Fuels

No.	Scopus Subject Categories	Sample Papers (%)	Population Papers (%)	Surplus (%)
1	Energy	43	27.6	15.4
2	Chemistry	36	23.7	12.3
3	Chemical engineering	35	24.5	10.5
4	Environmental science	33	36.5	–3.5
5	Engineering	17	42.4	–25.4
6	Earth and Planetary Sciences	9	5.2	3.8
7	Economics	9	4.0	5.0
8	Physics and Astronomy	5	5.6	–0.6
9	Multidisciplinary	5	0.8	4.2
10	Biochemistry, Genetics, and Molecular Biology	4	3.1	0.9
11	Social Sciences	3	4.1	–1.1

On the other hand, the country with the most citation impact is again the USA with 18.8% surplus. Turkey follows this top country with 6.6% surplus. Similarly, the country with the least citation impact is China with 5.0% deficit. Japan and Russia follow China with 2.9% and 0.3% deficit, respectively.

80.3.8 The Most Prolific Scopus Subject Categories in the Field of the Gasoline Fuels

Information about the most prolific 11 Scopus subject categories indexing at least 3% of the sample papers each is given in Table 80.7.

The most prolific Scopus subject category is 'Energy' with 43 sample papers. This top category is followed by 'Chemistry', 'Chemical Engineering', and 'Environmental Science' with at least 33 sample papers each.

On the other hand, the Scopus subject category with the most citation impact is again 'Energy' with 15.4% surplus. This top category is followed by 'Chemistry' and 'Chemical Engineering' with over 10.5% surplus each. Similarly, the Scopus subject category with the least citation impact is 'Engineering' with 25.4% deficit. The other categories with the least impact are 'Environmental Science', 'Social Sciences', and 'Physics and Astronomy' with at least 0.6% deficit each.

80.3.9 The Most Prolific Keywords in the Field of the Gasoline Fuels

Information about the keywords with at least 5% and 2% in the sample and population papers, respectively is given in Table 80.8. For this purpose, keywords related to the keyword set given in the appendix are selected from a list of the most prolific keyword set provided by Scopus database.

There are four keywords used related to the gasoline fuels: Gasoline, gasoline engines, gasoline direct injection engines, and gasoline direct injection. The first keyword is used around 82% in both paper sets.

The most prolific keywords related to the emissions are exhaust gases and hydrocarbons with 24% and 20% of the sample papers each. On the other hand, the most prolific keywords related to engines are diesel engines, engines, and SI engines with at least 10% of the sample papers each. The most prolific keywords related to the fuels are ethanol, fuels, and diesel fuels with at least 18% of the sample papers each. The most prolific keywords related to the combustion are combustion and internal combustion engines with 17% of the sample papers each.

The most prolific keywords related to the engine performance are engine performance, anti-knock rating, and brake-specific fuel consumption with at least 5% of the sample papers each.

On the other hand, there are some keywords with the significant surplus: These are exhaust gases, gasoline, and diesel fuels with at least 16% surplus each. Similarly, Direct injection, gasoline direct injection engines, and gasoline direct injection are some of the keywords with the least impact.

80.3.10 The Most Prolific Research Fronts in the Field of the Gasoline Fuels

Information about the most prolific research fronts for the sample papers in the field of the gasoline fuels is given in Table 80.9.

There are two primary research fronts in this field: The gasoline fuels and the gasoline-bioalcohol fuel blends. There are also eight and six research fronts in these primary research fronts, respectively.

The gasoline emissions are the most prolific research front with 25 papers for the primary field of gasoline fuels (Miguel et al., 1998; Schauer et al., 2002). The other research fronts in this research front are gasoline combustion (Kalghatgi et al., 2007; Mehl et al., 2011), gasoline economics (Borenstein et al., 1997; Newman and Kenworthy, 1989), gasoline production (Bjorgen et al., 2008; Meisel et al., 1976), gasoline fuels in the environment (Hill et al., 2009; Moldovan et al., 2002), gasoline fuel properties (Chica and Corma, 1999; Kelly et al., 1989), gasoline desulfurization (Ma et al., 2002; Song, 2003), and the health impact of the gasoline fuels (Hill et al., 2009; Jacobson, 2007).

The gasoline-bioethanol fuel blends are the most prolific research front in the field of gasoline-bioalcohol fuel blends with 15 papers (Hsieh et al., 2002; Najafi et al., 2009). The other research fronts in this field are the gasoline oxygenates (Aleiferis and van Romunde, 2013; Christensen et al., 2011), gasoline-methanol blends (Eyidogan et al., 2010; Liu et al., 2007), gasoline-butanol fuel blends (Gu et al., 2012), gasoline-biohydrogen blends (D'Andrea et al., 2004), and gasoline-compressed natural gas blends (Aslam et al., 2006).

80.4 DISCUSSION

80.4.1 Introduction

Crude oil-based gasoline fuels have been widely used in the transportation sector since the 1920s. However, there have been great public concerns over the adverse effects of these fuels on both

TABLE 80.8
The Most Prolific Keywords in the Field of the Gasoline Fuels

No.	Keywords	Sample Papers (%)	Population Papers (%)	Surplus (%)
1.		**Gasoline Fuels**		
	Gasoline	82	65.8	16.2
	Gasoline engines	8	8.3	–0.3
	Gasoline direct injection engines		4.1	–4.1
	Gasoline direct injection		3.8	–3.8
2.		**Environmental Issues**		
	Exhaust gases	24	6.4	17.6
	Hydrocarbons	20	7.8	12.2
	Exhaust emissions	12	3.2	8.8
	Nitrogen oxides	11	5.4	5.6
	Carbon monoxide	11	4.3	6.7
	Particulate emissions	11	3.8	7.2
	Air pollution	10	2.9	7.1
	Polycyclic aromatic hydrocarbons	10	0	10
	Aerosol	9	0	9
	Gas emissions	8	3.5	4.5
	Particulate matter	8	2.7	5.3
	Suspended particulate matter	8	0	8
	Emissions	7	3.4	3.6
	Vehicle emissions	7	2.2	4.8
	Carbon dioxide	6	3.4	2.6
	Desulfurization	6	2.1	3.9
	Traffic emissions	6	1.7	4.3
	Aromatic hydrocarbons	5	1.2	3.8
	Air pollutants	4	3.2	0.8
	Emission control		2.4	–2.4
3.		**Engines**		
	Diesel engines	20	9.8	10.2
	Engines	14	15.8	–1.8
	SI engines	10	1.1	8.9
	Gasoline engines	8	8.3	–0.3
	Engine cylinders	7	8.5	–1.5
	Direct injection spark ignition engines	5	0	5
	Gasoline direct injection engines		4.1	–4.1
	Internal combustion engines		3.7	–3.7
	Automobile engines		1.6	–1.6
4.		**Fuels**		
	Ethanol	23	9.6	13.4
	Fuels	20	9.1	10.9
	Diesel fuels	18	2	16
	Alternative fuels	12	2.3	9.7
	Fuel consumption	11	2.6	8.4

(Continued)

TABLE 80.8 (*Continued*)
The Most Prolific Keywords in the Field of the Gasoline Fuels

No.	Keywords	Sample Papers (%)	Population Papers (%)	Surplus (%)
	Alcohols	8	1.8	6.2
	Blending	6	3.8	2.2
	Ethanol fuels	6	3.2	2.8
	Fuel economy	5	4.9	0.1
	Methanol	4	3.6	0.4
	Biofuels	4	2.3	1.7
	Fuel injection	3	2.4	0.6
5.		**Combustion**		
	Combustion	17	10.6	6.4
	Internal combustion engines	17	3.7	13.3
	Ignition	7	9.5	−2.5
	Combustion chambers		2.7	−2.7
	Combustion characteristics		1.8	−1.8
	Compression ignition		1.7	−1.7
	Combustion knock		1.6	−1.6
	Combustion process		1.6	−1.6
6.		**Performance**		
	Engine performance	6	2.6	3.4
	Antiknock rating	5	1.8	3.2
	Brake-specific fuel consumption	5	1.1	3.9
	Direct injection	4	11.5	−7.5

TABLE 80.9
The Most Prolific Research Fronts in the Field of Gasoline Fuels and Their Blends

No.	Research Fronts	Sample Papers (%)
1.	Gasoline fuels	86
1.1.	Gasoline emissions	25
1.2.	Gasoline combustion	12
1.3.	Gasoline economics	11
1.4.	Gasoline production	10
1.5.	Gasoline in environment	9
1.6.	Gasoline properties	8
1.7.	Gasoline desulfurization	7
1.8.	Gasoline and health	4
2.	Gasoline-bioalcohol blends	23
2.1.	Gasoline-bioethanol blends	15
2.2.	Gasoline oxygenates	3
2.3.	Gasoline-methanol blends	2
2.4.	Gasoline-butanol blends	1
2.5.	Gasoline-biohydrogen blends	1
2.6.	Gasoline-compressed natural gas blends	1

environmental and the human. Hence, biomass-based bioalcoholic fuels have increasingly been used in blending gasoline fuels in recent years to reduce these adverse effects. In this context, there has been intense research in the research fronts of emissions, blends with bioethanol fuels and other bioalcohols, combustion, economics, production, and, to a lesser extent, the environmental impact, properties, desulfurization, and health impact of gasoline fuels. However, it is essential to develop efficient incentive structures for the primary stakeholders to enhance the research in this field as the funding ate for this field has been relatively very low.

The scientometric analysis has been traditionally used to inform the primary stakeholders about the current state of the research in a selected research field. However, there has been no scientometric study of the research in the field of the gasoline fuels. This book chapter presents a scientometric study of the research in this field. It examines the scientometric characteristics of the sample and population data presenting scientometric characteristics of these both datasets in the order of documents, authors, publication years, institutions, funding bodies, source titles, countries, Scopus subject categories, keywords, and research fronts.

The search for this study was carried out using Scopus database in August 2021. As a first step for the search of the relevant literature, the keywords were selected using the first most cited 500 papers. The selected keyword list was optimized to obtain a representative sample of papers for the searched research field. This keyword list was provided in the appendix for future replication studies. Additionally, the information about the most used keywords was given in the Section 80.3.9 to highlight the key research fronts in the Section 80.3.10.

As a second step, two sets of data were used for this study. First, a population sample of over 13,100 papers was used to examine the scientometric characteristics of the population data. Secondly, a sample of 100 most cited papers was used to examine the scientometric characteristics of these citation classics with over 180 citations each.

80.4.2 The Most Prolific Documents in the Field of the Gasoline Fuels

The review papers and articles are overrepresented in the sample dataset while conference papers are overrepresented in the population dataset as shown in Table 80.1. It is also interesting to note that all of the papers in the sample dataset were published in the journals, while only 95.8% of the papers were published in the journals for the population dataset. Furthermore, 3.1% and 1.0% of the population papers were published in books and book series, respectively.

Furthermore, 24% of the population papers are conference papers published in journals. Scopus differs from the Web of Science database in differentiating and showing both articles and conference papers published in the journals. Therefore, both articles and conference papers could be regrouped as articles. In such as a case, these regrouped articles from 91% and 93.5% of the sample and population papers, respectively.

The reviews and articles have a 5.3% and 17.4% surplus while conference papers have 17% deficit. These findings suggest that articles and conference papers have the most and least citation impact, respectively.

It is observed during the search process that there has been inconsistency in the classification of the documents in Scopus as well as in other databases such as Web of Science. This is especially relevant for the classification of papers as reviews or articles as the papers not involving a literature review may be erroneously classified as a review paper. In this context, it would be helpful to provide a classification note for the published papers in the books and journals at the first instance. It would also be helpful to use the document types listed in Table 80.1 for this purpose. Alternatively, the authors could insert the words, such as review or overview, denoting that the papers are a review paper in the titles or abstracts of their papers. This would be immensely helpful in differentiating the review papers from the ordinary papers such as articles or notes as it has been done in Song (2003).

80.4.3 The Most Prolific Authors in the Field of the Gasoline Fuels

There have been most prolific 23 authors with at least three papers in the sample dataset and with at least four population papers each as given in Table 80.2. These authors have shaped the development of the research in this field.

The most prolific authors are Robert A. Harley and Chunshan Song with four and three papers, separately. All the other prolific authors have two paper each (Table 80.2).

It is important to note the inconsistencies in indexing of the author names in Scopus and other databases. It is especially an issue for the names with more than two components such as 'Judge Alex de Camp Sirous'. The probable outcomes are 'Sirous, J.A.D.C.', 'de Camp Sirous, J.A.', or 'Camp Sirous, J.A.D.'. The first choice is the gold standard of the publishing sector where the last word in the name is taken as the last name. The second choice is a strong alternative while the last choice is an undesired outcome as it gives two words for the last name. It is notable that inconsistent indexing of the author names may cause substantial inefficiencies in the search process for the papers as well as allocating credit to the authors as there are often different author entries for each outcome in the databases.

There is also a difficulty in allowing credit for the authors especially for the authors with common names such as 'Wang, J.', or 'Huang, Z.', or 'Zhu, L.' in conducting scientometric studies. These difficulties strongly influence the efficiency of the scientometric studies as well as allocating credit to the authors as there are the same author entries for different authors with the same name, e.g. 'Wang J.' in the databases.

In this context, the coding of authors in Scopus database is a welcome innovation compared with the other databases such as Web of Science. In this process, Scopus allocates a unique number to each author in the database. However, there might still be substantial inefficiencies in this coding system especially for common names. It is possible that Scopus uses a number of software programs to differentiate the author names (Shin et al., 2014).

In this context, it does not help that author names are not given in full in some journals: For example, 'Wang, Y.' is given instead of 'Wang, Yuan'. This makes difficult to differentiate authors with common names and makes the scientometric studies further difficult in the author domain. Therefore, the author names should be given in full in all books and journals at the first instance.

There also inconsistencies in naming of the authors with more than two components in journal papers and book chapters. For example, 'Faaij, A.P.C.', 'Wyman C.E.', and 'Shuai, S.J.' might be given as 'Faaij, A.', 'Wyman C.', or 'Shuai S.' in the journals and books. This also makes the scientometric studies difficult in the author domain. Hence, contributing authors should use their name consistently in their publications.

The other critical issue regarding the author names is the spelling of the author names in the national spellings, 'Gonçales Sübadeye' rather than in the English spellings, 'Gonvales Subadeye', in Scopus database. Scopus differs from the Web of Science database in this respect where the author names are given in the English spellings. It is observed that the national spellings of the author names do not help in conducting scientometric studies as well in allocating credits to the authors as sometimes there are the different author entries for the English and National spellings for the same author in the Scopus database.

The most prolific institutions for the sample dataset are the Lawrence Livermore National Laboratory, University of California Berkeley, the Chinese Academy of Sciences, the Desert Research Institute, Gazi University, Gubkin Russian State University of Oil and Gas, and Pennsylvania State University. In total, ten institutions house these authors.

Similarly, the most prolific countries for the sample dataset are the USA, China, Turkey, Malaysia, and Russia. In total, seven countries house these authors.

The most prolific research fronts for the sample dataset are the emissions, combustion, properties, performance, and desulfurization of gasoline fuels. Additionally, 16 and 7 of these prolific papers are concerned with gasoline fuels and gasoline-bioalcohol fuel blends, respectively.

It is also notable that there is significant gender deficit for the sample dataset as surprisingly nearly all these top researchers are male. This finding is the most thought-provoking with strong public policy implications. Hence, institutions, funding bodies, and policy makers should take efficient measures to reduce the gender deficit in this field as well as other scientific fields with strong gender deficit.

80.4.4 The Most Prolific Research Output by Years in the Field of the Gasoline Fuels

The research output observed between 1970 and 2021 is illustrated in Figure 80.1. This figure clearly shows that the bulk of the research papers in the population dataset were published in the 2000s and 2010s. Similarly, the bulk of the research papers in the sample dataset were published in the 1990s, 2000s, and 2010s.

These data suggest that the most cited sample papers were primarily published in the 2000s and the early 2010s while the population papers were primarily published in the late 2000s and 2010s. These are the thought-provoking findings as there has been a persistent rise in the research output since the 1970s as seen from Figure 80.1, unlike the other studies in the field of bioethanol fuels. For example, 5.6% of the population papers has been published in the pre-1970s. This finding is not surprising since gasoline fuels have been produced and consumed since the 1920s (Botsford, 1920; Brooks, 1915).

In this context, the increasing public concerns about climate change (Change, 2007), greenhouse gas emissions (Carlson et al., 2017), and global warming (Kerr, 2007) have been certainly behind the boom in the research in this field in the last two decades.

The data in Figure 80.1 also suggest that the research in this field has boomed in the last two decades and the size of the population papers likely to more than double in the current decade, provided that the public concerns about climate change, greenhouse gas emissions, and global warming are translated efficiently to the research funding in this field.

80.4.5 The Most Prolific Institutions in the Field of the Gasoline Fuels

The most prolific 21 institutions publishing papers on the gasoline fuels with at least two sample papers and at least 0.3% of the population papers each given in Table 80.3 have shaped the development of the research in this field.

The most prolific institutions are University of California Berkeley, Pennsylvania State University, the Chinese Academy of Sciences, and Stanford University. The top countries for these most prolific institutions are the USA, China, and France.

On the other hand, the institutions with the most citation impact are University of California Berkeley, Pennsylvania State University, the Chinese Academy of Sciences, Lawrence Livermore National Laboratory, IFP New Energies, and Royal Dutch Shell. Similarly, the institutions with the least impact are Tianjin University, Tsinghua University, General Motors Co., and Exxon Mobil Co.

80.4.6 The Most Prolific Funding Bodies in the Field of the Gasoline Fuels

Information about the most prolific ten funding bodies funding at least two sample papers each is given in Table 80.4. The most prolific funding bodies are the National Natural Science Foundation of China, the U.S. Department of Energy, and the U.S. Environmental Protection Agency. It is notable that only 33% and 22.9% of the sample and population papers are funded, respectively. The most prolific country for these top funding bodies is the USA.

These findings on the funding of the research in this field suggest that the level of the funding is low and there is ample scope to increase the funding in this field to enhance the research further (Ebadi and Schiffauerova, 2016) above 22.9% for the population papers in the light of North's institutional framework (North, 1991). It is notable that despite the global initiatives to promote biofuels, gasoline fuels are still the most used crude oil-based fuels together with the petrodiesel fuels and liquefied natural gas in the vehicles.

80.4.7 The Most Prolific Source Titles in the Field of the Gasoline Fuels

The most prolific 16 source titles publishing at least two papers and 0.3% of the population papers each have shaped the development of the research in this field (Table 80.5). The most prolific source titles are 'Fuel', 'Atmospheric Environment', 'Renewable Energy', and 'Renewable and Sustainable Energy Reviews'.

On the other hand, the source titles with the most citation impact are 'Atmospheric Environment', 'Environmental Science and Technology', and 'Fuel'. Similarly, the source title with the least impact is the 'SAE Technical Papers'.

It is notable that bulk of these top source titles are indexed within the subject category of energy. This finding suggests that the energy and environment-related journals have significantly shaped the development of the research in this field. The research fronts of the combustion, performance, and emissions of the gasoline fuels and their gasoline-bioalcohol fuel blends have been primarily researched in recent years to increase the efficiency of these fuels and to reduce their adverse effects on the environment and human health.

80.4.8 The Most Prolific Countries in the Field of the Gasoline Fuels

Seventeen countries publishing at least two papers each have significantly shaped the development of the research in this field (Table 80.6). The most prolific countries are the USA, China, the UK, and Turkey. On the other hand, the countries with the most citation impact are Turkey and the USA. Similarly, the countries with the least citation impact are China, Japan, and Russia.

The close examination of these findings suggests that the USA, China, and Europe are the major producers of the research in this field. It is a fact that the USA has been a major player in science (Leydesdorff and Wagner, 2009; Leydesdorff et al., 2014). The USA has further developed a strong research infrastructure to support its gasoline fuel industry and consumption (An-Loh et al., 1985; Carolan, 2009).

However, China has been a rising star in scientific research in competition with the USA and Europe (Leydesdorff and Zhou, 2005). China is also a major player in this field as a major producer and consumer of gasoline fuels (Huo et al., 2012).

Next, the Europe has been a persistent player in the scientific research in competition with both the USA and China (Leydesdorff, 2000). The Europe has also been a persistent producer and consumer of gasoline fuels along with the USA and China (Pock, 2010; von Storch et al., 2003).

80.4.9 The Most Prolific Scopus Subject Categories in the Field of the Gasoline Fuels

The most prolific 11 Scopus subject categories indexing at least 3% of the sample papers each given in Table 80.7 have shaped the development of the research in this field.

The most prolific Scopus subject categories are 'Energy', 'Chemistry', 'Chemical Engineering', and 'Environmental Science'. On the other hand, the Scopus subject categories with the most citation impact are 'Energy', 'Chemical Engineering', and 'Chemistry'. Similarly, the Scopus subject categories with the least citation impact are 'Engineering', 'Environmental Science', 'Social Sciences', and 'Physics and Astronomy'.

These findings are thought-provoking suggesting that the primary subject categories are energy, chemical engineering, chemistry, environmental sciences, and engineering. All these research fronts are mostly related to the hard sciences, engineering, and environmental sciences. These subject categories hint the substantial weight of the research fronts related to the production and refining of gasoline fuels and their gasoline-bioalcohol fuel blends as well as those related to the impact of these fuels on both the environment and human health (Table 80.9).

The other key finding is that social sciences are not well represented in the sample and population papers, unlike the field of the evaluative studies in bioethanol fuels. However, it should be noted that the studies on the economics of gasoline fuels have been well developed over time as a primary research front as the taxation and pricing of these fuels impact heavily on the consumer choices to consume the fuels in relation to other types of fuels.

80.4.10 The Most Prolific Keywords in the Field of the Gasoline Fuels

A limited number of keywords have shaped the development of the research in this field as shown in Table 80.8 and the Appendix.

There are four keywords used related to the gasoline fuels: Gasoline, gasoline engines, gasoline direct injection engines, and gasoline direct injection. The first keyword is used around 82% in both paper sets.

The most prolific keywords related to the emissions are exhaust gases and hydrocarbons. On the other hand, the most prolific keywords related to engines are diesel engines, engines, and SI engines. The most prolific keywords related to the fuels are ethanol, fuels, and diesel fuels. The most prolific keywords related to the combustion are combustion and internal combustion engines. The most prolific keywords related to the engine performance are engine performance, antiknock rating, and brake-specific fuel consumption.

On the other hand, there are some keywords with the significant surplus: These are exhaust gases, gasoline, and diesel fuels with at least 16% surplus each. Similarly, direct injection, gasoline direct injection engines, and gasoline direct injection are some of the keywords with the least impact.

The keywords listed in Table 80.8 simply highlight the key research fronts for these fuels: Environmental issues covering fuel emissions, engines, fuels, combustion, and performance. For example, the most used keywords for the research front of the environmental issues are exhaust gases, hydrocarbons, exhaust emissions, nitrogen oxides, carbon monoxide, particulate emissions, air pollution, polycyclic aromatic hydrocarbons, aerosol, gas emissions, particulate matter, suspended particulate matter, emissions, vehicle emissions, carbon dioxide, desulfurization, traffic emissions, aromatic hydrocarbons, air pollutants, and emission control. In other words, these limited numbers of keywords shape the development of the research in the field of the environmental impact of the gasoline fuels and their gasoline-bioalcohol fuel blends.

80.4.11 The Most Prolific Research Fronts in the Field of the Gasoline Fuels

Information about the most prolific research fronts for the sample papers in the field of the gasoline fuels is given in Table 80.9. There are two primary research fronts in this field: The gasoline fuels and the gasoline-bioalcohol fuel blends. There are also eight and six research fronts in these primary research fronts, respectively. It is apparent the first research front has a long standing since the 1920s while the second research front has been a more recent innovative development to reduce the adverse impact of the gasoline fuels on both the environment and human health and to improve the combustion and performance of the gasoline fuels.

The gasoline emissions are the most prolific research front with 25 papers for the primary field of gasoline fuels. The other research fronts in the research front are gasoline combustion, gasoline economics, gasoline production, gasoline fuels in the environment, gasoline fuel properties, gasoline desulfurization, and the health impact of the gasoline fuels.

The gasoline-bioethanol fuel blends are the most prolific research front in the field of gasoline-bioalcohol fuel blends. The other research fronts in this field are the gasoline oxygenates, gasoline-methanol blends, gasoline-butanol fuel blends, gasoline-biohydrogen blends, and gasoline-compressed natural gas blends.

These findings are thought-provoking. First, the focus in on the properties, combustion, emissions, production, economics of gasoline fuels and their blends with bioalcohols. However, it is expected that the bulk of the research would focus on the emissions of bioethanol fuels and the removal of toxic compounds in the light of the public concerns about global warming, GHG emissions, and climate change in the future. The research on the adverse impact of gasoline fuels on the environment and human health has also great public importance.

Secondly, as gasoline fuels are increasingly used blending them with bioalcohols, the research focus is likely to be more on these blends rather than gasoline fuels alone in the future.

80.5 CONCLUSION AND FUTURE RESEARCH

The research on the gasoline fuels has been mapped through a scientometric study of both sample and population datasets.

The critical issue in this study has been to obtain a representative sample of the research as in any other scientometric study. Therefore, the keyword set has been carefully devised and optimized after a number of runs in the Scopus database. Apparently, the keyword set for the gasoline fuels has been relative simple as three keywords of gasoline*, biogasoline*, and gasohol* capture the researched studies well.

The other issue has been the selection of a multidisciplinary database to carry out the scientometric study of the research in this field. For this purpose, Scopus database has been selected. The journal coverage of this database has been wider than that of Web of Science. For example, SAE Technical Papers which form 19.7% of the population papers is covered by the Scopus database compared with the Web of Science database. Additionally, the Scopus database index the papers back to the 1900s. This issue is most relevant for this study as 5.6% of the population papers were published before 1970.

The key scientometric properties of the research in this field have been determined and discussed in this book chapter. It is evident that a limited number of documents, authors, institutions, publication periods, institutions, funding bodies, source titles, countries, Scopus subject categories, keywords, and research fronts have shaped the development of the research in this field.

There is ample scope to increase the efficiency of the scientometric studies in this field in the author and document domains by developing consistent policies and practices in both domains. In this respect, authors, journals, and academic databases have a lot to do. Further, the significant gender deficit as in most scientific fields emerges as a public policy issue.

The research in this field has boomed in the late 2000s, 2010s possibly promoted by the public concerns on global warming, greenhouse gas emissions, and climate change. However, there has been a persistent rise in the research output since the 1970s with a notable research output in the pre-1970s, unlike the studies in the field of bioethanol fuels where the research output has been more recent.

Institutions from the USA, China, Turkey, and the Europe have mostly shaped the research in this field. The low funding rate of 33% and 22.9% for sample and population papers, respectively, suggests that there is ample scope to increase funding in this field to enhance the research in this field, possibly more than doubling in the current decade. This is especially important as gasoline fuels are one of the most used fuels in cars together with petrodiesel and liquefied natural gas.

The most prolific journals have been mostly indexed by the subject category of energy. The USA, China, and Europe have been the major producers of the research in this field.

The primary subject categories have been Energy, Chemistry, Chemical Engineering, Engineering, and Environmental Sciences. Due the technological emphasis of this field, social sciences have not been fairly represented in both the sample and population papers, unlike the evaluative studies in bioethanol fuels. However, there has been a significant focus on the economics of gasoline fuels due to its public importance.

Gasoline is the primary keyword for the gasoline fuels. The primary keywords for the emissions of bioethanol fuels highlight the broad content of the emissions from gasoline fuels. Finally, the primary keywords related to engines, fuels, and combustion of gasoline highlight the fact that gasoline fuels are blended in the bioalcohol fuels.

The keywords listed in Table 80.8 simply highlight the key research fronts for these fuels: environmental issues covering fuel emissions, engines, fuels, combustion, and performance. For example, the most used keywords for the research front of the environmental issues are exhaust gases, hydrocarbons, exhaust emissions, nitrogen oxides, carbon monoxide, particulate emissions, air pollution, polycyclic aromatic hydrocarbons, aerosol, gas emissions, particulate matter, suspended particulate matter, emissions, vehicle emissions, carbon dioxide, desulfurization, traffic emissions, aromatic hydrocarbons, air pollutants, and emission control. In other words, these limited number of the keywords shape the development of the research in the field of the environmental impact of the gasoline fuels and their gasoline-bioalcohol fuel blends.

There are two primary research fronts in this field: The gasoline fuels and the gasoline-bioalcohol fuel blends. There are also eight and six research fronts in these primary research fronts, respectively. These findings are thought-provoking. First, the focus in on the properties, combustion, emissions, production, economics, desulfurization of gasoline fuels, and their blends with bioalcohol fuels. However, as it is expected, the bulk of the research would focus on the emissions of gasoline fuels and their adverse impact both on the environment and the human health in the future in the light of the public concerns about global warming, GHG emissions, and climate change. Secondly, as gasoline fuels are increasingly used blending them with bioalcohol fuels, the research focus is currently more on these blends rather than gasoline fuels alone.

Thus, the scientometric analysis has a great potential to gain valuable insights into the evolution of the research in this field as in other scientific fields.

It is recommended that further scientometric studies are carried out about the other aspects of both production and utilization of bioethanol fuels. It is further recommended that reviews of the most cited papers are carried out for each research front to complement these scientometric studies. Next, the scientometric studies of the hot papers in these primary fields are carried out.

ACKNOWLEDGMENTS

The contribution of the highly cited researchers in the field of the gasoline fuels has been gratefully acknowledged.

APPENDIX: THE KEYWORD SET FOR THE FIELD OF THE GASOLINE FUELS

(TITLE (gasoline* OR biogasoline* OR gasohol*) OR SRCTITLE (gasoline* OR biogasoline* OR gasohol*)) AND NOT (SRCTITLE (astronom*) OR TITLE (astronom*).

REFERENCES

Aleiferis, P. G. and Z. R. van Romunde. 2013. An analysis of spray development with iso-octane, n-pentane, gasoline, ethanol and n-butanol from a multi-hole injector under hot fuel conditions. *Fuel* 105:143–168.

An-loh, L., E. N. Botsas and S. A. Monroe. 1985. State gasoline consumption in the USA: An econometric analysis. *Energy Economics* 7:29–36.

Aslam, M. U., H. H. Masjuki and M. A. Kalam, et al. 2006. An experimental investigation of CNG as an alternative fuel for a retrofitted gasoline vehicle. *Fuel* 85:717–724.

Beaudry, C. and V. Lariviere. 2016. Which gender gap? Factors affecting researchers' scientific impact in science and medicine. *Research Policy* 45:1790–1817.

Bjorgen, M., F. Joensen and M. S. Holm, et al. 2008. Methanol to gasoline over zeolite H-ZSM-5: Improved catalyst performance by treatment with NaOH. *Applied Catalysis A: General* 345:43–50.

Borenstein, S., A. C. Cameron and R. Gilbert. 1997. Do gasoline prices respond asymmetrically to crude oil price changes? *Quarterly Journal of Economics* 112:304–339.

Botsford, H. 1920. Gasoline from natural gas. *Scientific American* 122:359–375.
Brooks, B. 1915. A statistical review of the question of gasoline supply. *Industrial & Engineering Chemistry* 7:176–179.
Burnham, J. F. 2006. Scopus database: A review. *Biomedical Digital Libraries* 3:1–8.
Carlson, K. M., J. S. Gerber and N. D. Mueller, et al. 2017. Greenhouse gas emissions intensity of global croplands. *Nature Climate Change* 7:63–68.
Carolan, M. S. 2009. Ethanol versus gasoline: The contestation and closure of a socio-technical system in the USA. *Social Studies of Science* 39:421–448.
Change, C. 2007. Climate change impacts, adaptation and vulnerability. *Science of the Total Environment* 326:95–112.
Chica, A. and A. Corma. 1999. Hydroisomerization of pentane, hexane, and heptane for improving the octane number of gasoline. *Journal of Catalysis* 187:167–176.
Christensen, E., J. Yanowitz, M. Ratcliff and R. L. McCormick. 2011. Renewable oxygenate blending effects on gasoline properties. *Energy and Fuels* 25:4723–4733.
D'Andrea, T., P. F. Henshaw and D. S. K. Ting. 2004. The addition of hydrogen to a gasoline-fuelled SI engine. *International Journal of Hydrogen Energy* 29:1541–1552.
Ebadi, A. and A. Schiffauerova. 2016. How to boost scientific production? A statistical analysis of research funding and other influencing factors. *Scientometrics* 106:1093–1116.
Eyidogan, M., A. N. Ozsezen, M. Canakci and A. Turkcan. 2010. Impact of alcohol-gasoline fuel blends on the performance and combustion characteristics of an SI engine. *Fuel* 89:2713–2720.
Garfield, E. 1955. Citation indexes for science. *Science* 122:108–111.
Gu, X., Z. Huang and J. Cai, et al. 2012. Emission characteristics of a spark-ignition engine fuelled with gasoline-n-butanol blends in combination with EGR. *Fuel* 93:611–617.
Hill, J., S. Polasky and E. Nelson, et al. 2009. Climate change and health costs of air emissions from biofuels and gasoline. *Proceedings of the National Academy of Sciences of the United States of America* 106:2077–2082.
Hsieh, W. D., R. H. Chen, T. L. Wu and T. H. Lin. 2002. Engine performance and pollutant emission of an SI engine using ethanol-gasoline blended fuels. *Atmospheric Environment* 36:403–410.
Huo, H., Z. Yao and Y. Zhang, et al. 2012. On-board measurements of emissions from light-duty gasoline vehicles in three mega-cities of China. *Atmospheric Environment* 49:371–377.
Jacobson, M. Z. 2007. Effects of ethanol (E85) versus gasoline vehicles on cancer and mortality in the United States. *Environmental Science and Technology* 41:4150–4157.
Kalghatgi, G. T., P. Risberg and H. E. Angstrom. 2007. Partially pre-mixed auto-ignition of gasoline to attain low smoke and low NO_x at high load in a compression ignition engine and comparison with a diesel fuel. SAE Technical Papers 2007-01-0006.
Kelly, J. J., C. H. Barlow, T. M. Jinguji and J. B. Callis. 1989. Prediction of gasoline octane numbers from near-infrared spectral features in the range 660-1215 nm. *Analytical Chemistry* 61:313–320.
Kerr, R. A. 2007. Global warming is changing the world. *Science* 316:188–190.
Konur, O. 2000. Creating enforceable civil rights for disabled students in higher education: An institutional theory perspective. *Disability & Society* 15:1041–1063.
Konur, O. 2002a. Access to nursing education by disabled students: Rights and duties of nursing programs. *Nurse Education Today* 22:364–374.
Konur, O. 2002b. Assessment of disabled students in higher education: Current public policy issues. *Assessment and Evaluation in Higher Education* 27:131–152.
Konur, O. 2002c. Access to employment by disabled people in the UK: Is the Disability Discrimination Act working? *International Journal of Discrimination and the Law* 5:247–279.
Konur, O. 2006a. Participation of children with dyslexia in compulsory education: Current public policy issues. *Dyslexia* 12:51–67.
Konur, O. 2006b. Teaching disabled students in higher education. *Teaching in Higher Education* 11:351–363.
Konur, O. 2007a. A judicial outcome analysis of the Disability Discrimination Act: A windfall for the employers? *Disability & Society* 22:187–204.
Konur, O. 2007b. Computer-assisted teaching and assessment of disabled students in higher education: The interface between academic standards and disability rights. *Journal of Computer Assisted Learning* 23:207–219.
Konur, O. 2011. The scientometric evaluation of the research on the algae and bio-energy. *Applied Energy* 88:3532–3540.
Konur, O. 2012a. Prof. Dr. Ayhan Demirbas' scientometric biography. *Energy Education Science and Technology Part A: Energy Science and Research* 28:727–738.

Konur, O. 2012b. The evaluation of the biogas research: A scientometric approach. *Energy Education Science and Technology Part A: Energy Science and Research* 29:1277–1292.
Konur, O. 2012c. The evaluation of the global energy and fuels research: A scientometric approach. *Energy Education Science and Technology Part A: Energy Science and Research* 30:613–628.
Konur, O. 2012d. The evaluation of the research on the biodiesel: A scientometric approach. *Energy Education Science and Technology Part A: Energy Science and Research* 28:1003–1014.
Konur, O. 2012e. The evaluation of the research on the bioethanol: A scientometric approach. *Energy Education Science and Technology Part A: Energy Science and Research* 28:1051–1064.
Konur, O. 2012f. The evaluation of the research on the biofuels: A scientometric approach. *Energy Education Science and Technology Part A: Energy Science and Research* 28:903–916.
Konur, O. 2012g. The evaluation of the research on the biohydrogen: A scientometric approach. *Energy Education Science and Technology Part A: Energy Science and Research* 29:323–338.
Konur, O. 2012h. The evaluation of the research on the microbial fuel cells: A scientometric approach. *Energy Education Science and Technology Part A: Energy Science and Research* 29:309–322.
Konur, O. 2012i. The scientometric evaluation of the research on the production of bioenergy from biomass. *Biomass and Bioenergy* 47:504–515.
Konur, O. 2015. Current state of research on algal bioethanol. In *Marine Bioenergy: Trends and Developments*, Eds. S. K. Kim and C. G. Lee, pp. 217–244. Boca Raton, FL: CRC Press.
Konur, O., Ed. 2018a. *Bioenergy and Biofuels*. Boca Raton, FL: CRC Press.
Konur, O. 2018b. Bioenergy and biofuels science and technology: Scientometric overview and citation classics. In *Bioenergy and Biofuels*, Ed. O. Konur, pp. 3–63. Boca Raton, FL: CRC Press.
Konur, O. 2019. Cyanobacterial bioenergy and biofuels science and technology: A scientometric overview. In *Cyanobacteria: From Basic Science to Applications*, Eds. A. K. Mishra, D. N. Tiwari and A. N. Rai, pp. 419–442. Amsterdam: Elsevier.
Konur, O. 2020a. The scientometric analysis of the research on the bioethanol production from green macroalgae. In *Handbook of Algal Science, Technology and Medicine*, Ed. O. Konur, pp. 385–401. London: Academic Press.
Konur, O., Ed. 2020b. *Handbook of Algal Science, Technology and Medicine*. London: Academic Press.
Konur, O., Ed. 2021a. *Handbook of Biodiesel and Petrodiesel Fuels: Science, Technology, Health, and Environment*. Boca Raton, FL: CRC Press.
Konur, O., Ed. 2021b. *Handbook of Biodiesel and Petrodiesel Fuels: Science, Technology, Health, and Environment. Volume 1. Biodiesel Fuels: Science, Technology, Health, and Environment*. Boca Raton, FL: CRC Press.
Konur, O., Ed. 2021c. *Handbook of Biodiesel and Petrodiesel Fuels: Science, Technology, Health, and Environment. Volume 2. Biodiesel Fuels Based on the Edible and Nonedible Feedstocks, Wastes, and Algae: Science, Technology, Health, and Environment*. Boca Raton, FL: CRC Press.
Konur, O., Ed. 2021d. *Handbook of Biodiesel and Petrodiesel Fuels: Science, Technology, Health, and Environment. Volume 3. Petrodiesel Fuels: Science, Technology, Health, and Environment*. Boca Raton, FL: CRC Press.
Konur, O. and F. L. Matthews. 1989. Effect of the properties of the constituents on the fatigue performance of composites: A review. *Composites* 20:317–328.
Leydesdorff, L. 2000. Is the European Union becoming a single publication system? *Scientometrics* 47:265–280.
Leydesdorff, L. and C. Wagner. 2009. Is the United States losing ground in science? A global perspective on the world science system. *Scientometrics* 78:23–36.
Leydesdorff, L. and P. Zhou. 2005. Are the contributions of China and Korea upsetting the world system of science? *Scientometrics* 63:617–630.
Leydesdorff, L., C. S. Wagner and L. Bornmann. 2014. The European Union, China, and the United States in the top-1% and top-10% layers of most-frequently cited publications: Competition and collaborations. *Journal of Informetrics* 8:606–617.
Liu, S., E. R. Cuty Clemente, T. Hu and Y. Wei. 2007. Study of spark ignition engine fueled with methanol/gasoline fuel blends. *Applied Thermal Engineering* 27:1904–1910.
Ma, X., L. Sun and C. Song. 2002. A new approach to deep desulfurization of gasoline, diesel fuel and jet fuel by selective adsorption for ultra-clean fuels and for fuel cell applications. *Catalysis Today* 77:107–116.
Mehl, M., W. J. Pitz, C. K. Westbrook and H. J. Curran. 2011. Kinetic modeling of gasoline surrogate components and mixtures under engine conditions. *Proceedings of the Combustion Institute* 33:193–200.
Meisel, S. L., J. P. McCullough, C. H. Lechthaler and P. B. Weisz. 1976. Gasoline from methanol in one step. *Chemische Technik* 6:86–89.

Miguel, A. H., T. W. Kirchstetter, R. A. Harley and S. V. Hering. 1998. On-road emissions of particulate polycyclic aromatic hydrocarbons and black carbon from gasoline and diesel vehicles. *Environmental Science and Technology* 32:450–455.

Moldovan, M., M. A. Palacios and M. M. Gomez, et al. 2002. Environmental risk of particulate and soluble platinum group elements released from gasoline and diesel engine catalytic converters. *Science of the Total Environment* 296:199–208.

Najafi, G., B. Ghobadian and T. Tavakoli, et al. 2009. Performance and exhaust emissions of a gasoline engine with ethanol blended gasoline fuels using artificial neural network. *Applied Energy* 86:630–639.

Newman, P. W. G. and J. R. Kenworthy. 1989. Gasoline consumption and cities: A comparison of U.S. cities with a global survey. *Journal of the American Planning Association* 55:24–37.

North, D. C. 1991. Institutions. *Journal of Economic Perspectives* 5:97–112.

Pock, M. 2010. Gasoline demand in Europe: New insights. *Energy Economics* 32:54–62.

Schauer, J. J., M. J. Kleeman, G. R. Cass and B. R. T. Simoneit. 2002. Measurement of emissions from air pollution sources. 5. C_1-C_{32} organic compounds from gasoline-powered motor vehicles. *Environmental Science and Technology*, 36:1169–1180.

Shin, D., T. Kim, J. Choi and J. Kim. 2014. Author name disambiguation using a graph model with node splitting and merging based on bibliographic information. *Scientometrics* 100:15–50.

Song, C. 2003. An overview of new approaches to deep desulfurization for ultra-clean gasoline, diesel fuel and jet fuel. *Catalysis Today* 86:211–263.

Von Storch, H., M. Costa-Cabral and C. Hagner, et al. 2003. Four decades of gasoline lead emissions and control policies in Europe: A retrospective assessment. *Science of the Total Environment* 311:151–176.

81 Gasoline Fuels
Review

Ozcan Konur
(Formerly) Ankara Yildirim Beyazit University

81.1 INTRODUCTION

Crude oil-based gasoline fuels (Ma et al., 2002; Newman and Kenworthy, 1989) have been widely used in the transportation sector since the 1920s. However, there have been great public concerns over the adverse environmental impact of these fuels (Hill et al., 2006, 2009; Konur, 2012, 2015, 2019, 2020a). Hence, biomass-based bioethanol fuels (Hsieh et al., 2002; Najafi et al., 2009) and other bioalcohol fuels (Aleiferis and van Romunde, 2013; Christensen et al., 2011) have increasingly been used in blending gasoline fuels.

In this context, there has been intense research in the research fronts of emissions (Miguel et al., 1998; Schauer et al., 2002), blends with bioethanol fuels and other bioalcohols, combustion (Kalghatgi et al., 2007; Mehl et al., 2011), economics (Borenstein et al., 1997; Newman and Kenworthy, 1989), production (Bjorgen et al., 2008; Meisel et al., 1976), and to a lesser extent the environmental impact (Hill et al., 2009; Moldovan et al., 2002), properties (Chica and Corma, 1999; Kelly et al., 1989), desulfurization (Ma et al., 2002; Song, 2003), and health impact (Hill et al., 2009; Jacobson, 2007) of gasoline fuels.

However, it is essential to develop efficient incentive structures (North, 1991, 1993; North and Weingast, 1989) for the primary stakeholders to enhance the research in this field (Konur, 2000, 2002a–c, 2006a,b, 2007a,b).

Although there have been a number of review papers (Battin-Leclerc, 2008; Brunet et al., 2005; Song, 2003), there has been no review of the research of the 25 most cited articles in the field of the gasoline blends and their bioalcohol blends.

This book chapter presents a review of the research of the 25 most cited articles in this field. Then, it discusses the key findings of these highly influential papers and comments on the future research priorities in this field.

81.2 MATERIALS AND METHODS

The search for this study was carried out using Scopus database (Burnham, 2006) in August 2021.

As a first step for the search of the relevant literature, the keywords were selected using the first 500 most cited papers. The selected keyword list was optimized to obtain a representative sample of papers for the searched research field. This keyword list was provided in the appendix for future replication studies.

As a second step, a sample data set was used for this study. The first 25 articles in the sample of 100 most cited papers with at least 283 citations each were selected for the review study. Key findings from each paper were taken from the abstracts of these papers and were discussed.

Additionally, a number of brief conclusions were drawn and a number of relevant recommendations were made to enhance the future research landscape.

DOI: 10.1201/9781003226567-109

81.3 RESULTS

The brief information about 25 most cited papers with at least 283 citations each in this field is given below under two major headings: Gasoline fuels and gasoline-bioalcohol fuel blends. There are additionally three research fronts for the first major research field of gasoline fuels: Gasoline engines, gasoline fuel economics, and gasoline fuel production. The research front of gasoline engines covers emissions, combustion, engine performance, and desulfurization, properties of gasoline fuels.

81.3.1 Gasoline Fuels

81.3.1.1 Gasoline Engines

Khalili et al. (1995) evaluate the chemical composition of airborne polycyclic aromatic hydrocarbons (PAHs) in both the particulate and gas phases for gasoline engines in a paper with 1061 citations. They find that two and three ring PAHs were responsible for 73% of the total concentration of measured 20 PAHs. Six ring PAHs such as indeno(1,2,3-cd)pyrene and benzo(ghi)perylene were mostly below the detection limit of this study.

Schauer et al. (2002) evaluate the chemical composition of the gas- and particle-phase organic compounds present in the tailpipe emissions from C_1 to C_{32} organic compounds from gasoline-powered motor vehicles in a paper with 834 citations. They identify six isoprenoids and two tricyclic terpanes as potential tracers for these emissions. They observe that the distribution of n-alkanes and isoprenoids emitted from the catalyst-equipped gasoline-powered vehicles is the same as the distribution of these compounds found in the gasoline used, whereas the distribution of these compounds in the emissions from the noncatalyst vehicles is very different from the distribution in the fuel. In contrast, the distribution of the PAHs and their methylated homologues in the gasoline is significantly different from the distribution of the PAH in the tailpipe emissions from both types of vehicles.

Mehl et al. (2011) analyze the combustion behavior of several components relevant to gasoline surrogate formulation in a paper with 813 citations. They focus on linear and branched saturated hydrocarbons (PRF mixtures), olefins (1-hexene) and aromatics (toluene).

Ma et al. (2002) develop a new desulfurization process by selective adsorption for removing sulfur (SARS) by developing an adsorbent for gasoline fuels in a paper with 465 citations. They observe that the transition metal-based adsorbent is effective for selectively adsorbing the sulfur compounds. The SARS process can effectively remove sulfur compounds in the liquid hydrocarbon fuels at ambient temperature under atmospheric pressure with low investment and operating cost. They propose a novel integrated process for deep desulfurization of the liquid hydrocarbon fuels in a future refinery, which combines a selective adsorption (SARS) of the sulfur compounds and a hydrodesulfurization process of the concentrated sulfur fraction (HDSCS). The SARS concept may be used for on-site or on-board removal of sulfur from fuels for fuel cell systems.

Miguel et al. (1998) evaluate the on-road emissions of particulate PAHs and fine particle black carbon from gasoline vehicles in a paper with 459 citations. They measure gas- and particle-phase pollutant concentrations in the USA in 1996 for light-duty vehicles. They find that light-duty vehicles emitted 30 mg of fine black carbon particles per kg of fuel burned. The light-duty gasoline vehicles were the dominant source of higher molecular weight PAHs such as benzo[a]pyrene and dibenz[a,h]anthracene. Gasoline engine-derived PAH emissions were almost entirely in the ultrafine mode.

Odum et al. (1997a) evaluate the atmospheric aerosol-forming potential of whole gasoline vapor in a paper with 455 citations. They observe that the atmospheric organic aerosol formation potential of whole gasoline vapor can be accounted for solely in terms of the aromatic fraction of the fuel. The total amount of secondary organic aerosol produced from the atmospheric oxidation of whole gasoline vapor is the sum of the contributions of the individual aromatic molecular constituents of the fuel. The urban atmospheric, anthropogenic hydrocarbon profile was approximated well by evaporated whole gasoline and it is possible to model atmospheric secondary organic aerosol formation.

Harris and Maricq (2001) evaluate the signature size distributions for gasoline engine exhaust particulate matter (PM) in a paper with 360 citations. They observe distinctly different characteristic distributions for direct injection and for port injection gasoline vehicles. They find that the coagulation models alone are incapable of reproducing the observed characteristic distributions. They discuss the possibility that soot oxidation also plays a role in determining the shape of the characteristic distributions, and the possibility that these signatures could be utilized to distinguish soot emissions from other aerosols.

Kalghatgi et al. (2007) evaluate combustion and emissions of gasoline fuels in a paper with 333 citations. They find that gasoline always gives much lower smoke compared with the petrodiesel fuel because of its higher ignition delay. This usually allows the heat release to be separated in time from the injection event. The nitrogen oxides (NO_x) can be controlled by exhaust gas recirculation (EGR). With gasoline, pilot injection helps reduce the maximum heat release rate for a given indicated mean effective pressure (IMEP) and enables heat release to occur later with low cyclic variation compared to single injection. This enables higher mean IMEP to be reached with lower smoke, NO_x and maximum heat release rate compared to single injection.

Benbrahim-Tallaa et al. (2012) evaluate the carcinogenicity of gasoline engine exhausts in a paper with 323 citations. The Working Group concluded that there was sufficient evidence in experimental animals for the carcinogenicity of condensates of gasoline engine exhaust. The Working Group further concluded that there is strong evidence for a genotoxic mechanism for the carcinogenicity of organic solvent extracts of particles from gasoline engine exhaust. In conclusion, the Working Group classified gasoline engine exhaust as possibly carcinogenic to humans (Group 23 2B).

Nriagu (1990) outlines the history of the lead (Pb) additives in gasoline in a paper in a paper with 314 citations. The leaded gasoline was first sold in the USA in 1923, and the lead was removed from gasoline in the 1980s.

Odum et al. (1997b) obtain secondary organic aerosol (SOA) yield curves for 17 individual aromatic species in a paper with 301 citations. They use these yield curves to quantitatively account for the SOA that is formed in a series of smog chamber experiments performed with the whole vapor of 12 different reformulated gasolines. They note that the total amount of SOA produced from the atmospheric oxidation of whole gasoline vapor can be represented as the sum of the contributions of the individual aromatic molecular constituents of the fuel.

Ressler et al. (2000) evaluate the chemical nature of Mn-bearing PM emitted from vehicles burning (methylcyclopentadienyl) manganese tricarbonyl-added gasolines in a paper with 295 citations. They observe that the average Mn valence in these particulates is ~2.2. The number and type of probable species contained in these particulates as three, consisting of Mn_3O_4, $MnSO_4$.H2O, and a divalent manganese phosphate. They distinguish two groups of Mn-bearing particulates.

Zielinska et al. (2004) evaluate the emission rates and chemical composition of gasoline-fueled vehicles in a paper with 289 citations. They find that the total PM emission rates ranged from below 3 mg/mi up to more than 700 mg/mi for the white smoker gasoline vehicle. The compositions of emissions were highly dependent on the fuel type (gasoline vs. diesel), the state of vehicle maintenance (low, average, or high emitters; white or black smokers), and ambient conditions (i.e., temperature) of the vehicles. Oil served as a repository for combustion byproducts (e.g., PAH), and oil-burning gasoline vehicles emitted PAH in higher concentrations than did other vehicles. These PAH emissions matched the PAH compositions observed in oil (Table 81.1).

81.3.1.2 Gasoline Fuel Economics

Newman and Kenworthy (1989) evaluate the gasoline consumption in a sample of US and global cities in a paper with 571 citations. They observe that gasoline consumption per capita in the US cities varies by up to 40%, primarily because of land use and transportation planning factors, rather than gasoline price or income variations. The same patterns, though more extreme, appear in a global sample of 32 cities. The average gasoline consumption in U.S. cities was nearly twice as high as in Australian cities, four times higher than in European cities, and ten times higher than in Asian cities. Allowing for variations in gasoline price, income, and vehicle efficiency explains only half of these differences. They introduce physical planning policies, particularly reurbanization and a reorientation of transportation priorities, as a means of reducing gasoline consumption and automobile dependence.

TABLE 81.1
The Gasoline Engines

No.	Papers	Issues	Research Fronts	Cits
1	Khalili et al. (1995)	Chemical composition of airborne PAHs in both the particulate and gas phases	Emissions	1061
2	Schauer et al. (2002)	Chemical composition of the gas- and particle-phase organic compounds present in the tailpipe emissions from C_1 to C_{32} organic compounds from gasoline-powered motor vehicles	Emissions	814
3	Mehl et al. (2011)	Combustion behavior of several components relevant to gasoline surrogate formulation	Combustion	813
4	Miguel et al. (1998)	On-road emissions of particulate PAHs and fine particle black carbon from gasoline vehicles	Emissions	459
5	Ma et al. (2002)	Desulfurization process development by selective adsorption for removing sulfur (SARS)	Desulfurization	465
6	Odum et al. (1997a)	Atmospheric aerosol-forming potential of whole gasoline vapor	Emissions	455
7	Harris and Maricq (2001)	Signature size distributions for gasoline engine exhaust PM	Emissions	360
8	Kalghatgi et al. (2007)	Combustion and emissions of gasoline fuels	Emissions, combustion	333
9	Benbrahim-Tallaa et al. (2012)	Carcinogenicity of gasoline engine exhausts	Emissions, health	323
10	Nriagu (1990)	History of the led additives in gasoline	Emissions	314
11	Odum et al. (1997b)	Secondary organic aerosol (SOA) yield curves	Emissions	301
12	Ressler et al. (2000)	Chemical nature of Mn-bearing particulates emitted from autos burning gasoline fuels	Emissions	295
13	Zielinska et al. (2004)	Emission rates and chemical composition of gasoline-fueled vehicles	Emissions	289

Borenstein et al. (1997) evaluate the gasoline prices-crude oil price relationships in a paper with 477 citations. They observe that retail gasoline prices respond more quickly to increases than to decreases in crude oil prices. Spot prices for generic gasoline show asymmetry in responding to crude oil price changes, which may reflect inventory adjustment effects. Asymmetry also appears in the response of retail prices to wholesale price changes, possibly indicating short-run market power among retailers.

Parry and Small (2005) study the determinants of the gasoline tax in the USA and the UK in a paper with 340 citations. They observe that the optimal gasoline tax for the USA is more than double its current rate, while that for the UK is about half its current rate. Leaving aside externalities, a heavy reliance on fuel taxes might be appropriate if gasoline is consumed disproportionately by high-income groups and they are given low distributional weights. However, budget shares for gasoline are either constant or mildly declining with income across U.S. households, especially with measures of lifetime income. Probably the gasoline tax is somewhat more progressive in the United Kingdom, where auto ownership is less widely distributed, but not to the extent of justifying such a high tax rate (Table 81.2).

81.3.1.3 Gasoline Fuel Production

Carlson et al. (2008) produce biogasoline by catalytic fast pyrolysis of solid biomass feedstocks in a single catalytic reactor at short residence times in a paper 437 citations. High heating rates and catalyst-to-feed ratios are needed to ensure that pyrolized biomass compounds enter the pores of the ZSM5 catalyst and that thermal decomposition is avoided. Product selectivity is a function of the active site and pore structure of the catalyst.

TABLE 81.2
The Gasoline Fuel Economics

No.	Papers	Issues	Cits
1	Newman and Kenworthy (1989)	Gasoline consumption in a sample of US and global cities	571
2	Borenstein et al. (1997)	Gasoline prices-crude oil price relationships	477
3	Parry and Small (2005)	Determinants of the gasoline tax in the USA and the UK	340

Bjorgen et al. (2008) produce gasoline from methanol over zeolite H-ZSM-5 focusing on the improved catalyst performance by treatment with NaOH in a paper with 412 citations. They find that the catalyst lifetime increased by a factor of 3.3 as a consequence of the most severe treatment. The procedure led to a moderate increase in the initial activities. Further, the product selectivities were altered dramatically. The selectivity toward the gasoline fraction (C_{5+}) was at best increased by a factor of 1.7. Hydrogen transfer reactions became faster and led to more aromatic and paraffinic compounds in the products. They finally observe increases in the propene/ethene ratios at moderate conversion.

Wei et al. (2017) develop a Na-Fe_3O_4/HZSM-5 catalyst for directly converting CO_2 to gasoline-range (C_5–C_{11}) hydrocarbons in a paper with 410 citations. The selectivity was up to 78% of all hydrocarbons while only 4% methane at a CO_2 conversion of 22% under industrial relevant conditions. This multifunctional catalyst provided three types of active sites (Fe_3O_4, Fe_5C_2, and acid sites), which cooperatively catalyze a tandem reaction. More significantly, the appropriate proximity of three types of active sites plays a crucial role in the successive and synergetic catalytic conversion of CO_2 to gasoline.

Meisel et al. (1976) develop a process for converting methanol to gasoline-range hydrocarbons and water using a zeolite cracking catalyst in a paper with 377 citations. The hydrocarbons produced are predominantly in the petrol boiling range, and the product resembles conventional petrol chemically. Since coal can be converted to methanol, the new process provides the final step in the conversion of coal to petrol. However, the product cannot compete economically at that time with petrol produced from crude oil (Table 81.3).

81.3.2 Gasoline-Bioalcohol Fuel Blends

Hsieh et al. (2002) evaluate the fuel properties, engine performance and pollutant emissions of an SI engine using bioethanol-gasoline fuel blends of E0, E5, E10, E20, and E30 in a paper with 484 citations. They observe that with increasing the bioethanol content, the heating value of the blended fuels is decreased, while the octane number of the blended fuels increases. They then find that with increasing the with ethanol content, the Reid vapor pressure of the blended fuels initially increases to a maximum at 10% ethanol addition, and then decreases. They next find that using these blends, torque output and fuel consumption of the engine slightly increase; carbon monoxide (CO) and hydrocarbon (HC) emissions decrease dramatically and carbon dioxide (CO_2) emission increases because of the improved combustion. Finally, NO_x emission depends on the engine operating condition rather than the ethanol content.

Al-Hasan (2003) evaluate the emissions and engine performance of bioethanol-unleaded gasoline fuel blends in a paper with 399 citations. They find that blending unleaded gasoline with bioethanol increases the brake power, torque, volumetric and brake thermal efficiencies, and fuel consumption, while it decreases the brake-specific fuel consumption (BSFC) and equivalence air-fuel ratio. The CO and HC emissions concentrations in the engine exhaust decrease, while the CO_2 concentration increases. The 20 vol.% bioethanol in fuel blend gave the best results for all measured parameters at all engine speeds.

TABLE 81.3
The Gasoline Fuel Production

No.	Papers	Issues	Cits.
1	Carlson et al. (2008)	Biogasoline production by catalytic fast pyrolysis of solid biomass feedstocks	437
2	Bjorgen et al. (2008)	Gasoline production from methanol over zeolite H-ZSM-5	412
3	Wei et al. (2017)	Gasoline production from CO_2 with Na-Fe_3O_4/HZSM-5 catalyst	410
4	Mesiel et al. (1976)	Gasoline production from methanol	377

Najafi et al. (2009) evaluate the performance and exhaust emissions of bioethanol-gasoline fuel blends of E5, E10, E15, and E20 using artificial neural network in a paper with 305 citations. They found that using these blends increased the power and torque output of the engine marginally. The BSFC decreased while the brake thermal efficiency (BTE) and volumetric efficiency (ηv) increased. The concentration of CO and HC emissions in the exhaust pipe decreased with these blends. This was due to the high oxygen percentage in the bioethanol. In contrast, the concentration of CO_2 and NO_x increased when bioethanol is introduced.

Yuksel and Yuksel (2004) evaluate the effect of the engine modification for the performance, fuel consumption, and exhaust emissions of bioethanol-gasoline fuel blend of E60 in a paper with 283 citations. They design a new carburetor to solve the phase problem, and the bioalcohol ratio in the total fuel was increased. They use E60 blend to test the performance, the fuel consumption, and the exhaust emissions.

Gu et al. (2012) evaluate the emission characteristics of a spark-ignition (SI) engine fueled with gasoline-n-butanol blends in combination with EGR in a paper with 283 citations. They observe that these blends decrease engine specific HC, CO, and NO_x emissions compared with those of gasoline. Pure n-butanol increases engine specific HC and CO emissions and decreases NO_x and particle number concentration compared with those of gasoline. n-Butanol addition can decrease particle number concentration emissions compared with those of gasoline. Advancing spark timing increases engine specific HC, NO_x emissions and particle number concentration while it decreases engine specific CO emissions. EGR can reduce engine specific NO_x emissions and particle number concentration simultaneously in spark-ignition engine fueled with gasoline and n-butanol blends (Table 81.4).

81.4 DISCUSSION

81.4.1 INTRODUCTION

Crude oil-based gasoline fuels have been widely used in the transportation sector. However, there have been great public concerns over the adverse environmental impact and sustainability of these fuels. Hence, biomass-based bioethanol fuels and other bioalcohol fuels have increasingly been used in blending gasoline fuels in recent years to reduce their adverse effects. In this context, there has been intense research in the research fronts of emissions, blends with bioethanol fuels and other bioalcohols, combustion, economics, production, and to a lesser extent the environmental impact, properties, desulfurization, and health impact of gasoline fuels. However, it is essential to develop efficient incentive structures for the primary stakeholders to enhance the research in this field.

Although there have been a number of review papers on the gasoline fuels, there has been no review of the research of the 25 most cited articles in this field. Hence, this book chapter presents a review of the research of the 25 most cited articles in this field. Then, it discusses the key findings of these highly influential papers and comments on the future research priorities in this field.

There are two major research fronts for this field: Gasoline fuels and gasoline-bioalcohol fuel blends with 20 and 5 studies, respectively.

TABLE 81.4
The Gasoline-Bioalcohol Fuel Blends

No.	Papers	Bioethanol Fuels	Issues	Research Fronts	Cits.
1	Hsieh et al. (2002)	Bioethanol-gasoline blends	Properties, engine performance, and emissions of bioethanol-gasoline fuel blends of E0, E5, E10, E20, E30	Emissions, performance, properties	484
2	Al-Hasan (2003)	Bioethanol-gasoline blends	Emissions and engine performance of bioethanol-unleaded gasoline fuel blends	Emissions, performance	399
3	Najafi et al. (2009)	Bioethanol-gasoline blends	Performance and exhaust emissions of bioethanol-gasoline fuel blends of E5, E10, E15, and E20	Emissions, performance	305
4	Yuksel and Yuksel (2004)	Bioethanol-gasoline blends	Effect of the engine modification for the performance, fuel consumption, and exhaust emissions of bioethanol-gasoline fuel blends	Performance, consumption, and emissions	283
5	Gu et al. (2012)	Butanol-gasoline blends	Emission characteristics of a spark-ignition engine fueled with gasoline-n-butanol blends in combination with EGR	Emissions	283

81.4.2 Gasoline Fuels

There are 13, 3, and 4 papers in the research fronts of gasoline engines, gasoline economics, and gasoline production, respectively.

81.4.2.1 Gasoline Engines

There are 13 papers on the gasoline engines. The bulk of these papers are related to the emissions from the gasoline fuels (Benbrahim-Tallaa et al., 2012; Harris and Maricq, 2001; Kalghatgi et al., 2007; Khalili et al., 1995; Miguel et al., 1998; Nriagu, 1990; Odum et al., 1997a,b; Ressler et al., 2000; Schauer et al., 2002; Zielinska et al., 2004). The other papers are concerned with the combustion of gasoline fuels (Kalghatgi et al., 2007; Mehl et al., 2011), desulfurization of gasoline fuels (Ma et al., 2002), and health impact of gasoline fuels (Benbrahim-Tallaa et al., 2012).

The studies on the emissions from the gasoline fuels fueling gasoline engines cover the composition and emissions of airborne PAHs (Khalili et al., 1995; Miguel et al, 1998; Schauer et al., 2002; Zielinska et al., 2004). The PAHs have very strong public policy implications with heavy impact on the environment and human health (Cerniglia, 1993).

Similarly, Miguel et al. (1998) evaluate the black carbon emissions from gasoline fuels. The black carbon in the soot has also had very strong public policy implications with a heavy impact on the environment and human health (Kuhlbusch, 1998).

Odum (1997a,b) study the aerosol formation (Pandis et al., 1992) in the atmosphere from gasoline fuels.

The emissions of these fuels are also of the most importance in the light of the public concerns about climate change (Change, 2007), greenhouse gas emissions (Carlson et al., 2017), and global warming (Kerr, 2007). As these concerns have been certainly behind the boom in the research in this field in the last two decades, it is not surprising that the bulk of the majority of the most cited papers sought to evaluate the environmental impact of gasoline fuels and their blends.

The emissions of CO_2 (Ang, 2007) are especially important and any potential reduction of these emissions by the bioethanol blends compared with gasoline or petrodiesel fuels would have significant implications for the environmental protection. The emissions of the PM are also important for the protection of the environment (Klimont et al., 2017).

Kalghatgi et al. (2007) and Mehl et al. (2011) study the combustion of gasoline fuels. These studies suggest there is ample room to optimize the combustion of gasoline fuels to reduce the emissions through the EGR (Abd-Alla, 2002).

The emissions of the NO_x from gasoline fuels are studied widely (Gu et al., 2012; Hsieh et al., 2002; Kalghatgi et al., 2007; Najafi et al., 2009). The emissions of the NO_x are also studied in the wider contexts (Carslaw, 2005).

The studies on the health impact of gasoline fuels have great public importance (Hoffmann et al., 1965; Jacobson, 2007). Benbrahim-Tallaa et al. (2012) summarize the research in this field and conclude that there was sufficient evidence in experimental animals for the carcinogenicity of condensates of gasoline engine exhaust. There was strong evidence for a genotoxic mechanism for the carcinogenicity of organic solvent extracts of particles from gasoline engine exhaust. In conclusion, the Working Group classified gasoline engine exhaust as possibly carcinogenic to humans (Group 23 2B).

Nriagu (1990) discusses the history of the incorporation and removal of lead in the gasoline fuels. The leaded gasoline is a major source of human lead exposure (Thomas 1995). Similarly, Ma et al. (2002) study the desulfurization (Brunet et al., 2005; Song, 2003) of gasoline fuels to reduce the adverse environmental impact of gasoline fuels and other fossil-based fuels.

In the end, these most cited papers in this field hint that the benefits sought from these fuels could be maximized using the structure, processing, and property relationships of these fuels (Formela et al., 2016; Konur, 2018a,b, 2020a,b, 2021a–d; Konur and Matthews, 1989).

81.4.2.2 Gasoline Fuel Economics

There are three most cited papers related to the economics of gasoline fuels (Borenstein et al., 1997; Newman and Kenworthy, 1989; Parry and Small, 2005). These studies focus on the consumption, taxation, and prices of gasoline fuels in relation to crude oil and other fossil-based fuels. The taxation and pricing of fuels are just two of the parameters impacting the fuel consumption as they are often used incentives or disincentives for the fuel consumption.

81.4.2.3 Gasoline Fuel Production

There are four papers on the production of gasoline fuels (Bjorgen et al., 2008; Carlson et al., 2008; Meisel et al., 1976; Wei et al., 2017). The feedstocks used for the gasoline production range from biomass to CO_2 and methanol. The recent studies focus more on the biogasoline production from the biomass.

81.4.3 Gasoline-Bioalcohol Fuel Blends

The second major research front is related to the blends of gasoline fuels with bioalcohols such as bioethanol, *n*-butanol, biodiesel, and biomethanol. The most cited papers are related to the bioethanol-gasoline fuel blends (Al-Hasan, 2003; Hsieh et al., 2002; Najafi et al., 2009; Yuksel and Yuksel, 2004), and to the gasoline-n-butanol blends (Gu et al., 2012).

The studied topics are the emissions, engine performance, properties, and fuel consumption. These studies hint that the emissions, engine performance, properties, and fuel consumption of these blends could be optimized using the bioalcohol content as well as the operating conditions of the gasoline engine.

81.5 CONCLUSION AND FUTURE RESEARCH

The brief information about the key research fronts covered by the 25 most cited papers with at least 283 citations each in the field of the gasoline fuels and their bioethanol and other bioalcohol blends is given in Table 81.5.

There are two major research fronts for this field: Gasoline fuels and gasoline fuel-bioalcohol fuel blends with 20 and 5 papers, respectively. There are three more research fronts for the first major section: Gasoline engines, gasoline fuel economics, and gasoline fuel production.

TABLE 81.5
The Most Prolific Research Fronts in the Field of the Gasoline Fuels and Their Bioalcohol Blends

No.	Research Fronts	Sample Papers (%)	Reviewed Papers (No.)	Reviewed Papers (%)
1		**Gasoline Fuels**		
1.1.	Gasoline fuel engines	56	13	56
1.2.	Gasoline fuel economics	11	3	12
1.3.	Gasoline fuel production	10	4	16
2.	Gasoline-bioalcohol fuel blends	22	5	20

The total number of reviewed papers = 25. Sample papers: the sample of the 100 most cited papers.

It is notable that there is similarity of these research fronts in the sample of reviewed papers and the sample of the 100 most cited papers (the first column data in Table 81.5) in this field.

The bulk of these papers on the gasoline fuels are related to the emissions from the gasoline fuels. The other papers are concerned with the combustion of gasoline fuels, desulfurization of gasoline fuels, and health impact of gasoline fuels.

The studies on the emissions from the gasoline fuels fueling gasoline engines cover the composition and emissions of airborne PAHs. The PAHs have very strong public policy implications with heavy impact on the environment and human health.

Similarly, a study evaluates the black carbon emissions from gasoline fuels. The black carbon in the soot also has very strong public policy implications with heavy impact on the environment and human health. Additionally, two papers study the aerosol formation in the atmosphere from gasoline fuels. Then, two papers study the combustion of gasoline fuels. These studies suggest there is ample room to optimize the combustion of gasoline fuels to reduce the emissions through the EGR.

The emissions of the NO_x from gasoline fuels are studied widely. The emissions of the NO_x are also studied in the wider contexts with strong public policy implications regarding their adverse impact on both the environment and human health.

The studies on the health impact of gasoline fuels have great public importance. A working group concludes that there was sufficient evidence in experimental animals for the carcinogenicity of condensates of gasoline engine exhaust (Benbrahim-Tallaa et al., 2012). There was strong evidence for a genotoxic mechanism for the carcinogenicity of organic solvent extracts of particles from gasoline engine exhaust. In conclusion, the Working Group classified gasoline engine exhaust as possibly carcinogenic to humans (Group 23 2B).

A paper discusses the history of the incorporation and removal of lead in the gasoline fuels. The leaded gasoline is a major source of human lead exposure. Similarly, a paper studies the desulfurization of gasoline fuels to reduce the adverse environmental impact of gasoline fuels and other fossil-based fuels.

The papers on the economics of gasoline fuels focus on the consumption, taxation, and prices of gasoline fuels in relation to crude oils and other fossil-based fuels. The taxation and pricing of fuels are just two of the parameters impacting the fuel consumption as they are often used incentives or disincentives for the fuel consumption.

The papers on the production of gasoline fuels use the feedstocks for the gasoline production range from biomass to CO_2 and methanol. The focus of the current research is more on the biogasoline production.

The papers in the second major research front focus on the blends of gasoline fuels with bioalcohols such as bioethanol, n-butanol, biodiesel, and methanol. The most cited papers are related to the bioethanol-gasoline fuel blends. The studied topics are the emissions, engine performance, properties, and fuel consumption. These studies hint that the emissions, engine performance, properties, and fuel consumption of these blends could be optimized using the bioalcohol content as well as the operating conditions of the gasoline engine. It is expected the research would focus more on these blends in the future.

The fuel properties are important as they determine the emission, combustion, and engine performance of these fuels. There is a room to optimize these properties by adjusting the composition and blend ratios of these fuels. The performance of these fuels in the blends is of the outmost importance as a function of the properties of these fuels as well as the proportions of the bioalcohol fuels in these blends. These studies hint that the blends have better fuel performance compared to gasoline fuels.

The emissions of these fuels are also of the most importance in the light of the public concerns about climate change, greenhouse gas emissions, and global warming. As these concerns have been certainly behind the boom in the research in this field in the last two decades, it is not surprising that the bulk of the majority of the most cited papers sought to evaluate the environmental impact of gasoline fuels and their blends.

The emissions of CO_2 are especially important and any potential reduction of these emissions by the blends compared with gasoline fuels would have significant implications for the environmental protection. The emissions of the PM are also important for the protection of the environment. These most cited studies highlight the determinants of the better emission performance for these blends. The operating conditions of the engines as well as the proportions of bioalcohol concentrations in the fuel blends play a key role in determining the emission performance of these blends. With the optimization of operating conditions of the engines as well as then proportions of bioethanol, these studies show that bioethanol blends might have better emission performance compared with both gasoline and petrodiesel fuels.

In the end, these most cited papers in this field hint that the benefits sought from these fuels could be maximized using the structure, processing, and property relationships of these fuels as in materials science and engineering.

These findings confirm that gasoline fuels have evolved since the 1920s through the removal of the toxic compounds such as lead and sulfur and more recently through their blending with bioalcohol fuels, thereby reducing their adverse impact on environment and the human health to a large extent and impacting favorably global warming, greenhouse gas emissions, and climate change.

It is recommended that such review studies should be performed for the other research fronts on both the production and utilization of bioethanol fuels complementing the corresponding scientometric studies.

ACKNOWLEDGMENTS

The contribution of the highly cited researchers in the field of the gasoline fuels and their bioalcohol fuel blends has been gratefully acknowledged.

APPENDIX: THE KEYWORD SET FOR THE FIELD OF THE GASOLINE FUELS

(TITLE (gasoline* OR biogasoline* OR gasohol*) OR SRCTITLE (gasoline* OR biogasoline* OR gasohol*)) AND NOT (SRCTITLE (astronom*) OR TITLE (astronom*).

REFERENCES

Abd-Alla, G. H. 2002. Using exhaust gas recirculation in internal combustion engines: A review. *Energy Conversion and Management* 43:1027–1042.

Aleiferis, P. G. and Z. R. van Romunde. 2013. An analysis of spray development with iso-octane, n-pentane, gasoline, ethanol and n-butanol from a multi-hole injector under hot fuel conditions. *Fuel* 105:143–168.

Al-Hasan, M. 2003. Effect of ethanol-unleaded gasoline blends on engine performance and exhaust emission. *Energy Conversion and Management* 44:1547–1561.

Ang, J. B. 2007. CO_2 emissions, energy consumption, and output in France. *Energy Policy* 35:4772–4778.

Battin-Leclerc, F. 2008. Detailed chemical kinetic models for the low-temperature combustion of hydrocarbons with application to gasoline and diesel fuel surrogates. *Progress in Energy and Combustion Science* 34:440–498.

Benbrahim-Tallaa, L., R. A. Baan, and Y. Grosse, et al. 2012. Carcinogenicity of diesel-engine and gasoline-engine exhausts and some nitroarenes. *Lancet Oncology* 13:663–664.

Bjorgen, M., F. Joensen and M. S. Holm, et al. 2008. Methanol to gasoline over zeolite H-ZSM-5: Improved catalyst performance by treatment with NaOH. *Applied Catalysis A: General* 345:43–50.

Borenstein, S., A. C. Cameron and R. Gilbert. 1997. Do gasoline prices respond asymmetrically to crude oil price changes? *Quarterly Journal of Economics* 112:305–339.

Brunet, S., D. Mey, G. Perot, C. Bouchy, F. Diehl, F. 2005. On the hydrodesulfurization of FCC gasoline: A review. *Applied Catalysis A: General* 278:143–172.

Burnham, J. F. 2006. Scopus database: A review. *Biomedical Digital Libraries* 3:1–8.

Carlson, K. M., J. S. Gerber and N. D. Mueller, et al. 2017. Greenhouse gas emissions intensity of global croplands. *Nature Climate Change* 7:63–68.

Carlson, T.R., T. P. Vispute and G. W. Huber. 2008. Green gasoline by catalytic fast pyrolysis of solid biomass derived compounds. *ChemSusChem* 1:397–400.

Carslaw, D. C. 2005. Evidence of an increasing NO_2/NO_x emissions ratio from road traffic emissions. *Atmospheric Environment* 39:4793–4802.

Cerniglia, C. E. 1993. Biodegradation of polycyclic aromatic hydrocarbons. *Current Opinion in Biotechnology* 4:331–338.

Change, C. 2007. Climate change impacts, adaptation and vulnerability. *Science of the Total Environment* 326:95–112.

Chica, A. and A. Corma. 1999. Hydroisomerization of pentane, hexane, and heptane for improving the octane number of gasoline. *Journal of Catalysis* 187:167–176.

Christensen, E., J. Yanowitz, M. Ratcliff and R. L. McCormick. 2011. Renewable oxygenate blending effects on gasoline properties. *Energy and Fuels* 25:4723–4733.

Formela, K., A. Hejna, L. Piszczyk, M. R. Saeb and X. Colom. 2016. Processing and structure-property relationships of natural rubber/wheat bran biocomposites. *Cellulose* 23:3157–3175.

Gu, X., Z. Huang, and J. Cai, et al. 2012. Emission characteristics of a spark-ignition engine fuelled with gasoline-n-butanol blends in combination with EGR. *Fuel* 93:611–617.

Harris, S. J. and M. M. Maricq. 2001. Signature size distributions for diesel and gasoline engine exhaust particulate matter. *Journal of Aerosol Science* 32:749–764.

Hill, J., E. Nelson, D. Tilman, S. Polasky and D. Tiffany. 2006. Environmental, economic, and energetic costs and benefits of biodiesel and ethanol biofuels. *Proceedings of the National Academy of Sciences of the United States of America* 103:11206–11210.

Hill, J., S. Polasky and E. Nelson, et al. 2009. Climate change and health costs of air emissions from biofuels and gasoline. *Proceedings of the National Academy of Sciences of the United States of America* 106:2077–2082.

Hoffmann, D., E. Theisz and E. L. Wynder. 1965. Studies on the carcinogenicity of gasoline exhaust. *Journal of the Air Pollution Control Association* 15:162–165.

Hsieh, W. D., R. H. Chen, T. L. Wu and T. H. Lin. 2002. Engine performance and pollutant emission of an SI engine using ethanol-gasoline blended fuels. *Atmospheric Environment* 36:403–410.

Jacobson, M. Z. 2007. Effects of ethanol (E85) versus gasoline vehicles on cancer and mortality in the United States. *Environmental Science & Technology* 41:4150–4157.

Kalghatgi, G. T., P. Risberg and H. E. Angstrom. 2007. Partially pre-mixed auto-ignition of gasoline to attain low smoke and low NO_x at high load in a compression ignition engine and comparison with a diesel fuel. SAE Technical Papers 2007-01-0006.

Kelly, J. J., C. H. Barlow, T. M. Jinguji and J. B. Callis. 1989. Prediction of gasoline octane numbers from near-infrared spectral features in the range 660-1215 nm. *Analytical Chemistry* 61:313–320.

Kerr, R. A. 2007. Global warming is changing the world. *Science* 316:188–190.

Khalili, N. R., P. A. Scheff and T. M. Holsen. 1995. PAH source fingerprints for coke ovens, diesel and, gasoline engines, highway tunnels, and wood combustion emissions. *Atmospheric Environment* 29:533–542.

Klimont, Z., K. Kupiainen and C. Heyes, et al. 2017. Global anthropogenic emissions of particulate matter including black carbon. *Atmospheric Chemistry and Physics* 17: 8681–8723.

Konur, O. 2000. Creating enforceable civil rights for disabled students in higher education: An institutional theory perspective. *Disability & Society* 15:1041–1063.

Konur, O. 2002a. Access to nursing education by disabled students: Rights and duties of nursing programs. *Nurse Education Today* 22:364–374.

Konur, O. 2002b. Assessment of disabled students in higher education: Current public policy issues. *Assessment and Evaluation in Higher Education* 27:131–52.

Konur, O. 2002c. Access to employment by disabled people in the UK: Is the Disability Discrimination Act working? *International Journal of Discrimination and the Law* 5:247–279.

Konur, O. 2006a. Participation of children with dyslexia in compulsory education: Current public policy issues. *Dyslexia* 12:51–67.

Konur, O. 2006b. Teaching disabled students in higher education. *Teaching in Higher Education* 11:351–363.

Konur, O. 2007a. A judicial outcome analysis of the Disability Discrimination Act: A windfall for the employers? *Disability & Society* 22:187–204.

Konur, O. 2007b. Computer-assisted teaching and assessment of disabled students in higher education: The interface between academic standards and disability rights. *Journal of Computer Assisted Learning* 23:207–219.

Konur, O. 2015. Current state of research on algal bioethanol. In *Marine Bioenergy: Trends and Developments*, Eds. S. K. Kim and C. G. Lee, pp. 217–244. Boca Raton, FL: CRC Press.

Konur, O., Ed. 2018a. *Bioenergy and Biofuels*. Boca Raton, FL: CRC Press.

Konur, O. 2018b. Bioenergy and biofuels science and technology: Scientometric overview and citation classics. In *Bioenergy and Biofuels*, Ed. O. Konur, pp. 3–63. Boca Raton, FL: CRC Press.

Konur, O. 2019. Cyanobacterial bioenergy and biofuels science and technology: A scientometric overview. In *Cyanobacteria: From Basic Science to Applications*, Eds. A. K. Mishra, D. N. Tiwari and A. N. Rai, pp. 419–442. Amsterdam: Elsevier.

Konur, O. 2020a. The scientometric analysis of the research on the bioethanol production from green macroalgae. In *Handbook of Algal Science, Technology and Medicine*, Ed. O. Konur, pp. 385–401. London: Academic Press.

Konur, O., Ed. 2020b. *Handbook of Algal Science, Technology and Medicine*. London: Academic Press.

Konur, O., Ed. 2021a. *Handbook of Biodiesel and Petrodiesel Fuels: Science, Technology, Health, and Environment*. Boca Raton, FL: CRC Press.

Konur, O., Ed. 2021b. *Handbook of Biodiesel and Petrodiesel Fuels: Science, Technology, Health, and Environment. Volume 1. Biodiesel Fuels: Science, Technology, Health, and Environment*. Boca Raton, FL: CRC Press.

Konur, O., Ed. 2021c. *Handbook of Biodiesel and Petrodiesel Fuels: Science, Technology, Health, and Environment. Volume 2. Biodiesel Fuels Based on the Edible and Nonedible Feedstocks, Wastes, and Algae: Science, Technology, Health, and Environment*. Boca Raton, FL: CRC Press.

Konur, O., Ed. 2021d. *Handbook of Biodiesel and Petrodiesel Fuels: Science, Technology, Health, and Environment. Volume 3. Petrodiesel Fuels: Science, Technology, Health, and Environment*. Boca Raton, FL: CRC Press.

Konur, O. and F. L. Matthews. 1989. Effect of the properties of the constituents on the fatigue performance of composites: A review. *Composites* 20:317–328.

Kuhlbusch, T. A. 1998. Black carbon and the carbon cycle. *Science* 280:1903–1904.

Ma, X., L. Sun and C. Song. 2002. A new approach to deep desulfurization of gasoline, diesel fuel and jet fuel by selective adsorption for ultra-clean fuels and for fuel cell applications. *Catalysis Today* 77:107–116.

Mehl, M., W. J. Pitz, C. K. Westbrook and H. J. Curran. 2011. Kinetic modeling of gasoline surrogate components and mixtures under engine conditions. *Proceedings of the Combustion Institute* 33:193–200.

Meisel, S. L., J. P. McCullough, C. H. Lechthaler and P. B. Weisz. 1976. Gasoline from methanol in one step. *Chemische Technik* 6:86–89.

Miguel, A. H., T. W. Kirchstetter, R. A., Harley and S. V. Hering. 1998. On-road emissions of particulate polycyclic aromatic hydrocarbons and black carbon from gasoline and diesel vehicles. *Environmental Science & Technology* 32:450–455.

Moldovan, M., M. A. Palacios and M. M. Gomez, et al. 2002. Environmental risk of particulate and soluble platinum group elements released from gasoline and diesel engine catalytic converters. *Science of the Total Environment* 296:199–208.

Najafi, G., B. Ghobadian and T. Tavakoli, et al. 2009. Performance and exhaust emissions of a gasoline engine with ethanol blended gasoline fuels using artificial neural network. *Applied Energy* 86:630–639.

Newman, P. W. G. and J. R. Kenworthy. 1989. Gasoline consumption and cities: A comparison of U.S. cities with a global survey. *Journal of the American Planning Association* 55:24–37.

North, D. C. 1991. Institutions. *Journal of Economic Perspectives* 5:97–112.

Nriagu, J. O. 1990. The rise and fall of leaded gasoline. *Science of the Total Environment C* 92:13–28.

Odum, J. R., T. P. W. Jungkamp, R. J. Griffin, R. C. Flagan and J. H. Seinfeld. 1997a. The atmospheric aerosol-forming potential of whole gasoline vapor. *Science* 276:96–99.

Odum, J.R., T. P. W. Jungkamp and R. J. Griffin, et al. 1997b. Aromatics, reformulated gasoline, and atmospheric organic aerosol formation. *Environmental Science and Technology* 31:1890–1897.

Pandis, S. N., R. A. Harley, G. R. Cass and J. H. Seinfeld. 1992. Secondary organic aerosol formation and transport. *Atmospheric Environment. Part A. General Topics*, 26:2269–2282.

Parry, I. W. H. and K. A. Small. 2005. Does Britain or the United States have the right gasoline tax? *American Economic Review* 95:1276–1289.

Ressler, T., J. Wong, J. Roos, and I. L. Smith. 2000. Quantitative speciation of Mn-bearing particulates emitted from autos burning (methylcyclopentadienyl) manganese tricarbonyl-added gasolines using XANES spectroscopy. *Environmental Science and Technology* 34:950–958.

Schauer, J. J., M. J. Kleeman, G. R. Cass and B. R. T. Simoneit. 2002. Measurement of emissions from air pollution sources. 5. C_1-C_{32} organic compounds from gasoline-powered motor vehicles. *Environmental Science and Technology* 36:1169–1180.

Song, C. 2003. An overview of new approaches to deep desulfurization for ultra-clean gasoline, diesel fuel and jet fuel. *Catalysis Today* 86:211–263.

Thomas, V. M. 1995. The elimination of lead in gasoline. *Annual Review of Energy and the Environment* 20:301–324.

Wei, J., Q. Ge and R. Yao, et al. 2017. Directly converting CO_2 into a gasoline fuel. *Nature Communications* 8:15174.

Yuksel, F. and B. Yuksel. 2004. The use of ethanol-gasoline blend as a fuel in an SI engine. *Renewable Energy* 29:1181–1191.

Zielinska, B., J. Sagebiel, J. D. Mc Donald, K. Whitney and D. R. Lawson. 2004. Emission rates and comparative chemical composition from selected in-use diesel and gasoline-fueled vehicles. *Journal of the Air and Waste Management Association* 54:1138–1150.

82 Nanotechnology Applications in Bioethanol Fuels

Scientometric Study

Ozcan Konur
(Formerly) Ankara Yildirim Beyazit University

82.1 INTRODUCTION

The crude oil-based gasoline fuels (Ma et al., 2002; Newman and Kenworthy, 1989) and petrodiesel fuels (Bosmann et al., 2001; Kim et al., 2006) have been widely used in the transportation sector since the 1920s. However, there have been great public concerns over the adverse environmental impact and sustainability of these fuels (Hill et al., 2009; Schauer et al., 2002).

Hence, biomass-based bioethanol fuels (Hill et al., 2009; Konur, 2012e, 2015, 2020) have increasingly been used in blending gasoline fuels (Hsieh et al., 2002; Najafi et al., 2009) as well as petrodiesel fuels in recent years (Hansen et al., 2005; Li et al., 2005).

In the meantime, the research in the nanomaterials and nanotechnology has intensified in recent years to become a major research field in the scientific research (Geim, 2009; Geim and Novoselov, 2007; Iijima, 1991; Novoselov et al., 2004). Thus, a large number of nanomaterials have been utilized nearly for nearly every research field. These materials offer an innovative way to increase the efficiency in the production and utilization of bioethanol fuels as in the other scientific fields (Konur, 2016a–f, 2017a–e, 2019b, 2021e,f).

In this context, there has been a significant focus on the nanotechnology applications in bioethanol fuels (Murdoch et al., 2011; Pan et al., 2007). The research in this field has particularly intensified the fields of sensing of bioethanol gas (Wan et al., 2004; Wang et al., 2012), oxidation of bioethanol fuels for the fuel cell applications (Dong et al., 2010; Xu et al., 2007), and to a lesser extent production of bioethanol fuels (Hoang et al., 2018; Pan et al., 2007), production of biohydrogen fuels from bioethanol fuels (Kugai et al., 2006; Murdoch et al., 2011), utilization of bioethanol fuels in the transport engines (Selvan et al., 2009; Shaafi and Velraj, 2015), and production of biochemicals from bioethanol fuels (Sun et al., 2011; Zhang et al., 2008).

However, it is essential to develop efficient incentive structures (North, 1991) for the primary stakeholders to enhance the research in this field (Konur, 2000, 2002a–c, 2006a,b, 2007a,b).

The scientometric analysis has been used in this context to inform the primary stakeholders about the current state of the research in a selected research field (Garfield, 1955; Konur, 2011, 2012a–i, 2015, 2018b, 2019a,b, 2020a).

Although there have been a number of scientometric studies on bioethanol fuels (Konur, 2012a–i, 2015, 2020a), there have been no scientometric study of the research in the field of the nanotechnology applications in bioethanol fuels.

This book chapter presents a scientometric study of the research in the field of the nanotechnology applications in bioethanol fuels. It examines the scientometric characteristics of both the sample and population data presenting scientometric characteristics of these both datasets in the order of documents, authors, publication years, institutions, funding bodies, source titles, countries, Scopus subject categories, keywords, and research fronts.

 DOI: 10.1201/9781003226567-110

82.2 MATERIALS AND METHODS

The search for this study was carried out using Scopus database (Burnham, 2006) in August 2021.

As a first step for the search of the relevant literature, the keywords were selected using the first most-cited 500 papers. The selected keyword list was optimized to obtain a representative sample of papers for the searched research field. This keyword list was provided in the appendix for future replication studies. Additionally, the information about the most-used keywords was given in Section 82.3.9 to highlight the key research fronts in Section 82.3.10.

As a second step, two sets of data were used for this study. First, a population sample of over 4,000 papers was used to examine the scientometric characteristics of the population data. Secondly, a sample of 100 most-cited papers was used to examine the scientometric characteristics of these citation classics with over 125 citations each.

The scientometric characteristics of these both sample and population datasets were presented in the order of documents, authors, publication years, institutions, funding bodies, source titles, countries, Scopus subject categories, keywords, and research fronts.

Lastly, the key scientometric findings for both datasets were discussed to highlight the research landscape for the nanotechnology applications in bioethanol fuels. Additionally, a number of brief conclusions were drawn, and a number of relevant recommendations were made to enhance the future research landscape.

82.3 RESULTS

82.3.1 The Most-Prolific Documents in the Field of the Nanotechnology Applications in Bioethanol Fuels

The information on the types of documents for both datasets is given in Table 82.1. The articles dominate both the sample and population datasets. The articles and conference papers have a surplus and deficit, respectively.

It is also interesting to note that all the papers in the sample dataset were published in the journals, while only 98.1% of the papers were published in the journals for the population dataset. Furthermore, 1.7% and 0.2% of the population papers were published in books and book series, respectively.

82.3.2 The Most-Prolific Authors in the Field of the Nanotechnology Applications in Bioethanol Fuels

The information about the most-prolific 29 authors with at least two sample papers and five population papers each in the sample dataset is given in Table 82.2.

The most-prolific authors are Yujin Chen, Chunling Zhu, and Taihong Wang with at least four sample papers each. Jong-Heun Lee, Yadong Li, Xiaoguang Wang, Zhonghua Zhang, Y.G. Wang, and X.Y. Xue follow these top authors with three sample papers each.

The most-prolific institutions for the sample dataset are the Harbin Engineering University and National Cheng Kung University with three authors each. The Chinese Academy of Sciences, Jilin University, Najran University, Shandong University, Sun Yat-Sen University, Tsinghua University, and Xiamen University follow these top institutions with two authors each. In total, 18 institutions house these authors.

The most-prolific country for the sample dataset is China with 20 authors. Saudi Arabia and Taiwan follow this top country with three authors each. In total, only five countries house these authors.

The most-prolific research fronts for the sample dataset are the oxidation of bioethanol fuels for the fuel cells and sensing of bioethanol gas with 14 and 15 authors, respectively.

TABLE 82.1
Documents in the Field of the Nanotechnology Applications in Bioethanol Fuels

Documents	Sample Dataset (%)	Population Dataset (%)	Surplus (%)
Article	99	95.9	3.1
Conference paper	0	3.2	–3.2
Book chapter	0	0.3	–0.3
Review	0	0.3	–0.3
Note	1	0.1	0.9
Letter	0	0.2	–0.2
Editorial	0	0.0	0.0
Book	0	0.0	0.0
Short Survey	0	0.0	0.0
Sample size	100	4,030	

On the other hand, there is a significant gender deficit (Beaudry and Lariviere, 2016) for the sample dataset as surprisingly nearly all of these top researchers are male.

82.3.3 The Most-Prolific Research Output by Years in the Field of the Nanotechnology Applications in Bioethanol Fuels

Information about papers published between 1970 and 2021 is given in Figure 82.1. This figure clearly shows that the bulk of the research papers in the population dataset were published primarily in the 2010s with 70.2% of the population dataset. This was followed by the early 2020s and 2000s with 16.1% and 13.1% of the population papers, respectively. The publication rates for the pre-2000s were negligible.

Similarly, the bulk of the research papers in the sample dataset were published in the 2000s and 2010s with 44% and 56% of the sample dataset, respectively. The publication rates for the pre-2000s were negligible.

The most-prolific publication years for the population dataset were after 2015 with at least 8% each of the dataset. Similarly, 39% of the sample papers were published between 2009 and 2011.

82.3.4 The Most-Prolific Institutions in the Field of the Nanotechnology Applications in Bioethanol Fuels

Information about the most-prolific 22 institutions publishing papers on the nanotechnology applications in bioethanol fuels with at least two sample papers and at least 0.3% of the population papers each is given in Table 82.3.

The most-prolific institution is the Chinese academy of Sciences with 14 papers. The Sun Yat-Sen University, Harbin Engineering University, Jilin University, Soochow University, and Beijing Institute of Technology follow this top institution with at least four sample papers each.

The top country for these most-prolific institutions is China with 13 institutions. Next, the USA and South Korea follow China with three and two institutions, respectively. In total, seven countries house these top institutions.

On the other hand, the institution with the most citation impact is the Chinese Academy of Sciences with 8.6% surplus. The Sun Yat-Sen University, Harbin Engineering University, and Beijing Institute of Technology follow this top institution with at least 3.5% surplus each. Similarly, the institutions with the least impact are Xiamen University, Beijing University of Chemical Technology, Chongqing University, and Nanyang Technological University with at least 0.4% surplus.

TABLE 82.2
The Most-Prolific Authors in the Field of the Nanotechnology Applications in Bioethanol Fuels

No.	Author Name	Author Code	Sample Papers	Population Papers	Institution	Country	Research Front
1	Chen, Yujin	7601437135	8	9	Harbin Inst. Technol.	China	Sensors
2	Zhu, Chunling	7403439505	5	6	Harbin Eng. Univ.	China	Sensors
3	Wang, Taihong	35241217600	4	10	Chinese Acad. Sci.	China	Sensors
4	Lee, Jong-Heun	26643283000	3	15	Korea Univ.	S. Korea	Sensors
5	Li Yadong	57192004602	3	6	Tsinghua Univ.	China	Oxidation
6	Wang, Xiaoguang	36138407600	3	6	Shandong Univ.	China	Oxidation
7	Zhang, Zhonghua	54913005200	3	6	Shandong Univ.	China	Oxidation
8	Wang, Y.G.	55954567600	3	5	Harbin Eng. Univ.	China	Sensors
9	Xue, X.Y.	13611568700	3	5	Harbin Eng. Univ.	China	Sensors
10	Sun, Shi-Gang	7404510197	2	11	Xiamen Univ.	China	Oxidation
11	Shen, Pei Kang	7201767641	2	10	Sun Yat-Sen Univ.	China	Oxidation
12	Chang, Shoou-Jinn	57221300233	2	9	Natl. Cheng Kung Univ.	Taiwan	Sensors
13	Hsueh, Ting-Jen	9336934600	2	9	Natl. Cheng Kung Univ.	Taiwan	Sensors
14	Wang, Hong	55963856800	2	9	Dongguan Univ. Technol.	China	Oxidation
15	Wang, Lili	57037822000	2	9	Jilin Univ.	China	Sensors
16	Guo, Shaojun	16202647200	2	8	Peking Univ.	China	Oxidation
17	Su, Dong	55154198900	2	8	Brookhaven Natl. Lab.	USA	Oxidation
18	Wang, Xun	56291267100	2	8	Tsinghua Univ.	China	Sensors
19	Zhou, Zhi-You	7406098551	2	8	Xiamen Univ.	China	Oxidation
20	Faisal, Muhammad	35617425400	2	7	Najran Univ.	S. Arabia	Sensors
21	Rahman, Muhammad M	56397398200	2	7	Najran Univ.	S. Arabia	Sensors
22	Wang, Huan	57207238487	2	7	Tianjin Univ.	China	Oxidation
23	Hsu, Cheng-Liang	7404946445	2	6	Natl. Cheng Kung Univ.	Taiwan	Sensors
24	Huang, Xiaoqing	56515570400	2	6	Soochow Univ.	China	Oxidation
25	Khan, Sher Bahadar	36059229400	2	6	King Abdulaziz Univ.	S. Arabia	Sensors
26	Wang, Erkang	34868964900	2	6	Chinese Acad. Sci.	China	Oxidation
27	Frenkel, Anatoly I	24404182600	2	5	Stony Brook Univ.	USA	Oxidation
28	Li Gao-Ren	7407053622	2	5	Sun Yat-Sen Univ.	China	Oxidation
29	Lou, Zheng	36739423500	2	5	Jilin Univ.	China	Sensors

Author code: the unique code given by Scopus to the authors. Sample papers: the number of papers authored in the sample dataset. Population papers: the number of papers authored in the population dataset.

82.3.5 The Most-Prolific Funding Bodies in the Field of the Nanotechnology Applications in Bioethanol Fuels

Information about the most-prolific 11 funding bodies funding at least two sample papers and with 0.4% of the population papers each is given in Table 82.4.

The most-prolific funding body is the National Natural Science Foundation of China with 20 sample papers. The Ministry of Science and Technology, Ministry of Education, and China Postdoctoral Science Foundation of China follow this top funding body with at least 4 sample papers each.

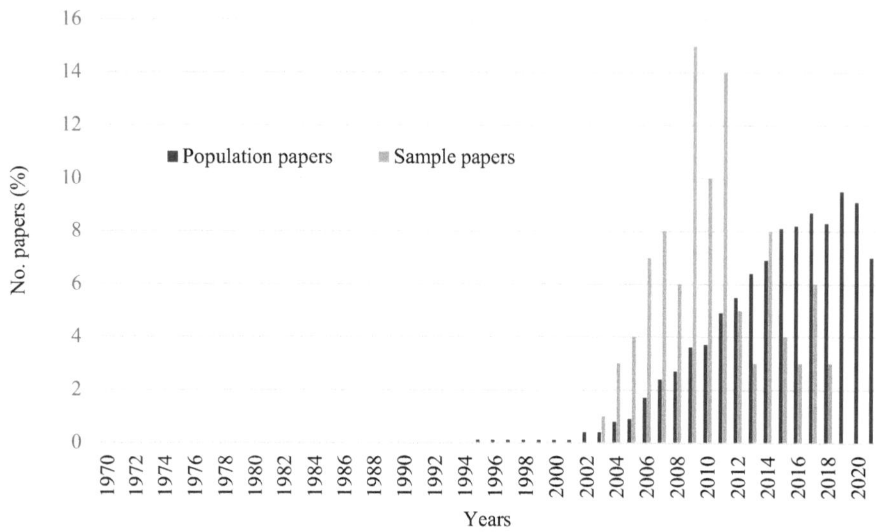

FIGURE 82.1 The research output by years regarding the nanotechnology applications in bioethanol fuels.

TABLE 82.3
The Most-Prolific Institutions in the Field of the Nanotechnology Applications in Bioethanol Fuels

No.	Institutions	Country	Sample Papers (%)	Population Papers (%)	Surplus (%)
1	Chinese Acad. Sci.	China	14	5.4	8.6
2	Sun Yat-Sen Univ.	China	6	0.5	5.5
3	Harbin Eng. Univ.	China	5	0.7	4.3
4	Jilin Univ.	China	4	2.0	2.0
5	Soochow Univ.	China	4	1.5	2.5
6	Beijing Inst. Technol.	China	4	0.5	3.5
7	Tianjin Univ.	China	3	1.5	1.5
8	Tsinghua Univ.	China	3	1.3	1.7
9	Shandong Univ.	China	3	1.0	2.0
10	Korea Univ.	S. Korea	3	0.7	2.3
11	Korea Adv. Inst. Sci. Technol.	S. Korea	3	0.4	2.6
12	Xiamen Univ.	China	2	1.6	0.4
13	Nanyang Technol. Univ.	Singapore	2	0.8	1.2
14	Beijing Univ. Chem. Technol.	China	2	0.8	1.2
15	Chongqing Univ.	China	2	0.8	1.2
16	Natl. Res. Counc.	Italy	2	0.6	1.4
17	Natl. Cheng Kung Univ.	Taiwan	2	0.5	1.5
18	Najran Univ.	S. Arabia	2	0.5	1.5
19	Brookhaven Natl. Univ.	USA	2	0.4	1.6
20	Georgia Inst. Technol.	USA	2	0.3	1.7
21	Nankai Univ.	China	2	0.3	1.7
22	Stony Brook Univ.	USA	2	0.3	1.7

It is notable that 99% and 58.8% of the sample and population papers are funded, respectively. The most-prolific country for these top funding bodies is China with seven funding bodies. Next, the USA follows China with three funding bodies. In total, China, the USA, and European Union house these top funding bodies.

TABLE 82.4
The Most-Prolific Funding Bodies in the Field of the Nanotechnology Applications in Bioethanol Fuels

No.	Funding Bodies	Country	Sample Papers No.	Population Paper No.	Surplus (%)
1	National Natural Science Foundation of China	China	20	23.4	–3.4
2	Ministry of Science and Technology	China	7	2.8	4.2
3	Ministry of Education	China	5	4.2	0.8
4	China Postdoctoral Science Foundation	China	4	1.4	2.6
5	Priority Academic Program Development of Jiangsu Higher Education Institutions	China	3	1.6	1.4
6	U.S. Department of Energy	USA	3	1.1	1.9
7	National Science Foundation	USA	2	1.6	0.4
8	European Commission	EU	2	1.6	0.4
9	National Basic Research Program of China	China	2	0.9	1.1
10	Specialized Research Fund for the Doctoral Program of Higher Education of China	China	2	0.4	1.6
11	U.S. Department of Defense	USA	2	0.4	1.6

TABLE 82.5
The Most-Prolific Source Titles in the Field of the Nanotechnology Applications in Bioethanol Fuels

No.	Source Titles	Sample Papers (%)	Population Papers (%)	Surplus (%)
1	Sensors and Actuators B Chemical	15	5.0	10.0
2	Journal of Power Sources	9	2.2	6.8
3	Journal of the American Chemical Society	7	0.3	6.7
4	ACS Applied Materials and Interfaces	6	1.7	4.3
5	Angewandte Chemie International Edition	6	0.3	5.7
6	Journal of Physical Chemistry C	4	1.5	2.5
7	Nanotechnology	4	1.2	2.8
8	Advanced Materials	4	0.3	3.7
9	Applied Physics Letters	4	0.2	3.8
10	International Journal of Hydrogen Energy	2	3.6	–1.6
11	Carbon	2	1.2	0.8
12	Journal of Catalysis	2	0.6	1.4
13	Journal of Materials Chemistry	2	0.5	1.5
14	Biosensors and Bioelectronics	2	0.4	1.6

82.3.6 The Most-Prolific Source Titles in the Field of the Nanotechnology Applications in Bioethanol Fuels

Information about the most-prolific 14 source titles publishing at least two sample papers and with 0.4% of the population papers each in the field of the nanotechnology applications in bioethanol fuels is given in Table 82.5.

The most-prolific source title is 'Sensors and Actuators B Chemical' with 20 sample papers. The 'Journal of Power Sources', 'Journal of the American Chemical Society', 'ACS Applied Materials and Interfaces', and 'Angewandte Chemie International Edition' follow this top journal with at least six papers each.

On the other hand, the source title with the most citation impact is 'Sensors and Actuators B Chemical' with 10% surplus. The 'Journal of Power Sources, 'Journal of the American Chemical Society', 'Angewandte Chemie International Edition', and 'ACS Applied Materials and Interfaces' follow this top journal with at least 6% surplus each. Similarly, the source titles with the least impact are the 'International Journal of Hydrogen Energy', 'Carbon', and 'Journal of Catalysis' with at least 1.8% deficit.

It is notable that six, three, and two of these top source titles are indexed within the subject categories of materials science, chemistry, and energy, respectively.

82.3.7 The Most-Prolific Countries in the Field of the Nanotechnology Applications in Bioethanol Fuels

Information about the most-prolific 12 countries publishing at least two papers and with 2% of the population papers each in the field of the nanotechnology applications in bioethanol fuels is given in Table 82.6.

The most-prolific country is China with 54 sample papers and the USA follows China with 22 sample papers. The other prolific countries are India, S. Korea, Spain, Japan, Taiwan, France, Italy, and the UK. Further, four European countries listed in Table 82.6 produce 13% and 9% of the sample and population papers, respectively.

On the other hand, the countries with the most citation impact are China and the USA with 10.8% and 14% publication surplus, respectively. Similarly, the countries with the least citation impact are Japan, Brazil, and India with at least 2.5% deficit.

82.3.8 The Most-Prolific Scopus Subject Categories in the Field of the Nanotechnology Applications in Bioethanol Fuels

Information about the most-prolific eight Scopus subject categories indexing at least 11% and 4% of the sample and population papers each is given in Table 82.7.

The most-prolific Scopus subject category in the field of the nanotechnology applications in bioethanol fuels is 'Chemistry' with 56 sample papers. This top category is followed by 'Materials Science', 'Engineering', and 'Chemical Engineering' with at least 32 sample papers each.

TABLE 82.6
The Most-Prolific Countries in the Field of the Nanotechnology Applications in Bioethanol Fuels

No.	Countries	Sample Papers (%)	Population Papers (%)	Surplus (%)
1	China	54	43.2	10.8
2	USA	22	8.0	14.0
3	India	9	9.3	–0.3
4	S. Korea	7	6.0	1.0
5	Spain	4	2.6	1.4
6	Japan	3	5.5	–2.5
7	Taiwan	3	2.8	0.2
8	France	3	2.4	0.6
9	Italy	3	2.2	0.8
10	UK	3	2.2	0.8
11	Brazil	2	3.7	–1.7
12	Canada	2	2.0	0.0

TABLE 82.7
The Most-Prolific Scopus Subject Categories in the Field of the Nanotechnology Applications in Bioethanol Fuels

No.	Scopus Subject Categories	Sample Papers (%)	Population Papers (%)	Surplus (%)
1	Chemistry	56	51.7	4.3
2	Materials Science	49	52.9	–3.9
3	Engineering	41	32.6	8.4
4	Chemical Engineering	32	32.3	–0.3
5	Physics and Astronomy	28	36.3	–8.3
6	Energy	20	15.0	5.0
7	Biochemistry, Genetics, and Molecular Biology	11	4.2	6.8

On the other hand, the Scopus subject categories with the most citation impact are 'Engineering', 'Biochemistry, Genetics and Molecular Biology', and 'Energy' with at least 5% surplus each. Similarly, the Scopus subject categories with the least citation impact are 'Physics and Astronomy', 'Materials Science', and 'Environmental Science' with at least 3.2% deficit each.

82.3.9 The Most-Prolific Keywords in the Field of the Nanotechnology Applications in Bioethanol Fuels

Information about the keywords used in at least 5% the sample and population papers each is given in Table 82.8. For this purpose, keywords related to the keyword set given in the Appendix are selected from a list of the most prolific keyword set provided by Scopus database.

These keywords are grouped under seven headings: bioethanol fuels, nanomaterials, sensors, oxidation, bioethanol production, materials, and catalysts.

There are eight keywords used related to the bioethanol fuels. The prolific keywords are ethanol, ethanol oxidation, and ethanol electrooxidation. It is notable that bioethanol keyword used only 3% and 2% of the sample and population papers, respectively.

The prolific keywords related to the nanomaterials are nanoparticles, nanostructured materials, nanowires, and nanostructures. The prolific keywords related to the sensors are gas sensors, chemical sensors, sensors, and gas detectors. The keywords related to the oxidation are oxidation and electrooxidation. The prolific keywords related to the bioethanol production are methanol and carbon dioxide. The most used materials are palladium, zinc oxide, and platinum. Finally, the prolific keywords related to the fuels cells and catalysts are fuel cells, electrocatalysts, catalysis, catalyst activity, and catalysts.

On the other hands, the keywords with the most citation impact are catalyst activity, fuel cells, gas sensors, palladium, nanostructured materials, chemical sensors, catalysis, oxidation, zinc oxide with at least 10% surplus each. Similarly, the keywords with the least citation impact are nanocomposites, metal nanoparticles, gas sensing, carbon nanotubes, graphene, and ethanol fuels.

82.3.10 The Most-Prolific Research Fronts in the Field of the Nanotechnology Applications in Bioethanol Fuels

Information about the most-prolific research fronts for the sample papers in the field of the nanotechnology applications in bioethanol fuels is given in Table 82.9.

There are two primary research fronts in this field: sensing of bioethanol gas (Wan et al., 2004; Wang et al., 2012) and oxidation of bioethanol fuels for the fuel cell applications (Dong et al., 2010; Xu et al., 2007) with 41 and 38 sample papers, respectively.

TABLE 82.8
The Most-Prolific Keywords in the Field of the Nanotechnology Applications in Bioethanol Fuels

No.	Keywords	Sample Papers (%)	Population Papers (%)	Surplus (%)
1.	**Bioethanol Fuels**			
	Ethanol	85	81.3	3.7
	Ethanol oxidation	15	10.5	4.5
	Ethanol electrooxidation	14	12.0	2.0
	Ethanol sensors	9	6.4	2.6
	Ethanol sensing	8	4.8	3.2
	Ethanol oxidation reaction	7	8.5	−1.5
	Ethanol fuels	5	7.0	−2
	Direct ethanol fuel cells	5	6.9	−1.9
	Bioethanol	3	2.0	1.0
2.	**Nanomaterials**			
	Nanoparticles	28	25.3	2.7
	Nanostructured materials	18	5.3	12.7
	Nanowires	11	4.7	6.3
	Nanostructures	9	6.1	2.9
	Graphene	7	9.2	−2.2
	Nanorods	7	4.7	2.3
	Nanotubes	6	4.4	1.6
	Nanotechnology	6	1.9	4.1
	Carbon nanotubes	5	8.0	−3
	Nanomaterials	5	0.0	5
	Nanocomposites	0	8.0	−8
	Metal nanoparticles	0	5.8	−5.8
3.	**Sensors**			
	Gas sensors	23	7.3	15.7
	Chemical sensors	22	9.3	12.7
	Sensors	17	7.3	9.7
	Gas detectors	10	10.0	0.0
	Electrochemical sensors	6	0.0	6.0
	Gas sensing	5	10.4	−5.4
4.	**Oxidation**			
	Oxidation	24	13.6	10.4
	Electrooxidation	20	12.3	7.7
	Catalytic oxidation	5	5.2	−0.2
	Alcohol oxidation	5	1.7	3.3
5.	**Bioethanol Production**			
	Dehydration	7	2.3	4.7
	Methanol	10	4.5	5.5
	Carbon dioxide	9	3.4	5.6
	Carbon monoxide	5	0.0	5

(Continued)

TABLE 82.8 (*Continued*)
The Most-Prolific Keywords in the Field of the Nanotechnology Applications in Bioethanol Fuels

No.	Keywords	Sample Papers (%)	Population Papers (%)	Surplus (%)
6.	**Materials**			
	Palladium	21	7.2	13.8
	Zinc oxide	18	7.8	10.2
	Platinum	11	9.6	1.4
	ZnO	7	2.6	4.4
	Gold	6	4.2	1.8
	Nickel	6	4.9	1.1
	Silver	6	2.3	3.7
	Platinum alloys	5	3.7	1.3
	Titanium dioxide	5	4.9	0.1
7.	**Catalysts**			
	Fuel cells	24	7.7	16.3
	Electrocatalysts	19	13.8	5.2
	Catalysis	18	5.4	12.6
	Catalyst activity	18	0.0	18
	Catalysts	17	10.4	6.6
	Electrocatalytic activity	11	7.7	3.3
	Electrocatalysis	8	5.8	2.2

TABLE 82.9
The Most-Prolific Research Fronts in the Field of the Nanotechnology Applications in Bioethanol Fuels

No.	Research Fronts	Sample Papers (%)
1	Sensing of bioethanol gas	41
2	Oxidation of bioethanol fuels for the fuel cell applications	38
3	Production of bioethanol fuels	11
4	Production of biohydrogen fuels from bioethanol fuels	5
5	Utilization of bioethanol fuels in the transport engines	3
6	Production of biochemicals from bioethanol fuels	2

Sample papers: the sample of the most-cited 100 papers.

There are also four other research fronts: production of bioethanol fuels (Hoang et al., 2018; Pan et al., 2007), production of biohydrogen fuels from bioethanol fuels for the fuel cells (Kugai et al., 2006; Murdoch et al., 2011), utilization of bioethanol fuels in the transport engines (Selvan et al., 2009; Shaafi and Velraj, 2015), and production of biochemicals from bioethanol fuels (Sun et al., 2011; Zhang et al., 2008) with 11, 5, 3, and 2 sample papers, respectively.

The focus of these most-cited 100 papers is the development of the nanomaterials for the most efficient processes for each of these research fronts.

82.4 DISCUSSION

82.4.1 Introduction

The crude oil-based gasoline and petrodiesel fuels have been widely used in the transportation sector. However, there have been great public concerns over the adverse environmental impact of these fuels. Hence, biomass-based bioethanol fuels have increasingly been used in blending gasoline fuels as well as petrodiesel fuels in recent years. There has been a significant focus on the nanotechnology applications in bioethanol fuels. However, it is essential to develop efficient incentive structures for the primary stakeholders to enhance the research in this field.

The scientometric analysis has been traditionally used to inform the primary stakeholders about the current state of the research in a selected research field. Although there have been a number of scientometric studies on bioethanol fuels, there has been no scientometric study of the research in the field of the nanotechnology applications in bioethanol fuels. This book chapter presents a scientometric study of the research in the field of the nanotechnology applications in bioethanol fuels. It examines the scientometric characteristics of the sample and population data presenting scientometric characteristics of these both datasets in the order of documents, authors, publication years, institutions, funding bodies, source titles, countries, Scopus subject categories, keywords, and research fronts.

The search for this study was carried out using Scopus database in August 2021. As a first step for the search of the relevant literature, the keywords were selected using the first most-cited 500 papers. The selected keyword list was optimized to obtain a representative sample of papers for the searched research field. This keyword list was provided in the appendix for future replication studies. Additionally, the information about the most-used keywords was given in Section 82.3.9 to highlight the key research fronts in Section 82.3.10.

As a second step, two sets of data were used for this study. First, a population sample of over 4,000 papers was used to examine the scientometric characteristics of the population data. Secondly, a sample of 100 most-cited papers was used to examine the scientometric characteristics of these citation classics with over 125 citations each.

82.4.2 The Most-Prolific Documents in the Field of the Nanotechnology Applications in Bioethanol Fuels

Articles and conference papers are overrepresented and underrepresented in the sample dataset, respectively, as shown in Table 82.1. It is highly notable that there are no review papers in the sample papers. Scopus differs from the Web of Science database in differentiating and showing articles and conference papers published in the journals.

It is observed during the search process that there has been inconsistency in the classification of the documents in Scopus as well as in other databases such as Web of Science. This is especially relevant for the classification of papers as reviews or articles as the papers not involving a literature review may be erroneously classified as a review paper. In this context, it would be helpful to provide a classification note for the published papers in the books and journals at the first instance. It would also be helpful to use the document types listed in Table 82.1 for this purpose. Book chapters may also be classified as articles or reviews as an additional classification to differentiate review chapters from the experimental chapters as it is done by the Web of Science.

82.4.3 The Most-Prolific Authors in the Field of the Nanotechnology Applications in Bioethanol Fuels

There have been most-prolific 29 authors with at least two sample papers and five population papers each in the sample dataset as given in Table 82.2. These authors have shaped the development of the research in this field.

The most-prolific authors are Yujin Chen, Chunling Zhu, Taihong Wang, Jong-Heun Lee, Yadong Li, Xiaoguang Wang, Zhonghua Zhang, Y.G. Wang, and X.Y. Xue (Table 82.2).

It is important to note the inconsistencies in indexing of the author names in Scopus and other databases. It is especially an issue for the names with more than two components such as 'Judge Alex de Camp Sirous'. The probable outcomes are 'Sirous J.A.D.C.', 'de Camp Sirous, J.A.', or 'Camp Sirous, J.A.D.'. The first choice is the gold standard of the publishing sector as the last word in the name is taken as the last name. The second choice is a strong alternative, while the last choice is an undesired outcome as two last words are taken as the last name. It is notable that inconsistent indexing of the author names may cause substantial inefficiencies in the search process for the papers as well as allocating credit to the authors as there are different author entries for each outcome in the databases.

There is also a difficulty in allowing credit for the authors especially for the authors with common names such as 'Wang, Y', or 'Huang, Y', or 'Zhu, Y.' in conducting scientometric studies. These difficulties strongly influence the efficiency of the scientometric studies as well as allocating credit to the authors as there are the same author entries for different authors with the same name, e.g., 'Wang Y' in the databases.

In this context, the coding of authors in Scopus database is a welcome innovation compared to the other databases such as Web of Science. In this process, Scopus allocates a unique number to each author in the database. However, there might still be substantial inefficiencies in this coding system especially for common names. It is possible that Scopus uses a number of software programs to differentiate the author names (Shin et al., 2014).

In this context, it does not help that author names are not given in full in some journals. This makes difficult to differentiate authors with common names and makes the scientometric studies further difficult in the author domain. Therefore, the author names should be given in all books and journals at the first instance.

There also inconsistencies in naming of the authors with more than two components in journal papers and book chapters. For example, 'Faaij, A.P.C.', 'Wyman CE', and 'Shuai, SJ' might be given as 'Faaij, A', 'Wyman C', or 'Shuai S.' in the journals and books. This also makes the scientometric studies difficult in the author domain. Hence, contributing authors should use their name consistently in their publications.

The other critical issue regarding the author names is the spelling of the author names in the national spellings (Gonçalves, Übeiro) rather than in the English spellings (Goncalves, Ubeiro) in Scopus database. Scopus differs from the Web of science database in this respect, where the author names are given in the English spellings. It is observed that national spellings of the author names do not help in conducting scientometric studies as well in allocating credits to the authors as sometimes there are the different author entries for the English and National spellings in the Scopus database.

The most-prolific institutions for the sample dataset are Harbin Engineering University, National Cheng Kung University, Chinese Academy of Sciences, Jilin University, Najran University, Shandong University, Sun Yat-Sen University, Tsinghua University, and Xiamen University. Similarly, the most-prolific countries for the sample dataset are China, Saudi Arabia, and Taiwan. The most-prolific research fronts for the sample dataset are the oxidation of bioethanol fuels for the fuel cells and the sensing of bioethanol gas.

It is also notable that there is a significant gender deficit for the sample dataset as surprisingly nearly all of these top researchers are male. This finding is the most thought-provoking with strong public policy implications. Hence, institutions, funding bodies, and policymakers should take efficient measures to reduce the gender deficit in this field as well as other scientific fields with strong gender deficit.

82.4.4 The Most-Prolific Research Output by Years in the Field of the Nanotechnology Applications in Bioethanol Fuels

The research output observed between 1970 and 2021 is illustrated in Figure 82.1. This figure clearly shows that the bulk of the research papers in the population dataset were published primarily in the

2010s. This was followed by the early 2020s and 2000s. The publication rates for the pre-2000s were negligible. Similarly, the bulk of the research papers in the sample dataset were published in the 2000s and 2010s. The publication rates for the pre-2000s were also negligible.

These data suggest that the most-cited sample papers were primarily published in the 2000s and 2010s, while the population papers were primarily published in the 2010s. These are the thought-provoking findings as there has been no significant research in this field in the pre-2000s, but there has been a significant research boom in the last two decades. In this context, the increasing public concerns about climate change (Change, 2007), greenhouse gas emissions (Lashof and Ahuja, 1990), and global warming (Kerr, 2007) have been certainly behind the boom in the research in this field in the last two decades.

The data in Figure 82.1 also suggest that the research in this field has boomed in the last two decades, and the size of the population papers is likely to more than double in the current decade, provided that the public concerns about climate change, greenhouse gas emissions, and global warming are translated efficiently to the research funding in this field.

82.4.5 The Most-Prolific Institutions in the Field of the Nanotechnology Applications in Bioethanol Fuels

The most-prolific 22 institutions publishing papers on the nanotechnology applications in bioethanol fuels with at least two sample papers and at least 0.3% of the population papers each given in Table 82.3 have shaped the development of the research in this field.

The most-prolific institutions are the Chinese Academy of Sciences, Sun Yat-Sen University, Harbin Engineering University, Jilin University, Soochow University, and Beijing Institute of Technology. The top countries for these most-prolific institutions are China, the USA, and South Korea.

On the other hand, the institutions with the most citation impact are the Chinese Academy of Sciences, Sun Yat-Sen University, Harbin Engineering University, and Beijing Institute of Technology. Similarly, the institutions with the least impact are Xiamen University, Beijing University of Chemical Technology, Chongqing University, and Nanyang Technological University.

82.4.6 The Most-Prolific Funding Bodies in the Field of the Nanotechnology Applications in Bioethanol Fuels

The most-prolific 11 funding bodies funding at least two sample papers and with 0.4% of the population papers each are given in Table 82.4.

The most-prolific funding bodies are the National Natural Science Foundation of China, the Ministry of Science and Technology, Ministry of Education, and China Postdoctoral Science Foundation of China. It is notable that 99% and 58.8% of the sample and population papers are funded, respectively. The most-prolific countries for these top funding bodies are China and the USA. In total, China, the USA, and European Union house these top funding bodies.

These findings on the funding of the research in this field suggest that the level of the funding, mostly in the 2010s, is highly intensive, and it has been largely instrumental in enhancing the research in this field (Ebadi and Schiffauerova, 2016) in the light of North's institutional framework (North, 1991).

82.4.7 The Most-Prolific Source Titles in the Field of the Nanotechnology Applications in Bioethanol Fuels

The most-prolific 14 source titles publishing at least two sample papers and with 0.4% of the population papers each in the field of the nanotechnology applications in bioethanol fuels have shaped the development of the research in this field (Table 82.5).

The most-prolific source titles are 'Sensors and Actuators B Chemical', 'Journal of Power Sources', 'Journal of the American Chemical Society', 'ACS Applied Materials and Interfaces', and 'Angewandte Chemie International Edition'.

On the other hand, the source titles with the most citation impact are 'Sensors and Actuators B Chemical', 'Journal of Power Sources, 'Journal of the American Chemical Society', 'Angewandte Chemie International Edition', and 'ACS Applied Materials and Interfaces'. Similarly, the source titles with the least impact are the 'International Journal of Hydrogen Energy', 'Carbon', and 'Journal of Catalysis'.

It is notable that six, three, and two of these top source titles are indexed within the subject categories of materials science, chemistry, and energy, respectively. This finding suggests that the journals in these fields have significantly shaped the development of the research in this field.

82.4.8 The Most-Prolific Countries in the Field of the Nanotechnology Applications in Bioethanol Fuels

The most-prolific 12 countries publishing at least two papers and with 2% of the population papers each have significantly shaped the development of the research in this field (Table 82.6).

The most-prolific countries are the USA and China. On the other hand, the countries with the most citation impact are again China and the USA. Similarly, the countries with the least citation impact are Japan, Brazil, and India.

The close examination of these findings suggests that the USA, China, and Europe are the major producers of the research in this field. It is a fact that the USA has been a major player in science (Leydesdorff and Wagner, 2009; Leydesdorff et al., 2014). The USA has further developed a strong research infrastructure to support its corn- and grass-based bioethanol industry (Vadas et al., 2008). The USA has also been very active in the nanotechnology research (Dong et al., 2016).

However, China has been a rising star in scientific research in competition with the USA and Europe (Leydesdorff and Zhou, 2005). China is also a major player in this field as a major producer of bioethanol (Li and Chan-Halbrendt, 2009). China has also been very active in the nanotechnology research (Dong et al., 2016).

Next, Europe has been a persistent player in the scientific research in competition with both the USA and China (Leydesdorff, 2000). Europe has also been a persistent producer of bioethanol along with the USA and Brazil (Gnansounou, 2010). The European Union has also been very active in the nanotechnology research (Schellekens, 2010).

India has also been a persistent player in scientific research at a moderate level (Karpagam et al., 2011). India, producing 11.4% of the population papers, has also developed a strong research infrastructure to support its biomass-based bioethanol industry (Sukumaran et al., 2010). India has also been very active in the nanotechnology research (Karpagam et al., 2011).

Like India, South Korea has also been a persistent player in scientific research at a moderate level (Leydesdorff and Zhou, 2005). S. Korea, producing 11.4% of the population papers, has also developed a strong research infrastructure to support its biomass-based bioethanol industry (Kim et al., 2010). S. Korea has also been very active in the nanotechnology research (Lee, 2002).

82.4.9 The Most-Prolific Scopus Subject Categories in the Field of the Nanotechnology Applications in Bioethanol Fuels

The most-prolific eight Scopus subject categories indexing at least 11% and 4% of the sample and population papers each given in Table 82.7 have shaped the development of the research in this field.

The most-prolific Scopus subject categories in the field of the nanotechnology applications in bioethanol fuels are 'Chemistry', 'Materials Science', 'Engineering', and 'Chemical Engineering'.

On the other hand, the Scopus subject categories with the most citation impact are 'Engineering', 'Biochemistry, Genetics and Molecular Biology', and 'Energy'. Similarly, the Scopus subject categories with the least citation impact are 'Physics and Astronomy', 'Materials Science', and 'Environmental Science'.

These findings are thought-provoking suggesting that the primary subject categories are chemical engineering, chemistry, engineering, and materials science. The other key finding is that social sciences are not well represented in both the sample and population papers, unlike the field of the evaluative studies in bioethanol fuels.

These finding are not surprising as the key research fronts in this field are the performance, combustion, and development of innovative nanomaterials for each research front in this field. All these research fronts are related to the hard sciences, engineering, and environmental sciences.

82.4.10 The Most-Prolific Keywords in the Field of the Nanotechnology Applications in Bioethanol Fuels

A limited number of keywords have shaped the development of the research in this field as shown in Table 82.8 and in Appendix.

These keywords are grouped under seven headings: bioethanol fuels, nanomaterials, sensors, oxidation, bioethanol production, materials, and catalysts.

The keywords with the most citation impact are catalyst activity, fuel cells, gas sensors, palladium, nanostructured materials, chemical sensors, catalysis, oxidation, and zinc oxide. Similarly, the keywords with the least citation impact are nanocomposites, metal nanoparticles, gas sensing, carbon nanotubes, graphene, and ethanol fuels.

These prolific keywords highlight the key research fronts in this field and reflect well he keywords used in the sample papers.

82.4.11 The Most-Prolific Research Fronts in the Field of the Nanotechnology Applications in Bioethanol Fuels

There are two primary research fronts in this field: sensing of bioethanol gas and oxidation of bioethanol fuels for the fuel cell applications with 41 and 38 sample papers, respectively.

There are also four other research fronts: Production of bioethanol fuels, production of biohydrogen fuels from bioethanol fuels for the fuel cells, utilization of bioethanol fuels in the transport engines, and production of biochemicals from bioethanol fuels.

The focus of these most-cited 100 papers is the development of the nanomaterials for the most efficient processes for each of these research fronts. Hence, there are strong structure–processing–property relationships for all the nanomaterials used for these research fronts. In the end, these most-cited papers in this field hint that the efficiency of the nanotechnology applications could be maximized using the structure, processing, and property relationships of the nanomaterials used in these applications (Formela et al., 2016; Konur, 2018a, 2020b, 2021a–d; Konur and Matthews, 1989).

82.5 CONCLUSION AND FUTURE RESEARCH

The research on the nanotechnology applications in bioethanol fuels has been mapped through a scientometric study of both sample and population datasets.

The critical issue in this study has been to obtain a representative sample of the research as in any other scientometric study. Therefore, the keyword set has been carefully devised and optimized after a number of runs in the Scopus database.

The other issue has been the selection of a multidisciplinary database to carry out the scientometric study of the research in this field. For this purpose, Scopus database has been selected. The journal coverage of this database has been wider than that of Web of Science.

The key scientometric properties of the research in this field have been determined and discussed in this book chapter. It is evident that a limited number of documents, authors, institutions, publication periods, institutions, funding bodies, source titles, countries, Scopus subject categories, keywords, and research fronts have shaped the development of the research in this field.

There is ample scope to increase the efficiency of the scientometric studies in this field in the author and document domains by developing consistent policies and practices in both domains. In this respect, authors, journals, and academic databases have a lot to do. Furthermore, the significant gender deficit as in most scientific fields emerges as a public policy issue.

The research in this field has boomed in the late 2010s possibly promoted by the public concerns on global warming, greenhouse gas emissions, and climate change. Institutions from the USA, China, and South Korea have mostly shaped the research in this field.

The extremely high funding rate of 99% and 59% for sample and population papers, respectively, suggests that this high funding rate significantly enhanced the research in this field primarily in the 2010s, possibly more than doubling in the current decade. The most-prolific journals have been mostly indexed by the subject categories of materials science, chemistry, and energy.

The USA, China, and Europe have been the major producers of the research in this field as the major producers of bioethanol fuels from different types of biomass such as corn, sugarcane, and grass as well as other types of biomass. India and South Korea have also contributed largely to the population papers.

The primary subject categories have been 'Chemistry', 'Materials Science', 'Engineering', and 'Chemical Engineering'. Due the technological emphasis of this field, social sciences have not been fairly represented in both the sample and population papers, unlike the evaluative studies in bioethanol fuels.

Ethanol is more popular than bioethanol as a keyword with strong implications for the search strategy. In other words, the search strategy using only bioethanol keyword would give only around 3% of the sample and population papers each.

These keywords are grouped under seven headings: bioethanol fuels, nanomaterials, sensors, oxidation, bioethanol production, materials, and catalysts. These groups of keywords highlight the potential primary research fronts for these fields.

There are two primary research fronts in this field: sensing of bioethanol gas and oxidation of bioethanol fuels for the fuel cell applications. There are also four other research fronts: production of bioethanol fuels, production of biohydrogen fuels from bioethanol fuels for the fuel cells, utilization of bioethanol fuels in the transport engines, and production of biochemical from bioethanol fuels.

These findings are thought-provoking. The focus of these most-cited 100 papers is the development of the nanomaterials for the most efficient processes for each of these research fronts. There are strong structure–processing–property relationships for all the nanomaterials used for these research fronts. In the end, these most-cited papers in this field hint that the efficiency of the nanotechnology applications could be maximized using the structure, processing, and property relationships of the nanomaterials used in the in these applications.

Thus, the scientometric analysis has a great potential to gain valuable insights into the evolution of the research in this field as in other scientific fields.

It is recommended that further scientometric studies are carried out about the other aspects of both production and utilization of bioethanol fuels. It is further recommended that reviews of the most-cited papers are carried out for each research front to complement these scientometric studies. Next, the scientometric studies of the hot papers in these primary fields are carried out.

ACKNOWLEDGMENTS

The contribution of the highly cited researchers in the field of the applications of nanotechnology in bioethanol fuels has been gratefully acknowledged.

APPENDIX: THE KEYWORD SET FOR THE FIELD OF THE NANOTECHNOLOGY APPLICATIONS IN BIOETHANOL FUELS

((TITLE (ethanol* OR bioethanol* OR {ethyl alcohol*} OR c2h5oh OR defc*) OR SRCTITLE (ethanol* OR bioethanol* OR gasohol)) AND (TITLE (*nano* OR {quantum dot*} OR {quantum-dot*} OR plasmon* OR {carbon dot*} OR dendrimer* OR {exfoliated graphite*} OR *fulleren* OR cnt OR mwnt* OR swcnt* OR mwcnt* OR mos2 OR *graphene* OR mose2 OR ws2 OR phosphorene OR {g-c3n4} OR mxene* OR {graphitic carbon nitride*} OR {atomic layer*} OR {inorganic-organic*} OR {organic-inorganic*} OR {metal-organic framework*} OR mof* OR mesoporous OR microporous OR {coordination polymer*} OR {zeolitic imidazolate framework*} OR {*molecular sieve*} OR mesocrystal* OR {single atom*} OR {atomic scale}) OR SRCTITLE (nano* OR mesoporous))) AND NOT (SUBJAREA (bioch OR phar OR medi OR neur OR immu OR dent OR heal OR psyc) OR TITLE (extract*) OR SRCTITLE (food*)) AND (LIMIT-TO (DOCTYPE, "ar") OR LIMIT-TO (DOCTYPE, "cp") OR LIMIT-TO (DOCTYPE, "re") OR LIMIT-TO (DOCTYPE, "ch") OR LIMIT-TO (DOCTYPE, "le") OR LIMIT-TO (DOCTYPE, "no")) AND (LIMIT-TO (LANGUAGE, "English")) AND (LIMIT-TO (SRCTYPE, "j") OR LIMIT-TO (SRCTYPE, "k") OR LIMIT-TO (SRCTYPE, "b")).

REFERENCES

Beaudry, C. and V. Lariviere. 2016. Which gender gap? Factors affecting researchers' scientific impact in science and medicine. *Research Policy* 45:1790–1817.

Bosmann, A., L. Datsevich and A. Jess, et al. 2001. Deep desulfurization of diesel fuel by extraction with ionic liquids. *Chemical Communications* 23:2494–2495.

Burnham, J. F. 2006. Scopus database: A review. *Biomedical Digital Libraries* 3:1–8.

Change, C. 2007. Climate change impacts, adaptation and vulnerability. *Science of the Total Environment* 326:95–112.

Dong, H., Y. Gao and P. J. Sinko, et al. 2016. The nanotechnology race between China and the United States. *Nano Today* 11:7–12.

Dong, L., R. R. S. Gari, Z. Li, M. M. Craig and S. Hou. 2010. Graphene-supported platinum and platinum-ruthenium nanoparticles with high electrocatalytic activity for methanol and ethanol oxidation. *Carbon* 48:781–787.

Ebadi, A. and A. Schiffauerova. 2016. How to boost scientific production? A statistical analysis of research funding and other influencing factors. *Scientometrics* 106:1093–1116.

Formela, K., A. Hejna, L. Piszczyk, M. R. Saeb and X. Colom. 2016. Processing and structure-property relationships of natural rubber/wheat bran biocomposites. *Cellulose* 23:3157–3175.

Garfield, E. 1955. Citation indexes for science. *Science* 122:108–111.

Geim, A. K. 2009. Graphene: Status and prospects. *Science* 324:1530–1534.

Geim, A. K. and K. S. Novoselov. 2007. The rise of graphene. *Nature Materials* 6:183–191.

Gnansounou, E. 2010. Production and use of lignocellulosic bioethanol in Europe: Current situation and perspectives. *Bioresource Technology* 101:4842–4850.

Hansen, A. C., Q. Zhang and P. W. L. Lyne. 2005. Ethanol-diesel fuel blends: A review. *Bioresource Technology* 96:277–285.

Hill, J., S. Polasky and E. Nelson, et al. 2009. Climate change and health costs of air emissions from biofuels and gasoline. *Proceedings of the National Academy of Sciences of the United States of America* 106:2077–2082.

Hoang, T. T. H., S. Verma and S. Ma, et al. 2018. Nanoporous copper-silver alloys by additive-controlled electrodeposition for the selective electroreduction of CO_2 to ethylene and ethanol. *Journal of the American Chemical Society* 140:5791–5797.

Hsieh, W. D., R. H. Chen, T. L. Wu and T. H. Lin. 2002. Engine performance and pollutant emission of an SI engine using ethanol-gasoline blended fuels. *Atmospheric Environment* 36:403–410.

Iijima, S. 1991. Helical microtubules of graphitic carbon. *Nature* 354:56–58.

Karpagam, R., S. Gopalakrishnan, M. Natarajan and B. R. Babu. 2011. Mapping of nanoscience and nanotechnology research in India: A scientometric analysis, 1990-2009. *Scientometrics* 89:501–522.

Kerr, R. A. 2007. Global warming is changing the world. *Science* 316:188–190.

Kim, J. H., X. L. Ma, A. N. Zhou and C. S. Song. 2006. Ultra-deep desulfurization and denitrogenation of diesel fuel by selective adsorption over three different adsorbents: A study on adsorptive selectivity and mechanism. *Catalysis Today* 111:74–83.

Kim, J. S., S. C. Park and J. W. Kim, et al. 2010. Production of bioethanol from lignocellulose: Status and perspectives in Korea. *Bioresource Technology* 101:4801–4805.

Konur, O. 2000. Creating enforceable civil rights for disabled students in higher education: An institutional theory perspective. *Disability & Society* 15:1041–1063.

Konur, O. 2002a. Access to nursing education by disabled students: Rights and duties of nursing programs. *Nurse Education Today* 22:364–374.

Konur, O. 2002b. Assessment of disabled students in higher education: Current public policy issues. *Assessment and Evaluation in Higher Education* 27:131–152.

Konur, O. 2002c. Access to employment by disabled people in the UK: Is the Disability Discrimination Act working? *International Journal of Discrimination and the Law* 5:247–279.

Konur, O. 2006a. Participation of children with dyslexia in compulsory education: Current public policy issues. *Dyslexia* 12:51–67.

Konur, O. 2006b. Teaching disabled students in higher education. *Teaching in Higher Education* 11:351–363.

Konur, O. 2007a. A judicial outcome analysis of the Disability Discrimination Act: A windfall for the employers? *Disability & Society* 22:187–204.

Konur, O. 2007b. Computer-assisted teaching and assessment of disabled students in higher education: The interface between academic standards and disability rights. *Journal of Computer Assisted Learning* 23:207–219.

Konur, O. 2011. The scientometric evaluation of the research on the algae and bio-energy. *Applied Energy* 88:3532–3540.

Konur, O. 2012a. Prof. Dr. Ayhan Demirbas' scientometric biography. *Energy Education Science and Technology Part A: Energy Science and Research* 28:727–738.

Konur, O. 2012b. The evaluation of the biogas research: A scientometric approach. *Energy Education Science and Technology Part A: Energy Science and Research* 29:1277–1292.

Konur, O. 2012c. The evaluation of the global energy and fuels research: A scientometric approach. *Energy Education Science and Technology Part A: Energy Science and Research* 30:613–628.

Konur, O. 2012d. The evaluation of the research on the biodiesel: A scientometric approach. *Energy Education Science and Technology Part A: Energy Science and Research* 28:1003–1014.

Konur, O. 2012e. The evaluation of the research on the bioethanol: A scientometric approach. *Energy Education Science and Technology Part A: Energy Science and Research* 28:1051–1064.

Konur, O. 2012f. The evaluation of the research on the biofuels: A scientometric approach. *Energy Education Science and Technology Part A: Energy Science and Research* 28:903–916.

Konur, O. 2012g. The evaluation of the research on the biohydrogen: A scientometric approach. *Energy Education Science and Technology Part A: Energy Science and Research* 29:323–338.

Konur, O. 2012h. The evaluation of the research on the microbial fuel cells: A scientometric approach. *Energy Education Science and Technology Part A: Energy Science and Research* 29:309–322.

Konur, O. 2012i. The scientometric evaluation of the research on the production of bioenergy from biomass. *Biomass and Bioenergy* 47:504–515.

Konur, O. 2015. Current state of research on algal bioethanol. In *Marine Bioenergy: Trends and Developments*, Eds. S. K. Kim and C. G. Lee, pp. 217–244. Boca Raton, FL: CRC Press.

Konur, O. 2016a. Scientometric overview in nanobiodrugs. In *Nanoarchitectonics for Smart Delivery and Drug Targeting*, Eds. A. M. Holban and A. M. Grumezescu, pp. 405–428. Amsterdam: Elsevier.

Konur, O. 2016b. Scientometric overview regarding nanoemulsions used in the food industry. In *Emulsions: Nanotechnology in the Agri-Food Industry*, Ed. A. M. Grumezescu, pp. 689–711. Amsterdam: Elsevier.

Konur, O. 2016c. Scientometric overview regarding the nanobiomaterials in antimicrobial therapy. In *Nanobiomaterials in Antimicrobial Therapy*, Ed. A. M. Grumezescu, pp. 511–535. Amsterdam: Elsevier.

Konur, O. 2016d. Scientometric overview regarding the nanobiomaterials in dentistry. In *Nanobiomaterials in Dentistry*, Ed. A. M. Grumezescu, pp. 425–453. Amsterdam: Elsevier.

Konur, O. 2016e. Scientometric overview regarding the surface chemistry of nanobiomaterials. In *Surface Chemistry of Nanobiomaterials*, Ed. A. M. Grumezescu, pp. 463–486. Amsterdam: Elsevier.

Konur, O. 2016f. The scientometric overview in cancer targeting. In *Nanoarchitectonics for Smart Delivery and Drug Targeting*, Eds. A. M. Holban and A. Grumezescu, pp. 871–895. Amsterdam: Elsevier.

Konur, O. 2017a. Recent citation classics in antimicrobial nanobiomaterials. In *Nanostructures for Antimicrobial Therapy*, Eds. A. Ficai and A. M. Grumezescu, pp. 669–685. Amsterdam: Elsevier.

Konur, O. 2017b. Scientometric overview in nanopesticides. In *New Pesticides and Soil Sensors*, Ed. A. M. Grumezescu, pp. 719–744. Amsterdam: Elsevier.

Konur, O. 2017c. Scientometric overview regarding oral cancer nanomedicine. In *Nanostructures for Oral Medicine*, Eds. E. Andronescu and A. M. Grumezescu, pp. 939–962. Amsterdam: Elsevier.

Konur, O. 2017d. Scientometric overview regarding water nanopurification. In *Water Purification*, Ed. A. M. Grumezescu, pp. 693–716. Amsterdam: Elsevier.

Konur, O. 2017e. Scientometric overview in food nanopreservation. In *Food Preservation,* Ed. A. M. Grumezescu, pp. 703–729. Amsterdam: Elsevier.

Konur, O., Ed. 2018a. *Bioenergy and Biofuels*. Boca Raton, FL: CRC Press.

Konur, O. 2018b. Bioenergy and biofuels science and technology: Scientometric overview and citation classics. In *Bioenergy and Biofuels*, Ed. O. Konur, pp. 3–63. Boca Raton, FL: CRC Press.

Konur, O. 2019a. Cyanobacterial bioenergy and biofuels science and technology: A scientometric overview. In *Cyanobacteria: From Basic Science to Applications*, Eds. A. K. Mishra, D. N. Tiwari and A. N. Rai, pp. 419–442. Amsterdam: Elsevier.

Konur, O. 2019b. Nanotechnology applications in food: A scientometric overview. In *Nanoscience for Sustainable Agriculture*, Eds. R. N. Pudake, N. Chauhan and C. Kole, pp. 683–711. Cham: Springer.

Konur, O. 2020a. The scientometric analysis of the research on the bioethanol production from green macroalgae. In *Handbook of Algal Science, Technology and Medicine,* Ed. O. Konur, pp. 385–401. London: Academic Press.

Konur, O., Ed. 2020b. *Handbook of Algal Science, Technology and Medicine.* London: Academic Press.

Konur, O., Ed. 2021a. *Handbook of Biodiesel and Petrodiesel Fuels: Science, Technology, Health, and Environment*. Boca Raton, FL: CRC Press.

Konur, O., Ed. 2021b. *Handbook of Biodiesel and Petrodiesel Fuels: Science, Technology, Health, and Environment. Volume 1. Biodiesel Fuels: Science, Technology, Health, and Environment.* Boca Raton, FL: CRC Press.

Konur, O., Ed. 2021c. *Handbook of Biodiesel and Petrodiesel Fuels: Science, Technology, Health, and Environment. Volume 2. Biodiesel Fuels Based on the Edible and Nonedible Feedstocks, Wastes, and Algae: Science, Technology, Health, and Environment.* Boca Raton, FL: CRC Press.

Konur, O., Ed. 2021d. *Handbook of Biodiesel and Petrodiesel Fuels: Science, Technology, Health, and Environment. Volume 3. Petrodiesel Fuels: Science, Technology, Health, and Environment.* Boca Raton, FL: CRC Press.

Konur, O. 2021e. Nanotechnology applications in diesel fuels and the related research fields: A review of the research. In *Handbook of Biodiesel and Petrodiesel Fuels: Science, Technology, Health, and Environment. Volume 1. Biodiesel Fuels: Science, Technology, Health, and Environment,* Ed. O. Konur, pp. 89–110. Boca Raton, FL: CRC Press.

Konur, O. 2021f. Nanobiosensors in agriculture and foods: A scientometric review. In *Nanobiosensors in Agriculture and Food*, Ed. R. N. Pudake, pp. 365–384. Cham: Springer.

Konur, O. and F. L. Matthews. 1989. Effect of the properties of the constituents on the fatigue performance of composites: A review. *Composites* 20:317–328.

Kugai, J., V. Subramani, C. Song, M. H. Engelhard and Y. H. Chin, 2006. Effects of nanocrystalline CeO_2 supports on the properties and performance of Ni-Rh bimetallic catalyst for oxidative steam reforming of ethanol. *Journal of Catalysi* 238:430–440.

Lashof, D. A. and D. R. Ahuja. 1990. Relative contributions of greenhouse gas emissions to global warming. *Nature* 344:529–531.

Lee, J. W. 2002. Overview of nanotechnology in Korea-10 years blueprint. *Journal of Nanoparticle Research* 4:473–476.

Leydesdorff, L. 2000. Is the European Union becoming a single publication system? *Scientometrics* 47:265–280.

Leydesdorff, L. and C. Wagner. 2009. Is the United States losing ground in science? A global perspective on the world science system. *Scientometrics* 78:23–36.

Leydesdorff, L. and P. Zhou. 2005. Are the contributions of China and Korea upsetting the world system of science? *Scientometrics* 63:617–630.

Leydesdorff, L., C. S. Wagner and L. Bornmann. 2014. The European Union, China, and the United States in the top-1% and top-10% layers of most-frequently cited publications: Competition and collaborations. *Journal of Informetrics* 8:606–617.

Li, D., Z. Huang, X. C. Lu, W. Zhang and J. Yang. 2005. Physico-chemical properties of ethanol-diesel blend fuel and its effect on performance and emissions of diesel engines. *Renewable Energy* 30:967–976.

Li, S. Z. and C. Chan-Halbrendt. 2009. Ethanol production in (the) People's Republic of China: Potential and technologies. *Applied Energy* 86:S162–S169.

Ma, X., L. Sun and C. Song. 2002. A new approach to deep desulfurization of gasoline, diesel fuel and jet fuel by selective adsorption for ultra-clean fuels and for fuel cell applications. *Catalysis Today* 77:107–116.

Murdoch, M., G. I. N. Waterhouse and M. A. Nadeem, et al. 2011. The effect of gold loading and particle size on photocatalytic hydrogen production from ethanol over Au/TiO_2 nanoparticles. *Nature Chemistry* 3:489–492.

Najafi, G., B. Ghobadian and T. Tavakoli, et al. 2009. Performance and exhaust emissions of a gasoline engine with ethanol blended gasoline fuels using artificial neural network. *Applied Energy* 86:630–639.

Newman, P. W. G. and J. R. Kenworthy. 1989. Gasoline consumption and cities: A comparison of U.S. cities with a global survey. *Journal of the American Planning Association* 55:24–37.

North, D. C. 1991. Institutions. *Journal of Economic Perspectives* 5:97–112.

Novoselov, K. S., A. K. Geim and S. V. Morozov, et al. 2004. Electric field in atomically thin carbon films. *Science* 306:666–669.

Pan, X., Z. Fan and W. Chen, et al. 2007. Enhanced ethanol production inside carbon-nanotube reactors containing catalytic particles. *Nature Materials* 6:507–511.

Schauer, J. J., M. J. Kleeman, G. R. Cass and B. R. T. Simoneit. 2002. Measurement of emissions from air pollution sources. 5. C_1-C_{32} organic compounds from gasoline-powered motor vehicles. *Environmental Science and Technology* 36:1169–1180.

Schellekens, M. 2010. Patenting nanotechnology in Europe: Making a good start? An analysis of issues in law and regulation. *Journal of World Intellectual Property* 13:47–76.

Selvan, V. A. M., R. B. Anand and M. Udayakumar. 2009. Effects of cerium oxide nanoparticle addition in diesel and diesel-biodiesel-ethanol blends on the performance and emission characteristics of a CI engine. *Journal of Engineering and Applied Sciences* 4:1–6.

Shaafi, T. and R. Velraj. 2015. Influence of alumina nanoparticles, ethanol and isopropanol blend as additive with diesel-soybean biodiesel blend fuel: Combustion, engine performance and emissions. *Renewable Energy* 80:655–663.

Shin, D., T. Kim, J. Choi and J. Kim. 2014. Author name disambiguation using a graph model with node splitting and merging based on bibliographic information. *Scientometrics* 100:15–50.

Sukumaran, R. K., V. J. Surender and R. Sindhu, et al. 2010. Lignocellulosic ethanol in India: Prospects, challenges and feedstock availability. *Bioresource Technology* 101:4826–4833.

Sun, J., K. Zhu and F. Gao, et al. 2011. Direct conversion of bio-ethanol to isobutene on nanosized $Zn_xZr_yO_z$ mixed oxides with balanced acid-base sites. *Journal of the American Chemical Society* 133:11096–11099.

Vadas, P. A., K. H. Barnett and D. J. Undersander 2008. Economics and energy of ethanol production from alfalfa, corn, and switchgrass in the Upper Midwest, USA. *Bioenergy Research* 1:44–55.

Wan, Q., Q. H. Li and Y. J. Chen, et al. 2004. Fabrication and ethanol sensing characteristics of ZnO nanowire gas sensors. *Applied Physics Letters* 84:3654–3656.

Wang, L., Y. Kang and X. Liu, et al. 2012. ZnO nanorod gas sensor for ethanol detection. *Sensors and Actuators, B: Chemical* 162:237–243.

Xu, C., H. Wang, P. K. Shen and S. P. Jiang. 2007. Highly ordered Pd nanowire arrays as effective electrocatalysts for ethanol oxidation in direct alcohol fuel cells. *Advanced Materials* 19:4256–4259.

Zhang, X., R. Wang, X. Yang and F. Zhang. 2008. Comparison of four catalysts in the catalytic dehydration of ethanol to ethylene. *Microporous and Mesoporous Materials* 116:210–215.

83 Nanotechnology Applications in Bioethanol Fuels
Review

Ozcan Konur
(Formerly) Ankara Yildirim Beyazit University

83.1 INTRODUCTION

The crude oil-based gasoline fuels (Ma et al., 2002; Newman and Kenworthy, 1989) have been widely used in the transportation sector since the 1920s. However, there have been great public concerns over the adverse environmental impact of these fuels (Hill et al., 2006, 2009). Hence, biomass-based bioethanol fuels (Hill et al., 2006, 2009; Konur, 2012, 2015, 2019a, 2020a) have increasingly been used in blending gasoline fuels (Hsieh et al., 2002; Najafi et al., 2009).

In the meantime, the research in the nanomaterials and nanotechnology has intensified in recent years to become a major research field in the scientific research (Geim, 2009; Geim and Novoselov, 2007; Novoselov et al., 2012). Thus, a large number of nanomaterials have been utilized nearly for nearly every research field. These materials offer an innovative way to increase the efficiency in the production and utilization of bioethanol fuels as in the other scientific fields (Konur, 2016a–f, 2017a–e, 2019b, 2021e,f).

In this context, there has been a significant focus on the nanotechnology applications in bioethanol fuels (Murdoch et al., 2011; Pan et al., 2007). The research in this field has particularly intensified the fields of sensing of bioethanol gas (Wan et al., 2004; Wang et al., 2012), oxidation of bioethanol fuels for the fuel cell applications (Dong et al., 2010; Xu et al., 2007), and to a lesser extent production of bioethanol fuels (Hoang et al., 2018; Pan et al., 2007), production of biohydrogen fuels from bioethanol fuels (Kugai et al., 2006; Murdoch et al., 2011), utilization of bioethanol fuels in the transport engines (Selvan et al., 2009; Shaafi and Velraj, 2015), and production of biochemicals from bioethanol fuels (Sun et al., 2011; Zhang et al., 2008).

However, it is essential to develop efficient incentive structures (North, 1991) for the primary stakeholders to enhance the research in this field (Konur, 2000, 2002a–c, 2006a,b, 2007a,b).

Although there have been a number of review papers on the nanotechnology applications in bioethanol fuels (Bion et al., 2012; Ju et al., 2014), there has been no review of the research of the most-cited 25 articles in this field.

This book chapter presents a review of the research of the most-cited 25 articles in the nanotechnology applications in bioethanol fuels. Then, it discusses the key findings of these highly influential papers and comments on the future research priorities in this field.

83.2 MATERIALS AND METHODS

The search for this study was carried out using Scopus database (Burnham, 2006) in August 2021.

As a first step for the search of the relevant literature, the keywords were selected using the first most-cited 500 papers. The selected keyword list was optimized to obtain a representative sample of papers for the searched research field. This keyword list was provided in the appendix of Konur (2023) for future replication studies.

DOI: 10.1201/9781003226567-111

As a second step, a sample dataset was used for this study. The first 25 articles in the sample of 100 most-cited papers with at least 217 citations each were selected for the review study. Key findings from each paper were taken from the abstracts of these papers and were discussed.

Additionally, a number of brief conclusions were drawn, and a number of relevant recommendations were made to enhance the future research landscape.

83.3 RESULTS

The brief information about 25 most-cited papers with at least 217 citations each in this field is given below under four headings: Nanotechnology applications in the sensing and oxidation of bioethanol fuels, in the production of bioethanol fuels, and in the biohydrogen production from bioethanol fuels with 13, 8, 2, and 2 papers, respectively.

83.3.1 The Nanotechnology Applications in the Sensing of Bioethanol Fuels

The brief information about 13 prolific studies with at least 217 citations each on the nanotechnology applications in the sensing of bioethanol fuels is given in Table 83.1. Furthermore, the brief notes on the contents of these studies are also given.

Wan et al. (2004) fabricate the zinc oxide (ZnO) nanowire gas sensors using the microelectromechanical system (MEMS) technology and study their ethanol sensing characteristics in a paper with 1,834 citations. They find that these sensors exhibited a very high sensitivity to ethanol gas and fast response time at 300°C.

Liu et al. (2005a) develop highly selective and stable bioethanol sensors based on single-crystalline divanadium pentoxide (V_2O_5) nanobelts in a paper with 505 citations. They obtain these nanobelts by a simple mild hydrothermal method with high yield. These gas sensors show great potential for the detection of bioethanol molecules at a relatively low temperature. The experiments with variations in relative humidity and tests with other gases indicated no problems of interference with bioethanol.

TABLE 83.1
The Nanotechnology Applications in the Sensing of Bioethanol Fuels

No.	Papers	Nanomaterials	Issues	Cits.
1	Wan et al. (2004)	ZnO nanowires	ZnO nanowire-based gas sensors	1,834
2	Liu et al. (2005a)	V_2O_5 nanobelts	V_2O_5 nanobelt-based gas sensors	505
3	Wang et al. (2012)	ZnO nanorods	ZnO nanorod-based gas sensors	359
4	Bie et al. (2007)	ZnO nanorods	ZnO nanorod-based gas sensors	296
5	Hsueh et al. (2007)	ZnO nanowires	ZnO nanowire-based gas sensors	292
6	Shan et al. (2010)	Graphene	Graphene-based gas sensors	285
7	Rout et al. (2006)	ZnO nanorods, nanowires and nanotubes	ZnO nanorods, nanowires, and nanotube-based gas sensors	247
8	Chen et al. (2006)	SnO_2 nanorods	SnO_2 nanorod-based gas sensors	247
9	Na et al. (2011)	Co_3O_4-decorated ZnO nanowires	Co_3O_4-decorated ZnO nanowire-based gas sensors	237
10	Chen et al. (2005)	SnO_2 nanorods	SnO_2 nanorod-based gas sensors	228
11	Choudhury (2009)	PANI/Ag nanocomposites	PANI/Ag nanocomposite-based gas sensors	228
12	Hwang et al. (2011)	Ag nanoclusters on SnO_2 nanowire networks	Ag nanoclusters on SnO_2 nanowire network-based gas sensors	220
13	Wang et al. (2016)	Al-doped NiO nanorod-flowers	Al-doped NiO nanorod-flower-based gas sensors	217

Wang et al. (2012) evaluate the ZnO nanorod gas sensor for ethanol detection in a paper with 359 citations. They fabricate these nanorods by a simple low-temperature hydrothermal process in high yield (about 85%), starting with $Zn(OH)_4^{2-}$ aqueous solution. They observe that this gas sensor exhibited a high, reversible and fast response to bioethanol, indicating its potential application as a gas sensor to detect ethanol.

Bie et al. (2007) evaluate the ZnO nanorod gas sensor for sensing bioethanol in a paper with 296 citations. They fabricate these aligned nanorods via a two-step solution approach on an Al_2O_3 tube. They observe that these nanorods were uniform with diameters of 10–30 nm and lengths about 1.4 µm. The response Sr (= Ra/Rg) of the aligned ZnO nanorod sensor reached 18.29 ppm ethanol, which was a 2-fold increase compared with that reported in literature.

Hsueh et al. (2007) evaluate the laterally grown ZnO nanowire ethanol gas sensors in a paper with 92 citations. They fabricate these nanowires on ZnO:Ga/glass templates. They observe that growth direction of the nanowires depends strongly on growth parameters. Resistivity of the fabricated sensor decreased upon ethanol gas injection. By introducing 1,500 ppm ethanol gas, they find that the device responses were around 20%, 35%, 58%, and 61% when the gas sensor was operated at 180°C, 230°C, 260°C, and 300°C, respectively. The device responses at 300°C were around 18%, 26%, 43%, 55%, and 61% when the concentration of injected ethanol gas was 50, 100, 500, 1,000, and 1,500 ppm, respectively.

Shan et al. (2010) carry out the low-potential determination of bioethanol based on ionic liquid-functionalized graphene (IL-graphene) in a paper with 285 citations. With alcohol dehydrogenase (ADH) as a model, they fabricate the ADH/IL-graphene/chitosan-modified electrode through a simple casting method. The resulting sensor showed rapid and highly sensitive amperometric response to bioethanol with a low detection limit (5 µm). Moreover, they use this sensor to determine bioethanol in real samples.

Rout et al. (2006) evaluate bioethanol sensors based on ZnO nanorods, nanowires, and nanotubes in a paper with 247 citations. They observe that the nanorods and nanowires impregnated with 1% Pt show high sensitivity for 1,000 ppm of bioethanol at or below 150°C, with short recovery and response times.

Chen et al. (2006) evaluate the linear ethanol sensing of tin oxide (SnO_2) nanorods with extremely high sensitivity in a paper with 247 citations. They fabricate these nanorods with a diameter down to 3 nm through a hydrothermal route. They observe that the sensitivity is up to 83.8 as the nanorod sensor is exposed to 300 ppm bioethanol vapor in air. Moreover, they observe the linear dependence of the sensitivity on the bioethanol concentration for each of the 20 sensors. Compared with the measured results of 80–180 nm SnO_2 particles, such linear dependence is related to the small size effect.

Na et al. (2011) evaluate the selective detection of bioethanol using a cobalt tetraoxide (Co_3O_4)-decorated ZnO nanowire network sensor in a paper with 239 citations. They explain the gas selectivity by the catalytic effect of nanocrystalline Co_3O_4 and the extension of the electron depletion layer via the formation of p–n junctions.

Chen et al. (2005) evaluate bioethanol sensing characteristics of single crystalline SnO_2 nanorods in a paper with 228 citations. They fabricate these nanorods with diameters of 4–15 nm and lengths of 100–200 nm using $SnCl_4$ as a precursor. They observe that these sensors exhibited the sensitivity of 31.4 for 300 ppm of bioethanol. Both the response and recovery time are short, around 1 s. Moreover, they observe a linear dependence of the sensitivity on the bioethanol concentration. They attribute these behaviors to the high surface-to-volume ratio of the nanorods.

Choudhury (2009) evaluate the dielectric properties and ethanol vapor sensitivity of the polyaniline (PANI)/Ag nanocomposites in a paper with 228 citations. They prepare these materials by in-situ oxidative polymerization of aniline monomer in the presence of different concentrations of Ag nanoparticles. They find that the particle size increased with increasing Ag concentration in the composite, owing to the aggregation effect. They observe higher conductivity, dielectric constant, and dielectric loss of PANI/Ag nanocomposites than those of pure PANI. The conductivity of the nanocomposites increased with increasing Ag concentration. Finally, they find that these materials possess superior bioethanol sensing capacity compared to pure PANI, and there are a linear relationship between the responses and the bioethanol and/or Ag concentration.

Hwang et al. (2011) evaluate the effect of Ag nanocluster decoration on the bioethanol sensing characteristics on SnO_2 nanowire networks (NWs) in a paper with 220 citations. They coat the Ag layers with thicknesses of 5–50 nm on the surface of SnO_2 NWs via e-beam evaporation. They observe that the SnO_2 NWs decorated by isolated Ag nano-islands displayed a 3.7-fold enhancement in gas response to 100 ppm bioethanol at 450°C compared to pristine SnO_2 NWs. In contrast, as the Ag decoration layers became continuous, the response to bioethanol decreased significantly. The enhancement and deterioration of the bioethanol sensing characteristics by the introduction of the Ag decoration layer were strongly governed by the morphological configurations of the Ag catalysts on SnO_2 NWs and their sensitization mechanism.

Wang et al. (2016) evaluate a superior bioethanol gas sensor based on Al-doped nickel monoxide (NiO) nanorod-flowers in a paper with 217 citations. They fabricate these nanorod-flowers with uniform sizes and well-defined morphologies by a facile solvothermal reaction. They find that the 2.15 at% Al-doped NiO nanorod-flowers showed improved gas sensing properties compared to those of pure NiO nanorod-flowers. The incorporation of Al ions with NiO nanocrystals adjusts the carrier concentration and induces the change of the oxygen deficiency and chemisorbed oxygen of NiO nanorod-flowers.

83.3.2 The Nanotechnology Applications in the Oxidation of Bioethanol Fuels

The brief information about eight prolific studies with at least 232 citations each on the nanotechnology applications in the oxidation of bioethanol fuels is given in Table 83.2. Furthermore, the brief notes on the contents of these studies are also given.

Dong et al. (2010) evaluate the graphene-supported platinum (Pt) and Pt-ruthenium (Ru) nanoparticles with high electrocatalytic activity for methanol and bioethanol oxidation in a paper with 523 citations. In comparison to the widely used carbon black catalyst supports, these nanoparticles demonstrate enhanced efficiency for both methanol and ethanol electrooxidation with regard to diffusion efficiency, oxidation potential, forward oxidation peak current density, and the ratio of the forward peak current density to the reverse peak current density. For instance, the forward peak current density of bioethanol oxidation for graphene- and carbon black-supported Pt nanoparticles is 16.2 and 13.8 mA/cm^2, respectively, and the ratios are 3.66 and 0.90, respectively.

Xu et al. (2007) evaluate the highly ordered palladium (Pd) nanowire arrays (NWA) as effective electrocatalysts for bioethanol oxidation in direct alcohol fuel cells (DAFC) in a paper with 449 citations. They fabricate these electrodes by the anodized aluminum oxide (AAO) template-electrodeposition method. They observe that Pd nanowires were highly ordered, with uniform diameter

TABLE 83.2
The Nanotechnology Applications in the Oxidation of Bioethanol Fuels

No.	Papers	Nanomaterials	Issues	Cits.
1	Dong et al. (2010)	Graphene-supported Pt and Pt-Ru nanoparticles	Graphene-supported Pt and Pt-Ru nanoparticles for bioethanol oxidation	523
2	Xu et al. (2007)	Pd nanowire arrays	Pd nanowire arrays for bioethanol oxidation	449
3	Tian et al. (2010)	Tetrahedral Pd nanocrystals	Tetrahedral Pd nanocrystals for bioethanol oxidation	393
4	Bambagioni et al. (2009)	Pd and Pt-Ru nanoparticles on MWCNTs	Pd and Pt-Ru anode electrocatalysts on MWCNTs for bioethanol oxidation	376
5	Wang et al. (2013)	Pd/PANI/Pd sandwich-structured nanotube	Pd/PANI/Pd sandwich-structured nanotube array (SNTA) catalysts for bioethanol oxidation	295
6	Zhou et al. (2010)	Pt nanocrystals	Pt nanocrystals for bioethanol oxidation	286
7	Hu et al. (2012)	Pt/PdCu nanoboxes anchored on 3D graphene framework	Pt/PdCu nanoboxes anchored on 3D graphene framework for bioethanol oxidation	278
8	Chen et al. (2017)	Pd-Ni-P ternary nanoparticles	Pd-Ni-P ternary nanoparticles for bioethanol oxidation	232

and length. They then observe that the NWs were uniform, well isolated, parallel to one another, and standing vertically to the electrode substrate surface. Furthermore, the Pd NWAs exhibited a face-centered cubic (FCC) lattice structure. This Pd NWA electrode had the high electrocatalytic activity by its superior performance for the electrooxidation reaction of ethanol in DAFC.

Tian et al (2010) evaluate the direct electrodeposition of tetrahexahedral (THH) Pd nanocrystals (NCs) with high-index facets and high catalytic activity for bioethanol electrooxidation in a paper with 393 citations. They fabricate that NCs with {730} high-index facets on a glassy carbon substrate in a dilute $PdCl_2$ solution by a programmed electrodeposition method. These NCs, due to their high density of surface atomic steps, exhibit four to six times higher catalytic activity than commercial Pd black catalyst toward bioethanol electrooxidation in alkaline solutions.

Bambagioni et al. (2009) evaluate the Pd and Pt-Ru anode electrocatalysts supported on multiwalled carbon nanotubes (MWCNTs) in passive and active DAFCs with an anion-exchange membrane in a paper with 376 citations. They find that this catalyst is very active for the oxidation of all alcohols, with ethanol showing the lowest onset potential. Pd/MWCNT exhibits unrivalled activity as anode electrocatalyst for alcohol oxidation. The ethanol is selectively oxidized to acetic acid, detected as acetate ion in the alkaline media of the reaction. The results obtained with Pt-Ru/MWCNT anodes in acid media are largely inferior to those provided by Pd/MWCNT electrodes in alkaline media.

Wang et al. (2013) fabricate hybrid Pd/PANI/Pd sandwich-structured nanotube array (SNTA) catalysts with special shape effects and synergistic effects for bioethanol electrooxidation for DAFCs in a paper with 295 citations. They observe that these SNTAs exhibit significantly improved electrocatalytic activity and durability compared with Pd NTAs and commercial Pd/C catalysts. The unique SNTAs provide fast transport and short diffusion paths for electroactive species and high utilization rate of catalysts. Besides the merits of nanotube arrays, they attribute the improved electrocatalytic activity and durability to the special Pd/PANI/Pd sandwich-like nanostructures, which results in electron delocalization between Pd d orbitals and PANI π-conjugated ligands and in electron transfer from Pd to PANI.

Zhou et al. (2010) evaluate the high-index faceted Pt nanocrystals supported on carbon black as highly efficient catalysts for ethanol electrooxidation in a paper with 282 citations. They fabricate Pt nanocrystals on carbon black by an electrochemical square-wave potential method. These nanocrystals have high-index facets and a high density of atomic steps. Due to this high density, the catalysts exhibit at least twice the activity and selectivity of commercial Pt/C catalysts for bioethanol electrooxidation into CO_2.

Hu et al. (2012) evaluate the complex ternary Pt/PdCu nanoboxes anchored on three-dimensional graphene framework (3DGF) for highly efficient ethanol oxidation in a paper with 278 citations. They fabricate these materials via a dual solvothermal strategy. This structurally well-defined system possesses an approximately 4-fold improvement in catalytic activity for bioethanol oxidation in alkaline media over the commercial 20% Pt/C catalyst as normalized by the total mass of active metals.

Chen et al. (2017) evaluate ultrasmall (~5 nm) Pd-nickel (Ni)-phosphorus (P) ternary nanoparticles for ethanol electrooxidation in a paper with 232 citations. They observe that the activity is improved up to 4.95 A per mg Pd, which is 6.88 times higher than commercial Pd/C (0.72 A per mg Pd), by shortening the distance between Pd and Ni active sites, achieved through shape transformation from Pd/Ni-P heterodimers into Pd-Ni-P nanoparticles and tuning the Ni/Pd atomic ratio to 1:1. The improved activity and stability stems from the promoted production of free OH radicals (on Ni active sites) which facilitate the oxidative removal of carbonaceous poison and combination with CH_3CO radicals on adjacent Pd active sites.

83.3.3 The Nanotechnology Applications in the Production of Bioethanol Fuels

The brief information about two prolific studies with at least 279 citations each on the nanotechnology applications in the production of bioethanol fuels is given in Table 83.3. Furthermore, the brief notes on the contents of these studies are also given.

TABLE 83.3
The Nanotechnology Applications in the Production of Bioethanol Fuels

No.	Papers	Nanomaterials	Issues	Cits.
1	Pan et al. (2007)	Carbon nanotubes	Bioethanol production from CO and H_2 inside CNT reactors containing the enhanced catalytic Rh particle	767
2	Hoang et al. (2018)	Nanoporous copper-silver alloys	Nanoporous copper-silver alloys for the selective electroreduction of CO_2 to bioethanol	279

Pan et al. (2007) evaluate the enhanced bioethanol production from CO and H_2 inside carbon nanotube (CNT) reactors containing the enhanced catalytic rhodium (Rh) particles in a paper with 767 citations. The observe that the overall formation rate of bioethanol (30.0 mol/mol Rhh) inside the nanotubes exceeds that on the outside of the nanotubes by more than an order of magnitude, although the latter is much more accessible.

Hoang et al. (2018) evaluate the nanoporous copper (Cu)-silver (Ag) alloys by additive-controlled electrodeposition for the selective electroreduction of CO_2 to bioethanol in a paper with 279 citations. They find that these alloy films were high surface area catalysts and homogeneously mixed. The alloy film containing 6% Ag exhibits the best CO_2 electroreduction performance, with the Faradaic efficiency for the bioethanol production reaching nearly 25%, at a cathode potential of just −0.7 V vs RHE and a total current density of approximately −300 mA/cm^2. The origin of the high selectivity toward C_2 products is a combined effect of the enhanced stabilization of the Cu_2O overlayer and the optimal availability of the CO intermediate due to the Ag incorporated in the alloy.

83.3.4 The Nanotechnology Applications in the Production of Biohydrogen Fuels from Bioethanol Fuels

The brief information about two prolific studies with at least 222 citations each on the nanotechnology applications in the production of biohydrogen from bioethanol fuels is given in Table 83.4. Furthermore, the brief notes on the contents of these studies are also given.

Murdoch et al. (2011) evaluate the effect of gold (Au) loading and particle size on photocatalytic biohydrogen production from bioethanol over Au/TiO_2 nanoparticles in a paper with 933 citations. They observe that Au particles in the size range 3–30 nm on TiO_2 are very active in biohydrogen production from bioethanol. Au particles of similar size on anatase nanoparticles delivered a rate two orders of magnitude higher than that recorded for Au on rutile nanoparticles. Surprisingly, Au particle size does not affect the photoreaction rate over the 3–12 nm range. The high hydrogen yield observed makes these catalysts promising materials for solar conversion.

Kugai et al. (2006) evaluate the effect of nanocrystalline cerium dioxide (CeO_2) support properties on the properties and performance of Ni-Rh/CeO_2 catalysts for oxidative steam reforming

TABLE 83.4
The Nanotechnology Applications in the Production of Biohydrogen from Bioethanol Fuels

No.	Papers	Nanomaterials	Issues	Cits.
1	Murdoch et al. (2011)	Au/TiO_2 nanoparticles	Effect of gold loading and particle size on photocatalytic hydrogen production	933
2	Kugai et al. (2006)	Nanocrystalline CeO_2	Properties and performance of Ni-Rh bimetallic catalyst for oxidative steam reforming of ethanol	222

(OSR) of ethanol for biohydrogen production in a paper with 222 citations. The surface areas of these supports increases in the order of CeO_2-I<CeO_2-II<CeO_2-III, but their crystallite sizes were about 10.2, 29.3, and 6.5 nm, respectively. They find that the Rh metal dispersion increased while the Ni metal dispersion decreased with decreasing crystallite sizes of CeO_2. There was a Rh–CeO_2 metal–support interaction as well as Ni–Rh interaction in the Ni–Rh bimetallic catalyst supported on CeO_2-III with a crystallite size of about 6.5 nm. The reduced Ni and Rh species were reversibly oxidized, suggesting the existence of Ni–Rh redox species rather than NiRh surface alloy in the present catalyst system. The Rh species became highly dispersed when the crystallite size of CeO_2 support was smaller. Both ethanol conversion and H_2 selectivity increased and the selectivity for undesirable byproducts decreased with increasing Rh metal dispersion. Best catalytic performance for OSR was achieved by supporting Ni–Rh bimetallic catalysts on the nanocrystalline CeO_2-III. The Ni–Rh/CeO_2-III catalyst exhibited stable activity and selectivity during on-stream operations at 450°C and as well as at 600°C.

83.4 DISCUSSION

83.4.1 Introduction

The crude oil-based gasoline fuels have been widely used in the transportation sector since the 1920s. However, there have been great public concerns over the adverse environmental impact of these fuels. Hence, biomass-based bioethanol fuels have increasingly been used in blending gasoline fuels.

In the meantime, the research in the nanomaterials and nanotechnology has intensified in recent years to become a major research field in the scientific research. Thus, a large number of nanomaterials have been developed nearly for nearly every research field. These materials offer an innovative way to increase the efficiency in the production and utilization of bioethanol fuels as in the other scientific fields.

The research in this field has particularly intensified the fields of sensing of bioethanol gas, oxidation of bioethanol fuels for the fuel cell applications, and to a lesser extent production of bioethanol fuels, production of biohydrogen fuels from bioethanol fuels, utilization of bioethanol fuels in the transport engines, and production of biochemicals from bioethanol fuels.

However, it is essential to develop efficient incentive structures for the primary stakeholders to enhance the research in this field. Although there have been a number of review papers on the nanotechnology applications in bioethanol fuels, there has been no review of the research of the most-cited 25 articles in this field.

This book chapter presents a review of the research of the most-cited 25 articles in the nanotechnology applications in bioethanol fuels. Then, it discusses the key findings of these highly influential papers and comments on the future research priorities in this field.

There are two major research fronts for this field: The nanotechnology applications regarding the sensing of bioethanol gas and oxidation of bioethanol fuels in the fuel cells with 13 and 8 papers, respectively. Additionally, there are two papers each in the fields of the production of bioethanol fuels and production of biohydrogen fuels from bioethanol fuels for the fuel cell applications.

83.4.2 The Nanotechnology Applications in the Sensing of Bioethanol Fuels

The brief information about 13 prolific studies with at least 217 citations each on the nanotechnology applications in the sensing of bioethanol fuels is given in Table 83.1. Furthermore, the brief notes on the contents of these studies are also given.

Six of these prolific studies use ZnO nanomaterials (Bie et al., 2007; Hsueh et al., 2007; Na et al., 2011; Rout et al., 2006; Wan et al., 2004; Wang et al., 2012). The studies hint that ZnO nanomaterials perform well in sensing bioethanol gas, although there are differences in fabricating them.

Three studies use SnO_2 nanomaterials in sensing bioethanol gas (Chen et al., 2005; 2006; Hwang et al., 2011). These studies hint that SnO_2 nanomaterials perform well in sensing bioethanol gas like ZnO nanomaterials.

The other nanomaterials used in sensing bioethanol gas are V_2O_5 nanobelts (Liu et al., 2005a), graphene (Shan et al., 2010), PANI/Ag nanocomposites (Choudhury, 2009), and NiO nanorod-flowers (Wang et al., 2016). These studies also hint that these nanomaterials perform well in sensing bioethanol gas.

In the end, these most-cited papers in this field hint that the efficiency of the sensing bioethanol gas could be maximized using the structure, processing, and property relationships of the nanomaterials used in the sensing process (Formela et al., 2016; Konur, 2018a,b, 2020b, 2021a–d; Konur and Matthews, 1989).

The research in sensors has intensified in recent years (Brolo, 2012; Turner, 2013). In parallel, the research on the nanomaterials and nanotechnology has also intensified in recent decades (Geim, 2009; Geim and Novoselov, 2007; Novoselov et al., 2012).

The studies presented in this section focus on the use of the nanomaterials to enhance the efficiency of the sensors. The studies on the use of ZnO nanomaterials expand the research on the use of ZnO nanomaterials in general (Liu and Zeng, 2003; Yang et al., 2002). Similarly, the studies on the use of the SnO_2 nanomaterials expand the research on the use of the SnO_2 nanomaterials in general (Cheng et al., 2004; Park et al., 2007). Similarly, there has been intense research on the V_2O_5 nanomaterials (Andre et al., 2011; Ostermann et al., 2006).

However, the development of two-dimensional materials such as graphene has been a great innovation in this field (Geim, 2009; Geim and Novoselov, 2007; Novoselov et al., 2012). As there is only one study concerned with graphene, it is expected that the research in this field would focus more on the graphene and graphene-like two dimensional nanomaterials in the future.

83.4.3 The Nanotechnology Applications in the Oxidation of Bioethanol Fuels

The brief information about eight prolific studies with at least 232 citations each on the nanotechnology applications in the oxidation of bioethanol fuels is given in Table 83.2.

There are two types of nanomaterials used in the oxidation of bioethanol fuels in the fuel cells: Pt- and Pd-based nanomaterials. The Pt-based nonomaterials include the graphene-supported Pt and Pt-Ru nanoparticles, Pt nanocrystals, and Pt/PdCu nanoboxes anchored on 3D graphene framework (Bambagioni et al., 2009; Dong et al., 2010; Hu et al., 2012; Zhou et al., 2010). These studies have built on the wider research front of Pt-based nanomaterials (Mostafa et al., 2010; Song et al., 2005).

On the other hand, the Pd-based nanomaterials include Pd nanowire arrays, tetrahedral Pd nanocrystals, Pd and Pt-Ru nanoparticles on MWCNTs, Pd/PANI/Pd sandwich-structured nanotube, Pt/PdCu nanoboxes anchored on 3D graphene framework, and Pd-Ni-P ternary nanoparticles (Bambagioni et al., 2009; Chen et al., 2017; Hu et al., 2012; Tian et al., 2010; Wang et al., 2013; Xu et al., 2007). These studies have built on the wider research front of Pd-based nanomaterials (Balanta et al., 2011; Lim et al., 2009).

All these studies hint that there are strong structure-processing-property relationships for the all of the nanomaterials used in the gas sensors sensing bioethanol. In the end, these most-cited papers in this field hint that the efficiency of the oxidation of bioethanol fuels could be maximized using the structure, processing, and property relationships of the nanomaterials used in the oxidation process (Formela et al., 2016; Konur, 2018a, 2020b, 2021a–d; Konur and Matthews, 1989).

In parallel with the studies on the nanomaterials, the research on the oxidation of bioethanol fuels in the fuel cells has also been intense (Klosek and Raftery, 2002; Liang et al., 2009). This field has been a research front of the wider fuel cell research (Dyer, 2002; Logan et al., 2006). It is expected that ethanol fuels would be increasingly used in the fuel cells to power electric and hybrid vehicles in the future (Chan, 2002; Ehsani et al., 1997).

83.4.4 The Nanotechnology Applications in the Production of Bioethanol Fuels

The brief information about two prolific studies with at least 279 citations each on the nanotechnology applications in the production of bioethanol fuels is given in Table 83.3. Both studies are concerned with the production of bioethanol from the syngas (Hoang et al., 2018; Pan et al., 2007). These studies use carbon nanotubes and nanoporous copper-silver alloys. Both studies hint that there are strong structure–processing–property relationships for all the nanomaterials used in the production of bioethanol fuels.

The applications of nanotechnology regarding the production of bioethanol fuels have been diverse in terms of both nanomaterials used and the processes of production including the production from the syngas (Bai et al., 2017; Spivey and Egbebi, 2007) and separation processes (Huang et al., 2006; Liu et al., 2005b).

These studies have built on the wider research front of nanomaterials as well as the wider research front of the production of bioethanol fuels (Hahn-Hagerdal et al., 2006).

In the end, these most-cited papers in this field hint that the efficiency of the production of bioethanol fuels could be maximized using the structure, processing, and property relationships of the nanomaterials used in the production process (Formela et al., 2016; Konur, 2018a, 2020b, 2021a–d; Konur and Matthews, 1989).

83.4.5 The Nanotechnology Applications in the Production of Biohydrogen Fuels from Bioethanol Fuels

The brief information about two prolific studies with at least 222 citations each on the nanotechnology applications in the production of biohydrogen from bioethanol fuels is given in Table 83.3.4. These studies use Au/TiO_2 nanoparticles and nanocrystalline CeO_2 for the biohydrogen production from bioethanol fuels (Murdoch et al., 2011; Kugai et al., 2006). Both studies hint that there are strong structure–processing–property relationships for all the nanomaterials used in the production of biohydrogen fuels from bioethanol fuels.

In the end, these most-cited papers in this fields hint that the efficiency of the biohydrogen production from bioethanol fuels could be maximized using the structure, processing, and property relationships of the nanomaterials used in the production process (Formela et al., 2016; Konur, 2018a, 2020b, 2021a–d; Konur and Matthews, 1989).

These studies have built on the wider research front of nanomaterials as well as the wider research front of the production of biohydrogen from bioethanol fuels (Haryanto et al., 2005; Ni et al., 2007). Like the bioethanol fuels which are used directly in the fuels cells, this field has also been a research front of the wider fuel cell research (Logan et al., 2006; Minh, 1993). It is expected that biohydrogen fuels from bioethanol fuels would be increasingly used in the fuel cells to power electric and hybrid vehicles in the future (Chan, 2002; Ehsani et al., 1997).

83.5 CONCLUSION AND FUTURE RESEARCH

The brief information about the key research fronts covered by the 25 most-cited papers with at least 217 citations each in the field of the nanotechnology applications in bioethanol fuels is given in Table 83.5.

There are two major research fronts for this field: the nanotechnology applications regarding the sensing of bioethanol gas and oxidation of bioethanol fuels in the fuel cells with 13 and 8 papers, respectively. Additionally, there are two papers each in the fields of the production of bioethanol fuels from syngas and production of biohydrogen fuels from bioethanol fuels for the fuel cell applications. However, there are no papers in the nanotechnology applications regarding the utilization of bioethanol fuels in the transport engines and the production of biochemicals from bioethanol fuels.

TABLE 83.5
The Most-Prolific Research Fronts in the Nanotechnology Applications in Bioethanol Fuels

No.	Research Fronts	Sample Papers (%)	Reviewed Papers (No.)	Reviewed Papers (%)
1	Sensing of bioethanol gas	41	13	52
2	Oxidation of bioethanol fuels for the fuel cell applications	38	8	32
3	Production of bioethanol fuels	11	2	8
4	Production of biohydrogen fuels from bioethanol fuels	5	2	8
5	Utilization of bioethanol fuels in the transport engines	3	0	0
6	Production of biochemical from bioethanol fuels	2	0	0

The total number of reviewed papers = 25. Sample papers: the sample of the most-cited 100 papers.

It is notable that there is similarity of these research fronts in the sample of reviewed papers and the sample of the most-cited 100 papers (the first column data in Table 83.5) in this field.

The first group of the prolific papers focus on the ZnO- and SnO_2-based nanomaterials for sensing bioethanol gas building on the wider research fronts of both the sensors and nanomaterials and nanotechnology. These studies show that all the studied nanomaterials perform well in sensing bioethanol gas. These studies also show that the efficiency of the sensing bioethanol gas could be maximized using the structure, processing, and property relationships of the nanomaterials used in the sensing process. It is notable that there is ample room to expand the nanotechnology applications in sensing bioethanol gas especially in using graphene and other two-dimensional nanomaterials.

Similarly, there are two types of nanomaterials used in the oxidation of bioethanol fuels in the fuel cells: Pt and Pd-based nanomaterials. All of these studies hint that there are strong structure–processing–property relationships for all the nanomaterials used in oxidation of bioethanol fuels. In the end, they hint that the efficiency of the oxidation of bioethanol fuels could be maximized using the structure, processing, and property relationships of the nanomaterials used in the oxidation process. Again, it is notable that there is ample room to expand the nanotechnology applications in the oxidation of bioethanol fuels in the fuel cells, especially in using graphene and other two-dimensional nanomaterials.

The applications of nanotechnology regarding the production of bioethanol fuels from syngas have been diverse in terms of both nanomaterials used and the processes of production including the production from the syngas and the separation of bioethanol fuels.

The studies on the production of biohydrogen fuels from biocthanol fuels built on the wider research front of nanomaterials as well as the wider research front of the production of biohydrogen from bioethanol fuels. Like the bioethanol fuels which are used directly in the fuels cells through the oxidation processes, this field has also been a research front of the wider fuel cell research. It is expected that bioethanol fuels and biohydrogen fuels from bioethanol fuels would be increasingly used in the fuel cells to power electric and hybrid vehicles in the future.

It is also expected that the research in the nanotechnology applications regarding the utilization of bioethanol fuels in the transport engines and the production of biochemical sfrom bioethanol fuels would intensify in the future due to their public importance.

These findings confirm that the application of nanomaterials and nanotechnology in the production and utilization of bioethanol fuels significantly improve the efficiency of these processes through the enhancement of the structure–processing–property relationships. This would make bioethanol fuels as a viable alternative to crude oil-based gasoline and petrodiesel fuels.

It is recommended that such review studies should be performed for the other research fronts on both the production and utilization of bioethanol fuels complementing the corresponding scientometric studies.

ACKNOWLEDGMENTS

The contribution of the highly cited researchers in the nanotechnology applications in bioethanol fuels has been gratefully acknowledged.

REFERENCES

Andre, R., F. Natalio and M. Humanes, et al. 2011. V_2O_5 nanowires with an intrinsic peroxidase-like activity. *Advanced Functional Materials* 21:501–509.

Bai, S. X., Q. Shao and P. T. Wang, et al. 2017. Highly active and selective hydrogenation of CO_2 to ethanol by ordered Pd-Cu nanoparticles. *Journal of the American Chemical Society* 139:6827–6830.

Balanta, A., C. Godard and C. Claver. 2011. Pd nanoparticles for C-C coupling reactions. *Chemical Society Reviews* 40:4973–4985.

Bambagioni, V., C. Bianchini and A. Marchionni, et al. 2009. Pd and Pt-Ru anode electrocatalysts supported on multi-walled carbon nanotubes and their use in passive and active direct alcohol fuel cells with an anion-exchange membrane (alcohol=methanol, ethanol, glycerol). *Journal of Power Sources* 190:241–251.

Bie, L. J., X. N. Yan, J. Yin, Y. Q. Duan and Z. H. Yuan. 2007. Nanopillar ZnO gas sensor for hydrogen and ethanol. *Sensors and Actuators, B: Chemical* 126:604–608.

Bion, N., D. Duprez and F. Epron. 2012. Design of nanocatalysts for green hydrogen production from bioethanol. *ChemSusChem* 5:76–84.

Brolo, A. G. 2012. Plasmonics for future biosensors. *Nature Photonics* 6:709–713.

Burnham, J. F. 2006. Scopus database: A review. *Biomedical Digital Libraries* 3:1–8.

Chan, C. C. 2002. The state of the art of electric and hybrid vehicles. *Proceedings of the IEEE* 90:247–275.

Chen, L., L. Lu and H. Zhu, et al. 2017. Improved ethanol electrooxidation performance by shortening Pd-Ni active site distance in Pd-Ni-P nanocatalysts. *Nature Communications* 8:14136.

Chen, Y. J., L. Nie, X. Y. Xue, Y. G. Wang and T. H. Wang. 2006. Linear ethanol sensing of SnO_2 nanorods with extremely high sensitivity. *Applied Physics Letters* 88:083105.

Chen, Y. J., X. Y. Xue, Y. G. Wang and T. H. Wang. 2005. Synthesis and ethanol sensing characteristics of single crystalline SnO_2 nanorods. *Applied Physics Letters* 87:1–3.

Cheng, B., J. M. Russell, W. Shi, L. Zhang and E. T. Samulski. 2004. Large-scale, solution-phase growth of single-crystalline SnO_2 nanorods. *Journal of the American Chemical Society* 126:5972–5973.

Choudhury, A. 2009. Polyaniline/silver nanocomposites: Dielectric properties and ethanol vapour sensitivity. *Sensors and Actuators, B: Chemical* 138:318–325.

Dong, L., R. R. S. Gari, Z. Li, M. M. Craig and S. Hou. 2010. Graphene-supported platinum and platinum-ruthenium nanoparticles with high electrocatalytic activity for methanol and ethanol oxidation. *Carbon* 48:781–787.

Dyer, C. K. 2002. Fuel cells for portable applications. *Journal of Power Sources* 106:31–34.

Ehsani, M., K. M. Rahman and H. A. Toliyat. 1997. Propulsion system design of electric and hybrid vehicles. *IEEE Transactions on Industrial Electronics* 44:19–27.

Formela, K., A. Hejna, L. Piszczyk, M. R. Saeb and X. Colom. 2016. Processing and structure-property relationships of natural rubber/wheat bran biocomposites. *Cellulose* 23:3157–3175.

Geim, A. K. 2009. Graphene: Status and prospects. *Science* 324:1530–1534.

Geim, A. K. and K. S. Novoselov. 2007. The rise of graphene. *Nature Materials* 6:183–191.

Hahn-Hagerdal, B., M. Galbe and M. F. Gorwa-Grauslund. 2006. Bio-ethanol: The fuel of tomorrow from the residues of today. *Trends in Biotechnology* 24:549–556.

Haryanto, A., S. Fernando, N. Murali and S. Adhikari. 2005. Current status of hydrogen production techniques by steam reforming of ethanol: A review. *Energy & Fuels* 19:2098–2106.

Hill, J., E. Nelson, D. Tilman, S. Polasky and D. Tiffany. 2006. Environmental, economic, and energetic costs and benefits of biodiesel and ethanol biofuels. *Proceedings of the National Academy of Sciences of the United States of America* 103:11206–11210.

Hill, J., S. Polasky and E. Nelson, et al. 2009. Climate change and health costs of air emissions from biofuels and gasoline. *Proceedings of the National Academy of Sciences of the United States of America* 106:2077–2082.

Hoang, T. T. H., S. Verma and S. Ma, et al. 2018. Nanoporous copper-silver alloys by additive-controlled electrodeposition for the selective electroreduction of CO_2 to ethylene and ethanol. *Journal of the American Chemical Society* 140:5791–5797.

Hsieh, W. D., R. H. Chen, T. L. Wu and T. H. Lin. 2002. Engine performance and pollutant emission of an SI engine using ethanol-gasoline blended fuels. *Atmospheric Environment* 36:403–410.

Hsueh, T. J., C. L. Hsu, S. J. Chang and I. C. Chen. 2007. Laterally grown ZnO nanowire ethanol gas sensors. *Sensors and Actuators, B: Chemical* 126:473–477.
Hu, C., H. Cheng and Y. Zhao, et al. 2012. Newly-designed complex ternary Pt/PdCu nanoboxes anchored on three-dimensional graphene framework for highly efficient ethanol oxidation. *Advanced Materials* 24:5493–5498.
Huang, Z., H. M. Guan, W. L. Tan, X. Y. Qiao and S. Kulprathipanja. 2006. Pervaporation study of aqueous ethanol solution through zeolite-incorporated multilayer poly(vinyl alcohol) membranes: Effect of zeolites. *Journal of Membrane Science* 276:260–271.
Hwang, I. S., J. K. Choi and H. S. Woo, et al. 2011. Facile control of C_2H_5OH sensing characteristics by decorating discrete Ag nanoclusters on SnO_2 nanowire networks. *ACS Applied Materials and Interfaces* 3:3140–3145.
Ju, D., H. Xu, J. Zhang, J. Guo and B. Cao. 2014. Direct hydrothermal growth of ZnO nanosheets on electrode for ethanol sensing. *Sensors and Actuators, B: Chemical*, 201:444–451.
Klosek, S. and D. Raftery. 2002. Visible light driven V-doped TiO_2 photocatalyst and its photooxidation of ethanol. *Journal of Physical Chemistry B* 105:2815–2819.
Konur, O. 2000. Creating enforceable civil rights for disabled students in higher education: An institutional theory perspective. *Disability & Society* 15:1041–1063.
Konur, O. 2002a. Access to nursing education by disabled students: Rights and duties of nursing programs. *Nurse Education Today* 22:364–374.
Konur, O. 2002b. Assessment of disabled students in higher education: Current public policy issues. *Assessment and Evaluation in Higher Education* 27:131–152.
Konur, O. 2002c. Access to employment by disabled people in the UK: Is the Disability Discrimination Act working? *International Journal of Discrimination and the Law* 5:247–279.
Konur, O. 2006a. Participation of children with dyslexia in compulsory education: Current public policy issues. *Dyslexia* 12:51–67.
Konur, O. 2006b. Teaching disabled students in higher education. *Teaching in Higher Education* 11:351–363.
Konur, O. 2007a. A judicial outcome analysis of the Disability Discrimination Act: A windfall for the employers? *Disability & Society* 22:187–204.
Konur, O. 2007b. Computer-assisted teaching and assessment of disabled students in higher education: The interface between academic standards and disability rights. *Journal of Computer Assisted Learning* 23:207–219.
Konur, O. 2012. The evaluation of the research on the bioethanol: A scientometric approach. *Energy Education Science and Technology Part A: Energy Science and Research* 28:1051–1064.
Konur, O. 2015. Current state of research on algal bioethanol. In *Marine Bioenergy: Trends and Developments*, Eds. S. K. Kim and C. G. Lee, pp. 217–244. Boca Raton, FL: CRC Press.
Konur, O. 2016a. Scientometric overview in nanobiodrugs. In *Nanoarchitectonics for Smart Delivery and Drug Targeting*, Eds. A. M. Holban and A. M. Grumezescu, pp. 405–428. Amsterdam: Elsevier.
Konur, O. 2016b. Scientometric overview regarding nanoemulsions used in the food industry. In *Emulsions: Nanotechnology in the Agri-Food Industry*, Ed. A. M. Grumezescu, pp. 689–711. Amsterdam: Elsevier.
Konur, O. 2016c. Scientometric overview regarding the nanobiomaterials in antimicrobial therapy. In *Nanobiomaterials in Antimicrobial Therapy*, Ed. A. M. Grumezescu, pp. 511–535. Amsterdam: Elsevier.
Konur, O. 2016d. Scientometric overview regarding the nanobiomaterials in dentistry. In *Nanobiomaterials in Dentistry*, Ed. A. M. Grumezescu, pp. 425–453. Amsterdam: Elsevier.
Konur, O. 2016e. Scientometric overview regarding the surface chemistry of nanobiomaterials. In *Surface Chemistry of Nanobiomaterials*, Ed. A. M. Grumezescu, pp. 463–486. Amsterdam: Elsevier.
Konur, O. 2016f. The scientometric overview in cancer targeting. In *Nanoarchitectonics for Smart Delivery and Drug Targeting*, Eds. A. M. Holban and A. Grumezescu, pp. 871–895. Amsterdam: Elsevier.
Konur, O. 2017a. Recent citation classics in antimicrobial nanobiomaterials. In *Nanostructures for Antimicrobial Therapy*, Eds. A. Ficai and A. M. Grumezescu, pp. 669–685. Amsterdam: Elsevier.
Konur, O. 2017b. Scientometric overview in nanopesticides. In *New Pesticides and Soil Sensors*, Ed. A. M. Grumezescu, pp. 719–744. Amsterdam: Elsevier.
Konur, O. 2017c. Scientometric overview regarding oral cancer nanomedicine. In *Nanostructures for Oral Medicine*, Eds. E. Andronescu and A. M. Grumezescu, pp. 939–962. Amsterdam: Elsevier.
Konur, O. 2017d. Scientometric overview regarding water nanopurification. In *Water Purification*, Ed. A. M. Grumezescu, pp. 693–716. Amsterdam: Elsevier.
Konur, O. 2017e. Scientometric overview in food nanopreservation. In *Food Preservation*, Ed. A. M. Grumezescu, pp. 703–729. Amsterdam: Elsevier.
Konur, O., Ed. 2018a. *Bioenergy and Biofuels*. Boca Raton, FL: CRC Press.

Konur, O. 2018b. Bioenergy and biofuels science and technology: Scientometric overview and citation classics. In *Bioenergy and Biofuels*, Ed. O. Konur, pp. 3–63. Boca Raton, FL: CRC Press.

Konur, O. 2019a. Cyanobacterial bioenergy and biofuels science and technology: A scientometric overview. In *Cyanobacteria: From Basic Science to Applications*, Eds. A. K. Mishra, D. N. Tiwari and A. N. Rai, pp. 419–442. Amsterdam: Elsevier.

Konur, O. 2019b. Nanotechnology applications in food: A scientometric overview. In *Nanoscience for Sustainable Agriculture*, Eds. R. N. Pudake, N. Chauhan and C. Kole, pp. 683–711. Cham: Springer.

Konur, O. 2020a. The scientometric analysis of the research on the bioethanol production from green macroalgae. In *Handbook of Algal Science, Technology and Medicine,* Ed. O. Konur, pp. 385–401. London: Academic Press.

Konur, O., Ed. 2020b. *Handbook of Algal Science, Technology and Medicine.* London: Academic Press.

Konur, O., Ed. 2021a. *Handbook of Biodiesel and Petrodiesel Fuels: Science, Technology, Health, and Environment.* Boca Raton, FL: CRC Press.

Konur, O., Ed. 2021b. *Handbook of Biodiesel and Petrodiesel Fuels: Science, Technology, Health, and Environment. Volume 1. Biodiesel Fuels: Science, Technology, Health, and Environment.* Boca Raton, FL: CRC Press.

Konur, O., Ed. 2021c. *Handbook of Biodiesel and Petrodiesel Fuels: Science, Technology, Health, and Environment. Volume 2. Biodiesel Fuels Based on the Edible and Nonedible Feedstocks, Wastes, and Algae: Science, Technology, Health, and Environment.* Boca Raton, FL: CRC Press.

Konur, O., Ed. 2021d. *Handbook of Biodiesel and Petrodiesel Fuels: Science, Technology, Health, and Environment. Volume 3. Petrodiesel Fuels: Science, Technology, Health, and Environment.* Boca Raton, FL: CRC Press.

Konur, O. 2021e. Nanotechnology applications in diesel fuels and the related research fields: A review of the research. In *Handbook of Biodiesel and Petrodiesel Fuels: Science, Technology, Health, and Environment. Volume 1. Biodiesel Fuels: Science, Technology, Health, and Environment,* Ed. O. Konur, pp. 89–110. Boca Raton, FL: CRC Press.

Konur, O. 2021f. Nanobiosensors in agriculture and foods: A scientometric review. In *Nanobiosensors in Agriculture and Food*, Ed. R. N. Pudake, pp. 365–384. Cham: Springer.

Konur, O. 2023. Nanotechnology applications in bioethanol fuels: Scientometric study. In *Evaluation and Utilization of Bioethanol Fuels. I.: Evaluation of Bioethanol Fuels, Transport Engines, and Bioethanol Sensors. Handbook of Bioethanol Fuels Volume 5*, Ed. O. Konur, pp. 120–139. Boca Raton, FL: CRC Press.

Konur, O. and F. L. Matthews. 1989. Effect of the properties of the constituents on the fatigue performance of composites: A review. *Composites* 20:317–328.

Kugai, J., V. Subramani, C. Song, et al. 2006. Effects of nanocrystalline CeO_2 supports on the properties and performance of Ni-Rh bimetallic catalyst for oxidative steam reforming of ethanol. *Journal of Catalysis* 238:430–440.

Liang, Z. X., T. S. Zhao, J. B. Xu and L. D. Zhu. 2009. Mechanism study of the ethanol oxidation reaction on palladium in alkaline media. *Electrochimica Acta* 54:2203–2208.

Lim, B., M. Jiang and J. Tao, et al. 2009. Shape-controlled synthesis of Pd nanocrystals in aqueous solutions. *Advanced Functional Materials* 19:189–200.

Liu, B. and H. C. Zeng. 2003. Hydrothermal synthesis of ZnO nanorods in the diameter regime of 50 nm. *Journal of the American Chemical Society* 125:4430–4431.

Liu, J., X. Wang, Q. Peng and Y. Li. 2005a. Vanadium pentoxide nanobelts: Highly selective and stable ethanol sensor materials. *Advanced Materials* 17:764–767.

Liu, Y. L., C. Y. Hsu, Y. H. Su and J. Y. Lai. 2005b. Chitosan-silica complex membranes from sulfonic acid functionalized silica nanoparticles for pervaporation dehydration of ethanol-water solutions. *Biomacromolecules* 6:368–373.

Logan, B. E., B. Hamelers and R. Rozendal, et al. 2006. Microbial fuel cells: Methodology and technology. *Environmental Science & Technology* 40:5181–5192.

Ma, X., L. Sun and C. Song. 2002. A new approach to deep desulfurization of gasoline, diesel fuel and jet fuel by selective adsorption for ultra-clean fuels and for fuel cell applications. *Catalysis Today* 77:107–116.

Minh, N. Q. 1993. Ceramic fuel cells. *Journal of the American Ceramic Society* 76:563–588.

Mostafa, S., F. Behafarid and J. R. Croy, et al. 2010. Shape-dependent catalytic properties of Pt nanoparticles. *Journal of the American Chemical Society* 132:15714–15719.

Murdoch, M., G. I. N. Waterhouse and M. A. Nadeem, et al. 2011. The effect of gold loading and particle size on photocatalytic hydrogen production from ethanol over Au/TiO_2 nanoparticles. *Nature Chemistry* 3:489–492.

Na, C. W., H. S. Woo, I. D. Kim and J. H. Lee. 2011. Selective detection of NO_2 and C_2H_5OH using a Co_3O_4-decorated ZnO nanowire network sensor. *Chemical Communications* 47:5148–5150.

Najafi, G., B. Ghobadian and T. Tavakoli, et al. 2009. Performance and exhaust emissions of a gasoline engine with ethanol blended gasoline fuels using artificial neural network. *Applied Energy* 86:630–639.

Newman, P. W. G. and J. R. Kenworthy. 1989. Gasoline consumption and cities: A comparison of U.S. cities with a global survey. *Journal of the American Planning Association* 55:24–37.

Ni, M., D. Y. C. Leung and M. K. H. Leung. 2007. A review on reforming bio-ethanol for hydrogen production. *International Journal of Hydrogen Energy* 32:3238–3247.

North, D. C. 1991. Institutions. *Journal of Economic Perspectives* 5:97–112.

Novoselov, K. S., V. I. Fal and L. Colombo, et al. 2012. A roadmap for graphene. *Nature* 490:192–200.

Ostermann, R., D. Li, Y. Yin, J. T. McCann and Y. Xia. 2006. V_2O_5 nanorods on TiO_2 nanofibers: A new class of hierarchical nanostructures enabled by electrospinning and calcination. *Nano Letters* 6:1297–1302.

Pan, X., Z. Fan and W. Chen, et al. 2007. Enhanced ethanol production inside carbon-nanotube reactors containing catalytic particles. *Nature Materials* 6:507–511.

Park, M. S., G. X. Wang and Y. M. Kang, et al. 2007. Preparation and electrochemical properties of SnO_2 nanowires for application in lithium-ion batteries. *Angewandte Chemie International Edition* 46:750–753.

Rout, C. S., S. H. Krishna, S. R. C. Vivekchand, A. Govindaraj and C. N. R. Rao. 2006. Hydrogen and ethanol sensors based on ZnO nanorods, nanowires and nanotubes. *Chemical Physics Letters* 418:586–590.

Selvan, V. A. M., R. B. Anand and M. Udayakumar. 2009. Effects of cerium oxide nanoparticle addition in diesel and diesel-biodiesel-ethanol blends on the performance and emission characteristics of a CI engine. *Journal of Engineering and Applied Sciences* 4:1–6.

Shaafi, T. and R. Velraj. 2015. Influence of alumina nanoparticles, ethanol and isopropanol blend as additive with diesel-soybean biodiesel blend fuel: Combustion, engine performance and emissions. *Renewable Energy* 80:655–663.

Shan, C., H. Yang and D. Han, et al. 2010. Electrochemical determination of NADH and ethanol based on ionic liquid-functionalized graphene. *Biosensors and Bioelectronics* 25:1504–1508.

Song, H., F. Kim, S. Connor, G. A. Somorjai and P. Yang. 2005. Pt nanocrystals: Shape control and Langmuir-Blodgett monolayer formation. *Journal of Physical Chemistry B* 109:188–193.

Spivey, J. J. and A. Egbebi. 2007. Heterogeneous catalytic synthesis of ethanol from biomass-derived syngas. *Chemical Society Reviews* 36:1514–1528.

Sun, J., K. Zhu and F. Gao, et al. 2011. Direct conversion of bio-ethanol to isobutene on nanosized $Zn_xZr_yO_z$ mixed oxides with balanced acid-base sites. *Journal of the American Chemical Society* 133:11096–11099.

Tian, N., Z. Y. Zhou, N. F. Yu, L. Y. Wang and S. G. Sun. 2010. Direct electrodeposition of tetrahexahedral Pd nanocrystals with high-index facets and high catalytic activity for ethanol electrooxidation. *Journal of the American Chemical Society* 132:7580–7581.

Turner, A. P. 2013. Biosensors: Sense and sensibility. *Chemical Society Reviews* 42:3184–3196.

Wan, Q., Q. H. Li and Y. J. Chen, et al. 2004. Fabrication and ethanol sensing characteristics of ZnO nanowire gas sensors. *Applied Physics Letters* 84:3654–3656.

Wang, A. L., H. Xu and J. X. Feng, et al. 2013. Design of Pd/PANI/Pd sandwich-structured nanotube array catalysts with special shape effects and synergistic effects for ethanol electrooxidation. *Journal of the American Chemical Society* 135:10703–10709.

Wang, C., X. Cui and J. Liu, et al. 2016. Design of superior ethanol gas sensor based on al-doped NiO nanorod-flowers. *ACS Sensors* 1:131–136.

Wang, L., Y. Kang and X. Liu, et al. 2012. ZnO nanorod gas sensor for ethanol detection. *Sensors and Actuators, B: Chemical* 162:237–243.

Xu, C., H. Wang, P. K. Shen and J. P. Jiang. 2007. Highly ordered Pd nanowire arrays as effective electrocatalysts for ethanol oxidation in direct alcohol fuel cells. *Advanced Materials* 19:4256–4259.

Yang, P., H. Yan and S. Mao, et al. 2002. Controlled growth of ZnO nanowires and their optical properties. *Advanced Functional Materials* 12:323–331.

Zhang, X., R. Wang, X. Yang and F. Zhang. 2008. Comparison of four catalysts in the catalytic dehydration of ethanol to ethylene. *Microporous and Mesoporous Materials* 116:210–215.

Zhou, Z. Y., Z. Z. Huang and D. J. Chen, et al. 2010. High-index faceted platinum nanocrystals supported on carbon black as highly efficient catalysts for ethanol electrooxidation. *Angewandte Chemie: International Edition* 49:411–414.

Part 25

Utilization of Bioethanol Fuels in the Transport Engines

84 Utilization of Bioethanol Fuels in the Transport Engines
Scientometric Study

Ozcan Konur
(Formerly) Ankara Yildirim Beyazit University

84.1 INTRODUCTION

The crude oil-based gasoline fuels (Ma et al., 2002; Newman and Kenworthy, 1989) and petrodiesel fuels (Bosmann et al., 2001; Ma et al., 1994) have been widely used in the transportation sector for a long time. However, there have been great public concerns over the adverse environmental impact and sustainability of these fuels (Hill et al., 2009; Schauer et al., 2002).

Hence, biomass-based bioethanol fuels (Hill et al., 2006, 2009; Konur, 2012e, 2015, 2019, 2020a) have increasingly been used in blending gasoline fuels (Hsieh et al., 2002; Najafi et al., 2009) as well as petrodiesel fuels in recent years (Hansen et al., 2005; Li et al., 2005). In this context, the research has intensified in the research fronts of engine emissions (65%) (Hsieh et al., 2002; Najafi et al., 2009), fuel combustion (43%) (Hulwan and Joshi, 2011; Xing-Cai et al., 2004), engine performance (35%) (Hsieh et al., 2002; Najafi et al., 2009), fuel properties (17%) (Hansen et al., 2005; Kwanchareon et al. 2007), and to a lesser extent gasoline engine (4%) (Anderson et al., 2012; Christensen et al., 1997) and health impact of bioethanol fuels (1%) (Jacobson, 2007). Thus, the focus has been on the utilization of bioethanol fuels in blending with both gasoline fuels (Hsieh et al., 2002; Najafi et al., 2009) and petrodiesel fuels (Hansen et al., 2005; Li et al., 2005) in gasoline and diesel engines, respectively.

However, it is essential to develop efficient incentive structures (North, 1991) for the primary stakeholders to enhance the research in this field (Konur, 2000, 2002a–c, 2006a,b, 2007a,b). The scientometric analysis has been used in this context to inform the primary stakeholders about the current state of the research in a selected research field (Garfield, 1955; Konur, 2011, 2012a–i, 2015, 2018b, 2019, 2020a).

Although there have been a number of scientometric studies on bioethanol fuels (Konur, 2012e, 2015, 2019, 2020a), there has been no scientometric study of the research in the field of the utilization of bioethanol fuels in the transport engines.

This book chapter presents a scientometric study of the research in the field of the utilization of bioethanol fuels in the transport engines. It examines the scientometric characteristics of both the sample and population data presenting scientometric characteristics of these both datasets in the order of documents, authors, publication years, institutions, funding bodies, source titles, countries, Scopus subject categories, keywords, and research fronts.

84.2 MATERIALS AND METHODS

The search for this study was carried out using Scopus database (Burnham, 2006) in August 2021.

As a first step for the search of the relevant literature, the keywords were selected using the first most-cited 500 papers. The selected keyword list was optimized to obtain a representative sample of papers for the searched research field. This keyword list was provided in the appendix for

DOI: 10.1201/9781003226567-113

future replication studies. Additionally, the information about the most-used keywords was given in Section 84.3.9 to highlight the key research fronts in Section 84.3.10.

As a second step, two sets of data were used for this study. First, a population sample of over 3,400 papers was used to examine the scientometric characteristics of the population data. Secondly, a sample of 100 most-cited papers was used to examine the scientometric characteristics of these citation classics with over 119 citations each.

The scientometric characteristics of these both sample and population datasets were presented in the order of documents, authors, publication years, institutions, funding bodies, source titles, countries, Scopus subject categories, keywords, and research fronts.

Lastly, the key scientometric findings for both datasets were discussed to highlight the research landscape for the utilization of bioethanol fuels in the transport engines. Additionally, a number of brief conclusions were drawn and a number of relevant recommendations were made to enhance the future research landscape.

84.3 RESULTS

84.3.1 The Most-Prolific Documents in the Field of the Utilization of Bioethanol Fuels in the Transport Engines

The information on the types of documents for both datasets is given in Table 84.1. The articles, review papers, and conference papers published in the journals dominate the sample dataset while articles and conference papers dominate the population dataset.

Review papers and articles are overrepresented in the sample dataset while conference papers are overrepresented in the population dataset as shown in Table 84.1. Reviews and articles have a surplus over 7% and 11%, respectively, while conference papers have over 17% deficit.

It is also interesting to note that all of the papers in the sample dataset were published in the journals, while only 96.4% of the papers were published in the journals for the population dataset. Furthermore, 3.1% and 0.5% of the population papers were published in books and book series, respectively.

84.3.2 The Most-Prolific Authors in the Field of the Utilization of Bioethanol Fuels in the Transport Engines

The information about the most-prolific 16 authors with at least three papers each in the sample dataset is given in Table 84.2.

TABLE 84.1
Documents in the Field of the Utilization of Bioethanol Fuels in the Transport Engines

Documents	Sample Dataset (%)	Population Dataset (%)	Surplus (%)
Article	84	72.8	11.2
Conference paper	6	23.4	−17.4
Book chapter	0	1.0	−1.0
Review	9	1.4	7.6
Note	1	0.9	0.1
Letter	0	0.2	−0.2
Editorial	0	0.0	0.0
Book	0	0.0	0.0
Short survey	0	0.3	−0.3
Sample size	100	3,447	

TABLE 84.2
Most-Prolific Authors in the Field of the Utilization of Bioethanol Fuels in the Transport Engines

No.	Author Name	Author Code	Sample Papers	Population Papers	Institution	Country	Research Front
1	Shuai, Shijing	6603356005	7	10	Tsinghua Univ.	China	Emissions
2	Rakopoulos, Constantine D.	35570765900	6	13	Natl. Tech. Univ. Athens	Greece	Emissions
3	Rakopoulos, Dimitrios C.	6603012578	6	13	Natl. Tech. Univ. Athens	Greece	Emissions
4	Wang, Jianxin	35254238800	5	14	Tsinghua Univ.	China	Emissions
5	He, Hong	55229497700	4	10	Chinese Acad. Sci.	China	Emissions
6	Giakoumis, Evangelos G.	6602269983	4	7	Natl. Tech. Univ. Athens	Greece	Emissions
7	Johansson, Bengt	35366669100	3	24	Lund Univ.	Sweden	Combustion
8	Xu, Hongming	16199134500	3	23	Univ. Birmingham	UK	Combustion
9	Masjuki, Haji Hassan	57175108000	3	16	Int. Islamic Univ.	Malaysia	Emissions, performance
10	Huang, Zhen	57205480286	3	13	Shanghai Jiao Tıng Univ.	China	Combustion
11	Aleiferis, Pavlos G.	9243666900	3	11	Imperial Coll.	UK	Combustion
12	Kalam, Mohammad Abul	55103352400	3	9	Univ. Malaya	Malaysia	Emissions, performance
13	Li, Rulong	12797581500	3	6	Tsinghua Univ.	China	Emissions
14	Shi, Xiaoyan	8043874200	3	6	Chinese Acad. Sci.	China	Emissions
15	Yilmaz, Nadir	20736043400	3	6	New Mexico Inst. Min. Technol.	USA	Performance
16	Zhen, Huang	57205480286	3	6	Shanghai Jiao Tong Univ.	China	Performance, emissions

Author code: the unique code given by Scopus to the authors. Sample papers: the number of papers authored in the sample dataset. Population papers: the number of papers authored in the population dataset.

The most-prolific authors are Shijing Shuhai, Constantine D. Rakopoulos, Dimitrios C. Rakopoulos, Jianxin Wang, Hong He, and Evangelos G. Giakoumis with at least four papers each.

The most-prolific institutions for the sample dataset are National Technical University of Athens, Tsinghua University, Shanghai Jiao Tong University, and Chinese Academy of Sciences with at least two authors each. In total, nine institutions house these authors.

The most-prolific country for the sample dataset is China with seven authors. This is followed by Greece, the UK, and Malaysia with at least two authors each. In total, six countries house these authors.

The most-prolific research front for the sample dataset is the emissions of bioethanol fuels in the transport engines with 11 authors. This is followed by the performance and combustion of bioethanol fuels in the transport engines with four authors each.

On the other hand, there is significant gender deficit (Beaudry and Lariviere, 2016) for the sample dataset as surprisingly all of these top researchers are male.

84.3.3 The Most-Prolific Research Output by Years in the Field of the Utilization of Bioethanol Fuels in the Transport Engines

Information about papers published between 1970 and 2021 is given in Figure 84.1. This figure clearly shows that the bulk of the research papers in the population dataset were published in the 2000s, 2010s, and 2020s with 17.9%, 61.7%, and 14.0% of the population dataset, respectively. The publication rates were 0.4%, 2.4%, and 3.6% for the 1970s, 1980s, and 1990s, respectively. There was a rising trend for the population papers between 2003 and 2013, and thereafter, it steadied around 7% of the population papers for each year, losing its momentum.

Similarly, the bulk of the research papers in the sample dataset were published in the 2000s and 2010s with 45% and 50% of the sample dataset, respectively. Furthermore, 1% and 5% of the sample papers were published in the 1970s and 1990s, respectively.

The most-prolific publication years for the population dataset were after 2007 with at least 3% each of the dataset. Additionally, 7.5% of the population papers were published in 2020. Similarly, 13% and 10% of the sample papers were published in 2010 and 2011, respectively. Additionally, there were 9 papers each published in 2007 and 2009.

84.3.4 The Most-Prolific Institutions in the Field of the Utilization of Bioethanol Fuels in the Transport Engines

Information about the most-prolific 16 institutions publishing papers on the utilization of bioethanol fuels in the transport engines with at least three sample papers and at least 0.1% of the population papers is given in Table 84.3.

The most-prolific institution is Tsinghua University with nine papers. Shanghai Jiao Tong University, Lund University, and the National Technical University of Athens follow this top institution with six, five, and five papers, respectively.

The top country for these most-prolific institutions is the USA with four institutions. Next, the UK and China follow the USA with three institutions each. In total, nine countries house these top institutions.

On the other hand, the institution with the most citation impact is Tsinghua University with 7.4% surplus. Shanghai Jiao Tong University, the National Technical University of Athens, and

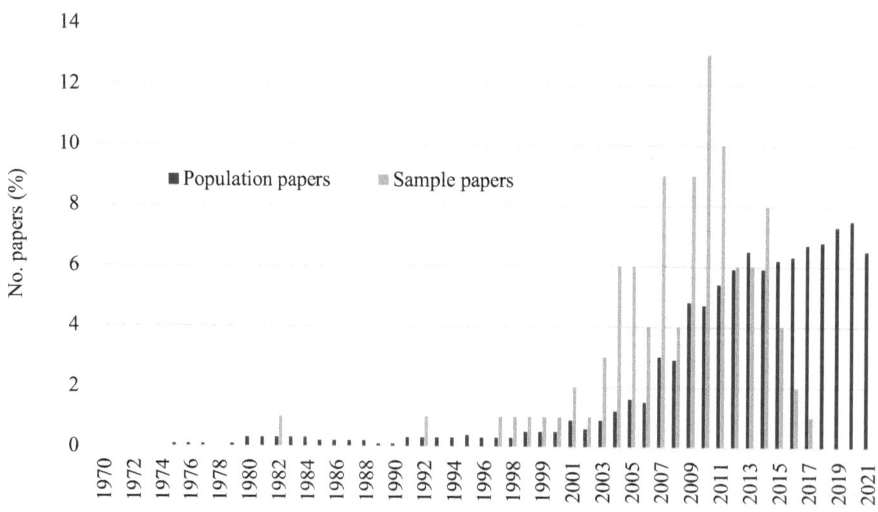

FIGURE 84.1 The research output by years regarding the utilization of bioethanol fuels in the transport engines.

TABLE 84.3
The Most-Prolific Institutions in the Field of the Utilization of Bioethanol Fuels in the Transport Engines

No.	Institutions	Country	Sample Papers (%)	Population Papers (%)	Surplus (%)
1	Tsinghua Univ.	China	9	1.6	7.4
2	Shanghai Jiao Tong Univ.	China	6	1.3	4.7
3	Lund Univ.	Sweden	5	0.9	4.1
4	Natl. Tech. Univ. Athens	Greece	5	0.6	4.4
5	Univ. Illinois U. C.	USA	4	1.5	2.5
6	Univ. Castilla la Mancha	Spain	4	0.9	3.1
7	Chinese Acad. Sci.	China	4	0.7	3.3
8	Ford Motor Co.	USA	3	1.4	1.6
9	Anna Univ.	India	3	1.0	2.0
10	Univ. Birmingham	UK	3	0.8	2.2
11	Univ. Malaya	Malaysia	3	0.7	2.3
12	Lawrence Livermore Natl. Lab.	USA	3	0.6	2.4
13	Univ. Coll. London	UK	3	0.4	2.6
14	Gazi Univ.	Turkey	3	0.3	2.7
15	Jaguar Landrover PLC	UK	3	0.3	2.7
16	New Mexico Inst. Min. Technol.	USA	3	0.1	2.9

Lund University follow this top institution with at least 4.1% surplus each. Similarly, the institution with the least impact is Ford Motor Co. with 1.6% surplus. University of Malaya, University of Birmingham, and Anna University follow this top institution with at least 2.1% surplus each.

84.3.5 The Most-Prolific Funding Bodies in the Field of the Utilization of Bioethanol Fuels in the Transport Engines

Information about the most-prolific 12 funding bodies funding at least two sample papers is given in Table 84.4.

The most-prolific funding bodies are the Engineering and Physical Sciences Research Council of the UK and the UK Research and Innovation funding five sample papers each.

It is notable that only 41% and 33.9% of the sample and population papers are funded, respectively.

The most-prolific country for these top funding bodies is China with four funding bodies. Next, the USA and the UK follow China with three and two funding bodies, respectively. In total, five countries and Europe house these top funding bodies.

84.3.6 The Most-Prolific Source Titles in the Field of the Utilization of Bioethanol Fuels in the Transport Engines

Information about the most-prolific 14 source titles publishing at least two papers each in the field of the utilization of bioethanol fuels in the transport engines is given in Table 84.5.

The most-prolific source title is 'Fuel' with 27 sample papers. The 'Atmospheric Environment', 'Energy', and 'Renewable Energy' follow this top journal with eight, seven, and seven papers, respectively. Next, 'Combustion and Flame', 'Applied Thermal Engineering', and 'Renewable and Sustainable Energy Reviews' publish six papers each.

On the other hand, the source title with the most citation impact is 'Fuel' with 16.7% surplus. The 'Atmospheric Environment', 'Renewable Energy', and 'Renewable and Sustainable Energy

TABLE 84.4
The Most-Prolific Funding Bodies in the Field of the Utilization of Bioethanol Fuels in the Transport Engines

No.	Funding Bodies	Country	Sample Papers No.	Population Paper No.
1	Engineering and Physical Sciences Research Council	UK	5	34
2	UK Research and Innovation	UK	5	30
3	U.S. Department of Energy	USA	3	84
4	National Science Foundation	USA	3	52
5	Univ. Malaya	Malaysia	3	11
6	Chinese Academy of Sciences	China	3	5
7	National Natural Science Foundation of China	China	2	257
8	European Commission	EU	2	45
9	U.S. Department of Defense	USA	2	22
10	Hong Kong Polytechnic University	China	2	14
11	Shanghai Jiao Tong University	China	2	8
12	National Science Council	Taiwan	2	7

TABLE 84.5
The Most-Prolific Source Titles in the Field of the Utilization of Bioethanol Fuels in the Transport Engines

No.	Source Titles	Sample Papers (%)	Population Papers (%)	Surplus (%)
1	Fuel	27	10.3	16.7
2	Atmospheric Environment	8	0.8	7.2
3	Energy	7	2.3	4.7
4	Renewable Energy	7	1.1	5.9
5	Combustion and Flame	6	2.6	3.4
6	Applied Thermal Engineering	6	1.7	4.3
7	Renewable and Sustainable Energy Reviews	6	0.4	5.6
8	Energy and Fuels	5	3.3	1.7
9	Applied Energy	4	1.9	2.1
10	Energy Conversion and Management	4	1.4	2.6
11	SAE Technical Papers	3	19.8	−16.8
12	Proceedings of the Combustion Institute	2	1.3	0.7
13	Environmental Science and Technology	2	1.0	1.0
14	Symposium (International) on Combustion	2	0.1	1.9

Reviews' follow this top journal with at least 5.6% surplus each. Similarly, the source title with the least impact is the 'SAE Technical Papers' with 16.8% deficit. The 'Proceedings of the Combustion Institute', 'Environmental Science and Technology', and 'Energy and Fuels' follow this top journal with at least 0.7% surplus each. It is notable that 11 of these top source titles are indexed within the subject category of 'Energy'.

84.3.7 The Most-Prolific Countries in the Field of the Utilization of Bioethanol Fuels in the Transport Engines

Information about the most-prolific 16 countries publishing at least two papers each and with at least 1.2% of the population papers each in the field of the utilization of bioethanol fuels in the transport engines is given in Table 84.6.

TABLE 84.6
The Most-Prolific Countries in the Field of the Utilization of Bioethanol Fuels in the Transport Engines

No.	Countries	Sample Papers (%)	Population Papers (%)	Surplus (%)
1	USA	23	21.0	2.0
2	China	22	16.7	5.3
3	UK	9	4.6	4.4
4	Turkey	9	2.8	6.2
5	India	8	11.4	–3.4
6	Greece	7	1.0	6.0
7	Sweden	6	2.3	3.7
8	France	5	2.4	2.6
9	Germany	4	3.2	0.8
10	Canada	4	2.9	1.1
11	Australia	4	2.6	1.4
12	Spain	4	2.4	1.6
13	Malaysia	4	2.2	1.8
14	Brazil	3	10.1	–7.1
15	Iran	2	1.9	0.1
16	Thailand	2	1.2	0.8

The most-prolific countries are the USA, China, and to a lesser extent the UK, Turkey, India, Greece, and Sweden. Further, seven European countries listed in Table 84.6 produce 44% and 19% of the sample and population papers, respectively.

On the other hand, the countries with the most citation impact are Turkey and Greece with 6.2% and 6.0% publication surplus, respectively. Similarly, the countries with the least citation impact are Brazil and India with 7.1% and 3.4% deficit, respectively.

84.3.8 The Most-Prolific Scopus Subject Categories in the Field of the Utilization of Bioethanol Fuels in the Transport Engines

Information about the most-prolific eight Scopus subject categories indexing at least 2% of the sample papers each is given in Table 84.7.

The most-prolific Scopus subject category in the field of the utilization of bioethanol fuels in the transport engines is 'Energy' with 73 sample papers. This top category is followed by 'Chemical Engineering', 'Chemistry', 'Environmental Science', and 'Engineering' with at least 27 sample papers each.

On the other hand, the Scopus subject category with the most citation impact is again 'Energy' with 30.8% surplus. This top category is followed by 'Chemical Engineering' and 'Chemistry' with over 15% surplus each. Similarly, the Scopus subject category with the least citation impact is 'Engineering' with 22.6% deficit. The other categories with the least impact are 'Environmental Science', 'Physics and Astronomy', and 'Materials Science' with at least 4.0% deficit each.

84.3.9 The Most-Prolific Keywords in the Field of the Utilization of Bioethanol Fuels in the Transport Engines

Information about the keywords used in both the sample and population papers is given in Table 84.8. For this purpose, keywords related to the keyword set given in the appendix are selected from a list of the most-prolific keyword set provided by Scopus database.

TABLE 84.7
The Most-Prolific Scopus Subject Categories in the Field of the Utilization of Bioethanol Fuels in the Transport Engines

No.	Scopus Subject Categories	Sample Papers (%)	Population Papers (%)	Surplus (%)
1	Energy	73	42.2	30.8
2	Chemical Engineering	47	30.9	16.1
3	Chemistry	41	25.7	15.3
4	Environmental Science	30	36.6	–6.6
5	Engineering	27	49.6	–22.6
6	Earth and Planetary Sciences	8	3.2	4.8
7	Physics and Astronomy	7	11.1	–4.1
8	Materials Science	2	6.1	–4.1

TABLE 84.8
The Most-Prolific Keywords in the Field of the Utilization of Bioethanol Fuels in the Transport Engines

No.	Keywords	Sample Papers (%)	Population Papers (%)	Surplus (%)
1.	**Bioethanol Fuels**			
	Ethanol	95	83.9	11.1
	Ethanol fuels	23	24.9	–1.9
	Ethanol blend	7	3.0	4
	Bioethanol	7	8.4	–1.4
2.	**Environmental Issues**			
	Emissions	41	9.8	31.2
	Carbon monoxide	28	9.6	18.4
	Nitrogen oxides	22	10.9	11.1
	Particulate emissions	21	7.0	14
	PM emissions	21	1.5	19.5
	Hydrocarbons	20	7.1	12.9
	Exhaust emissions	18	5.8	12.2
	Gas emissions	10	6.0	4
	Air pollution	10	2.3	7.7
	Particulate matter	9	3.1	5.9
	Exhaust gases	9	5.5	3.5
	Smoke	9	3.4	5.6
	Acetaldehyde	7	2.2	4.8
3.	**Engines**			
	Diesel engine	70	31.5	38.5
	Engines	27	17.5	9.5
	Engine cylinders	22	12.9	9.1
	Engine performance	20	6.1	13.9
	Internal combustion engines	12	5.3	6.7
	Direct injection	12	12.2	–0.2
4.	**Fuels**		0.0	0
	Gasoline	38	33.0	5
	Diesel fuels	34	24.9	9.1

(*Continued*)

TABLE 84.8 (*Continued*)
The Most-Prolific Keywords in the Field of the Utilization of Bioethanol Fuels in the Transport Engines

No.	Keywords	Sample Papers (%)	Population Papers (%)	Surplus (%)
	Fuels	30	21.0	9
	Biodiesel	25	8.0	17
	Alcohols	24	7.4	16.6
	Alternative fuels	23	10.8	12.2
	Biofuels	20	10.2	9.8
	Fuel consumption	17	3.9	13.1
	Blends	11	1.7	9.3
	Methanol	10	4.4	5.6
5.	**Combustion**			
	Combustion	35	25.6	9.4
	Ignition	20	15.7	4.3
	laminar burning	7	1.7	5.3
	Heat release rate	6	2.8	3.2
6.	**Performance**			
	Engine performance	20	6.1	13.9
	Brake thermal efficiency	10	4.4	5.6
	Brakes	9	3.9	5.1
	Antiknock rating	9	1.7	7.3
	Brake-specific fuel consumption	8	2.8	5.2
	Performance	7	3.9	3.1
	Compression ratio	5	3.3	1.7

There are two keywords used related to the bioethanol fuels: bioethanol and ethanol. The former keyword is used over 90% in both paper sets.

The most-prolific keywords related to the emissions in the field of the utilization of bioethanol fuels in the transport engines are emissions, carbon monoxide, nitrogen oxides, particulate emissions, particulate matter emissions, and hydrocarbons with at least 20% of the sample papers each.

On the other hand, the most-prolific keywords related to engines in the field of the utilization of bioethanol fuels in the transport engines are diesel engines, engines, engine cylinders, and engine performance with at least 12% of the sample papers each.

The most-prolific keywords related to the fuels in the field of the utilization of bioethanol fuels in the transport engines are gasoline, diesel fuels, fuels, biodiesel, and alcohol with at least 20% of the sample papers each.

The most-prolific keywords related to the combustion in the field of the utilization of bioethanol fuels in the transport engines are combustion, ignition, combustion characteristics, and laminar burning with at least 7% of the sample papers each.

The most-prolific keywords related to the engine performance in the field of the utilization of bioethanol fuels in the transport engines are engine performance, brake thermal efficiency, brakes, antiknock rating, and brake-specific fuel consumption with at least 7% of the sample papers each.

On the other hand, there are some keywords with the significant surplus. These are biofuels, engines, combustion, blends, engine cylinders, diesel fuels, and fuels. Similarly, ethanol fuels, bioethanol, and direct injection are some of the keywords with the least impact.

84.3.10 The Most-Prolific Research Fronts in the Field of the Utilization of Bioethanol Fuels in the Transport Engines

Information about the most-prolific research fronts for the sample papers in the field of the utilization of bioethanol fuels in the transport engines is given in Table 84.9.

Biomass-based bioethanol fuels (Hill et al., 2006, 2009; Konur, 2012e, 2015, 2019, 2020a) have increasingly been used in blending gasoline fuels (Hsieh et al., 2002; Najafi et al., 2009) as well as petrodiesel fuels in recent years (Hansen et al., 2005; Li et al., 2005). In this context, the research has intensified in the research fronts of engine emissions (65%) (Hsieh et al., 2002; Najafi et al., 2009), fuel combustion (43%) (Hulwan and Joshi, 2011; Xing-Cai et al., 2004), engine performance (35%) (Hsieh et al., 2002; Najafi et al., 2009), fuel properties (17%) (Hansen et al., 2005; Kwanchareon et al. 2007), and to a lesser extent gasoline engines (4%) (Anderson et al., 2012; Christensen et al., 1997) and health impact of bioethanol fuels (1%) (Jacobson, 2007). The focus has been on the utilization of bioethanol fuels in blending with both gasoline fuels (Hsieh et al., 2002; Najafi et al., 2009) and petrodiesel fuels (Hansen et al., 2005; Li et al., 2005) in gasoline and diesel engines, respectively.

The focus on the first group is on the fuel properties of bioethanol–gasoline fuel blends in gasoline engines (13%). In the second group of papers, the focus is on both bioethanol–gasoline fuel blends (14%) and bioethanol–petrodiesel fuel blends (20%). The focus is again on both

TABLE 84.9
The Most-Prolific Research Fronts in the Field of the Utilization of Bioethanol Fuels in the Transport Engines

No.	Research Fronts	Sample Papers (%)
1.	**Fuel Properties**	**17**
	Bioethanol fuels	1
	Bioethanol–gasoline fuel blends	3
	Bioethanol–petrodiesel fuel blends	13
2.	**Engine Performance**	**35**
	Bioethanol fuels	1
	Bioethanol–gasoline fuel blends	14
	Bioethanol–petrodiesel fuel blends	20
3.	**Engine Emissions**	**65**
	Bioethanol fuels	8
	Bioethanol–gasoline fuel blends	20
	Bioethanol–petrodiesel fuel blends	37
4.	**Fuel Combustion**	**43**
	Bioethanol fuels	19
	Bioethanol–gasoline fuel blends	6
	Bioethanol–petrodiesel fuel blends	18
5.	**Engine**	**4**
	Bioethanol fuels	2
	Bioethanol–gasoline fuel blends	2
	Bioethanol–petrodiesel fuel blends	0
6.	**Fuel Health Impact**	**1**
	Bioethanol fuels	0
	Bioethanol–gasoline fuel blends	1
	Bioethanol–petrodiesel fuel blends	0

bioethanol–gasoline fuel blends (20%) and bioethanol–petrodiesel fuel blends (37%) for the third group of papers. However, the focus is on both bioethanol fuels (19%) and bioethanol–petrodiesel fuel blends (18%) in the fourth group of papers.

84.4 DISCUSSION

84.4.1 Introduction

The crude oil-based gasoline and petrodiesel fuels have been widely used in the transportation sector. However, there have been great public concerns over the adverse environmental impact of these fuels. Hence, biomass-based bioethanol fuels have increasingly been used in blending gasoline fuels as well as petrodiesel fuels in recent years.

In this context, there has been a significant focus on the utilization of bioethanol fuels in the transport engines in the research fronts of engine emissions, fuel combustion, engine performance, fuel properties, and to a lesser extent gasoline engine and health impact of bioethanol fuels.

There are also three secondary research fronts for each group: bioethanol fuels, bioethanol–gasoline fuel blends, and bioethanol–petrodiesel fuel blends. The focus has also been on the utilization of bioethanol fuels in blending with both gasoline fuels and petrodiesel fuels in gasoline and diesel engines, respectively.

However, it is essential to develop efficient incentive structures for the primary stakeholders to enhance the research in this field. The scientometric analysis has been traditionally used to inform the primary stakeholders about the current state of the research in a selected research field. Although there have been a number of scientometric studies on bioethanol fuels, there has been no scientometric study of the research in the field of the utilization of bioethanol fuels in the transport engines.

Thus, this book chapter presents a scientometric study of the research in the field of the utilization of bioethanol fuels in the transport engines. It examines the scientometric characteristics of the sample and population data presenting scientometric characteristics of these both datasets in the order of documents, authors, publication years, institutions, funding bodies, source titles, countries, Scopus subject categories, keywords, and research fronts.

The search for this study was carried out using Scopus database in August 2021. As a first step for the search of the relevant literature, the keywords were selected using the first most-cited 500 papers. The selected keyword list was optimized to obtain a representative sample of papers for the searched research field. This keyword list was provided in the appendix for future replication studies. Additionally, the information about the most-used keywords is given in Section 84.3.9 to highlight the key research fronts in Section 84.3.10.

As a second step, two sets of data were used for this study. First, a population sample of over 3,400 papers was used to examine the scientometric characteristics of the population data. Secondly, a sample of 100 most-cited papers was used to examine the scientometric characteristics of these citation classics with over 119 citations each.

84.4.2 The Most-Prolific Documents in the Field of the Utilization of Bioethanol Fuels in the Transport Engines

Review papers and articles are overrepresented in the sample dataset while conference papers and articles are overrepresented in the population dataset as shown in Table 84.1. Scopus differs from the Web of Science database in differentiating and showing articles and conference papers published in the journals.

It is observed during the search process that there has been inconsistency in the classification of the documents in Scopus as well as in other databases such as Web of Science. This is especially relevant for the classification of papers as reviews or articles as the papers not involving a literature review may be erroneously classified as a review paper. In this context, it would be helpful to provide a classification note for the published papers in the books and journals at the first instance. It would also be helpful to use the document types listed in Table 84.1 for this purpose.

84.4.3 The Most-Prolific Authors in the Field of the Utilization of Bioethanol Fuels in the Transport Engines

There have been most-prolific 16 authors with at least three papers each in the sample dataset as given in Table 84.2. These authors have shaped the development of the research in this field.

The most-prolific authors are Shijing Shuhai, Constantine D. Rakopoulos, Dimitrios C. Rakopoulos, Jianxin Wang, Hong He, and Evangelos G. Giakoumis (Table 84.2).

It is important to note the inconsistencies in indexing of the author names in Scopus and other databases. It is especially an issue for the names with more than two components such as 'Judge Alex de Camp Sirous'. The probable outcomes are 'Sirous J.A.D.C.', 'de Camp Sirous, J.A.', or 'Camp Sirous, J.A.D.'. The first choice is the gold standard of the publishing sector. The second choice is a strong alternative while the last choice is an undesired outcome. It is notable that inconsistent indexing of the author names may cause substantial inefficiencies in the search process for the papers as well as allocating credit to the authors as there are different author entries for each outcome in the databases.

There is also a difficulty in allowing credit for the authors especially for the authors with common names such as 'Wang, J.' or 'Huang, Z.' or 'Zhu, L.' in conducting scientometric studies. These difficulties strongly influence the efficiency of the scientometric studies as well as allocating credit to the authors as there are the same author entries for different authors with the same name, for example, 'Wang J.' in the databases.

In this context, the coding of authors in Scopus database is a welcome innovation compared to the other databases such as Web of Science. In this process, Scopus allocates a unique number to each author in the database. However, there might still be substantial inefficiencies in this coding system especially for common names. It is possible that Scopus uses a number of software programs to differentiate the author names (Shin et al., 2014).

In this context, it does not help that author names are not given full in some journals. This makes difficult to differentiate authors with common names and makes the scientometric studies further difficult in the author domain. Therefore, the author names should be given in all books and journals at the first instance.

There are also inconsistencies in naming of the authors with more than two components in journal papers and book chapters. For example, 'Faaij, A.P.C.', 'Wyman C.E.', and 'Shuai, S.J.' might be given as 'Faaij, A.', 'Wyman C.', or 'Shuai S.' in the journals and books. This also makes the scientometric studies difficult in the author domain. Hence, contributing authors should use their name consistently in their publications.

The other critical issue regarding the author names is the spelling of the author names in the national spellings rather than in the English spellings in Scopus database. Scopus differs from the Web of Science database in this respect where the author names are given in the English spellings. It is observed that national spellings of the author names do not help in conducting scientometric studies as well as in allocating credits to the authors as sometimes there are the different author entries for the English and National spellings in the Scopus database.

The most-prolific institutions for the sample dataset are National Technical University of Athens, Tsinghua University, Shanghai Jiao Tong University, and Chinese Academy of Sciences with at least two authors each. Similarly, the most-prolific countries for the sample dataset are China, Greece, the UK, and Malaysia. The most-prolific research fronts for the sample dataset are emissions, performance, and the combustion of bioethanol fuels in the transport engines.

It is also notable that there is significant gender deficit for the sample dataset as surprisingly all these top researchers are male. This finding is the most thought-provoking with strong public policy implications. Hence, institutions, funding bodies, and policy-makers should take efficient measures to reduce the gender deficit in this field as well as other scientific fields with strong gender deficit.

84.4.4 The Most-Prolific Research Output by Years in the Field of the Utilization of Bioethanol Fuels in the Transport Engines

The research output observed between 1970 and 2021 is illustrated in Figure 84.1. This figure clearly shows that the bulk of the research papers in the population dataset were published in the 2000s, 2010s, and 2020s. Similarly, the bulk of the research papers in the sample dataset were published in the 2000s and 2010s.

These data suggest that the most-cited sample papers were primarily published in the 2000s and the early 2010s while the population papers were primarily published in the late 2000s, 2010s, and the early 2020s. These are the thought-provoking findings as there has been no significant research in this field in the 1970s, 1980s, 1990s, and the early 2000s, but there has been significant research boom in the last two decades. In this context, the increasing public concerns about climate change (Change, 2007), greenhouse gas emissions (Carlson et al., 2017), and global warming (Kerr, 2007) have been certainly behind the boom in the research in this field in the last two decades.

The data in Figure 84.1 also suggest that the research in this field has boomed in the last two decades and the size of the population papers likely to more than double in the current decade, provided that the public concerns about climate change, greenhouse gas emissions, and global warming are translated efficiently to the research funding in this field. However, it should be noted that the research output for the population papers lost its momentum after 2013.

84.4.5 The Most-Prolific Institutions in the Field of the Utilization of Bioethanol Fuels in the Transport Engines

The most-prolific 16 institutions publishing papers on the utilization of bioethanol fuels in the transport engines with at least three sample papers each and at least 0.1% of the population papers each given in Table 84.3 have shaped the development of the research in this field.

The most-prolific institutions are Tsinghua University, Shanghai Jiao Tong University, Lund University, and the National Technical University of Athens. The top countries for these most prolific institutions are the USA, China, and Brazil.

On the other hand, the institutions with the most citation impact are Tsinghua University, Shanghai Jiao Tong University, the National Technical University of Athens, and Lund University. Similarly, the institutions with the least impact are Ford Motor Co., University of Malaya, University of Birmingham, and Anna University.

84.4.6 The Most-Prolific Funding Bodies in the Field of the Utilization of Bioethanol Fuels in the Transport Engines

The most-prolific funding bodies are the Engineering and Physical Sciences Research Council of the UK and the UK Research and Innovation (Table 84.4). It is notable that only 41% and 34% of the sample and population papers are funded, respectively. The most-prolific countries for these top funding bodies are China, the USA, and the UK.

These findings on the funding of the research in this field suggest that the level of the funding is not intensive and there is ample scope to increase the funding in this field to enhance the research further (Ebadi and Schiffauerova, 2016) in light of North's institutional framework (North, 1991).

84.4.7 The Most-Prolific Source Titles in the Field of the Utilization of Bioethanol Fuels in the Transport Engines

The most-prolific 14 source titles publishing at least two papers each have shaped the development of the research in this field (Table 84.5). The most-prolific source titles are 'Fuel', the 'Atmospheric Environment', 'Energy', 'Renewable Energy', 'Combustion and Flame', 'Applied Thermal Engineering', and 'Renewable and Sustainable Energy Reviews'.

On the other hand, the source titles with the most citation impact are 'Fuel', 'Atmospheric Environment', 'Renewable Energy', and 'Renewable and Sustainable Energy Reviews'. Similarly, the source titles with the least impact are the 'SAE Technical Papers', 'Proceedings of the Combustion Institute', 'Environmental Science and Technology', and 'Energy and Fuels'.

It is notable that 11 of these top source titles are indexed within the subject category of 'Energy'. This finding suggests that the energy-related journals have significantly shaped the development of the research in this field.

84.4.8 The Most-Prolific Countries in the Field of the Utilization of Bioethanol Fuels in the Transport Engines

Sixteen countries publishing at least two papers each and with at least 1.2% of the population papers each have significantly shaped the development of the research in this field (Table 84.6). The most-prolific countries are the USA, China, the UK, Turkey, India, Greece, and Sweden.

On the other hand, the countries with the most citation impact are Turkey and Greece. Similarly, the countries with the least citation impact are Brazil and India.

The close examination of these findings suggests that the USA, China, and Europe are the major producers of the research in this field. It is a fact that the USA has been a major player in science (Leydesdorff and Wagner, 2009; Leydesdorff et al., 2014). The USA has further developed a strong research infrastructure to support its corn- and grass-based bioethanol industry (Vadas et al., 2008).

However, China has been a rising star in scientific research in competition with the USA and Europe (Leydesdorff and Zhou, 2005). China is also a major player in this field as a major producer of bioethanol (Li and Chan-Halbrendt, 2009).

Next, Europe has been a persistent player in the scientific research in competition with both the USA and China (Leydesdorff, 2000). Europe has also been a persistent producer of bioethanol along with the USA and Brazil (Gnansounou, 2010).

Brazil has also been a persistent player in scientific research at a moderate level (Glanzel et al., 2006). Brazil, producing 10.1% of the population papers, has also developed a strong research infrastructure to support its sugarcane-based bioethanol industry (Soccol et al., 2010).

Like Brazil, India has also been a persistent player in scientific research at a moderate level (Karpagam et al., 2011). India, producing 11.4% of the population papers, has also developed a strong research infrastructure to support its biomass-based bioethanol industry (Sukumaran et al., 2010).

84.4.9 The Most-Prolific Scopus Subject Categories in the Field of the Utilization of Bioethanol Fuels in the Transport Engines

The most-prolific eight Scopus subject categories indexing at least 2% of the sample papers each given in Table 84.7 have shaped the development of the research in this field.

The most-prolific Scopus subject categories in the field of the utilization of bioethanol fuels in the transport engines are 'Energy', 'Chemical Engineering', 'Chemistry', 'Environmental Science', and 'Engineering'.

On the other hand, the Scopus subject categories with the most citation impact are 'Energy', 'Chemical Engineering', and 'Chemistry'. Similarly, the Scopus subject categories with the least citation impact are 'Engineering', 'Environmental Science', 'Physics and Astronomy', and 'Materials Science'.

These findings are thought-provoking suggesting that the primary subject categories are energy, chemical engineering, chemistry, environmental sciences, and engineering. The other key finding is that social sciences are not well represented in both the sample and population papers, unlike the field of the evaluative studies in bioethanol fuels.

These finding are not surprising as the key research fronts in this field are the performance, combustion, and emissions of bioethanol fuels alone or as gasoline and petrodiesel blends in the diesel or gasoline engines. All these research fronts are related to the hard sciences, engineering, and environmental sciences.

84.4.10 The Most-Prolific Keywords in the Field of the Utilization of Bioethanol Fuels in the Transport Engines

A limited number of keywords have shaped the development of the research in this field as shown in Table 84.8 and Appendix.

There are two keywords used related to the bioethanol fuels: ethanol and bioethanol. The former keyword is used over 90% in both paper sets. It seems that ethanol is more popular than bioethanol with strong implications for the search strategy. In other words, the search strategy using only bioethanol keyword would give only 8% and 7.4% of the sample and population papers, respectively. Hence, the search strategy should cover both keywords as it is done in this study.

There are some keywords with the significant surplus: These are biofuels, engines, combustion, blends, engine cylinders, diesel fuels, and fuels. Similarly, ethanol fuels, bioethanol, and direct injection are some of the keywords with the least impact.

84.4.11 The Most-Prolific Research Fronts in the Field of the Utilization of Bioethanol Fuels in the Transport Engines

There are four primary research fronts in this field: fuel properties (17%), engine performance (35%), engine emissions (65%), and fuel combustion (43%). There are also two other research fronts: engine (4%), and fuel health impact (1%). There are also three secondary research fronts for each group: bioethanol fuels, bioethanol–gasoline fuel blends, and bioethanol–diesel fuel blends.

The focus on the first group is on the fuel properties of bioethanol–gasoline fuel blends in gasoline engines (13%). In the second group of papers, the focus is on both bioethanol–gasoline fuel blends (14%) and bioethanol–diesel fuel blends (20%). The focus is again on both bioethanol–gasoline fuel blends (20%) and bioethanol–petrodiesel fuel blends (37%) for the third group of papers. However, the focus is on both bioethanol fuels (19%) and bioethanol–petrodiesel fuel blends (18%) in the fourth group of papers.

These findings are thought-provoking. First, the focus is on the properties, combustion, and emissions of bioethanol fuels and their blends with gasoline, petrodiesel, and biodiesel fuels. However, it is expected that the bulk of the research focuses on the emissions of bioethanol fuels in light of the public concerns about global warming, GHG emissions, and climate change.

Secondly, as bioethanol fuels are used blending them with gasoline, petrodiesel, and biodiesel fuels, the research focus is more on these blends rather than bioethanol fuels alone.

There are strong structure–processing–property relationships for all of the nanomaterials used for these research fronts. In the end, these most-cited papers in this field hint that the efficiency of the utilization of bioethanol fuels in both gasoline and diesel engines could be maximized using the structure, processing, and property relationships of the bioethanol fuels used in these engines (Formela et al., 2016; Konur, 2018a, 2020b, 2021a–d; Konur and Matthews, 1989).

84.5 CONCLUSION AND FUTURE RESEARCH

The research on the utilization of bioethanol fuels in the transport engines has been mapped through a scientometric study of both sample and population datasets.

The critical issue in this study has been to obtain a representative sample of the research as in any other scientometric study. Therefore, the keyword set has been carefully devised and optimized after a number of runs in the Scopus database.

The other issue has been the selection of a multidisciplinary database to carry out the scientometric study of the research in this field. For this purpose, Scopus database has been selected. The journal coverage of this database has been wider than that of Web of Science.

The key scientometric properties of the research in this field have been determined and discussed in this book chapter. It is evident that a limited number of documents, authors, institutions, publication periods, institutions, funding bodies, source titles, countries, Scopus subject categories, keywords, and research fronts have shaped the development of the research in this field.

There is ample scope to increase the efficiency of the scientometric studies in this field in the author and document domains by developing consistent policies and practices in both domains. In this respect, authors, journals, and academic databases have a lot to do.

The significant gender deficit as in most scientific fields emerges as a public policy issue.

The research in this field has boomed in the late 2000s, 2010s, and early 2020s possibly promoted by the public concerns on global warming, greenhouse gas emissions, and climate change.

Institutions from the USA, China, and the UK have mostly shaped the research in this field. The low funding rate of 41% and 34% for sample and population papers, respectively, suggests that there is ample scope to increase funding in this field to enhance the research in this field, possibly more than doubling in the current decade.

The most-prolific journals have been mostly indexed by the subject category of energy. The USA, China, and Europe have been the major producers of the research in this field as the major producers of bioethanol fuels from different types of biomass such as corn, sugarcane, and grass as well as other types of biomass. India and Brazil have also contributed largely to the population papers.

The primary subject categories have been energy, chemistry, chemical engineering, engineering, and environmental sciences. Due to the technological emphasis of this field, social sciences have not been fairly represented in both the sample and population papers, unlike the evaluative studies in bioethanol fuels.

Ethanol is more popular than bioethanol as a keyword with strong implications for the search strategy. In other words, the search strategy using only bioethanol keyword would give only around 8% of the sample and population papers. The primary keywords for the emissions of bioethanol fuels highlight the broad content of the emissions from bioethanol fuels in the gasoline or diesel engines. Finally, the primary keywords related to engines and fuels in the field of the utilization of bioethanol fuels in the transport engines highlight the fact that bioethanol fuels are blended in the gasoline or diesel engines.

There are four primary research fronts in this field: fuel properties, engine performance, engine emissions, and fuel combustion. There are also two other research fronts: engine and fuel health impact. There are further three secondary research fronts for each group: bioethanol fuels, bioethanol–gasoline fuel blends, and bioethanol–petrodiesel fuel blends.

These findings are thought-provoking. First, the focus is on the properties, combustion, and emissions of bioethanol fuels and their blends with gasoline, petrodiesel, and biodiesel fuels. However, it is expected that the bulk of the research focuses on the emissions of bioethanol fuels in light of the public concerns about global warming, GHG emissions, and climate change. Secondly, as bioethanol fuels are used blending them with gasoline, petrodiesel, and biodiesel fuels, the research focus is more on these blends rather than bioethanol fuels alone.

Thus, the scientometric analysis has a great potential to gain valuable insights into the evolution of the research in this field as in other scientific fields.

It is recommended that further scientometric studies are carried out about the other aspects of both production and utilization of bioethanol fuels. It is further recommended that reviews of the most-cited papers are carried out for each research front to complement these scientometric studies. Next, the scientometric studies of the hot papers in these primary fields are carried out.

ACKNOWLEDGMENTS

The contribution of the highly cited researchers in the field of the utilization of bioethanol fuels in transport engines has been gratefully acknowledged.

APPENDIX: THE KEYWORD SET FOR THE FIELD OF THE UTILIZATION OF BIOETHANOL FUELS IN THE TRANSPORT ENGINES

(((TITLE (ethanol OR bioethanol OR e85 OR e10) OR SRCTITLE (ethanol* OR bioethanol*)) AND (TITLE (blend* OR vehicle* OR vehicular OR engine* OR cars OR car OR automobile* OR automotive OR exhaust* OR flame* OR combust* OR *ignit* OR burning OR spray* OR *octane OR laminar OR *gasoline OR gasohol OR diesehol OR brake* OR {heat release*}) OR SRCTITLE ({journal of combust*} OR {combustion and flame} OR {combustion explosion } OR {combustion theory} OR {combustion institute} OR {combustion dynamics} OR automotive OR engine OR automobile OR sae)))) AND NOT (SUBJAREA (bioc* OR dent* OR heal OR immu OR medi OR neur OR nurs OR phar OR psyc OR vete) OR TITLE (tolerance OR metabolic OR engineer* OR reforming OR sensing OR membrane* OR demand OR btex OR OR btx mixing OR lignite OR price* OR *fiber* OR nitrate OR solar OR water)) AND (LIMIT-TO (DOCTYPE, 'ar') OR LIMIT-TO (DOCTYPE, 'cp') OR LIMIT-TO (DOCTYPE, 're') OR LIMIT-TO (DOCTYPE, 'no') OR LIMIT-TO (DOCTYPE, 'ch') OR LIMIT-TO (DOCTYPE, 'sh') OR LIMIT-TO (DOCTYPE, 'le') OR LIMIT-TO (DOCTYPE, 'ed') OR LIMIT-TO (DOCTYPE, 'bk') OR LIMIT-TO (DOCTYPE, 'cr')) AND (LIMIT-TO (LANGUAGE, 'English')) AND (LIMIT-TO (SRCTYPE, 'j') OR LIMIT-TO (SRCTYPE, 'k') OR LIMIT-TO (SRCTYPE, 'b'))

REFERENCES

Anderson, J. E., D. M. Dicicco and J. M. Ginder, et al. 2012. High octane number ethanol-gasoline blends: Quantifying the potential benefits in the United States. *Fuel* 97:585–594.

Beaudry, C. and V. Lariviere. 2016. Which gender gap? Factors affecting researchers' scientific impact in science and medicine. *Research Policy* 45:1790–1817.

Bosmann, A., L. Datsevich and A. Jess, et al. 2001. Deep desulfurization of diesel fuel by extraction with ionic liquids. *Chemical Communications* 23:2494–2495.

Burnham, J. F. 2006. Scopus database: A review. *Biomedical Digital Libraries* 3:1–8.

Carlson, K. M., J. S. Gerber and N. D. Mueller, et al. 2017. Greenhouse gas emissions intensity of global croplands. *Nature Climate Change* 7:63–68.

Change, C. 2007. Climate change impacts, adaptation and vulnerability. *Science of the Total Environment* 326:95–112.

Christensen, M., B. Johansson and P. Einewall. 1997. Homogeneous charge compression ignition (HCCI) using isooctane, ethanol and natural gas: A comparison with spark ignition operation. SAE Technical Papers 1997:972874.

Ebadi, A. and A. Schiffauerova. 2016. How to boost scientific production? A statistical analysis of research funding and other influencing factors. *Scientometrics* 106:1093–1116.

Formela, K., A. Hejna, L. Piszczyk, M. R. Saeb and X. Colom. 2016. Processing and structure-property relationships of natural rubber/wheat bran biocomposites. Cellulose 23:3157–3175.

Garfield, E. 1955. Citation indexes for science. *Science* 122:108–111.

Glanzel, W., J. Leta and B. Thijs. 2006. Science in Brazil. Part 1: A macro-level comparative study. *Scientometrics* 67:67–86.

Gnansounou, E. 2010. Production and use of lignocellulosic bioethanol in Europe: Current situation and perspectives. *Bioresource Technology* 101:4842–4850.

Hansen, A. C., Q. Zhang and P. W. L. Lyne. 2005. Ethanol-diesel fuel blends: A review. *Bioresource Technology* 96:277–285.

Hill, J., E. Nelson, D. Tilman, S. Polasky and D. Tiffany. 2006. Environmental, economic, and energetic costs and benefits of biodiesel and ethanol biofuels. *Proceedings of the National Academy of Sciences of the United States of America* 103:11206–11210.

Hill, J., S. Polasky and E. Nelson, et al. 2009. Climate change and health costs of air emissions from biofuels and gasoline. *Proceedings of the National Academy of Sciences of the United States of America* 106:2077–2082.

Hsieh, W. D., R. H. Chen, T. L. Wu and T. H. Lin. 2002. Engine performance and pollutant emission of an SI engine using ethanol-gasoline blended fuels. *Atmospheric Environment* 36:403–410.

Hulwan, D. B. and S. V. Joshi. 2011. Performance, emission and combustion characteristic of a multicylinder DI diesel engine running on diesel-ethanol-biodiesel blends of high ethanol content. *Applied Energy* 88:5042–5055.

Jacobson, M. Z. 2007. Effects of ethanol (E85) versus gasoline vehicles on cancer and mortality in the United States. *Environmental Science and Technology* 41:4150–4157.

Karpagam, R., S. Gopalakrishnan, M. Natarajan and B. Ramesh Babu. 2011. Mapping of nanoscience and nanotechnology research in India: A scientometric analysis, 1990-2009. *Scientometrics* 89:501–522.

Kerr, R. A. 2007. Global warming is changing the world. *Science* 316:188–190.

Konur, O. 2000. Creating enforceable civil rights for disabled students in higher education: An institutional theory perspective. *Disability & Society* 15:1041–1063.

Konur, O. 2002a. Access to nursing education by disabled students: Rights and duties of nursing programs. *Nurse Education Today* 22:364–374.

Konur, O. 2002b. Assessment of disabled students in higher education: Current public policy issues. *Assessment and Evaluation in Higher Education* 27:131–152.

Konur, O. 2002c. Access to employment by disabled people in the UK: Is the Disability Discrimination Act working? *International Journal of Discrimination and the Law* 5:247–279.

Konur, O. 2006a. Participation of children with dyslexia in compulsory education: Current public policy issues. *Dyslexia* 12:51–67.

Konur, O. 2006b. Teaching disabled students in higher education. *Teaching in Higher Education* 11:351–363.

Konur, O. 2007a. A judicial outcome analysis of the Disability Discrimination Act: A windfall for the employers? *Disability & Society* 22:187–204.

Konur, O. 2007b. Computer-assisted teaching and assessment of disabled students in higher education: The interface between academic standards and disability rights. *Journal of Computer Assisted Learning* 23:207–219.

Konur, O. 2011. The scientometric evaluation of the research on the algae and bio-energy. *Applied Energy* 88:3532–3540.

Konur, O. 2012a. Prof. Dr. Ayhan Demirbas' scientometric biography. *Energy Education Science and Technology Part A: Energy Science and Research* 28:727–738.

Konur, O. 2012b. The evaluation of the biogas research: A scientometric approach. *Energy Education Science and Technology Part A: Energy Science and Research* 29:1277–1292.

Konur, O. 2012c. The evaluation of the global energy and fuels research: A scientometric approach. *Energy Education Science and Technology Part A: Energy Science and Research* 30:613–628.

Konur, O. 2012d. The evaluation of the research on the biodiesel: A scientometric approach. *Energy Education Science and Technology Part A: Energy Science and Research* 28:1003–1014.

Konur, O. 2012e. The evaluation of the research on the bioethanol: A scientometric approach. *Energy Education Science and Technology Part A: Energy Science and Research* 28:1051–1064.

Konur, O. 2012f. The evaluation of the research on the biofuels: A scientometric approach. *Energy Education Science and Technology Part A: Energy Science and Research* 28:903–916.

Konur, O. 2012g. The evaluation of the research on the biohydrogen: A scientometric approach. *Energy Education Science and Technology Part A: Energy Science and Research* 29:323–338.

Konur, O. 2012h. The evaluation of the research on the microbial fuel cells: A scientometric approach. *Energy Education Science and Technology Part A: Energy Science and Research* 29:309–322.

Konur, O. 2012i. The scientometric evaluation of the research on the production of bioenergy from biomass. *Biomass and Bioenergy* 47:504–515.

Konur, O. 2015. Current state of research on algal bioethanol. In *Marine Bioenergy: Trends and Developments*, Eds. S. K. Kim and C. G. Lee, pp. 217–244. Boca Raton, FL: CRC Press.

Konur, O., Ed. 2018a. *Bioenergy and Biofuels*. Boca Raton, FL: CRC Press.

Konur, O. 2018b. Bioenergy and biofuels science and technology: Scientometric overview and citation classics. In *Bioenergy and Biofuels*, Ed. O. Konur, pp. 3–63. Boca Raton, FL: CRC Press.

Konur, O. 2019. Cyanobacterial bioenergy and biofuels science and technology: A scientometric overview. In *Cyanobacteria: From Basic Science to Applications*, Eds. A. K. Mishra, D. N. Tiwari and A. N. Rai, pp. 419–442. Amsterdam: Elsevier.

Konur, O. 2020a. The scientometric analysis of the research on the bioethanol production from green macroalgae. In *Handbook of Algal Science, Technology and Medicine*, Ed. O. Konur, pp. 385–401. London: Academic Press.

Konur, O., Ed. 2020b. *Handbook of Algal Science, Technology and Medicine.* London: Academic Press.

Konur, O., Ed. 2021a. *Handbook of Biodiesel and Petrodiesel Fuels: Science, Technology, Health, and Environment.* Boca Raton, FL: CRC Press.

Konur, O., Ed. 2021b. *Handbook of Biodiesel and Petrodiesel Fuels: Science, Technology, Health, and Environment. Volume 1. Biodiesel Fuels: Science, Technology, Health, and Environment.* Boca Raton, FL: CRC Press.

Konur, O., Ed. 2021c. *Handbook of Biodiesel and Petrodiesel Fuels: Science, Technology, Health, and Environment. Volume 2. Biodiesel Fuels Based on the Edible and Nonedible Feedstocks, Wastes, and Algae: Science, Technology, Health, and Environment.* Boca Raton, FL: CRC Press.

Konur, O., Ed. 2021d. Handbook of Biodiesel and Petrodiesel Fuels: Science, Technology, Health, and Environment. Volume 3. Petrodiesel Fuels: Science, Technology, Health, and Environment. Boca Raton, FL: CRC Press.

Konur, O. and F. L. Matthews. 1989. Effect of the properties of the constituents on the fatigue performance of composites: A review. *Composites* 20:317–328.

Kwanchareon, P., A. Luengnaruemitchai, A. and S. Jai-In, S. 2007. Solubility of a diesel-biodiesel-ethanol blend, its fuel properties, and its emission characteristics from diesel engine. *Fuel* 86:1053–1061.

Leydesdorff, L. 2000. Is the European Union becoming a single publication system? *Scientometrics* 47:265–280.

Leydesdorff, L. and C. Wagner. 2009. Is the United States losing ground in science? A global perspective on the world science system. *Scientometrics* 78:23–36.

Leydesdorff, L. and P. Zhou. 2005. Are the contributions of China and Korea upsetting the world system of science? *Scientometrics* 63:617–630.

Leydesdorff, L., C. S. Wagner and L. Bornmann. 2014. The European Union, China, and the United States in the top-1% and top-10% layers of most-frequently cited publications: Competition and collaborations. *Journal of Informetrics* 8:606–617.

Li, D., Z. Huang, X. C. Lu, W. Zhang and J. Yang. 2005. Physico-chemical properties of ethanol-diesel blend fuel and its effect on performance and emissions of diesel engines. *Renewable Energy* 30:967–976.

Li, S. Z. and C. Chan-Halbrendt. 2009. Ethanol production in (the) People's Republic of China: Potential and technologies. *Applied Energy* 86:S162–S169.

Ma, X., L. Sun and C. Song. 2002. A new approach to deep desulfurization of gasoline, diesel fuel and jet fuel by selective adsorption for ultra-clean fuels and for fuel cell applications. *Catalysis Today* 77:107–116.

Ma, X. L., K. Y. Sakanish and I. Mochida. 1994. Hydrodesulfurization reactivities of various sulfur-compounds in diesel fuel. *Industrial & Engineering Chemistry Research* 33:218–222.

Najafi, G., B. Ghobadian and T. Tavakoli, et al. 2009. Performance and exhaust emissions of a gasoline engine with ethanol blended gasoline fuels using artificial neural network. *Applied Energy* 86:630–639.

Newman, P. W. G. and J. R. Kenworthy. 1989. Gasoline consumption and cities: A comparison of U.S. cities with a global survey. *Journal of the American Planning Association* 55:24–37.

North, D. C. 1991. Institutions. *Journal of Economic Perspectives* 5:97–112.

Schauer, J. J., M. J. Kleeman, G. R. Cass and B. R. T. Simoneit. 2002. Measurement of emissions from air pollution sources. 5. C_1-C_{32} organic compounds from gasoline-powered motor vehicles. *Environmental Science and Technology* 36:1169–1180.

Shin, D., T. Kim, J. Choi and J. Kim. 2014. Author name disambiguation using a graph model with node splitting and merging based on bibliographic information. *Scientometrics* 100:15–50.

Soccol, C. R., L. P. de Souza Vandenberghe and A. B. P. Medeiros, et al. 2010. Bioethanol from lignocelluloses: Status and perspectives in Brazil. *Bioresource Technology* 101:4820–4825.

Sukumaran, R. K., V. J. Surender and R. Sindhu, et al. 2010. Lignocellulosic ethanol in India: Prospects, challenges and feedstock availability. *Bioresource Technology* 101:4826–4833.

Vadas, P. A., K. H. Barnett and D. J. Undersander. 2008. Economics and energy of ethanol production from alfalfa, corn, and switchgrass in the Upper Midwest, USA. *Bioenergy Research* 1:44–55.

Xing-Cai, L., Y. Jian-Guang, Z. Wu-Gao and H. Zhen. 2004. Effect of cetane number improver on heat release rate and emissions of high speed diesel engine fueled with ethanol-diesel blend fuel. *Fuel* 83:2013–2020.

85 Utilization of Bioethanol Fuels in the Transport Engines

Review

Ozcan Konur
(Formerly) Ankara Yildirim Beyazit University

85.1 INTRODUCTION

The crude oil-based gasoline fuels (Ma et al., 2002; Newman and Kenworthy, 1989) and petrodiesel fuels (Bosmann et al., 2001; Ma et al., 1994) have been widely used in the transportation sector for a long time. However, there have been great public concerns over the adverse environmental impact of these fuels (Hill et al., 2006, 2009; Schauer et al., 2002).

Hence, biomass-based bioethanol fuels (Hill et al., 2006, 2009; Konur, 2012, 2015, 2019, 2020a) have increasingly been used in blending gasoline fuels (Hsieh et al., 2002; Najafi et al., 2009) as well as petrodiesel fuels in recent years (Hansen et al., 2005; Li et al., 2005). In this context, the research has intensified in the research fronts of engine emissions (76%) (Hsieh et al., 2002; Najafi et al., 2009), fuel combustion (24%) (Hulwan and Joshi, 2011; Xing-Cai et al., 2004), engine performance (12%) (Hsieh et al., 2002; Najafi et al., 2009), fuel properties (16%) (Hansen et al., 2005; Kwanchareon et al. 2007), and to a lesser extent gasoline engine (8%) (Anderson et al., 2012; Christensen et al., 1997) and health impact of bioethanol fuels (4%) (Jacobson, 2007). The focus has been on the utilization of bioethanol fuels in blending with both gasoline fuels (Hsieh et al., 2002; Najafi et al., 2009) and petrodiesel fuels (Hansen et al., 2005; Li et al., 2005) in gasoline and diesel engines, respectively.

However, it is essential to develop efficient incentive structures (North, 1991) for the primary stakeholders to enhance the research in this field (Konur, 2000, 2002a–c, 2006a,b, 2007a,b).

Although there has been a number of review papers on the bioethanol fuels in the transport engines (Giakoumis et al., 2013; Hansen et al., 2005; Masum et al., 2013; Niven, 2005), there has been no review of the research of the most-cited 25 articles in the field of the utilization of bioethanol fuels in the transport engines.

This book chapter presents a review of the research of the most-cited 25 articles in the field of the utilization of bioethanol fuels in the transport engines. Then, it discusses the key findings of these highly influential papers and comments on the future research priorities in this field.

85.2 MATERIALS AND METHODS

The search for this study was carried out using Scopus database (Burnham, 2006) in August 2021.

As a first step for the search of the relevant literature, the keywords were selected using the first most-cited 500 papers. The selected keyword list was optimized to obtain a representative sample of papers for the searched research field. This keyword list was provided in the appendix of Konur (2023) for future replication studies.

As a second step, a sample dataset was used for this study. The first 25 articles in the sample of 100 most-cited papers with at least 215 citations each were selected for the review study. Key findings from each paper were taken from the abstracts of these papers and were discussed.

 DOI: 10.1201/9781003226567-114

Additionally, a number of brief conclusions were drawn and a number of relevant recommendations were made to enhance the future research landscape.

85.3 RESULTS

The brief information about 25 most-cited papers with at least 215 citations each in this field is given below under five headings: properties, performance, emissions, combustion, and engine modification and health impact of bioethanol fuels and their bioethanol–gasoline and bioethanol–petrodiesel fuel blends in gasoline or diesel engines.

85.3.1 The Fuel Properties of Bioethanol Fuels

The brief information about four prolific studies with at least 318 citations each on the fuel properties of bioethanol–gasoline and bioethanol–petrodiesel fuel blends is given in Table 85.1. Furthermore, the brief notes on the contents of these studies are also given.

Hsieh et al. (2002) evaluate the engine performance of bioethanol–gasoline fuel blends with various blending rates (E0, E5, E10, E20, E30) in a paper with 590 citations. They find that with increasing the bioethanol content, the heating value of the blended fuels is decreased, while the octane number of the blended fuels increases. With increasing the bioethanol content, the Reid vapor pressure of the blended fuels initially increases to a maximum at 10% bioethanol addition, and then decreases.

Kwanchareon et al. (2007) evaluate the fuel properties of bioethanol–petrodiesel–biodiesel blends in a paper with 485 citations. They find that the fuel properties were close to the standard limit for petrodiesel fuel. However, the flash point of blends containing bioethanol was quite different from that of petrodiesel. The high cetane value of biodiesel could compensate for the decrease of the cetane number (CN) of the blends caused by the presence of bioethanol. They advise that a blend of 80% petrodiesel, 15% biodiesel, and 5% bioethanol (D80-B15-E5) was the most suitable ratio for diesohol production because of the acceptable fuel properties (except flash point).

de Caro et al. (2001) evaluate the impact of additives on the properties of bioethanol–petrodiesel fuel blends in a paper with 355 citations. They employ a blend of 2% additive and bioethanol contents between 10% and 20% in volume (E10-E20) in relation to the petrodiesel fuel. They find that the engine behavior improved in the presence of additives with a reduction of pollutant emissions in exhaust gas, cyclic irregularities, and ignition delay. There was no trouble shooting, knocking, or vapor-lock phenomenon.

He et al. (2003) evaluate the fuel properties of bioethanol–petrodiesel fuel blends in a paper with 318 citations. They find that the addition of bioethanol to petrodiesel fuel simultaneously decreases CN, high heating value, aromatics fractions and kinematic viscosity of these blends, and changes distillation temperatures. An additive has favorable effects on the physicochemical properties related to ignition and combustion of E10 and E30 blends.

TABLE 85.1
The Fuel Properties of Bioethanol Fuels

No.	Papers	Bioethanol Fuels	Issues	Research Fronts	Cits.
1	Hsieh et al. (2002)	Bioethanol–gasoline blends	Properties of bioethanol–gasoline fuel blends of E0, E5, E10, E20, E30	Emissions, performance, properties	483
2	Kwanchareon et al. (2007)	Bioethanol–petrodiesel blends	Properties of bioethanol–petrodiesel fuel blends	Emissions, properties	405
3	de Caro et al. (2001)	Bioethanol–petrodiesel fuel blends	Impact of additives on the properties of bioethanol–petrodiesel fuel blends	Properties	355
4	He et al. (2003)	Bioethanol–petrodiesel fuel blends	Properties of bioethanol–petrodiesel fuel blends	Emissions, properties	318

85.3.2 The Engine Performance of Bioethanol Fuels

The brief information about 12 prolific studies with at least 219 citations each on the performance of bioethanol–gasoline and bioethanol–petrodiesel fuel blends in gasoline or diesel engines is given in Table 85.2. Furthermore, the brief notes on the contents of these studies are also given.

85.3.2.1 The Engine Performance of Bioethanol–Gasoline Fuel Blends

Hsieh et al. (2002) evaluate the engine performance of bioethanol–gasoline fuel blends with various blending rates (E0, E5, E10, E20, E30) in a paper with 590 citations. They find that using these blended fuels, torque output and fuel consumption of the engine slightly increase.

Al-Hasan (2003) evaluates the engine performance of bioethanol–unleaded gasoline fuel blends in a paper with 399 citations. He finds that blending unleaded gasoline with bioethanol increases the brake power, torque, volumetric efficiency (ηv), brake thermal efficiencies (BTEs), and fuel consumption, while it decreases the brake-specific fuel consumption (BSFC) and equivalence air–fuel ratio (λ).

Najafi et al. (2009) evaluate the performance of bioethanol–gasoline fuel blends of E5, E10, E15, and E20 with the aid of artificial neural network (ANN) in a paper with 306 citations. They find that using these blends increased the power and torque output of the engine marginally. The BSFC decreased while the BTE and the volumetric efficiency (ηv) increased.

Koc et al. (2009) evaluate the performance of bioethanol–gasoline blends of unleaded E0, E50, and E85 in a paper with 239 citations. They find that bioethanol addition to unleaded gasoline

TABLE 85.2
The Engine Performance of Bioethanol Fuels

No.	Papers	Bioethanol Fuels	Issues	Research Fronts	Cits.
1	Hsieh et al. (2002)	Bioethanol–gasoline blends	Performance of bioethanol–gasoline fuel blends of E0, E5, E10, E20, E30	Emissions, performance, properties	483
2	Al-Hasan (2003)	Bioethanol–gasoline blends	Performance of bioethanol–gasoline fuel blends	Emissions, performance	399
3	Li et al. (2005)	Bioethanol–petrodiesel blends	Performance of bioethanol–gasoline fuel blends of E5, E10, E15, E20	Emissions, performance	340
4	Najafi et al. (2009)	Bioethanol–gasoline blends	Performance of bioethanol–gasoline fuel blends of E5, E10, E15, and E20	Emissions performance	306
5	Hulwan and Joshi (2011)	Bioethanol–petrodiesel blends	Performance of bioethanol–petrodiesel fuel blends	Emissions, engine performance, combustion	285
6	Sayin (2010)	Bioethanol–petrodiesel blends	Performance of bioethanol–petrodiesel blends of E5 and E10	Performance, emissions	279
7	Zhu et al. (2011)	Bioethanol–petrodiesel blends	Performance of bioethanol–petrodiesel blends	Performance, combustion, emissions	279
8	Can et al. (2004)	Bioethanol–petrodiesel blends	Performance of bioethanol–petrodiesel blends of E10 and E15	Performance, emissions	258
9	Koc et al. (2009)	Bioethanol–gasoline blends	Performance of bioethanol–gasoline blends of unleaded E0, E50, and E85	Performance, emissions	239
10	Huang et al. (2009)	Bioethanol–petrodiesel blends	Performance of bioethanol–petrodiesel blends	Performance, emissions	226
11	Bayraktar (2005)	Bioethanol–gasoline blends	Performance of bioethanol–gasoline blends	Performance, emissions	225
12	Ajav et al. (1999)	Bioethanol–petrodiesel blends	Performance of bioethanol–petrodiesel blends of E5, E10, E15, and E20	Performance, emissions	219

Cits., Number of citations received by the papers.

increases the engine torque, power, and fuel consumption. They also find that these blends allow increasing compression ratio (CR) without knock occurrence.

Bayraktar (2005) evaluates the performance of bioethanol–gasoline fuel blends in a paper with 225 citations. They employ the blends containing 1.5, 3, 4.5, 6, 7.5, 9, 10.5, and 12 vol% bioethanol (E1–E11). They find that among the various blends, the blend of E7.5 was the most suitable one from the engine performance point of view. However, theoretical comparisons have shown that the blend containing 16.5% bioethanol was the most suited for SI engines.

85.3.2.2 The Engine Performance of Bioethanol–Petrodiesel Fuel Blends

Li et al. (2005) evaluate the engine performance of bioethanol–petrodiesel fuel blends of E5, E10, E15, and E20 in a paper with 340 citations. They find that the BSFC and BTE increased with an increase of bioethanol contents in the blended fuel at overall operating conditions.

Hulwan and Joshi (2011) evaluate the performance of petrodiesel–bioethanol–biodiesel fuel blends in a paper with 285 citations. They find that advancing injection timing increased peak firing pressure. The BSFC increased considerably and BTE improved slightly remarkably at high loads compared with petrodiesel fuels. Blend B which replaced 50% petrodiesel and having oxygen content up to 12.21% by weight has given satisfactory performance for steady-state running mode up to 1,600 revolutions per minute (rpm).

Sayin (2010) evaluates the performance of bioethanol–petrodiesel fuel blends in a paper with 279 citations. They employ E5 and E10 blends as well as methanol–petrodiesel blends of M5 and M10. They find that the BSFC increased while the BTE decreased with for both blends.

Zhu et al. (2011) evaluate the performance of bioethanol–petrodiesel blends compared to both petrodiesel and biodiesel fuels in a paper with 279 citations. They find that compared with biodiesel, the engine performance has improved slightly with E5 blend. In comparison with petrodiesel fuel, the biodiesel and blends have higher BTE.

Can et al. (2004) evaluate the performance of bioethanol–petrodiesel fuel blends of E10 and E15 in a paper with 258 citations. They add 1% isopropanol to maintain homogeneity and prevent phase separation. They find that bioethanol addition caused ~12.5% (for E10) and 20% (for E15) power reductions. They find that increasing the injection pressure of the engine running with these blends caused some reduction in power.

Huang et al. (2009) evaluate the performance of bioethanol–petrodiesel fuel blends in a paper with 226 citations. They find that the thermal efficiencies of the engine were comparable with those fueled by petrodiesel, with some increase of fuel consumptions, which is due to the lower heating value of bioethanol.

Ajav et al. (1999) evaluate the performance of bioethanol–petrodiesel fuel blends of E5, E10, E15, and E20 in a paper with 219 citations. They find no significant power reduction in the engine operation on blends (up to 20%). The exhaust gas temperature and lubricating oil temperatures were lower with operations on these blends compared to petrodiesel.

85.3.3 The Emissions from Bioethanol Fuels in the Transport Engines

The brief information about 19 prolific studies with at least 215 citations each on the emissions of bioethanol–gasoline and bioethanol–petrodiesel fuel blends in gasoline or diesel engine is given in Table 85.3. Furthermore, the brief notes on the contents of these studies are also given.

85.3.3.1 Emissions from Bioethanol–Gasoline Fuel Blends

Hsieh et al. (2002) evaluate the emissions of bioethanol–gasoline fuel blends with various blended rates (E0, E5, E10, E20, and E30) in a paper with 590 citations. They find that carbon monoxide (CO) and hydrocarbon (HC) emissions decrease dramatically, and carbon dioxide (CO_2) emission increases. Finally, nitrogen oxide (NO_x) emissions depend on the engine operating condition rather than the bioethanol content.

TABLE 85.3
The Emissions from Bioethanol Fuels in the Transport Engines

No.	Papers	Bioethanol Fuels	Issues	Research Fronts	Cits.
1	Hsieh et al. (2002)	Bioethanol–gasoline blends	Emissions of a commercial SI engine using bioethanol–gasoline blended fuels of E0, E5, E10, E20, E30	Emissions, performance	483
2	Kwanchareon et al. (2007)	Bioethanol–petrodiesel blends	Emissions of bioethanol–petrodiesel blends	Emissions, properties	405
3	Al-Hasan (2003)	Bioethanol–gasoline blends	Emissions of gasoline–bioethanol blends	Emissions, performance	399
4	Li et al. (2005)	Bioethanol–petrodiesel blends	Emissions of petrodiesel–bioethanol blends of E5, E10, E15, E20	Emissions, performance	340
5	He et al. (2003)	Bioethanol–petrodiesel blends	Emissions of petrodiesel–bioethanol blends	Emissions, properties	318
6	Najafi et al. (2009)	Bioethanol–gasoline blends	Emissions of gasoline–bioethanol blends of E5, E10, E15, and E20	Emissions performance	306
7	Xing-Cai et al. (2004)	Bioethanol–petrodiesel blends	Impact of the CN improver on the emissions of bioethanol–petrodiesel blends	Emissions, performance, combustion	306
8	Hulwan and Joshi (2011)	Bioethanol–petrodiesel blends	Emission of bioethanol–petrodiesel blends of E30, E40, and E50	Emissions, performance, combustion	285
9	Sayin (2010)	Bioethanol–petrodiesel blends	Emissions of bioethanol–petrodiesel blends of E5 and E10	Performance, emissions	279
10	Zhu et al. (2011)	Bioethanol–petrodiesel blends	Emissions of bioethanol–petrodiesel blends	Performance, combustion, emissions	279
11	Lapuerta et al. (2008)	Bioethanol–petrodiesel blends	Emissions of bioethanol–petrodiesel fuel blends of E10	Emissions	273
12	Can et al. (2004)	Bioethanol–petrodiesel blends	Emissions of bioethanol–petrodiesel blends of E10 and E15	Performance, emissions	258
13	Koc et al. (2009)	Bioethanol–gasoline blends	Emissions of bioethanol–gasoline blends of unleaded E0, E50, and E85	Performance, emissions	239
14	Shi et al. (2006)	Bioethanol–petrodiesel–biodiesel blends	Emissions of bioethanol–petrodiesel–biodiesel blends (E5-B20-P75)	Emissions	229
15	Huang et al. (2009)	Bioethanol–petrodiesel blends	Emissions of bioethanol–petrodiesel blends	Performance, emissions	226
16	Bayraktar (2005)	Bioethanol–gasoline blends	Emissions of bioethanol–gasoline blends	Performance, emissions	225
17	Ajav et al. (1999)	Bioethanol–petrodiesel blends	Emissions of bioethanol–petrodiesel blends of E5, E10, E15, and E20	Performance, emissions	219
18	Zhu et al. (2010)	Bioethanol–biodiesel blends	Bioethanol–petrodiesel–biodiesel blends of E5, E10, E15	Emissions	216
19	Poulopoulos et al. (2001)	Bioethanol–gasoline blends	Emissions of bioethanol–gasoline blends	Emissions	215

Al-Hasan (2003) evaluates the emissions of bioethanol–unleaded gasoline blends in a paper with 399 citations. He finds that the CO and unburned HC emission concentrations decrease, while the CO_2 concentration increases. The E20 gave the best results for all measured parameters at all engine speeds.

Najafi et al. (2009) evaluate the emissions of bioethanol–gasoline blends of E5, E10, E15, and E20 with the aid of ANN in a paper with 306 citations. They find that the concentration of CO and HC emissions decreased when bioethanol blends were introduced. In contrast, the concentration of CO_2 and NO_x was increased when bioethanol is introduced.

Koc et al. (2009) evaluate the emissions of bioethanol–unleaded gasoline blends of E0, E50, and E85 in a paper with 239 citations. They find that bioethanol addition to unleaded gasoline reduces CO, NO_x, and HC emissions.

Bayraktar (2005) evaluates the emissions of bioethanol–gasoline fuel blends in a paper with 225 citations. They employ the blends containing 1.5, 3, 4.5, 6, 7.5, 9, 10.5, and 12 vol% bioethanol. They find that among the various blends, the blend of E7.5 was the most suitable one from the CO emissions point of view. However, theoretical comparisons have shown that the blend containing 16.5% bioethanol was the most suited for SI engines.

Poulopoulos et al. (2001) evaluate the emissions from bioethanol–gasoline fuel blends in a paper with 215 citations. They find that the addition of bioethanol in the fuel up to 10% w/w had as a result an increase in the Reid vapor pressure of the fuel, which indicates indirectly increased evaporative emissions, while CO tailpipe emissions were decreased. For the blends, acetaldehyde emissions were appreciably increased (up to 100%), especially for E3 blend. In contrast, aromatics emissions were decreased by bioethanol addition to gasoline.

85.3.3.2 Emissions from Bioethanol–Petrodiesel Blends

Kwanchareon et al. (2007) evaluate the emissions of petrodiesel–biodiesel–bioethanol blends in a paper with 485 citations. They find that CO and HC were reduced significantly at high engine load, whereas NO_x increased, compared to petrodiesel. They advise that a blend of 80% petrodiesel, 15% biodiesel, and 5% bioethanol (D80-B15-E5) was the most suitable ratio for the fuel production because of the reduction of emissions.

Li et al. (2005) evaluate the emissions of bioethanol–petrodiesel fuel blends of E5, E10, E15, and E20, in a paper with 340 citations. They find that the smoke emissions decreased with blending, especially with E10 and E15. CO and NO_x emissions reduced for these blends, but HC increased significantly compared to petrodiesel fuel.

He et al. (2003) evaluate the emissions of bioethanol–petrodiesel fuel blends in a paper with 318 citations. They find that at high loads, the blends reduce smoke significantly with a small penalty on CO, acetaldehyde, and HC emissions compared to petrodiesel fuel. However, NO_x and CO_2 emissions of the blends are decreased. At low loads, the blends have slight effects on smoke reduction. They assert that with the aid of additive and ignition improver, CO, HC, and acetaldehyde emissions of the blends can be decreased moderately, even total HC emissions are less than those of petrodiesel fuel.

Xing-Cai et al. (2004) evaluate the impact of the CN improver on the emissions of bioethanol–petrodiesel blends in a paper with 306 citations. They find that the NO_x and smoke emissions decreased simultaneously when diesel engine fueled with these blends. NO_x and smoke emissions further reduced when CN improver was added to blends.

Hulwan and Joshi (2011) evaluate the emissions of petrodiesel–bioethanol–biodiesel blends in a paper with 285 citations. They test the blends tested D70/E20/B10 (blend A), D50/E30/B20 (blend B), D50/E40/B10 (blend C), and petrodiesel (D100). They find that advancing injection timing almost doubled the NO_x emissions and increased peak firing pressure. NO_x variation depended on operating conditions while CO emissions drastically increased at low loads. Blend B did not show any benefit on peak smoke emission during free acceleration test.

Sayin (2010) evaluates the emissions of bioethanol–petrodiesel fuel blends in a paper with 279 citations. They employ E5 and E10 blends as well as methanol–petrodiesel blends of M5 and M10. They find that the emissions of NO_x increased while smoke opacity, emissions of CO, and total HC decreased with for both blends.

Zhu et al. (2011) evaluate the emissions of bioethanol–petrodiesel fuel blends compared to both petrodiesel and biodiesel fuels in a paper with 279 citations. They find that compared with petrodiesel fuel, the blends could lead to reduction of both NO_x and particulate matter (PM) emissions of the diesel engine. The effectiveness of NO_x and PM reductions increases with increasing bioethanol in the blends. With high percentage of bioethanol in the blends, the HC and CO emissions could increase. However, the use of E5 blend could reduce the HC and CO emissions as well.

Lapuerta et al. (2008) evaluate the emissions of bioethanol–petrodiesel fuel blends of E10 in a paper with 273 citations. They find that these blends provide a significant reduction on PM emissions, with no substantial increase in other gaseous emissions.

Can et al. (2004) evaluate the emissions of bioethanol–petrodiesel fuel blends of E10 and E15 in a paper with 258 citations. They add 1% isopropanol to maintain homogeneity and prevent phase separation. They find that bioethanol addition reduces CO, soot, and sulfur dioxide (SO_2) emissions, although it caused an increase in NO_x emissions. They find that increasing the injection pressure of the engine running with these blends decreased CO and smoke emissions, especially between 1,500 and 2,500 rpm, with respect to petrodiesel fuel, while it caused some reduction in power.

Shi et al. (2006) evaluate the emissions of bioethanol–petrodiesel–biodiesel blends (E5-B20-D75) in a paper with 229 citations. They find a significant reduction in PM emissions and 2%–14% increase of NO_x emissions. The change of CO emission was not conclusive and depended on operating conditions. Total HC from this blend was lower compared to petrodiesel fuel. This blend led to a slight increase of acetaldehyde, propionaldehyde, and acetone emissions.

Huang et al. (2009) evaluate the emissions of bioethanol–petrodiesel fuel blends in a paper with 226 citations. They find that the smoke emissions from the engine fueled by the blends were all lower than those fueled by petrodiesel. The CO was reduced when the engine ran at and above its half loads, but was increased at low loads and low speed. The HC emissions were all higher except for the top loads at high speed. Finally, the NO_x emissions were different for different speeds, loads, and blends.

Ajav et al. (1999) evaluate the emissions of bioethanol-petrodiesel fuel blends of E5, E10, E15, and E20 in a paper with 219 citations. They find that the exhaust gas temperature, lubricating oil temperatures and exhaust emissions (CO and NO_x) were lower with operations on these blends compared petrodiesel.

Zhu et al. (2010) evaluate the emissions of bioethanol–petrodiesel–biodiesel blends (E5, E10, and E15) in a paper with 216 citations. They find that compared with petrodiesel fuel, the blended fuels could lead to reduction of both NO_x and PM, with the biodiesel–methanol blends being more effective than the biodiesel–bioethanol blends. The effectiveness of NO_x and PM reductions is more effective with increase of alcohol in the blends. With high percentage of alcohol in the blends, the HC and CO emissions could increase and the BTE might be slightly reduced, but the use of 5% blends could reduce the HC and CO emissions as well. With the diesel oxidation catalyst (DOC), the HC, CO, and PM emissions can be further reduced.

85.3.4 The Combustion of Bioethanol Fuels in the Transport Engines

The brief information about four prolific studies with at least 215 citations each on the performance of bioethanol fuels, bioethanol–gasoline and bioethanol–petrodiesel fuel blends in gasoline or diesel engines is given in Table 85.4. Furthermore, the brief notes on the contents of these studies are also given.

Veloo et al. (2010) evaluate the laminar flame speeds and extinction strain rates of premixed methanol, bioethanol, and n-butanol flames in the counterflow configuration at atmospheric pressure and elevated unburned mixture temperatures in a paper with 322 citations. They find that laminar flame speeds of bioethanol/air and n-butanol/air flames are similar to those of their n-alkane/air counterparts, and that methane/air flames have consistently lower laminar flame speeds than methanol/air flames. The laminar flame speeds of methanol/air flames are considerably higher compared to both bioethanol/air and n-butanol/air flames under fuel-rich conditions.

TABLE 85.4
The Combustion of Bioethanol Fuels

No.	Papers	Bioethanol Fuels	Issues	Research Fronts	Cits.
1	Veloo et al. (2010)	Bioethanol fuels	Laminar flame speeds and extinction strain rates of bioethanol fuels	Combustion	322
2	Xing-Cai et al. (2004)	Bioethanol–petrodiesel fuel blends	Impact of the CN improver on the combustion of bioethanol–petrodiesel fuel blends	Combustion, emissions, performance	302
3	Hulwan and Joshi (2011)	Bioethanol–petrodiesel blends	Combustion of bioethanol–petrodiesel fuel blends	Emissions, performance, combustion	285
4	Zhu et al. (2011)	Bioethanol–petrodiesel blends	Combustion of bioethanol–petrodiesel blends	Performance, combustion, emissions	279
5	Rakopoulos et al. (2011)	Bioethanol–petrodiesel blends	Combustion of bioethanol–petrodiesel fuel blends of E5 and E10	Combustion	278
6	Turner et al. (2011)	Bioethanol–gasoline blends	Combustion of bioethanol–gasoline blends	Combustion	215

Xing-Cai et al. (2004) study the impact of the CN improver on the combustion of bioethanol–petrodiesel fuel blends in a paper with 306 citations. They find that ignition delay prolonged, and the total combustion duration shortened for these blends compared to petrodiesel fuel. The combustion characteristics of these blends at large load may be resumed to petrodiesel fuel by CN improver, but a large difference exists at lower load yet.

Hulwan and Joshi (2011) evaluate the combustion of bioethanol–petrodiesel fuel blends in a paper with 285 citations. They find that advancing injection timing increased peak firing pressure. The combustion process of these blends delayed at low loads but approached to the petrodiesel fuel at high loads. The comparison of blend results with petrodiesel showed that thermal efficiency improved slightly and smoke opacity reduced remarkably at high loads.

Zhu et al. (2011) evaluate the combustion of bioethanol–petrodiesel blends compared to both petrodiesel and biodiesel fuels in a paper with 279 citations. They find that compared with biodiesel, the combustion characteristics of these blends changed and the engine performance has improved slightly with E5 blend.

Rakopoulos et al. (2011) evaluate the combustion of bioethanol–petrodiesel fuel blends of E5 and E10 compared with n-butanol-petrodiesel blends of B8 and B16 in a paper with 278 citations. They find that with the use of these blends, fuel injection pressure diagrams are very slightly displaced (delayed), ignition delay is increased, maximum cylinder pressures are slightly reduced, and cylinder temperatures are reduced during the first part of combustion.

Turner et al. (2011) evaluate the combustion of bioethanol–gasoline fuel blends in a paper with 215 citations. They find that the benefits of adding bioethanol into gasoline are reduced engine-out emissions and increased efficiency, and the impact changes with the blend ratio following a certain pattern. They attribute these benefits to the fact that the addition of bioethanol modifies the evaporation properties of the fuel blend which increases the vapor pressure for low blends and reduces the heavy fractions for high blends. This is furthermore coupled with the presence of oxygen within the bioethanol fuel molecule and the contribution of its faster flame speed, leading to enhanced combustion initiation and stability and improved engine efficiency.

85.3.5 The Engine Modification and Health Impact of Bioethanol Fuels

The brief information about two prolific studies with at least 239 citations each on the engine modification and health impact of bioethanol–gasoline fuel blends in gasoline is given in Table 85.5. Furthermore, the brief notes on the contents of these studies are also given.

TABLE 85.5
The Engines and Health Impact of Bioethanol Fuels

No.	Papers	Bioethanol Fuels	Issues	Research Fronts	Cits.
1	Yuksel and Yuksel (2004)	Bioethanol–gasoline fuel blends	Use of bioethanol–gasoline fuel blends of E60 by designing a new carburetor in a SI engine	Engines	283
2	Jacobson (2007)	Bioethanol–gasoline blends of E85	Health impact of bioethanol–gasoline blends of E85	Health impact	239

Yuksel and Yuksel (2004) evaluate the use of bioethanol–gasoline fuel blends by designing a new carburetor in a SI engine in a paper with 283 citations. They design a new carburetor to overcome the realization of a stable homogeneous liquid phase. They find that this phase problem was solved and the alcohol ratio in the total fuel was increased. They use E60 blend to test the performance, the fuel consumption, and the exhaust emissions.

Jacobson (2007) evaluates the health impact of bioethanol–gasoline fuel blends of E85 in a paper with 239 citations. He finds that E85 may increase ozone-related mortality, hospitalization, and asthma by about 9% in Los Angeles and 4% in the United States as a whole relative to 100% gasoline. Ozone increases in Los Angeles and the northeast were partially offset by decreases in the southeast. E85 also increased peroxyacetyl nitrate in the USA but was estimated to cause little change in cancer risk. He asserts that due to its ozone effects, future E85 may be a greater overall public health risk than gasoline and E85 is unlikely to improve air quality over future gasoline vehicles. Unburned bioethanol emissions from E85 may result in a global-scale source of acetaldehyde larger than that of direct emissions.

85.4 DISCUSSION

85.4.1 Introduction

The crude oil-based gasoline and petrodiesel fuels have been widely used in the transportation sector. However, there have been great public concerns over the adverse environmental impact and sustainability of these fuels. Hence, biomass-based bioethanol fuels have increasingly been used in blending gasoline fuels as well as petrodiesel fuels in recent years. In this context, the research has intensified in the research fronts of engine emissions, fuel combustion, engine performance, fuel properties, and to a lesser extent gasoline engine and health impact of bioethanol fuels. The focus has also been on the utilization of bioethanol fuels in blending with both gasoline fuels and petrodiesel fuels in gasoline and diesel engines, respectively. However, it is essential to develop efficient incentive structures for the primary stakeholders to enhance the research in this field.

Although there have been a number of review papers on the bioethanol fuels in the transport engines, there has been no review of the research of the most-cited 25 articles in this field. Hence, this book chapter presents a review of the research of the most-cited 25 articles in this field. Then, it discusses the key findings of these highly influential papers and comments on the future research priorities in this field.

There are four major research fronts for this field: emissions, performance, combustion, and properties of these fuels of the sample of the reviewed papers. Additionally, there is one paper each in the field of the engine modification and health impact of these fuels.

85.4.2 The Fuel Properties of Bioethanol Fuels

There are one and three papers on the fuel properties (Dooley et al., 2010; Schwab et al., 1987) of bioethanol–gasoline (Hsieh et al., 2002) and bioethanol–petrodiesel fuel blends (de Caro et al., 2001; He et al., 2003; Kwanchareon et al., 2007), respectively.

The most-studied fuel properties are the octane number, Reid vapor pressure, flash point, CN, higher heating value (HHV), and kinematic viscosity. These studies highlight that the bioethanol fuels have better fuel properties in general, compared to both gasoline and petrodiesel fuels. There is certainly a room to optimize these properties by adjusting the composition and blend ratios of these fuels to maximize the benefits related to the engine performance, combustion, and emissions of these fuels.

In the end, these most-cited papers in the field of the utilization of bioethanol fuels and their blends hint that the benefits sought from these fuels could be maximized using the structure, processing, and property relationships of these fuels (Formela et al., 2016; Konur, 2018, 2020b, 2021a–d; Konur and Matthews, 1989).

It should be noted here that the fuel properties of the bioethanol fuels are largely determined by the biomass used in their production (Lynd et al., 1991; Wyman, 1994). There is a persistent trend for the bioethanol research to focus more on the second- (Lynd et al., 1991; Wyman, 1994) and third generation (Ho et al., 2013; John et al., 2011) feedstocks and to move away from the first generation bioethanol feedstocks (Wang et al., 2007, 2011). These feedstocks would certainly impact on the properties, engine performance, emissions, and combustion of the resulting bioethanol fuels and their blends. This trend should be followed closely by the major stakeholders in the future.

85.4.3 The Engine Performance of Bioethanol Fuels

There are five and seven papers on the engine performance (Elnajjar et al., 2013; Haas et al., 2001) of bioethanol–gasoline (Al-Hasan, 2003; Bayraktar, 2005; Hsieh et al., 2002; Koc et al., 2009; Najafi et al., 2009) and bioethanol–petrodiesel fuel blends (Ajav et al., 1999; Can et al., 2004; Huang et al., 2009; Hulwan and Joshi, 2011; Li et al., 2005; Sayin, 2010; Zhu et al., 2011), respectively.

The most-studied engine parameters are torque output, fuel consumption, brake power, ηv, BTE, power output, and BSFC. These studies highlight that the bioethanol fuels have better fuel performance in both gasoline and diesel engines compared to both gasoline and petrodiesel fuels in general.

85.4.4 The Emissions from Bioethanol Fuels in the Transport Engines

There are 6 and 13 papers on the emissions of bioethanol–gasoline (Al-Hasan, 2003; Bayraktar, 2005; Hsieh et al., 2002; Koc et al., 2009; Najafi et al., 2009; Poulopoulos et al., 2001) and bioethanol–petrodiesel fuel blends (Ajav et al., 1999; Can et al., 2004; He et al., 2003; Huang et al., 2009; Hulwan and Joshi, 2011; Kwanchareon et al., 2007; Lapuerta et al., 2008; Li et al., 2005; Sayin, 2010; Shi et al., 2006; Xing-Cai et al., 2004; Zhu et al., 2010, 2011), respectively.

The most-studied emissions are CO, unburned HC, CO_2, NO_x, acetaldehyde, aromatics, smoke, PM, and SO_2. These studies highlight the determinants of the better emission performance for these blends. These studies hint that the operating conditions of the engines as well as the proportions of bioethanol concentrations play a key role in determining the emission performance of these blends. With the optimization of operating conditions of the engines as well as the proportions of bioethanol, these studies show that bioethanol blends might have better emission performance compared to both gasoline and petrodiesel fuels.

The emissions (Lapuerta et al., 2008; Searchinger et al., 2008) of these fuels are also of the most importance in light of the public concerns about climate change (Change, 2007; Karl and Trenberth, 2003), greenhouse gas emissions (Carlson et al., 2017; Riahi et al., 2017), and global warming (Kerr, 2007; Peters et al., 2013). As these concerns have been certainly behind the boom in the research in this field in the last two decades, it is not surprising that the bulk of the majority of the most-cited papers sought to evaluate the environmental impact of bioethanol fuels during its utilization in both gasoline and diesel engines.

The emissions of CO_2 (Ang, 2007; Holtz-Eakin and Selden, 1995) are especially important, and any potential of the reduction of these emissions by the bioethanol blends compared to gasoline or petrodiesel fuels would have significant implications for the environmental protection. The emissions of the PM are also important for the protection of the environment (Abdullahi et al., 2013; Pant and Harrison, 2013).

85.4.5 The Combustion of Bioethanol Fuels in the Transport Engines

There are one, one and four papers on the combustion of bioethanol (Veloo et al., 2010), bioethanol–gasoline fuel blends (Turner et al., 2011), and bioethanol–petrodiesel fuel blends (Rakopoulos et al., 2011; Zhu et al., 2011), respectively.

The most-studied combustion parameters are laminar flame speeds, extinction strain rates, ignition delay, total combustion duration, peak firing pressure, fuel injection pressure, maximum cylinder pressures, cylinder temperatures, and combustion initiation and stability.

The studies highlight the determinants of the better combustion performance (Demirbas, 2004; Spalding, 1950) of bioethanol fuels and their blends, and suggest that it is possible to optimize the combustion parameters to obtain better combustion performance for bioethanol fuels and their blends.

85.4.6 The Engine Modification and Health Impact of Bioethanol Fuels

There are two papers on the engine modification for bioethanol–gasoline fuel blends (Yuksel and Yuksel, 2004) and the health impact (Jacobson, 2007) of the bioethanol fuels. The first study focuses on the impact of a new carburetor design while the second study evaluates the impact of bioethanol–gasoline blend of E85 on ozone-related mortality, hospitalization, and asthma in the USA. He asserts that due to its ozone effects, future E85 may be a greater overall public health risk than gasoline. However, it should be noted that the optimal concentrations of bioethanol blends rarely exceed 10% in the field.

85.5 CONCLUSION AND FUTURE RESEARCH

The brief information about the key research fronts covered by the 25 most-cited papers with at least 215 citations each in the field of the utilization of bioethanol fuels and their bioethanol–gasoline and bioethanol–petrodiesel fuel blends in the gasoline or diesel engines is given in Table 85.6.

There are four major research fronts for this field: emissions, performance, combustion, and properties of these fuels. The other minor research fronts are the engine modification and health impact of these fuels.

It is notable that there is similarity of these research fronts in the sample of reviewed papers and the sample of the most-cited 100 papers (the first column data in Table 85.6) in this field.

The fuel properties are important as they determine the emission, combustion, and engine performance of these fuels. The most-studied fuel properties are the octane number, Reid vapor pressure, flash point, CN, HHV, and kinematic viscosity. These studies hint that the bioethanol fuels have better fuel properties compared to gasoline or petrodiesel fuels in general. There is a room to optimize these properties by adjusting the composition and blend ratios of these fuels.

There is a persistent trend for the bioethanol research to focus more on the second- and third generation feedstocks and to move away from the first generation bioethanol feedstocks. These feedstocks would certainly impact on the properties, engine performance, emissions, and combustion of the resulting bioethanol fuels and their blends. This trend should be followed closely by the major stakeholders in the future. For this reason, the impact of source of the biomass on the properties, engine performance emissions, and combustion of bioethanol fuels and their gasoline or petrodiesel

TABLE 85.6
The Most-Prolific Research Fronts in the Field of the Utilization of Bioethanol Fuels in the Transport Engines

No.	Research Fronts	Sample Papers (%)	Reviewed Papers (No.)	Reviewed Papers (%)
1	Fuel properties	18	4	16
	Bioethanol fuels	1	0	0
	Bioethanol–gasoline fuel blends	3	1	4
	Bioethanol–petrodiesel fuel blends	13	3	12
2.	Engine performance	35	12	48
	Bioethanol fuels	1	0	0
	Bioethanol–gasoline fuel blends	14	5	20
	Bioethanol–petrodiesel fuel blends	21	7	28
3.	Engine emissions	65	19	76
	Bioethanol fuels	8	0	0
	Bioethanol–gasoline fuel blends	20	6	24
	Bioethanol–petrodiesel fuel blends	37	13	52
4.	Fuel combustion	43	6	24
	Bioethanol fuels	19	1	4
	Bioethanol–gasoline fuel blends	6	1	4
	Bioethanol–petrodiesel fuel blends	18	4	16
5.	Engine modification	4	2	8
	Bioethanol fuels	2	1	4
	Bioethanol–gasoline fuel blends	2	1	4
	Bioethanol–petrodiesel fuel blends	0	0	0
6.	Fuel health impact	1	1	4
	Bioethanol fuels	0	1	4
	Bioethanol–gasoline fuel blends	1	0	0
	Bioethanol–petrodiesel fuel blends	0	0	0

The total number of reviewed papers = 25. Sample papers: the sample of the most-cited 100 papers.

blends in the diesel or gasoline engines should be focused on to establish structure–property–processing relationships for these bioethanol fuels in the future studies as it has been done in the other research streams of bioethanol production and utilization

The performance of these fuels in the gasoline or diesel engines is of the outmost importance as a function of the properties of these fuels as well as the proportions of the bioethanol in these blends. The most-studied engine parameters are torque output, fuel consumption, brake power, ηv, BTE, power output, and BSFC. These studies hint that the bioethanol fuels have better fuel performance compared to both gasoline and petrodiesel fuels in both gasoline and diesel engines in general.

The emissions of these fuels are also of the most importance in light of the public concerns about climate change, greenhouse gas emissions, and global warming. As these concerns have been certainly behind the boom in the research in this field in the last two decades, it is not surprising that the bulk of the majority of the most-cited papers sought to evaluate the environmental impact of bioethanol fuels during its utilization in both gasoline and diesel engines. The most-studied emissions are CO, unburned HC, CO_2, NO_x, acetaldehyde, aromatics, smoke, PM, and SO_2.

The emissions of CO_2 are especially important, and any potential of the reduction of these emissions by the bioethanol blends compared to gasoline or petrodiesel fuels would have significant implications for the environmental protection. The emissions of the PM are also important for the protection of the environment.

These most-cited studies highlight the determinants of the better emission performance for these blends. The operating conditions of the engines as well as the proportions of bioethanol concentrations in the fuel blends play a key role in determining the emission performance of these blends. With the optimization of operating conditions of the engines as well as the proportions of bioethanol, these studies show that bioethanol blends might have better emission performance compared to both gasoline and petrodiesel fuels.

The combustion of these fuels is also an important research front. The most-studied combustion parameters are laminar flame speeds, extinction strain rates, ignition delay, total combustion duration, peak firing pressure, fuel injection pressure, maximum cylinder pressures, cylinder temperatures, and combustion initiation and stability. These most-cited studies highlight the determinants of the better combustion performance of bioethanol fuels and their blends, and suggest that it is possible to optimize the combustion parameters to obtain better combustion performance for bioethanol fuels and their blends.

In the end, these most-cited papers in the field of the utilization of bioethanol fuels and their blends hint that the benefits sought from these fuels could be maximized using the structure, processing, and property relationships of these fuels as in materials science and engineering.

These findings confirm that bioethanol fuels are a viable alternative to crude oil-based gasoline and petrodiesel fuels and have environmental benefits impacting favorably global warming, greenhouse gas emissions, and climate change.

It is recommended that such review studies should be performed for the other research fronts on both the production and utilization of bioethanol fuels complementing the corresponding scientometric studies.

ACKNOWLEDGMENTS

The contribution of the highly cited researchers in the field of the utilization of bioethanol fuels in the transport engines has been gratefully acknowledged.

REFERENCES

Abdullahi, K. L., J. M. Delgado-Saborit and R. M. Harrison. 2013. Emissions and indoor concentrations of particulate matter and its specific chemical components from cooking: A review. *Atmospheric Environment* 71:260–294.

Ajav, E. A., B. Singh and T. K. Bhattacharya. 1999. Experimental study of some performance parameters of a constant speed stationary diesel engine using ethanol-diesel blends as fuel. *Biomass and Bioenergy* 17:357–365.

Al-Hasan, M. 2003. Effect of ethanol-unleaded gasoline blends on engine performance and exhaust emission. *Energy Conversion and Management* 44:1547–1561.

Anderson, J. E., D. M. Dicicco and J. M. Ginder, et al. 2012. High octane number ethanol-gasoline blends: Quantifying the potential benefits in the United States. *Fuel* 97:585–594.

Ang, J. B. 2007. CO_2 emissions, energy consumption, and output in France. *Energy Policy* 35:4772–4778.

Bayraktar, H. 2005. Experimental and theoretical investigation of using gasoline-ethanol blends in spark-ignition engines. *Renewable Energy* 30:1733–1747.

Bosmann, A., L. Datsevich and A. Jess, et al. 2001. Deep desulfurization of diesel fuel by extraction with ionic liquids. *Chemical Communications* 23:2494–2495.

Burnham, J. F. 2006. Scopus database: A review. *Biomedical Digital Libraries* 3:1–8.

Can, O., I. Celikten and N. Usta. 2004. Effects of ethanol addition on performance and emissions of a turbocharged indirect injection diesel engine running at different injection pressures. *Energy Conversion and Management* 45:2429–2440.

Carlson, K. M., J. S. Gerber and N. D. Mueller, et al. 2017. Greenhouse gas emissions intensity of global croplands. *Nature Climate Change* 7:63–68.

Change, C. 2007. Climate change impacts, adaptation and vulnerability. *Science of the Total Environment* 326:95–112.

Christensen, M., B. Johansson and P. Einewall. 1997. Homogeneous charge compression ignition (HCCI) using isooctane, ethanol and natural gas: A comparison with spark ignition operation. SAE Technical Papers 1997:972874.

de Caro, P. S., Z. Mouloungui, G. Vaitilingom and J. C. Berge. 2001. Interest of combining an additive with diesel-ethanol blends for use in diesel engines. *Fuel* 80:565–574.

Demirbas, A. 2004. Combustion characteristics of different biomass fuels. *Progress in Energy and Combustion Science* 30(2):219–230.

Dooley, S., S. H. Won and M. Chaos, et al. 2010. A jet fuel surrogate formulated by real fuel properties. *Combustion and Flame* 157:2333–2339.

Elnajjar, E., M. Y. Selim and M. O. Hamdan. 2013. Experimental study of dual fuel engine performance using variable LPG composition and engine parameters. *Energy Conversion and Management* 76:32–42.

Formela, K., A. Hejna, L. Piszczyk, M. R. Saeb and X. Colom. 2016. Processing and structure-property relationships of natural rubber/wheat bran biocomposites. *Cellulose* 23:3157–3175.

Giakoumis, E. G., C. D. Rakopoulos, A. M. Dimaratos and D. C. Rakopoulos. 2013. Exhaust emissions with ethanol or n-butanol diesel fuel blends during transient operation: A review. *Renewable and Sustainable Energy Reviews* 17:170–190.

Haas, M. J., K. M. Scott, T. L. Alleman and R. L. McCormick. 2001. Engine performance of biodiesel fuel prepared from soybean soapstock: A high quality renewable fuel produced from a waste feedstock. *Energy & Fuels* 15:1207–1212.

Hansen, A. C., Q. Zhang and P. W. L. Lyne. 2005. Ethanol-diesel fuel blends: A review. *Bioresource Technology* 96:277–285.

He, B. Q., S. J. Shuai, J. X. Wang and H. He. 2003. The effect of ethanol blended diesel fuels on emissions from a diesel engine. *Atmospheric Environment* 37:4965–4971.

Hill, J., E. Nelson, D. Tilman, S. Polasky and D. Tiffany. 2006. Environmental, economic, and energetic costs and benefits of biodiesel and ethanol biofuels. *Proceedings of the National Academy of Sciences of the United States of America* 103:11206–11210.

Hill, J., S. Polasky and E. Nelson, et al. 2009. Climate change and health costs of air emissions from biofuels and gasoline. *Proceedings of the National Academy of Sciences of the United States of America* 106:2077–2082.

Ho, S. H., S. W. Huang and C. Y. Chen, et al. 2013. Bioethanol production using carbohydrate-rich microalgae biomass as feedstock. *Bioresource Technology* 135:191–198.

Holtz-Eakin, D. and T. M. Selden. 1995. Stoking the fires? CO_2 emissions and economic growth. *Journal of Public Economics* 57:85–101.

Hsieh, W. D., R. H. Chen, T. L. Wu and T. H. Lin. 2002. Engine performance and pollutant emission of an SI engine using ethanol-gasoline blended fuels. *Atmospheric Environment* 36:403–410.

Huang, J., Y. Wang and S. Li, et al. 2009. Experimental investigation on the performance and emissions of a diesel engine fuelled with ethanol-diesel blends. *Applied Thermal Engineering* 29:2484–2490.

Hulwan, D. B. and S. V. Joshi. 2011. Performance, emission and combustion characteristic of a multicylinder DI diesel engine running on diesel-ethanol-biodiesel blends of high ethanol content. *Applied Energy* 88:5042–5055.

Jacobson, M. Z. 2007. Effects of ethanol (E85) versus gasoline vehicles on cancer and mortality in the United States. *Environmental Science and Technology* 41:4150–4157.

John, R. P., G. S. Anisha, K. M. Nampoothiri and A. Pandey. 2011. Micro and macroalgal biomass: A renewable source for bioethanol. *Bioresource Technology* 102:186–193.

Karl, T. R. and K. E. Trenberth. 2003. Modern global climate change. *Science* 302:1719–1723.

Kerr, R. A. 2007. Global warming is changing the world. *Science* 316:188–190.

Koc, M., Y. Sekmen, T. Topgul and H. S. Yucesu. 2009. The effects of ethanol-unleaded gasoline blends on engine performance and exhaust emissions in a spark-ignition engine. *Renewable Energy* 34:2101–2106.

Konur, O. 2000. Creating enforceable civil rights for disabled students in higher education: An institutional theory perspective. *Disability & Society* 15:1041–1063.

Konur, O. 2002a. Access to nursing education by disabled students: Rights and duties of nursing programs. *Nurse Education Today* 22:364–374.

Konur, O. 2002b. Assessment of disabled students in higher education: Current public policy issues. *Assessment and Evaluation in Higher Education* 27:131–152.

Konur, O. 2002c. Access to employment by disabled people in the UK: Is the Disability Discrimination Act working? *International Journal of Discrimination and the Law* 5:247–279.

Konur, O. 2006a. Participation of children with dyslexia in compulsory education: Current public policy issues. *Dyslexia* 12:51–67.

Konur, O. 2006b. Teaching disabled students in higher education. *Teaching in Higher Education* 11:351–363.

Konur, O. 2007a. A judicial outcome analysis of the Disability Discrimination Act: A windfall for the employers? *Disability & Society* 22:187–204.

Konur, O. 2007b. Computer-assisted teaching and assessment of disabled students in higher education: The interface between academic standards and disability rights. *Journal of Computer Assisted Learning* 23:207–219.

Konur, O. 2012. The evaluation of the research on the bioethanol: A scientometric approach. *Energy Education Science and Technology Part A: Energy Science and Research* 28:1051–1064.

Konur, O. 2015. Current state of research on algal bioethanol. In *Marine Bioenergy: Trends and Developments*, Eds. S. K. Kim and C. G. Lee, pp. 217–244. Boca Raton, FL: CRC Press.

Konur, O., Ed. 2018. *Bioenergy and Biofuels*. Boca Raton, FL: CRC Press.

Konur, O. 2019. Cyanobacterial bioenergy and biofuels science and technology: A scientometric overview. In *Cyanobacteria: From Basic Science to Applications*, Eds. A. K. Mishra, D. N. Tiwari and A. N. Rai, pp. 419–442. Amsterdam: Elsevier.

Konur, O. 2020a. The scientometric analysis of the research on the bioethanol production from green macroalgae. In *Handbook of Algal Science, Technology and Medicine*, Ed. O. Konur, pp. 385–401. London: Academic Press.

Konur, O., Ed. 2020b. *Handbook of Algal Science, Technology and Medicine*. London: Academic Press.

Konur, O., Ed. 2021a. *Handbook of Biodiesel and Petrodiesel Fuels: Science, Technology, Health, and Environment*. Boca Raton, FL: CRC Press.

Konur, O., Ed. 2021b. *Handbook of Biodiesel and Petrodiesel Fuels: Science, Technology, Health, and Environment. Volume 1. Biodiesel Fuels: Science, Technology, Health, and Environment*. Boca Raton, FL: CRC Press.

Konur, O., Ed. 2021c. *Handbook of Biodiesel and Petrodiesel Fuels: Science, Technology, Health, and Environment. Volume 2. Biodiesel Fuels Based on the Edible and Nonedible Feedstocks, Wastes, and Algae: Science, Technology, Health, and Environment*. Boca Raton, FL: CRC Press.

Konur, O., Ed. 2021d. *Handbook of Biodiesel and Petrodiesel Fuels: Science, Technology, Health, and Environment. Volume 3. Petrodiesel Fuels: Science, Technology, Health, and Environment*. Boca Raton, FL: CRC Press.

Konur, O. 2023. Utilization of bioethanol fuels in the transport engines: Scientometric study. In *Evaluation and Utilization of Bioethanol Fuels. I.: Evaluation of Bioethanol Fuels, Transport Engines, and Bioethanol Sensors. Handbook of Bioethanol Fuels Volume 5*, Ed. O. Konur, pp. 157–175. Boca Raton, FL: CRC Press.

Konur, O. and F. L. Matthews. 1989. Effect of the properties of the constituents on the fatigue performance of composites: A review. *Composites* 20:317–328.

Kwanchareon, P., A. Luengnaruemitchai, A. and S. Jai-In. 2007. Solubility of a diesel-biodiesel-ethanol blend, its fuel properties, and its emission characteristics from diesel engine. *Fuel* 86:1053–1061.

Lapuerta, M., O. Armas and J. M. Herreros. 2008. Emissions from a diesel-bioethanol blend in an automotive diesel engine. *Fuel* 87:25–31.

Lapuerta, M., O. Armas and J. Rodriguez-Fernandez. 2008. Effect of biodiesel fuels on diesel engine emissions. *Progress in Energy and Combustion Science* 34:198–223.

Li, D., Z. Huang, X. C. Lu, W. Zhang and J. Yang. 2005. Physico-chemical properties of ethanol-diesel blend fuel and its effect on performance and emissions of diesel engines. *Renewable Energy* 30:967–976.

Lynd, L. R., J. H. Cushman, R. J. Nichols and C. E. Wyman. 1991. Fuel ethanol from cellulosic biomass. *Science* 251:1318–1323.

Ma, X., L. Sun and C. Song. 2002. A new approach to deep desulfurization of gasoline, diesel fuel and jet fuel by selective adsorption for ultra-clean fuels and for fuel cell applications. *Catalysis Today* 77:107–116.

Ma, X. L., K. Y. Sakanish and I. Mochida. 1994. Hydrodesulfurization reactivities of various sulfur-compounds in diesel fuel. *Industrial & Engineering Chemistry Research* 33:218–222.

Masum, B. M., H. H. Masjuki and M. A. Kalam, et al. 2013. Effect of ethanol-gasoline blend on NO_x emission in SI engine. *Renewable and Sustainable Energy Reviews* 24:209–222.

Najafi, G., B. Ghobadian and T. Tavakoli, et al. 2009. Performance and exhaust emissions of a gasoline engine with ethanol blended gasoline fuels using artificial neural network. *Applied Energy* 86:630–639.

Newman, P. W. G. and J. R. Kenworthy. 1989. Gasoline consumption and cities: A comparison of U.S. cities with a global survey. *Journal of the American Planning Association* 55:24–37.

Niven, R. K. 2005. Ethanol in gasoline: Environmental impacts and sustainability review article. *Renewable and Sustainable Energy Reviews* 9:535–555.

North, D. C. 1991. Institutions. *Journal of Economic Perspectives* 5:97–112.

Pant, P. and R. M. Harrison. 2013. Estimation of the contribution of road traffic emissions to particulate matter concentrations from field measurements: A review. *Atmospheric Environment* 77:78–97.

Peters, G. P., R. M. Andrew and T. Boden, et al. 2013. The challenge to keep global warming below 2°C. *Nature Climate Change* 3:4–6.

Poulopoulos, S. G., D. P. Samaras and C. J. Philippopoulos. 2001. Regulated and unregulated emissions from an internal combustion engine operating on ethanol-containing fuels. *Atmospheric Environment* 35:4399–4406.

Rakopoulos, D. C., C. D. Rakopoulos, R. G. Papagiannakis and D. C. Kyritsis 2011. Combustion heat release analysis of ethanol or n-butanol diesel fuel blends in heavy-duty DI diesel engine. *Fuel* 90:1855–1867.

Riahi, K., D. P. Van Vuuren and E. Kriegler, et al. 2017. The shared socioeconomic pathways and their energy, land use, and greenhouse gas emissions implications: An overview. *Global Environmental Change* 42:153–168.

Sayin, C. 2010. Engine performance and exhaust gas emissions of methanol and ethanol-diesel blends. *Fuel* 89:3410–3415.

Schauer, J. J., M. J. Kleeman, G. R. Cass and B. R. T. Simoneit. 2002. Measurement of emissions from air pollution sources. 5. C_1-C_{32} organic compounds from gasoline-powered motor vehicles. *Environmental Science and Technology* 36:1169–1180.

Schwab, A. W., M. O. Bagby and B. Freedman. 1987. Preparation and properties of diesel fuels from vegetable oils. *Fuel* 66:1372–1378.

Searchinger, T., R. Heimlich and R. A. Houghton, et al. 2008. Use of US croplands for biofuels increases greenhouse gases through emissions from land-use change. *Science* 319:1238–1240.

Shi, X., X. Pang and Y. Mu, et al. 2006. Emission reduction potential of using ethanol-biodiesel-diesel fuel blend on a heavy-duty diesel engine. *Atmospheric Environment* 40:2567–2574.

Spalding, D. B. 1950. Combustion of liquid fuels. *Nature* 165:160–160.

Turner, D., X. Xu, R. F. Cracknell, V. Natarajan and X. Chen. 2011. Combustion performance of bio-ethanol at various blend ratios in a gasoline direct injection engine. *Fuel* 90:1999–2006.

Veloo, P. S., Y. L. Wang, F. N. Egolfopoulos and C. K. Westbrook. 2010. A comparative experimental and computational study of methanol, ethanol, and n-butanol flames. *Combustion and Flame* 157:1989–2004.

Wang, M., M. Wu and H. Huo. 2007. Life-cycle energy and greenhouse gas emission impacts of different corn ethanol plant types. *Environmental Research Letters* 2:024001.

Wang, M. Q., J. Han and Z. Haq, et al. 2011. Energy and greenhouse gas emission effects of corn and cellulosic ethanol with technology improvements and land use changes. *Biomass and Bioenergy* 35:1885–1896.

Wyman, C. E. 1994. Ethanol from lignocellulosic biomass: Technology, economics, and opportunities. *Bioresource Technology* 50:3–15.

Xing-Cai, L., Y. Jian-Guang, Z. Wu-Gao and H. Zhen. 2004. Effect of cetane number improver on heat release rate and emissions of high speed diesel engine fueled with ethanol-diesel blend fuel. *Fuel* 83:2013–2020.

Yuksel, F. and B. Yuksel. 2004. The use of ethanol-gasoline blend as a fuel in an SI engine. *Renewable Energy* 29:1181–1191.

Zhu, L., C. S. Cheung, W. G. Zhang and Z. Huang. 2011. Combustion, performance and emission characteristics of a DI diesel engine fueled with ethanol-biodiesel blends. *Fuel* 90:1743–1750.

Zhu, L., C. S. Cheung., W. G. Zhang and Z. Huang. 2010. Emissions characteristics of a diesel engine operating on biodiesel and biodiesel blended with ethanol and methanol. *Science of the Total Environment* 408:914–921.

Part 26

Evaluation of Bioethanol Fuels

86 Evaluation of Bioethanol Fuels

Scientometric Study

Ozcan Konur
(Formerly) Ankara Yildirim Beyazit University

86.1 INTRODUCTION

Crude oil-based gasoline fuels (Ma et al., 2002; Newman and Kenworthy, 1989) and petrodiesel fuels (Bosmann et al., 2001; Ma et al., 1994) have been widely used in the transportation sector for a long time. However, there have been great public concerns over the adverse environmental impact and sustainability of these fuels (Hill et al., 2009; Schauer et al., 1999, 2002).

Hence, biomass-based bioethanol fuels (Hill et al., 2006, 2009; Konur, 2012e, 2015, 2109, 2020a) have been increasingly used in blending gasoline fuels (Hsieh et al., 2002; Najafi et al., 2009) and petrodiesel fuels in recent years (Hansen et al., 2005; Li et al., 2005). In this context, there has been a significant focus on the evaluative studies in bioethanol fuels (Farrell et al., 2006; Hamelinck et al., 2005; Hill et al., 2006).

There have been three primary research fronts in this field: environmental impact, technoeconomics (Hamelinck et al., 2005; Hill et al., 2006), and technoassessment (Hsieh et al., 2002; Zhu and Pan, 2010). There have also been three other research fronts: policy (Lynd, 1996; Martinelli and Filoso, 2008), economics (Serra et al., 2011; Solomon et al., 2007), and business (Chen and Fan, 2012; Dal-Mas et al., 2011).

Furthermore, there have been five secondary research fronts for the research on environmental impact: emissions (Lynd, 1996; Macedo et al., 2008), life cycle assessment (Lynd, 1996; Sheehan et al., 2003), land use (Leite et al., 2009; Wang et al., 2011), water consumption (Chiu et al., 2009; Petersen et al., 2009), and other issues (Farrell et al., 2006; Goldemberg et al., 2008).

However, it is essential to develop efficient incentive structures (North, 1991) for the primary stakeholders to enhance the research in this field (Konur, 2000, 2002a–c, 2006a,b, 2007a,b).

The scientometric analysis has been used in this context to inform the primary stakeholders about the current state of the research in a selected research field (Garfield, 1955; Konur, 2011, 2012a–i, 2015, 2018b, 2019, 2020a).

Although there have been a number of scientometric studies on bioethanol fuels (Konur, 2012e, 2015, 2019, 2020a), there has been no scientometric study of the research in the field of the evaluative studies in bioethanol fuels.

This book chapter presents a scientometric study of the research in the field of the evaluative studies in bioethanol fuels. It examines the scientometric characteristics of the sample and population data presenting scientometric characteristics of these both datasets in the order of documents, authors, publication years, institutions, funding bodies, source titles, countries, Scopus subject categories, keywords, and research fronts.

86.2 MATERIALS AND METHODS

The search for this study was carried out using the Scopus database (Burnham, 2006) in August 2021.

As a first step for the search of the relevant literature, the keywords were selected using the first most-cited 500 papers. The selected keyword list was optimized to obtain a representative sample

DOI: 10.1201/9781003226567-116

of papers for the searched research field. This keyword list is provided in the appendix for future replication studies. Additionally, the information about the most-used keywords is given in Section 86.3.9 to highlight the key research fronts in Section 86.3.10.

As a second step, two sets of data were used for this study. First, a population sample of over 3,700 papers was used to examine the scientometric characteristics of the population data. Second, a sample of 100 most-cited papers was used to examine the scientometric characteristics of these citation classics with over 136 citations each.

The scientometric characteristics of these both sample and population datasets are presented in the order of documents, authors, publication years, institutions, funding bodies, source titles, countries, Scopus subject categories, keywords, and research fronts.

Lastly, the key scientometric findings for both datasets are discussed to highlight the research landscape for the evaluative studies in bioethanol fuels. Additionally, a number of brief conclusions were drawn, and a number of relevant recommendations are made to enhance the future research landscape.

86.3 RESULTS

86.3.1 The Most-Prolific Documents in the Field of the Evaluative Studies in Bioethanol Fuels

The information on the types of documents for both datasets is given in Table 86.1. The articles, review papers, and conference papers published in the journals dominate the sample dataset, while, articles, conference papers, book chapters, reviews, and notes dominate the population dataset.

Review papers are overrepresented in the sample dataset, while conference papers, articles, and book chapters are overrepresented in the population dataset, as shown in Table 86.1.

It is also interesting to note that all of the papers in the sample dataset were published in journals, while only 93% of the papers were published in journals for the population dataset. Furthermore, 3.8% and 3.2% of the population papers were published in books and book series, respectively.

86.3.2 The Most-Prolific Authors in the Field of the Evaluative Studies in Bioethanol Fuels

The information about the most-prolific 22 authors with at least two papers each in the sample dataset and five papers each in the population dataset is given in Table 86.2.

TABLE 86.1
Documents in the Field of Evaluative Studies in Bioethanol Fuels

Document	Sample Dataset (%)	Population Dataset (%)	Surplus (%)
Article	85	80.1	4.9
Conference paper	3	8.2	−5.2
Book chapter	0	4.5	−4.5
Review	11	3.3	7.7
Note	1	2.3	−1.3
Letter	0	0.7	−0.7
Editorial	0	0.3	−0.3
Book	0	0.3	−0.3
Short survey	0	0.3	−0.3
Sample size	100	3,778	

TABLE 86.2
Most-Prolific Authors in the Field of Evaluative Studies in Bioethanol Fuels

No.	Authors	Author Code	Sample Papers	Population Papers	Institution	Country	Research Front
1	Rakopulos, Dimitrios C.	6603012578	5	11	Natl. Tech. Univ. Athens	Greece	Emissions
2	Rakopoulos, Constantine D.	35570765900	5	11	Natl. Tech. Univ. Athens	Greece	Emissions
3	Aden, Andy	35324090200	5	8	Natl. Renew. Energ. Lab.	USA	Technoeconomics
4	Faaij, Andre	6701681600	4	7	Univ. Groningen	Netherlands	Technoeconomics
5	Giakoumis, Evangelos G.	6602269983	4	6	Natl. Tech. Univ. Athens	Greece	Emissions
6	Wyman, Charles E.	7004396809	4	6	Univ. Calif.	USA	Technoeconomics
7	Bezzo, Fabrizio	6602450288	3	20	Univ. Padua	Italy	Technoeconomics
8	Wang, Jianxin	35254238800	3	7	Tsinghua Univ.	China	Emissions
9	He, Hong	55229497700	3	6	Chinese acad. Sci.	China	Emissions
10	Shuai, Shijing	6603356005	3	6	Tsinghua Univ.	China	Emissions
11	Cheung, Chun Sun	57191305782	2	14	Hong Kong Polytech. Univ.	China	Emissions
12	Huang, Zhen	57205480286	2	11	Shanghai Jiao Tong Univ.	China	Emissions
13	Walter, Arnaldo	7003397572	2	11	State Univ. Campinas	Brazil	Technoeconomics
14	Ghobadian, Barat	22937734400	2	9	Tarbiat Modares Univ.	Iran	Emissions
15	Najafi, Gholamhassan	22938428000	2	9	Tarbiat Modares Univ.	Iran	Emissions
16	Zamboni, Andrea	24438630900	2	9	Leonardo Co.	Italy	Business
17	Masjuki, Haji Hassan	57175108000	2	7	Int. Islamic Univ.	Malaysia	Emissions
18	Pimentel, David	7005471319	2	7	Cornell Univ.	USA	Technoeconomics
19	Tao, Ling	34769458000	2	7	Natl. Renew. Energ. Lab.	USA	Technoeconomics
20	He, Bang-Quan	7402047862	2	6	Tianjin Univ.	China	Emissions
21	Zhu, Lei	56325990300	2	6	Shanghai Jiao Tong Univ.	China	Emissions
22	Herreros, Jose Martin	16642541300	2	5	Univ. Birmingham	UK	Emissions

Author code: the unique code given by Scopus to the authors. Sample papers: the number of papers authored in the sample dataset. Population papers: the number of papers authored in the population dataset.

The most-prolific authors are Andy Aden, Dimitrios C. Rakopoulos, and Constantine D. Rakopoulos, with five papers each. Next, Andre Faaij, Evangelos G. Giakoumis, and Charles E. Wyman contribute four papers each. Furthermore, Fabrizio Bezzo, He Hong, Shijing Shuai, and Jianxin Wang contribute three papers each.

The most-prolific institutions for the sample dataset are the National Technical University of Athens, the National Renewable Energy Laboratory (NREL), Shanghai Jiao Tong University, Tarbiat Modares University, and Tsinghua University, with at least two authors each.

The most-prolific country for the sample dataset is China, with seven authors. This is followed by the USA, Greece, Italy, and Iran, with four, three, two, and two authors each, respectively. In total, nine countries contribute to the sample dataset.

The most-prolific research fronts for the sample dataset are the emissions of bioethanol fuels and the technoeconomics of bioethanol fuels, with 13 and 7 authors, respectively. Furthermore, one author deals with the business management of bioethanol fuels.

On the other hand, there is significant gender deficit (Beaudry and Lariviere, 2016) for the sample dataset as surprisingly all of these top researchers are male.

86.3.3 The Most-Prolific Research Output by Years in the Field of the Evaluative Studies in Bioethanol Fuels

Information about papers published between 1970 and 2021 is given in Figure 86.1. This figure clearly shows that the bulk of the research papers in the population dataset were published in the 2000s, 2010s, and 2020s, with 17.7%, 63.5%, and 13.4% of the population dataset, respectively. The publication rates were 0.2%, 0.6%, 2.2%, and 2.8% for the pre-1970s, 1970s, 1980s, and 1990s, respectively. The number of population papers rose between 2004 and 2012, and there was a second rising trend for the research output between 2015 and 2020.

Similarly, the bulk of the research papers in the sample dataset were published in the 2000s and 2010s, with 57% and 36% of the sample dataset, respectively. Furthermore, 6% and 1% of the sample papers were published in the 1990s and 1970s, respectively.

The most-prolific publication years for the population dataset were after 2008, with at least 5% each of the dataset. Nearly 8% of the population papers were published in 2020. Similarly, 11%, 15%, and 13% of the sample papers were published in 2008, 2010, and 2009, respectively. Additionally, there were seven papers each published in 2005, 2007, and 2011.

86.3.4 The Most-Prolific Institutions in the Field of the Evaluative Studies in Bioethanol Fuels

Information about the most-prolific 25 institutions publishing papers on the evaluative studies in bioethanol fuels with at least two sample papers each and at least 0.4% of the population papers each is given in Table 86.3.

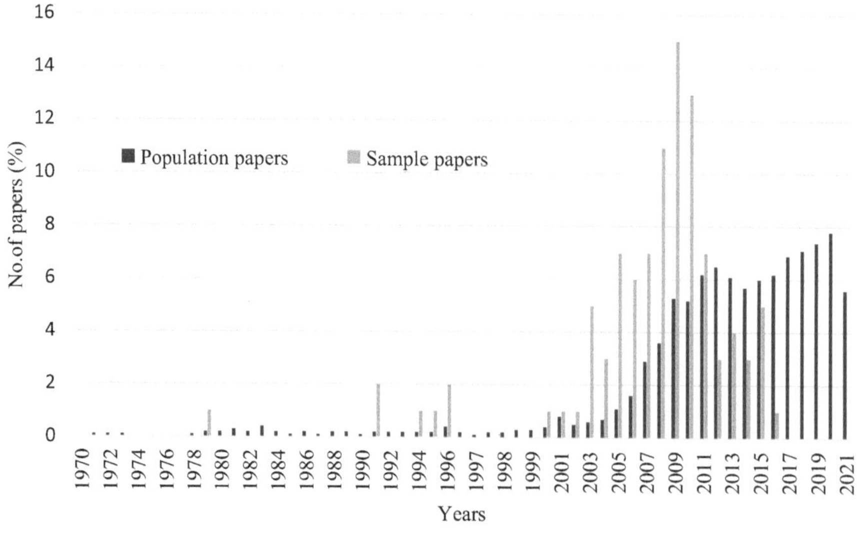

FIGURE 86.1 The research output regarding the evaluative studies in bioethanol fuels.

TABLE 86.3
The Most-Prolific Institutions in the Field of the Evaluative Studies in Bioethanol Fuels

No.	Institutions	Country	Sample Papers (%)	Population Papers (%)	Surplus (%)
1	Tsinghua Univ.	China	6	0.8	5.2
2	NREL	USA	6	0.7	5.3
3	State Univ. Campinas	Brazil	5	2.5	2.5
4	Shanghai Jiao Tong Univ.	China	4	0.9	3.1
5	Natl. Tech. Univ. Athens	Greece	4	0.6	3.4
6	Chinese Acad. Sci.	China	3	1.0	2.0
7	Purdue Univ.	USA	3	0.8	2.2
8	Univ. Padova	Italy	3	0.7	2.3
9	Univ. Minnesota	USA	3	0.7	2.3
10	Univ. Utrecht	Netherlands	3	0.4	2.6
11	Gazi Univ.	Turkey	3	0.3	2.7
12	Univ. Sao Paulo	Brazil	2	3.2	−1.2
13	Cornell Univ.	USA	2	0.7	1.3
14	Univ. Nebraska Lincoln	USA	2	0.7	1.3
15	Kansas State Univ.	USA	2	0.6	1.4
16	USDA Agr. Res. Serv.	USA	2	0.6	1.4
17	Univ. Calif. Riverside	USA	2	0.6	1.4
18	Michigan State Univ.	USA	2	0.6	1.4
19	Univ. Castilla la Mancha	Spain	2	0.5	1.5
20	Natl. Cheng Kung Univ.	Taiwan	2	0.5	1.5
21	Univ. Calif. Berkeley	USA	2	0.5	1.5
22	Univ. Malaya	Malaysia	2	0.5	1.5
23	Lund Univ.	Sweden	2	0.5	1.5
24	Hong Kong Polytech. Univ.	China	2	0.4	1.6
25	Tarbiat Modares Univ.	Iran	2	0.4	1.6

NREL, National Renewable Energy Laboratory.

The most-prolific institutions are Tsinghua University and the NREL, with six sample papers each. Next, the State University of Campinas, Shanghai Jiao Tong University, and the National Technical University of Athens publish five, four, and four papers, respectively. Furthermore 6 and 14 institutions publish three and two papers, respectively.

The top country for these most-prolific institutions is the USA with ten institutions. Next, China and Brazil cover four and two institutions, respectively. In total, ten countries house these top institutions.

On the other hand, the institutions with the most citation impact are the NREL and Tsinghua University, with 5.3% and 5.2% surplus, respectively. The National Technical university of Athens and Shanghai Jiao Tong University follow these top institutions, with 3.4% and 3.1% surplus, respectively.

Similarly, the institution with the least impact is the University of Sao Paulo, with 1.2% deficit. Cornell University and the University of Nebraska Lincoln follow this top institution, with 1.3% surplus each.

86.3.5 The Most-Prolific Funding Bodies in the Field of the Evaluative Studies in Bioethanol Fuels

Information about the most-prolific 12 funding bodies funding at least two sample papers each is given in Table 86.4.

TABLE 86.4
The Most-Prolific Funding Bodies in the Field of the Evaluative Studies in Bioethanol Fuels

No.	Funding Bodies	Country	Sample Papers No.	Population Paper No.
1	US Department of Energy	USA	3	76
2	Ministry of Economic Affairs	Netherlands	3	4
3	National Science Foundation	USA	2	62
4	European Commission	Europe	2	54
5	U.S. Environmental Protection Agency	USA	2	15
6	Hong Kong Polytechnic University	China	2	13
7	Chinese Academy of Sciences	China	2	11
8	National Nuclear Security Administration	USA	2	10
9	National Science Council	Taiwan	2	10
10	Los Alamos National Laboratory	USA	2	9
11	University of Malaya	Malaysia	2	7
12	NREL	USA	2	6

NREL, National Renewable Energy Laboratory.

The most-prolific funding bodies are the US Department of Energy and the Ministry of Economic Affairs of the Netherlands funding three sample papers each.

It is notable that only 34% of both sample and population papers are funded.

The most-prolific country for these top funding bodies is the USA, with six funding bodies. Next, China houses two funding bodies. In total, five countries and Europe houses these top funding bodies.

86.3.6 The Most-Prolific Source Titles in the Field of the Evaluative Studies in Bioethanol Fuels

Information about the most-prolific source titles publishing at least two papers each in the field of the evaluative studies in bioethanol fuels is given in Table 86.5.

The most-prolific source titles are Fuel and Biomass and Bioenergy, publishing 12 papers each. Next, Bioresource Technology, Renewable and Sustainable Energy Reviews, and Atmospheric Environment publish seven papers each. Then, Renewable Energy, Energy, Biofuels Bioproducts and Biorefining, and Energy Conversion and Management publish five, four, four, and four papers, respectively.

On the other hand, the source titles with the most citation impact are Biofuels Bioproducts and Biorefining, Renewable and Sustainable Energy Reviews, and Proceedings of the National Academy of Sciences of the United States of America, with 1.7%, 1.5%, and 1.5% publication surplus, respectively. Next, Atmospheric Environment, Applied Thermal Engineering, and International Journal of Life Cycle Assessment follow these top institutions with 1.3% publication surplus each.

It is notable that ten of these top source titles are indexed within the subject category of Energy.

86.3.7 The Most-Prolific Countries in the Field of the Evaluative Studies in Bioethanol Fuels

Information about the most-prolific countries publishing at least two papers each in the field of the evaluative studies in bioethanol fuels is given in Table 86.6.

The most-prolific two countries are the USA and China, publishing 37% and 12% of the sample papers, respectively. Brazil, the Netherlands, Greece, India, and Turkey follow these top countries,

TABLE 86.5
The Most-Prolific Source Titles in the Field of the Evaluative Studies in Bioethanol Fuels

No.	Source Titles	Sample Papers (%)	Population Papers (%)	Surplus (%)
1	Fuel	12	3.4	−1.4
2	Biomass and Bioenergy	12	2.4	−0.4
3	Bioresource Technology	7	2.7	−0.7
4	Renewable and Sustainable Energy Reviews	7	1.5	1.5
5	Atmospheric Environment	7	0.7	1.3
6	Renewable Energy	5	1.7	0.3
7	Energy	4	2.4	−0.4
8	Biofuels Bioproducts and Biorefining	4	1.3	1.7
9	Energy Conversion and Management	4	1.1	0.9
10	Applied Energy	3	2.4	−0.4
11	Science	3	2.4	−0.4
12	International Journal of Hydrogen Energy	3	1.0	1.0
13	Applied Thermal Engineering	3	0.7	1.3
14	Bioscience	2	3.4	−1.4
15	Journal of Industrial Ecology	2	2.7	−0.7
16	Proceedings of the National Academy of Sciences of the United States of America	2	1.5	1.5
17	International Journal of Life Cycle Assessment	2	0.7	1.3

TABLE 86.6
The Most-Prolific Countries in the Field of the Evaluative Studies in Bioethanol Fuels

No.	Countries	Sample Papers (%)	Population Papers (%)	Surplus (%)
1	USA	37	27.3	9.7
2	China	12	10.4	1.6
3	Brazil	8	12.1	−4.1
4	Netherlands	6	2.0	4.0
5	Greece	6	1.1	4.9
6	India	5	8.4	3.4
7	Turkey	5	2.1	2.9
8	United Kingdom	3	4.3	−1.3
9	Italy	3	3.1	−0.1
10	Spain	3	2.4	0.6
11	Malaysia	3	2.0	1
12	Australia	3	1.6	1.4
13	Iran	2	2.2	−0.2
14	Sweden	2	2.2	−0.2
15	Taiwan	2	1.1	0.9
16	Switzerland	2	1.0	1

with six, six, five, and five papers, respectively. Next, the United Kingdom, Italy, Spain, Malaysia, and Australia publish three papers each. Further, eight European countries listed in Table 86.6 produce 30% and 18% of the sample and population papers, respectively.

On the other hand, the countries with the most citation impact are the USA, Greece, and the Netherlands, with 9.7%, 4.9%, and 4.0% publication surplus, respectively. Similarly, the countries with the least citation impact are Brazil, India, and the United Kingdom, with 4.1, 3.4, and 1.3 publication deficit, respectively.

86.3.8 The Most-Prolific Scopus Subject Categories in the Field of the Evaluative Studies in Bioethanol Fuels

Information about the most-prolific 12 Scopus subject categories indexing at least 2% of the sample papers each is given in Table 86.7.

The most-prolific Scopus subject categories in the field of the evaluative studies in bioethanol fuels are 'Energy' and 'Environmental Science' with 66% and 46% of the sample papers, respectively. 'Chemical Engineering', 'Engineering', 'Chemistry', and 'Agricultural and Biological Sciences' further index 27%, 17%, 16%, and 16% of the sample papers, respectively.

On the other hand, the Scopus subject categories with the most citation impact are 'Energy' and 'Chemical Engineering', with 22.3% and 7.3% publication surplus, respectively. Similarly, the Scopus subject categories with the least citation impact are 'Engineering' 'Business, Management and Accounting', and 'Social Sciences', with 13.9%, 6.8%, and 6.6% publication deficit, respectively.

86.3.9 The Most-Prolific Keywords in the Field of the Evaluative Studies in Bioethanol Fuels

Information about the keywords used in the population papers is given in Table 86.8. For this purpose, keywords related to the keyword set given in the appendix are selected from a list of the most-prolific keyword set provided by the Scopus database.

There are two keywords used related to the bioethanol fuels: Ethanol (74.4%) and bioethanol (20.6%). There are also other keywords related to ethanol: Ethanol production (14.2%) and ethanol fuels (11.2%).

The most-prolific keywords related to the environmental issues in the field of the evaluative studies in bioethanol fuels are greenhouse gases, carbon dioxide, gas emissions, emissions, life cycle assessment, carbon monoxide, and nitrogen oxides, with over 7.2% of the population papers each.

TABLE 86.7
The Most-Prolific Scopus Subject Categories in the Field of the Evaluative Studies in Bioethanol Fuels

No.	Scopus Subject Categories	Sample Papers (%)	Population Papers (%)	Surplus (%)
1	Energy	66	43.7	22.3
2	Environmental Science	46	42.7	3.3
3	Chemical Engineering	27	19.7	7.3
4	Engineering	17	30.9	−13.9
5	Chemistry	16	13.3	2.7
6	Agricultural and Biological Sciences	16	12.2	3.8
7	Earth and Planetary Sciences	7	3.0	4.0
8	Multidisciplinary	5	1.3	3.7
9	Social Sciences	3	9.6	−6.6
10	Physics and Astronomy	3	4.5	−1.5
11	Materials Science	3	3.9	−0.9
12	Business Management and Accounting	2	8.8	−6.8

TABLE 86.8
The Most-Prolific Keywords in the Field of the Evaluative Studies in Bioethanol Fuels

No.	Keywords	Population Papers (%)
1.	**Bioethanol Fuels**	
	Ethanol	74.4
	Bioethanol	20.6
	Ethanol production	14.2
	Ethanol fuels	11.3
2.	**Environmental Issues**	
	Greenhouse gases	12.4
	Carbon dioxide	8.6
	Gas emissions	7.8
	Emissions	7.6
	Life cycle assessment	7.5
	Carbon monoxide	7.3
	Nitrogen oxides	7.2
	Life cycle	6.8
	Exhaust emissions	6.0
	Environmental impact	5.9
	Particulate emissions	5.4
	Emission control	4.4
	Global warming	4.0
	Life cycle analysis	3.8
	Sustainability	3.7
	Exhaust gas	3.6
	Land use	3.1
	Air pollution	2.9
	Particulate matter	2.4
	Climate change	1.9
	Carbon emission	1.8
	Pollution	1.7
	Water	1.6
3.	**Other Issues**	
	Costs	7.9
	Economic analysis	5.7
	Economics	4.5
	Energy policy	4.0
	Energy efficiency	3.9
	Fuel consumption	3.7
	Sustainability	3.7
	Investments	3.0
	Cost benefit analysis	2.8
	Brake thermal efficiency	2.8
	Supply chains	2.5
	Performance assessment	2.2

(Continued)

TABLE 86.8 (*Continued*)
The Most-Prolific Keywords in the Field of the Evaluative Studies in Bioethanol Fuels

No.	Keywords	Population Papers (%)
	Energy balance	1.9
	Exergy	1.9
	Brake specific fuel consumption	1.8
	Thermodynamics	1.8
	Cost-benefit analysis	1.8
	Techno-economic analysis	1.7
	Energy market	1.6
	Production cost	1.5

On the other hand, the most-prolific keywords related to the other issues such as economic, technoeconomic, and technoassessment in the field of the evaluative studies in bioethanol fuels are costs, economic analysis, sustainable development, economics, and energy policy, with over 4.0% of the population papers each.

86.3.10 The Most-Prolific Research Fronts in the Field of the Evaluative Studies in Bioethanol Fuels

Information about the most-prolific research fronts for the sample papers in the field of the evaluative studies in bioethanol fuels is given in Table 86.9.

There are three primary research fronts in this field: environmental impact (74%), technoeconomics (Hamelinck et al., 2005; Hill et al., 2006), and technoassessment (Hsieh et al., 2002; Zhu and Pan, 2010). There are also three other research fronts: policy (Lynd, 1996; Martinelli and Filoso, 2008; Solomon et al., 2007), economics (Serra et al., 2011; Solomon et al., 2007), and business (Chen and Fan, 2012; Dal-Mas et al., 2011).

Furthermore, there are five secondary research fronts for the research front of environmental issues: emissions (Lynd, 1996; Macedo et al., 2008), life cycle assessment (Lynd, 1996; Sheehan et al., 2003), land use (Leite et al., 2009; Wang et al., 2011), water consumption (Chiu et al., 2009; Petersen et al., 2009), and other issues (Farrell et al., 2006; Goldemberg et al., 2008; Hill et al., 2009).

86.4 DISCUSSION

86.4.1 Introduction

Crude oil-based gasoline and petrodiesel fuels have been widely used in the transportation sector. However, there have been great public concerns over the adverse environmental impact of these fuels. Hence, biomass-based bioethanol fuels have been increasingly used in blending gasoline fuels and petrodiesel fuels in recent years.

Hence, biomass-based bioethanol fuels (Hill et al., 2006, 2009; Konur, 2012e, 2015, 2109, 2020a) have been increasingly used in blending gasoline fuels (Hsieh et al., 2002; Najafi et al., 2009) and petrodiesel fuels in recent years (Hansen et al., 2005; Li et al., 2005). In this context, there has been a significant focus on the evaluative studies in bioethanol fuels (Farrell et al., 2006; Hamelinck et al., 2005; Hill et al., 2006).

TABLE 86.9
The Most-Prolific Research Fronts in the Field of the Evaluative Studies in Bioethanol Fuels

No.	Research Fronts	Sample Papers (%)
1.	Environmental issues	74
	Emissions	59
	Life cycle assessment	9
	Other issues	8
	Land use	2
	Water consumption	1
2.	Technoeconomics	33
3.	Technoassessments	42
4.	Policy	3
5.	Economics	1
6.	Business	3

There have been three primary research fronts in this field: environmental issues, technoeconomics (Hamelinck et al., 2005; Hill et al., 2006), and technoassessment (Hsieh et al., 2002; Zhu and Pan, 2010). There have also been three other research fronts: policy (Lynd, 1996; Martinelli and Filoso, 2008), economics (Serra et al., 2011; Solomon et al., 2007), and business (Chen and Fan, 2012; Dal-Mas et al., 2011).

However, it is essential to develop efficient incentive structures for the primary stakeholders to enhance the research in this field. The scientometric analysis has been traditionally used to inform primary stakeholders about the current state of the research in a selected research field. Although there have been a number of scientometric studies on bioethanol fuels, there has been no scientometric study of the research in the field of the evaluative studies in bioethanol fuels.

This book chapter presents a scientometric study of the research in the field of the evaluative studies in bioethanol fuels. It examines the scientometric characteristics of the sample and population data presenting scientometric characteristics of these both datasets in the order of documents, authors, publication years, institutions, funding bodies, source titles, countries, Scopus subject categories, keywords, and research fronts.

The search for this study was carried out using the Scopus database in August 2021. As a first step for the search of the relevant literature, the keywords were selected using the first most-cited 500 papers. The selected keyword list was optimized to obtain a representative sample of papers for the searched research field. This keyword list is provided in the appendix for future replication studies. Additionally, the information about the most-used keywords is given in Section 86.3.9 to highlight the key research fronts in Section 86.3.10.

As a second step, two sets of data were used for this study. First, a population sample of over 3,700 papers was used to examine the scientometric characteristics of the population data. Second, a sample of 100 most-cited papers was used to examine the scientometric characteristics of these citation classics with over 136 citations each.

86.4.2 The Most-Prolific Documents in the Field of the Evaluative Studies in Bioethanol Fuels

Review papers are overrepresented in the sample dataset while conference papers published in journals, articles, and book chapters are overrepresented in the population dataset, as shown in Table 86.1. It is notable that only 85% and 80% of the sample and population papers are articles, respectively.

It is observed during the search process that there has been inconsistency in the classification of the documents in Scopus and in other databases such as Web of Science. This is especially relevant for the classification of papers as reviews or articles as the papers not involving a literature review may be erroneously classified as a review paper. In this context, it would be helpful to provide a classification note for published papers in books and journals. It would also be helpful to use the document types listed in table 86.1 for this purpose.

86.4.3 The Most-Prolific Authors in the Field of the Evaluative Studies in Bioethanol Fuels

There have been most-prolific 22 authors with at least two papers each in the sample dataset and five papers each in the population dataset, as given in Table 86.2. These authors have shaped the development of the research in this field.

The top most-prolific authors are Andy Aden, Dimitrios C. Rakopoulos, Constantine D. Rakopoulos, Andre Faaij, Evangelos G. Giakoumis, Charles E. Wyman, Fabrizio Bezzo, He Hong, Shijing Shuai, and Jianxin Wang (Table 86.2).

It is important to note the inconsistencies in indexing of the author names in Scopus and other databases. It is especially an issue for the names with more than two components such as 'Judge Alex de Camp Sirous'. The probable outcomes are 'Sirous J.A.D.C.', 'de Camp Sirous, J.A.' OR 'Camp Sirous, J.A.D.'. It is notable that inconsistent indexing of the author names may cause substantial inefficiencies in the search process for papers.

There is also a difficulty in allowing credit for authors, especially for authors with common names such as 'Wang, J', or 'Huang, Z' or 'Zhu, L' in conducting scientometric studies. These difficulties strongly influence the efficiency of scientometric studies.

In this context, the coding of authors in the Scopus database is a welcome innovation compared to the other databases such as Web of Science. In this process, Scopus allocates a unique number to each author in the database. However, there might be still substantial inefficiencies in this coding system, especially for common names. It is possible that Scopus uses software programs to differentiate author names (Shin et al., 2014).

In this context, it does not help that author names are not given full in some journals. This makes it difficult to differentiate authors with common names and makes the scientometric studies more difficult in the author domain. Therefore, author names should be given in all books and journals.

There also inconsistencies in naming of authors with more than two components in journal papers and book chapters. For example, 'Faaij, A.P.C', 'Wyman C.E.', and 'Shuai, S.J.' might be given as 'Faaij, A.', 'Wyman C.', or 'Shuai S.' in journals and books. This also makes scientometric studies difficult in the author domain. Hence, contributing authors should use their name consistently in their publications.

The most-prolific institutions for the sample dataset are the National Technical University of Athens, the NREL, Shanghai Jiao Tong University, Tarbiat Modares University, and Tsinghua University. Similarly, the most-prolific countries for the sample dataset are China, the USA, Greece, Italy, and Iran. In total, nine countries contribute to the sample dataset.

The most-prolific research fronts for the sample dataset are emissions of bioethanol fuels and technoeconomics of bioethanol fuels.

It is also notable that there is significant gender deficit for the sample dataset as surprisingly all these top researchers are male. This finding is the most thought-provoking with strong public policy implications. Hence, institutions, funding bodies, and policy makers should take measures to reduce gender deficit in this field as well as in other scientific fields with strong gender deficit.

86.4.4 The Most-Prolific Research Output by Years in the Field of the Evaluative Studies in Bioethanol Fuels

The research output observed between 1970 and 2021 is illustrated in Figure 86.1. This figure clearly shows that the bulk of the research papers in the population dataset were published in the 2000s, 2010s, and 2020s. Similarly, the bulk of the research papers in the sample dataset were published in the 2000s and 2010s.

These data suggest that the most-cited sample papers were primarily published in the 2000s, and population papers were primarily published in the 2000s, 2010s, and the early 2020s. These are the thought-provoking findings as there has been no significant research in this field in the 1970s, 1980s, and 1990s, but there has been significant research boom in the last two decades. In this context, the increasing public concerns about climate change (Change, 2007), greenhouse gas emissions (Carlson et al., 2017), and global warming (Kerr, 2007) have been certainly behind the boom in the research in this field in the last two decades.

The data in Figure 86.1 also suggest that the research in this field has been saturated in the last decade, and the size of population papers is likely to double in the current decade, provided that the public concerns about climate change, greenhouse gas emissions, and global warming continue.

86.4.5 The Most-Prolific Institutions in the Field of the Evaluative Studies in Bioethanol Fuels

The most-prolific 25 institutions publishing papers on evaluative studies in bioethanol fuels with at least two sample papers each and at least 0.4% of the population papers each given in Table 86.3 have shaped the development of the research in this field.

The most-prolific institutions are Tsinghua University, the NREL, the State University of Campinas, Shanghai Jiao Tong University, and the National Technical University of Athens. The top country for these most-prolific institutions are the USA, China, and Brazil.

On the other hand, the institutions with the most impact are the NREL, Tsinghua University, the National Technical University of Athens, and Shanghai Jiao Tong University. Similarly, the institutions with the least impact are the University of Sao Paulo, Cornell University, and the University of Nebraska Lincoln.

86.4.6 The Most-Prolific Funding Bodies in the Field of the Evaluative Studies in Bioethanol Fuels

The most-prolific funding bodies are the US Department of Energy and the Ministry of Economic Affairs of the Netherlands (Table 86.4). It is notable that only 34% of both sample and population papers are funded. The most-prolific country for these top funding bodies are the USA, with six funding bodies, and China, with two funding bodies.

These findings on the funding of the research in this field suggest that the level of the funding is low, and there is ample scope to increase funding in this field to enhance the research further (Ebadi and Schiffauerova, 2016) above 34% in the light of North's institutional framework (North, 1991).

86.4.7 The Most-Prolific Source Titles in the Field of the Evaluative Studies in Bioethanol Fuels

Seventeen source titles publishing at least two papers each have shaped the development of the research in this field (Table 86.5). The most-prolific source titles are Fuel, Biomass and Bioenergy, Bioresource Technology, Renewable and Sustainable Energy Reviews, Atmospheric Environment, Renewable Energy, Energy, Biofuels Bioproducts and Biorefining, and Energy Conversion and Management.

On the other hand, the source titles with the most citation impact are Biofuels Bioproducts and Biorefining, Renewable and Sustainable Energy Reviews, Proceedings of the National Academy of Sciences of the United States of America, Atmospheric Environment, Applied Thermal Engineering, and International Journal of Life Cycle Assessment. It is notable that 10 of these top source titles are indexed within the subject category of Energy.

86.4.8 The Most-Prolific Countries in the Field of the Evaluative Studies in Bioethanol Fuels

Sixteen countries publishing at least two papers each have significantly shaped the development of the research in this field (Table 86.6). The most-prolific countries are the USA, China, Brazil, the Netherlands, Greece, India, Turkey, the United Kingdom, Italy, Spain, Malaysia, and Australia.

On the other hand, the countries with the most citation impact are the USA, Greece, and the Netherlands. Similarly, the countries with the least citation impact are Brazil, India, and the United Kingdom.

The close examination of these findings suggest that the USA, China, Europe, and Brazil are the major producers of the research in this field. It is a fact that the USA has been a major player in science (Leydesdorff and Wagner, 2009; Leydesdorff et al., 2014). The USA has further developed a strong research infrastructure to support its corn- and grass-based bioethanol industry (Vadas et al., 2008).

However, China has been a rising star in scientific research in competition with the USA and Europe (Leydesdorff and Zhou, 2005). China is also a major player in this field as a major producer of bioethanol (Li and Chan-Halbrendt, 2009).

Next, Europe has been a persistent player in the scientific research in competition with both the USA and China (Leydesdorff, 2000). Europe has also been a persistent producer of bioethanol along with the USA and Brazil (Gnansounou, 2010).

Brazil has also been a persistent player in scientific research at a moderate level (Glanzel et al., 2006). Brazil has also developed a strong research infrastructure to support its sugarcane-based bioethanol industry (Soccol et al., 2010).

86.4.9 The Most-Prolific Scopus Subject Categories in the Field of the Evaluative Studies in Bioethanol Fuels

The most-prolific 12 Scopus subject categories indexing at least 2% of the sample papers each given in Table 86.7 have shaped the development of the research in this field.

The most-prolific Scopus subject categories in the field of the evaluative studies in bioethanol fuels are 'Energy', 'Environmental Science', 'Chemical Engineering', 'Engineering', 'Chemistry', and 'Agricultural and Biological Sciences'.

On the other hand, the Scopus subject categories with the most citation impact are 'Energy' and 'Chemical Engineering'. Similarly, the Scopus subject categories with the least citation impact are 'Engineering' 'Business, Management and Accounting', and 'Social Sciences'.

These findings are thought-provoking, suggesting that the primary subject categories are energy and environmental sciences. The other key finding is that although social sciences including business management are fairly represented in the population papers, they are underrepresented significantly in the sample papers.

86.4.10 The Most-Prolific Keywords in the Field of the Evaluative Studies in Bioethanol Fuels

A limited number of keywords have shaped the development of the research in this field, as shown in Table 86.8 and the appendix.

There are two keywords used related to the bioethanol fuels: ethanol (74.4%) and bioethanol (20.6%). There are also keywords related to ethanol: ethanol production and ethanol fuels. It seems that ethanol is more popular than bioethanol, with strong implications for the search strategy. In other words, the search strategy using only bioethanol keyword would give only around 20% of the population papers. Hence, the search strategy should cover both keywords as it is done in this study.

The most-prolific keywords related to the environmental issues in the field of the evaluative studies in bioethanol fuels are greenhouse gases, carbon dioxide, gas emissions, emissions, life cycle assessment, carbon monoxide, and nitrogen oxides. These findings point out the key research fronts for the environmental impact of bioethanol fuels: greenhouse gases, emissions, and life cycle assessment.

On the other hand, the most-prolific keywords related to the other issues such as economic, technoeconomic, and technoassessment in the field of the evaluative studies in bioethanol fuels are costs, economic analysis, sustainable development, economics, and energy policy. These keyword point out the key research fronts for technoeconomics, technoassessment, economics, business, and policy.

86.4.11 The Most-Prolific Research Fronts in the Field of the Evaluative Studies in Bioethanol Fuels

There are three primary research fronts in this field: environmental issues, technoeconomics, and technoassessment. The other research fronts are policy, economics, and business. It is notable that the first three research fronts have strong public policy implications.

Furthermore, there are five secondary research fronts for the research front of environmental issues: emissions and, to a lesser extent, life cycle assessment, land use, water consumption, and other issues. It is notable that the field of emissions has a strong public policy implications. It appears that land use and water consumption also play a key role in assessing the environmental impact of bioethanol fuels.

There are strong structure–processing–property relationships for the all of the bioethanol fuels. In the end, these most-cited papers in this field hint that the efficiency of the production and utilization of bioethanol fuels could be maximized using the structure, processing, and property relationships of bioethanol fuels (Formela et al., 2016; Konur, 2018a, 2020b, 2021a–d; Konur and Matthews, 1989).

86.5 CONCLUSION AND FUTURE RESEARCH

Research on the evaluative studies in bioethanol fuels has been mapped through a scientometric study of both sample and population datasets.

The critical issue in this study has been to obtain a representative sample of the research as in any other scientometric study. Therefore, the keyword set has been carefully devised and optimized after a number of runs in the Scopus database.

The other issue has been the selection of a multidisciplinary database to carry out the scientometric study of the research in this field. For this purpose, the Scopus database has been selected. The journal coverage of this database has been wider than that of Web of Science.

The key scientometric properties of the research in this field have been determined and are discussed in this book chapter. It is evident that a limited number of documents, authors, institutions, publication periods, institutions, funding bodies, source titles, countries, Scopus subject categories, keywords, and research fronts have shaped the development of the research in this field.

There is ample scope to increase the efficiency of the scientometric studies in this field in the author and document domains by developing consistent policies and practices in both domains. In this respect, authors, journals, and academic databases have a lot to do.

The significant gender deficit as in most scientific fields emerges as a public policy issue. The research in this field has boomed in the 2000s, 2010s, and early 2020s possibly promoted by the public concerns on global warming, greenhouse gas emissions, and climate change.

Institutions from the USA, China, and Brazil have mostly shaped the research in this field. The funding rate of 34% for both sample and population papers suggests that there is ample scope to increase funding in this field to enhance the research in this field, possibly more than doubling in the current decade.

The most-prolific journals have been mostly indexed by the subject category of energy. The USA, China, Europe, and Brazil have been the major producers of the research in this field as the major producers of bioethanol fuels from different types of biomass such as corn, sugarcane, and grass as well as other types of biomass.

The primary subject categories have been energy and environmental sciences. Although social sciences including business management have been fairly represented in the population papers, but they have been underrepresented significantly in the sample papers.

Ethanol is more popular than bioethanol as a keyword with strong implications for the search strategy. In other words, the search strategy using only bioethanol as a keyword would give only around 20% of the population papers. The primary keywords for the environmental impact of bioethanol fuels emerge as greenhouse gases, emissions, and life cycle assessment.

There are three primary research fronts in this field: environmental issues, technoeconomics, and technoassessment related to bioethanol fuels. The other fronts are policy, economics, and business. Furthermore, there are five secondary research fronts for the research front of environmental issues: emissions and, to a lesser extent life cycle, assessment, land use water consumption, and other issues. It is notable that the field of emissions has a strong public policy implications. It appears that land use and water consumption also play a key role in assessing the environmental impact of bioethanol fuels.

Thus, the scientometric analysis has a great potential to gain valuable insights into the evolution of the research in this field as in other scientific fields.

It is recommended that further scientometric studies should be carried out on the other aspects of both production and utilization of bioethanol fuels. It is further recommended that reviews of the most-cited papers should be carried out for each research front to complement these scientometric studies. Next, the scientometric studies of the hot papers in these primary fields should be carried out.

ACKNOWLEDGMENTS

The contribution of the highly cited researchers in the field of the evaluative studies in bioethanol fuels has been gratefully acknowledged.

APPENDIX: THE KEYWORD SET FOR THE FIELD OF THE EVALUATIVE STUDIES IN BIOETHANOL FUELS

((((TITLE (ethanol OR bioethanol OR {ethyl alcohol*} OR c2h5oh OR gasohol OR diesohol OR e85 OR e10 OR defc*) OR SRCTITLE (ethanol OR bioethanol OR gasohol)) AND (TITLE ({life cycle*} OR {life-cycle*} OR lifecycle* OR econ* OR exergy OR technoecon* OR {techno-econ*} OR {techno econ*} OR {thermodynamic analysis} OR sustainability OR {Input-output analysis} OR environment* OR emission* OR {greenhouse gas*} OR {global warming} OR sustainability OR land OR ecol* OR ghg OR {water consumption} OR footprint* OR policy OR policies OR social OR societal OR security OR socio* OR food OR logist* OR public* OR history OR perception* OR climat* OR lca OR {air quality} OR pollut* OR groundwater* OR energetic* OR cost* OR investment* OR supply OR price* OR demand OR market* OR {net energy} OR {energy consumption} OR exergetic* OR trading OR trade OR exergy OR tax* OR subsid* OR financ* OR {energy balance} OR climat* OR lca) OR SRCTITLE ({life cycle*} OR {life-cycle*} OR lifecycle* OR technoecon* OR {techno-econ*} OR emission* OR {greenhouse gas*} OR climat* OR land OR warming OR policy OR social OR policies OR econ*) OR SUBJAREA (arts OR busi OR deci OR soci OR econ))) AND NOT SUBJAREA (medi OR neur OR bioc OR immu OR phar OR heal OR dent OR nurs OR vete OR psyc) AND NOT TITLE (extract* OR droplet* OR dibutylamino OR antioxidant* OR lignin OR *toxin* OR receptor* OR chronic))) AND (LIMIT-TO (SRCTYPE, "j") OR LIMIT-TO (SRCTYPE, "k") OR LIMIT-TO (SRCTYPE, "b")) AND (LIMIT-TO (DOCTYPE, "ar") OR LIMIT-TO (DOCTYPE, "cp") OR LIMIT-TO (DOCTYPE, "ch") OR LIMIT-TO (DOCTYPE, "re") OR LIMIT-TO (DOCTYPE, "no") OR LIMIT-TO (DOCTYPE, "le") OR LIMIT-TO (DOCTYPE, "ed") OR LIMIT-TO (DOCTYPE, "sh") OR LIMIT-TO (DOCTYPE, "bk") OR LIMIT-TO (DOCTYPE, "cr")) AND (LIMIT-TO (LANGUAGE, "English")).

REFERENCES

Beaudry, C. and V. Lariviere. 2016. Which gender gap? Factors affecting researchers' scientific impact in science and medicine. *Research Policy* 45:1790–1817.

Bosmann, A., L. Datsevich and A. Jess, et al. 2001. Deep desulfurization of diesel fuel by extraction with ionic liquids. *Chemical Communications* 23:2494–2495.

Burnham, J. F. 2006. Scopus database: A review. *Biomedical Digital Libraries* 3:1–8.

Carlson, K. M., J. S. Gerber and N. D. Mueller, et al. 2017. Greenhouse gas emissions intensity of global croplands. *Nature Climate Change* 7:63–68.

Change, C. 2007. Climate change impacts, adaptation and vulnerability. *Science of the Total Environment* 326:95–112.

Chen, C. W. and Y. Fan. 2012. Bioethanol supply chain system planning under supply and demand uncertainties. *Transportation Research Part E: Logistics and Transportation Review* 48:150–164.

Chiu, Y. W., B. Walseth and S. Suh. 2009. Water embodied in bioethanol in the United States. *Environmental Science and Technology* 43:2688–2692.

Dal-Mas, M., S. Giarola, A. Zamboni and F. Bezzo. 2011. Strategic design and investment capacity planning of the ethanol supply chain under price uncertainty. *Biomass and Bioenergy* 35:2059–2071.

Ebadi, A. and A. Schiffauerova. 2016. How to boost scientific production? A statistical analysis of research funding and other influencing factors. *Scientometrics* 106:1093–1116.

Farrell, A. E., R. J. Plevin and B. T. Turner, et al. 2006. Ethanol can contribute to energy and environmental goals. *Science* 311:506–508.

Formela, K., A. Hejna, L. Piszczyk, M. R. Saeb and X. Colom. 2016. Processing and structure-property relationships of natural rubber/wheat bran biocomposites. *Cellulose* 23:3157–3175.

Garfield, F. 1955. Citation indexes for science. *Science* 122:108–111.

Glanzel, W., J. Leta and B. Thijs. 2006. Science in Brazil. Part 1: A macro-level comparative study. *Scientometrics* 67:67–86.

Gnansounou, E. 2010. Production and use of lignocellulosic bioethanol in Europe: Current situation and perspectives. *Bioresource Technology* 101:4842–4850.

Goldemberg, J., S. T. Coelho and P. Guardabassi. 2008. The sustainability of ethanol production from sugarcane. *Energy Policy* 36:2086–2097.

Hamelinck, C. N., G. van Hooijdonk and A. P. C. Faaij. 2005. Ethanol from lignocellulosic biomass: Techno-economic performance in short-, middle- and long-term. *Biomass and Bioenergy* 28:384–410.

Hansen, A. C, Q. Zhang and P. W. L. Lyne. 2005. Ethanol-diesel fuel blends: A review. *Bioresource Technology* 96:277–285.

Hill, J., E. Nelson, D. Tilman, S. Polasky and D. Tiffany. 2006. Environmental, economic, and energetic costs and benefits of biodiesel and ethanol biofuels. *Proceedings of the National Academy of Sciences of the United States of America* 103:11206–11210.

Hill, J., S. Polasky and E. Nelson, et al. 2009. Climate change and health costs of air emissions from biofuels and gasoline. *Proceedings of the National Academy of Sciences of the United States of America* 106:2077–2082.

Hsieh, W. D., R. H. Chen, T. L. Wu and T. H. Lin. 2002. Engine performance and pollutant emission of an SI engine using ethanol-gasoline blended fuels. *Atmospheric Environment* 36:403–410.

Kerr, R. A. 2007. Global warming is changing the world. *Science* 316:188–190.
Konur, O. 2000. Creating enforceable civil rights for disabled students in higher education: An institutional theory perspective. *Disability & Society* 15:1041–1063.
Konur, O. 2002a. Access to nursing education by disabled students: Rights and duties of nursing programs. *Nurse Education Today* 22:364–374.
Konur, O. 2002b. Assessment of disabled students in higher education: Current public policy issues. *Assessment and Evaluation in Higher Education* 27:131–152.
Konur, O. 2002c. Access to employment by disabled people in the UK: Is the Disability Discrimination Act working? *International Journal of Discrimination and the Law* 5:247–279.
Konur, O. 2006a. Participation of children with dyslexia in compulsory education: Current public policy issues. *Dyslexia* 12:51–67.
Konur, O. 2006b. Teaching disabled students in higher education. *Teaching in Higher Education* 11:351–363.
Konur, O. 2007a. A judicial outcome analysis of the Disability Discrimination Act: A windfall for the employers? *Disability & Society* 22:187–204.
Konur, O. 2007b. Computer-assisted teaching and assessment of disabled students in higher education: The interface between academic standards and disability rights. *Journal of Computer Assisted Learning* 23:207–219.
Konur, O. 2011. The scientometric evaluation of the research on the algae and bio-energy. *Applied Energy* 88:3532–3540.
Konur, O. 2012a. Prof. Dr. Ayhan Demirbas' scientometric biography. *Energy Education Science and Technology Part A: Energy Science and Research* 28:727–738.
Konur, O. 2012b. The evaluation of the biogas research: A scientometric approach. *Energy Education Science and Technology Part A: Energy Science and Research* 29:1277–1292.
Konur, O. 2012c. The evaluation of the global energy and fuels research: A scientometric approach. *Energy Education Science and Technology Part A: Energy Science and Research* 30:613–628.
Konur, O. 2012d. The evaluation of the research on the biodiesel: A scientometric approach. *Energy Education Science and Technology Part A: Energy Science and Research* 28:1003–1014.
Konur, O. 2012e. The evaluation of the research on the bioethanol: A scientometric approach. *Energy Education Science and Technology Part A: Energy Science and Research* 28:1051–1064.
Konur, O. 2012f. The evaluation of the research on the biofuels: A scientometric approach. *Energy Education Science and Technology Part A: Energy Science and Research* 28:903–916.
Konur, O. 2012g. The evaluation of the research on the biohydrogen: A scientometric approach. *Energy Education Science and Technology Part A: Energy Science and Research* 29:323–338.
Konur, O. 2012h. The evaluation of the research on the microbial fuel cells: A scientometric approach. *Energy Education Science and Technology Part A: Energy Science and Research* 29:309–322.
Konur, O. 2012i. The scientometric evaluation of the research on the production of bioenergy from biomass. *Biomass and Bioenergy* 47:504–515.
Konur, O. 2015. Current state of research on algal bioethanol. In *Marine Bioenergy: Trends and Developments*, Eds. S. K. Kim and C. G. Lee, pp. 217–244. Boca Raton, FL: CRC Press.
Konur, O., Ed. 2018a. *Bioenergy and Biofuels.* Boca Raton, FL: CRC Press.
Konur, O. 2018b. Bioenergy and biofuels science and technology: Scientometric overview and citation classics. In *Bioenergy and Biofuels*, Ed. O. Konur, pp. 3–63. Boca Raton, FL: CRC Press.
Konur, O. 2019. Cyanobacterial bioenergy and biofuels science and technology: A scientometric overview. In *Cyanobacteria: From Basic Science to Applications*, Eds. A. K. Mishra, D. N. Tiwari and A. N. Rai, pp. 419–442. Amsterdam: Elsevier.
Konur, O. 2020a. The scientometric analysis of the research on the bioethanol production from green macroalgae. In *Handbook of Algal Science*, Technology and Medicine, Ed. O. Konur, pp. 385–401. London: Academic Press.
Konur, O., Ed. 2020b. *Handbook of Algal Science, Technology and Medicine.* London: Academic Press.
Konur, O., Ed. 2021a. *Handbook of Biodiesel and Petrodiesel Fuels: Science, Technology, Health, and Environment.* Boca Raton, FL: CRC Press.
Konur, O., Ed. 2021b. *Handbook of Biodiesel and Petrodiesel Fuels: Science, Technology, Health, and Environment. Volume 1. Biodiesel Fuels: Science, Technology, Health, and Environment.* Boca Raton, FL: CRC Press.
Konur, O., Ed. 2021c. *Handbook of Biodiesel and Petrodiesel Fuels: Science, Technology, Health, and Environment. Volume 2. Biodiesel Fuels Based on the Edible and Nonedible Feedstocks, Wastes, and Algae: Science, Technology, Health, and Environment.* Boca Raton, FL: CRC Press.
Konur, O., Ed. 2021d. *Handbook of Biodiesel and Petrodiesel Fuels: Science, Technology, Health, and Environment. Volume 3. Petrodiesel Fuels: Science, Technology, Health, and Environment.* Boca Raton, FL: CRC Press.

Konur, O. and F. L. Matthews. 1989. Effect of the properties of the constituents on the fatigue performance of composites: A review. *Composites* 20:317–328.

Leite, R. C. D. C., M. R. L. V. Leal, L. A. B. Cortez, M. W. Griffin and M. I. G. Scandiffio. 2009. Can Brazil replace 5% of the 2025 gasoline world demand with ethanol? *Energy* 34:655–661.

Leydesdorff, L. 2000. Is the European Union becoming a single publication system? *Scientometrics* 47:265–280.

Leydesdorff, L. and C. Wagner. 2009. Is the United States losing ground in science? A global perspective on the world science system. *Scientometrics* 78:23–36.

Leydesdorff, L. and P. Zhou. 2005. Are the contributions of China and Korea upsetting the world system of science? *Scientometrics* 63:617–630.

Leydesdorff, L., C. S. Wagner and L. Bornmann. 2014. The European Union, China, and the United States in the top-1% and top-10% layers of most-frequently cited publications: Competition and collaborations. *Journal of Informetrics* 8:606–617.

Li, D., Z. Huang, X. C. Lu, W. Zhang and J. Yang. 2005. Physico-chemical properties of ethanol-diesel blend fuel and its effect on performance and emissions of diesel engines. *Renewable Energy* 30:967–976.

Li, S. Z. and C. Chan-Halbrendt. 2009. Ethanol production in (the) People's Republic of China: Potential and technologies. *Applied Energy* 86:S162–S169.

Lynd, L. R. 1996. Overview and evaluation of fuel ethanol from cellulosic biomass: Technology, economics, the environment, and policy. *Annual Review of Energy and the Environment* 21:403–465.

Ma, X., L. Sun and C. Song. 2002. A new approach to deep desulfurization of gasoline, diesel fuel and jet fuel by selective adsorption for ultra-clean fuels and for fuel cell applications. *Catalysis Today* 77:107–116.

Ma, X. L., K. Y. Sakanish and I. Mochida. 1994. Hydrodesulfurization reactivities of various sulfur-compounds in diesel fuel. *Industrial & Engineering Chemistry Research* 33:218–222.

Macedo, I. C., J. E. A. Seabra and J. E. A. R. Silva 2008. Green house gases emissions in the production and use of ethanol from sugarcane in Brazil: The 2005/2006 averages and a prediction for 2020. *Biomass and Bioenergy* 32:582–595.

Martinelli, L. A. and S. Filoso. 2008. Expansion of sugarcane ethanol production in Brazil: Environmental and social challenges. *Ecological Applications* 18:885–898.

Najafi, G., B. Ghobadian and T. Tavakoli, et al. 2009. Performance and exhaust emissions of a gasoline engine with ethanol blended gasoline fuels using artificial neural network. *Applied Energy* 86:630–639.

Newman, P. W. G. and J. R. Kenworthy. 1989. Gasoline consumption and cities: A comparison of U.S. cities with a global survey. *Journal of the American Planning Association* 55:24–37.

North, D. C. 1991. Institutions. *Journal of Economic Perspectives* 5:97–112.

Petersen, M. O., J. Larsen and M. H. Thomsen. 2009. Optimization of hydrothermal pretreatment of wheat straw for production of bioethanol at low water consumption without addition of chemicals. *Biomass and Bioenergy* 33:834–840.

Schauer, J. J., M. J. Kleeman, G. R. Cass and B. R. T. Simoneit. 2002. Measurement of emissions from air pollution sources. 5. C_1-C_{32} organic compounds from gasoline-powered motor vehicles. *Environmental Science and Technology* 36:1169–1180.

Schauer, J. J., M. J. Kleeman, G. R. Cass and B. R. T. Simoneit. 1999. Measurement of emissions from air pollution sources. 2. C_1 through C_{30} organic compounds from medium duty diesel trucks. *Environmental Science & Technology* 33:1578–1587.

Serra, T., D. Zilberman, J. M. Gil and B. K. Goodwin. 2011. Nonlinearities in the U.S. corn-ethanol-oil-gasoline price system. *Agricultural Economics* 42:35–45.

Sheehan, J., A. Aden and K. Paustian, et al. 2003. Energy and environmental aspects of using corn stover for fuel ethanol. *Journal of Industrial Ecology* 7:117–146.

Shin, D., T. Kim, J. Choi and J. Kim. 2014. Author name disambiguation using a graph model with node splitting and merging based on bibliographic information. *Scientometrics* 100:15–50.

Soccol, C. R., L. P. de Souza Vandenberghe and A. B. P. Medeiros, et al. 2010. Bioethanol from lignocelluloses: Status and perspectives in Brazil. *Bioresource Technology* 101:4820–4825.

Solomon, B. D., J. R. Barnes and K. E. Halvorsen. 2007. Grain and cellulosic ethanol: History, economics, and energy policy. *Biomass and Bioenergy* 31:416–425.

Vadas, P. A., K. H. Barnett and D. J. Undersander 2008. Economics and energy of ethanol production from alfalfa, corn, and switchgrass in the Upper Midwest, USA. *Bioenergy Research* 1:44–55.

Wang, M. Q., J. Han and Z. Haq, et al. 2011. Energy and greenhouse gas emission effects of corn and cellulosic ethanol with technology improvements and land use changes. *Biomass and Bioenergy* 35:1885–1896.

Zhu, J. Y. and J. X. Pan. 2010. Woody biomass pretreatment for cellulosic ethanol production: Technology and energy consumption evaluation. *Bioresource Technology* 101:4992–5002.

87 Evaluation of Bioethanol Fuels
Review

Ozcan Konur
(Formerly) Ankara Yildirim Beyazit University

87.1 INTRODUCTION

Crude oil-based gasoline fuels (Ma et al., 2002; Newman and Kenworthy, 1989) and petrodiesel fuels (Bosmann et al., 2001) have been widely used in the transportation sector for a long time. However, there have been great public concerns over the adverse environmental impact of these fuels (Hill et al., 2006; 2009; Schauer et al., 1999, 2002).

Hence, biomass-based bioethanol fuels (Hill et al., 2006, 2009; Konur, 2012, 2015, 2019, 2020a) have increasingly been used in blending gasoline fuels (Hsieh et al., 2002; Najafi et al., 2009) and petrodiesel fuels in recent years (Hansen et al., 2005; Li et al., 2005). In this context, there has been a significant focus on the evaluative studies in bioethanol fuels (Farrell et al., 2006; Hamelinck et al., 2005; Hill et al., 2006).

There have been two primary research fronts in this field: environmental impact of bioethanol fuels (100%) and technoeconomics of bioethanol fuels (44%) (Farrell et al., 2006; Hamelinck et al., 2005; Hill et al., 2006; Schmer et al., 2008; Sheehan et al., 2003). There have also been two other research fronts: energy assessments of bioethanol fuels (Macedo et al., 2008; Zhu and Pan, 2010) and policy and economic issues in bioethanol production (Solomon et al., 2007).

Furthermore, there have been four secondary research fronts for the research front of environmental issues: emissions during bioethanol production (Goldemberg et al., 2008; Hertel et al., 2010; Hill et al., 2006; Macedo et al., 2008; Pimentel, 2003; Schmer et al., 2008), emissions during bioethanol utilization (Al-Hasan, 2003; Hsieh et al., 2002; Kwanchareon et al., 2007; Macedo et al., 2008; Najafi et al., 2009; Sheehan et al., 2003), life cycle assessment (LCA) of bioethanol fuels (Farrell et al., 2006; Hill et al., 2006; Sheehan et al., 2003), and other environmental issues (de Oliveira et al., 2005; Goldemberg et al., 2008; Hertel et al., 2010; Pimentel, 2003).

However, it is essential to develop efficient incentive structures (North, 1991) for the primary stakeholders to enhance the research in this field (Konur, 2000, 2002a–c, 2006a,b, 2007a,b).

Although there have been a number of reviews on the evaluative studies in bioethanol fuels (Giakoumis et al., 2013; Gnansounou and Dauriat, 2010; Hammerschlag, 2006; Lynd, 1996), there has been no review of the research of the most-cited 25 articles in the field of the evaluative studies in bioethanol fuels.

This book chapter presents a review of the research of the most-cited 25 articles in the field of the evaluative studies in bioethanol fuels. Then, it discusses the key findings of these highly influential papers and comments on the future research priorities in this field.

87.2 MATERIALS AND METHODS

The search for this study was carried out using the Scopus database (Burnham, 2006) in August 2021.

As a first step for the search of the relevant literature, the keywords were selected using the first most-cited 500 papers. The selected keyword list was optimized to obtain a representative sample of papers for the searched research field. This keyword list is provided in the appendix of Konur (2023)

DOI: 10.1201/9781003226567-117

for future replication studies. Additionally, the information about the most-used keywords is given in Section 87.3.9 to highlight the key research fronts in Section 87.3.10.

As a second step, a sample dataset was used for this study. The first 25 articles in the sample of 100 most-cited papers were selected for the review study. Key findings from each paper were taken from the abstracts of these papers and were discussed.

Additionally, a number of brief conclusions were drawn, and a number of relevant recommendations are made to enhance the future research landscape.

87.3 RESULTS

87.3.1 Environmental Impact of Bioethanol Fuels

87.3.1.1 Emissions from Bioethanol Fuels

In this study, the studies on the emissions emanating from the bioethanol production and bioethanol utilization in gasoline and diesel engines are differentiated and given separately.

87.3.1.1.1 Emissions from Bioethanol Fuel Production

The brief information about eight most-prolific papers with at least 343 citations each in the field of the emissions from bioethanol fuel production is given in Table 87.1. Furthermore, the brief notes about each paper are given.

Hill et al. (2006) evaluate the emissions from first generation corn grain-based bioethanol production compared to soybean-based biodiesel production through LCA in a paper with 1970 citations. They find that biodiesel fuels release 1.0%, 8.3%, and 13% of the agricultural nitrogen (N), phosphorus (P), and pesticide pollutants, respectively, per net energy gain compared to bioethanol fuels. Bioethanol and biodiesel fuels have 12% and 41% reduced greenhouse gas (GHG) emissions, respectively. Biodiesel fuels also release less air pollutants per net energy gain than

TABLE 87.1
The Emissions from the Production of Bioethanol Fuels

No.	Papers	Biomass	Issues	Research Fronts	Cits.
1	Hill et al. (2006)	Corn grains	Emissions of corn grain-based bioethanol production compared to biodiesel	Technoeconomics, LCA, emissions	1970
2	Schmer et al. (2008)	Switchgrass	Emissions of switchgrass-based bioethanol fuels	Technoeconomics, emissions	796
3	Macedo et al. (2008)	Sugarcane	GHG emissions from the production and use of sugarcane-based bioethanol fuels in Brazil	Emissions, energy assessments	601
4	Sheehan et al. (2003)	Corn stover	Emissions of both the production and use of corn stover-based bioethanol fuels of E85 blend	Emissions, LCA, technoeconomics	433
5	Goldemberg et al. (2008)	Sugarcane	Emissions of sugarcane-based bioethanol production in Brazil	Emissions, other environmental issues	404
6	Pimentel (2003)	Corn grains	Emissions of corn grain-based bioethanol production	Technoeconomics, emissions, other environmental issues	403
7	Hertel et al. (2010)	Corn grains	Emissions of corn-based bioethanol production in the USA	Emissions other environmental issues	353
8	de Oliveira et al. (2005)	Corn grains, sugarcane	Emissions and ecological footprints of bioethanol production in Brazil and the USA	Emissions, other environmental issues	343

LCA, life cycle assessment.
Cits.: The number of citations received by each paper.

bioethanol. They assert that these advantages of biodiesel fuels over bioethanol fuels are due to lower agricultural inputs and more efficient conversion of feedstocks to biofuel.

Schmer et al. (2008) evaluate the emissions of switchgrass-based bioethanol fuels in field trials in a paper with 796 citations. They find that estimated average GHG emissions from these fuels were 94% lower than estimated GHG from gasoline fuels. They assert that improved genetics and agronomics may further enhance energy sustainability and bioethanol yield of switchgrass.

Macedo et al. (2008) evaluate the GHG emissions in the production and use of first generation sugarcane-based bioethanol fuels in Brazil in a paper with 601 citations. They find that for bioethanol production, the total GHG emission was 436 kg CO_2 eq/m^3 bioethanol for 2005/2006, decreasing to 345 kg CO_2 eq/m^3 in the 2020 scenario. They find the high impact of sugarcane productivity and bioethanol yield variation on the energy and emission balances and of sugarcane bagasse and bioelectricity surpluses on GHG emissions avoidance based on a sensitivity analysis.

Sheehan et al. (2003) evaluate emissions of both the production and use of first generation corn stover-based bioethanol fuels of E85 blend through the LCA in a paper with 433 citations. This assessment incorporates results from individual models for soil carbon dynamics, soil erosion, agronomics of corn stover collection and transport, and bioconversion of stover to bioethanol. The GHG emissions such as fossil carbon dioxide (CO_2), nitrogen oxides (NO_x), and methane (CH_4) on a life cycle basis are 113% lower. Emissions of carbon monoxide (CO), NO_x, and sulfur oxides (SO_x) increase, whereas hydrocarbon (HC) ozone precursors are reduced.

Goldemberg et al. (2008) discuss the emissions of first generation sugarcane-based bioethanol production in Brazil in a paper with 404 citations. They note that the positive impacts are the elimination of lead compounds from gasoline and the reduction of noxious emissions. There is also the reduction of CO_2 emissions. These positive impacts are particularly noticeable in metropolitan areas showing air quality improvement and also in rural areas where mechanized harvesting of sugarcane is being introduced, eliminating the burning of sugarcane.

Pimentel (2003) discusses the emissions of first generation corn grain-based bioethanol production in a paper with 403 citations. He argues that bioethanol production increases environmental degradation as corn production causes more total soil erosion than any other crop. Corn production also uses more insecticides, herbicides, and nitrogen fertilizers than any other crop. All these factors degrade the agricultural and natural environment and contribute to water pollution and air pollution.

Hertel et al. (2010) evaluate the emissions from the first generation corn grain-based bioethanol production in the USA in a paper with 353 citations. The associated estimated GHG release is 800 g of carbon dioxide per megajoule (MJ); 27 g per MJ per year, over 30 years of bioethanol production. They argue that 800 of carbon dioxide g is enough to cancel out the benefits that corn-based bioethanol has on global warming.

de Oliveira et al. (2005) evaluate the emissions and ecological footprints of bioethanol production in Brazil and the USA compared to gasoline fuels in a paper with 343 citations. They note that the use of bioethanol as a substitute for gasoline is neither a sustainable nor an environmentally friendly option, considering ecological footprint values, and both net energy and CO_2 offset considerations are relatively unimportant compared to the ecological footprint.

87.3.1.2 Emissions from Bioethanol Fuel Utilization in Gasoline and Diesel Engines

The brief information about ten most-prolific papers with at least 285 citations each in the field of the emissions from bioethanol fuel utilization in gasoline and diesel engines is given in Table 87.2. Furthermore, the brief notes about each paper are given.

87.3.1.2.1 Emissions from Neat Bioethanol Fuels

Macedo et al. (2008) evaluate the GHG emissions in the production and use of sugarcane-based bioethanol fuels in Brazil in a paper with 601 citations. They find that for E100 use in Brazil, the avoided emissions were 2,181 kg CO_2 eq/m^3 bioethanol, and for E25, they were 2,323 kg CO_2 eq/m^3 bioethanol in 2005/2006. Both values would increase about 26% for the conditions assumed for

TABLE 87.2
The Emissions from the Utilization of Bioethanol Fuels

No.	Papers	Biomass	Bioethanol Fuels	Issues	Research Fronts	Cits.
1	Macedo et al. (2008)	Sugarcane	Bioethanol	GHG emissions from the production and use of sugarcane-based bioethanol fuels in Brazil	Emissions, energy assessments	601
2	Hsieh et al. (2002)	Biomass	Bioethanol–gasoline blends	Emission from bioethanol–gasoline fuel blends	Emissions, engine performance	483
3	Sheehan et al. (2003)	Corn stover	Bioethanol–gasoline blends	Emissions of both the production and use of corn stover-based bioethanol fuels of E85 blend though the LCA	Emissions, LCA, technoeconomics	433
4	Kwanchareon et al. (2007)	Biomass	Bioethanol–petrodiesel blends	Emissions of bioethanol–petrodiesel blends	Emissions, engine performance	405
5	Al-Hasan (2003)	Biomass	Bioethanol–gasoline blends	Emissions of bioethanol–unleaded gasoline blends	Emissions, engine performance	399
6	Li et al. (2005)	Biomass	Bioethanol–petrodiesel blends	Emissions of petrodiesel–bioethanol blends	Emissions, engine performance	340
7	He et al. (2003)	Biomass	Bioethanol–petrodiesel blends	Emissions of petrodiesel–bioethanol blends	Emissions, engine performance	318
8	Najafi et al. (2009)	Biomass	Bioethanol–gasoline blends	Emissions of gasoline–bioethanol blends	Emissions engine performance	306
9	Xing-Cai et al. (2004)	Biomass	Bioethanol–petrodiesel blends	Emissions of petrodiesel–bioethanol blends	Emissions, engine performance	306
10	Hulwan and Joshi (2011)	Biomass	Bioethanol–petrodiesel blends	Emissions of bioethanol–petrodiesel–biodiesel blends	Emissions, engine performance	285

LCA, life cycle assessment.

2020. They find the high impact of sugarcane productivity and bioethanol yield variation on the energy and emission balances and of sugarcane bagasse and bioelectricity surpluses on GHG emissions avoidance.

87.3.1.2.2 Emissions from Bioethanol–Gasoline Blends

Hsieh et al. (2002) evaluate the emissions of bioethanol–gasoline fuel blends with various blended rates (E0, E5, E10, E20, and E30) in a paper with 590 citations. They find that CO and HC emissions decrease dramatically and CO_2 emissions increase. Finally, NO_x emissions depend on the engine operating condition rather than the bioethanol content.

Sheehan et al. (2003) evaluate the emissions of both the production and use of first generation corn stover-based bioethanol fuels of E85 blend through the LCA in a paper with 433 citations. They find that for each kilometer fueled by the bioethanol portion of E85, the vehicle uses 95% less petroleum than gasoline. The GHG emissions (fossil CO_2, N_2O, and CH_4) on a life cycle basis are 113% lower. Emissions of CO, NO_x, and SO_x increase, whereas HC ozone precursors are reduced.

Al-Hasan (2003) evaluates the emissions of bioethanol–unleaded gasoline blends in a paper with 399 citations. He finds that the CO and unburned HC emission concentrations decrease, while the CO_2 concentration increases. The E20 gave the best results for all measured parameters at all engine speeds.

Najafi et al. (2009) evaluate the emissions of bioethanol–gasoline blends of E5, E10, E15, and E20 with the aid of artificial neural network in a paper with 306 citations. They find that the concentration of CO and HC emissions decreases when bioethanol blends are introduced. In contrast, the concentration of CO_2 and NO_x increases when bioethanol is introduced.

87.3.1.2.3 Emissions from Bioethanol–Petrodiesel Blends

Kwanchareon et al. (2007) evaluate the fuel properties of petrodiesel–biodiesel–bioethanol blends in a paper with 485 citations. They find that CO and HC reduces significantly at high engine load, whereas NO_x increases compared to petrodiesel. They advise that a blend of 80% petrodiesel, 15% biodiesel, and 5% bioethanol (D80-B15-E5) is the most suitable ratio for fuel production because of the acceptable fuel properties (except flash point) and the reduction of emissions.

Li et al. (2005) evaluate the emissions of bioethanol–petrodiesel fuel blends (E5, E10, E15, and E20) in a paper with 340 citations. They find that the smoke emissions decreased with blending, especially with E10 and E15. CO and NO_x emissions reduced for these blends, but HC increased significantly compared to neat petrodiesel fuel.

He et al. (2003) evaluate the emissions of bioethanol–petrodiesel fuel blends in a paper with 318 citations. They find that at high loads, the blends reduce smoke significantly with a small penalty on CO, acetaldehyde, and HC emissions compared to petrodiesel fuel. However, NO_x and CO_2 emissions of the blends are decreased. At low loads, the blends have slight effects on smoke reduction due to overall leaner mixture. With the aid of additive and ignition improver, CO, HC, and acetaldehyde emissions of the blends can be decreased moderately, even total HC emissions are less than those of petrodiesel fuel.

Xing-Cai et al. (2004) evaluate the emissions of bioethanol–petrodiesel fuel blends in a paper with 306 citations. They find that the NO_x and smoke emissions decrease simultaneously when diesel engine is fueled with these blends. NO_x and smoke emissions further reduce when cetane number improver is added to blends.

Hulwan and Joshi (2011) evaluate the emissions of bioethanol–biodiesel–petrodiesel blends in a paper with 285 citations. They test the blends tested D70/E20/B10 (blend A), D50/E30/B20 (blend B) D50/E40/B10 (blend C), and petrodiesel (D100). They find that advancing injection timing almost doubles the NO emissions and increases peak firing pressure. NO variation depends on operating conditions, while CO emissions drastically increase at low loads. Blend B did not show any benefit on peak smoke emission during the free acceleration test.

87.3.1.3 LCAs of Bioethanol Fuels

The brief information about three most-prolific papers with at least 433 citations each in the field of the LCA of bioethanol fuels is given in Table 87.3. Furthermore, the brief notes about each paper are given.

Farrell et al. (2006) carry out the LCA of bioethanol fuels compared to gasoline fuels in a paper with 2,066 citations. There are many important environmental effects of biofuel production that are poorly understood. They develop new metrics that measure specific resource inputs and recommend that further research into environmental metrics should be carried out.

Hill et al. (2006) carry out LCA of first generation corn grain-based bioethanol production compared to soybean-based biodiesel production in a paper with 1,970 citations. They find that

TABLE 87.3
The Life Cycle Assessment of Bioethanol Fuels

No.	Papers	Biomass	Issues	Research Fronts	Cits.
1	Farrell et al. (2006)	Corn grains, cellulosic biomass	LCA of bioethanol fuels compared to gasoline fuels	Technoeconomics, LCA	2,066
2	Hill et al. (2006)	Corn grains	LCA of corn grain-based bioethanol production compared to soybean-based biodiesel production	Technoeconomics, LCA, emissions	1,970
3	Sheehan et al. (2003)	Corn stover	LCA of corn stover-based E85 bioethanol–gasoline blends	Emissions, LCA, technoeconomics	433

LCA, life cycle assessment.

bioethanol has 25% net energy gains compared to 93% for biodiesel fuels. Bioethanol and biodiesel fuels cannot replace gasoline or petrodiesel fuels without impacting food supplies. Even dedicating all US corn and soybean production to bioethanol and biodiesel fuels, respectively, could meet only 12% of gasoline demand and 6% of petrodiesel fuel demand. Until recent increases in petroleum prices, high production costs made both biofuels unprofitable without subsidies. They assert that biodiesel provides sufficient environmental advantages to merit subsidy.

Sheehan et al. (2003) carry out the LCA of corn stover-based E85 bioethanol–gasoline blends in a paper with 433 citations. They assert that Iowa alone could produce almost 8 billion liters per year of E100 bioethanol at competitive prices. They find that for each kilometer fueled by the E85, the vehicle uses 95% less petroleum than gasoline. Total fossil energy use (coal, oil, and natural gas) and GHG emissions (fossil CO_2, N_2O, and CH_4) on a life cycle basis are 102% and 113% lower, respectively. Emissions of CO, NO_x, and SO_x increase, whereas HC ozone precursors are reduced.

87.3.1.4 Other Environmental Issues for Bioethanol Fuels

The brief information about four most-prolific papers with at least 343 citations each in the field of the environmental issues other than emissions for bioethanol fuels is given in Table 87.4. Furthermore, the brief notes about each paper are given.

Goldemberg et al. (2008) discuss the sustainability of first generation sugarcane-based bioethanol production in Brazil in a paper with 404 citations. They note that the negative impacts such as future large-scale bioethanol production might lead to the destruction or damage of high-biodiversity areas, deforestation, degradation or damaging of soils through the use of chemicals and soil decarbonization, water resource contamination or depletion, competition between food and fuel production decreasing food security, and worsening of labor conditions on the fields.

Pimentel (2003) discusses the environmental impact of first generation corn grain-based bioethanol production in a paper with 403 citations. He argues that bioethanol production increases environmental degradation as corn production causes more total soil erosion than any other crop. Corn production also uses more insecticides, herbicides, and nitrogen fertilizers than any other crop. All these factors degrade the agricultural and natural environment and contribute to water pollution and air pollution. He argues that increasing the cost of food and diverting human food resources to the costly inefficient production of bioethanol fuel raise major ethical questions. He advises that the ethical priority for corn and other food crops should be for food and feed.

Hertel et al. (2010) evaluate the effects of US first generation corn-based bioethanol production on global land use in a paper with 353 citations. Factoring market-mediated responses and by-product use into the analysis reduces cropland conversion by 72% from the land used for the bioethanol feedstock. Consequently, the associated estimated GHG release is 800 g of carbon dioxide per megajoule (MJ); 27 g per MJ per year, over 30 years of bioethanol production. However, 800 g are enough to cancel out the benefits that corn bioethanol has on global warming.

TABLE 87.4
The Other Environmental Issues in Bioethanol Fuels

No.	Papers	Biomass	Issues	Research Fronts	Cits.
1	Goldemberg et al. (2008)	Sugarcane	Sustainability of bioethanol production from sugarcane in Brazil	Emissions, other environmental issues	404
2	Pimentel (2003)	Corn grains	Adverse environmental impact of bioethanol production from corn grains	Technoeconomics, emissions, other environmental issues	403
3	Hertel et al. (2010)	Corn grains	Effects of US corn-based bioethanol production on global land use	Emissions other environmental issues	353
4	de Oliveira et al. (2005)	Corn grains, sugarcane	Ecological footprints of bioethanol production in Brazil and the USA	Emissions, other environmental issues	343

de Oliveira et al. (2005) evaluate the ecological footprints of bioethanol production in Brazil and the USA in a paper with 343 citations. They note that the use of bioethanol as a substitute for gasoline is neither a sustainable nor an environmentally friendly option, considering ecological footprint values, and both net energy and CO_2 offset considerations are relatively unimportant compared to the ecological footprint. The direct and indirect environmental impacts of growing, harvesting, and converting biomass to bioethanol far exceed any value in developing this alternative energy resource on a large scale.

87.3.2 Technoeconomics of Bioethanol Fuels

The brief information about 11 most-prolific papers with at least 285 citations each in the field of the technoeconomics of bioethanol fuels is given in Table 87.5. Furthermore, the brief notes about each paper are given.

Farrell et al. (2006) carry out the technoeconomic assessment of bioethanol fuels, compared to gasoline fuels, in a paper with 2,066 citations. They find that studies that reported negative net energy incorrectly ignored coproducts and used some obsolete data. Current first generation corn grain-based bioethanol technologies are much less crude oil-intensive than gasoline. They also recommend the use of second generation lignocellulosic biomass-based bioethanol production for the large-scale production of bioethanol and biodiesel fuels.

Hill et al. (2006) evaluate the technoeconomics of first generation corn grain-based bioethanol production compared to soybean-based biodiesel production in a paper with 1,970 citations. They find that bioethanol has 25% net energy gains compared to 93% for biodiesel fuels. These fuels

TABLE 87.5
The Technoeconomic Assessment of Bioethanol Fuels

No.	Papers	Biomass	Issues	Research Fronts	Cits.
1	Farrell et al. (2006)	Corn grains, cellulosic biomass	Technoeconomics of bioethanol fuels compared to gasoline fuels	Technoeconomics, LCA	2066
2	Hill et al. (2006)	Corn grains	Technoeconomics of corn grain-based bioethanol fuels compared to biodiesel fuels	Technoeconomics, LCA, emissions	170
3	Hamelinck et al. (2005)	Lignocellulosic biomass	Technoeconomics of lignocellulosic biomass-based bioethanol production	Technoeconomics	1162
4	Schmer et al. (2008)	Switchgrass	Technoeconomics of switchgrass-based bioethanol fuels	Technoeconomics, emissions	796
5	Sheehan et al. (2003)	Corn stover	Technoeconomics of E85 bioethanol–gasoline fuel blends	Emissions, LCA, technoeconomics	433
6	Pimentel (2003)	Corn grains	Technoeconomics of corn grain-based bioethanol production	Technoeconomics, emissions, other environmental issues	403
7	Wyman (1994)	Lignocellulosic biomass	Technoeconomics of lignocellulosic biomass-based bioethanol production	Technoeconomics	367
8	Kazi et al. (2010)	Corn stover	Technoeconomics of the process technologies for corn stover-based bioethanol production	Technoeconomics	359
9	Sassner et al. (2008)	Spruce, salix, corn stover	Technoeconomics of lignocellulosic biomass-based bioethanol production	Technoeconomics	344
10	Kwiatkowski et al. (2006)	Corn grains	Technoeconomics of corn grain-based bioethanol production	Technoeconomics	330
11	Gnansounou et al. (2005)	Sweet sorghum, sorghum bagasse	Technoeconomics of sweet sorghum-based bioethanol production in China	Technoeconomics	285

LCA, life cycle assessment.

cannot replace gasoline or petrodiesel fuels without impacting food supplies. Even dedicating all US corn and soybean production to bioethanol and biodiesel fuels, respectively, could meet only 12% of gasoline demand and 6% of petrodiesel fuel demand. Until recent increases in petroleum prices, high production costs made both biofuels unprofitable without subsidies. They assert that bioethanol fuels do not provide sufficient environmental advantages to merit subsidy. They recommend that second generation lignocellulosic biomass-based bioethanol fuels could provide much greater supplies and environmental benefits than first generation food-based bioethanol and biodiesel fuels.

Hamelinck et al. (2005) carry out the technoeconomic analysis of lignocellulosic biomass-based bioethanol production in a paper with 1,162 citations. They estimate current investment costs at 2.1 k€/kWHHV at 400 MWHHV input (a nominal 2,000 ton dry/day input). A future technology in a five times larger plant (2 GWHHV) could have investments of 900 k€/kWHHV. They assert that under the conditions of higher hydrolysis–fermentation efficiency, lower specific capital investments, increase in scale and cheaper biomass feedstock costs (from 3 to 2 €/GJHHV), could bring the bioethanol production costs from 22 €/GJHHV in the next 5 years to 13 €/GJ over the 10–15 year time scale, and down to 8.7 €/GJ in 20 or more years.

Schmer et al. (2008) evaluate the technoeconomics of switchgrass-based bioethanol fuels in field trials in a paper with 796 citations. They observe annual biomass yields averaged 5.2–11.1 Mg/ha, with an average estimated net energy yield (NEY) of 60 GJ/ha·year. Switchgrass-based bioethanol had 540% net energy gains. Switchgrass monocultures managed for high yield produce 93% more biomass yield and an equivalent estimated NEY than previous estimates from human-made prairies that received low agricultural inputs.

Sheehan et al. (2003) evaluate the technoeconomics of corn stover-based E85 bioethanol–gasoline blends in a paper with 433 citations. They assert that Iowa alone could produce almost 8 billion liters per year of E100 at competitive prices. They find that for each kilometer fueled by the bioethanol portion of E85, the vehicle uses 95% less petroleum than a kilometer driven in the same vehicle on gasoline.

Pimentel (2003) discusses technoeconomics of first generation corn grain-based bioethanol production in a paper with 403 citations. He notes that the $1.4 billion in government subsidies are encouraging the bioethanol fuel production without substantial benefits to the US economy where corn farmers receive minimal profits. Bioethanol has negative 29% net energy. He asserts that increasing subsidized bioethanol production takes more feed from livestock production, and cost consumers an additional $1 billion per year. He argues that increasing the cost of food and diverting human food resources to the costly inefficient production of bioethanol fuel raise major ethical questions. He advises that the ethical priority for first generation corn and other food crops should be for food and feed.

Wyman (1994) discusses the technoeconomics of lignocellulosic biomass-based bioethanol production in a paper with 367 citations. He notes that production of bioethanol from lignocellulosic biomass could improve energy security, reduce trade deficits, decrease urban air pollution, and contribute little net carbon dioxide accumulation to the atmosphere. He argues that developments in conversion technology have reduced the projected gate price of bioethanol from about US$0.95/L (US$3.60/gallon) in 1980 to only about US$0.32/L (US$1.22/gallon) in 1994. He identifies thee technical targets to bring the selling price down to about US$0.18/L (US$0.67/gallon), a level that is competitive when oil prices exceed US$25/barrel. However, at the projected costs, bioethanol from lignocellulosic biomass could be competitive with bioethanol from corn.

Kazi et al. (2010) carry out the technoeconomic evaluation of the process technologies for second generation corn stover-based bioethanol production in a paper with 359 citations. They find that the dilute acid pretreatment process has the lowest product value (PV) among all process scenarios, which is estimated to be $1.36/L of gasoline equivalent [LGE] ($5.13/gal of gasoline equivalent [GGE]). The PV is most sensitive to feedstock cost, enzyme cost, and installed equipment costs. A significant fraction of capital costs is related to producing heat and power from lignin in the biomass. The estimated value of PV for the pioneer plant is substantially larger than that for the nth plant. The PV for the pioneer plant model with dilute acid pretreatment is \$2.30/LGE (\$8.72/GGE) for the most probable scenario, and the estimated total capital investment was more than double the nth plant cost.

Sassner et al. (2008) carry out the technoeconomic evaluation of bioethanol production from spruce, salix, and corn stover in a paper with 344 citations. They show the importance of a high bioethanol yield and the necessity of utilizing the pentose fraction for bioethanol production to obtain good process economy, especially when using salix or corn stover. Furthermore, a less energy-demanding process, mainly achieved by increasing the dry matter content in SSF, reduces the capital cost and results in higher coproduct credit, and therefore has a significant effect on the overall process economy.

Kwiatkowski et al. (2006) evaluate the technoeconomics of first generation corn grain-based bioethanol in a paper with 330 citations. In one sensitivity analysis, they find that the cost of producing bioethanol increased from US\$ 0.235/L to US\$ 0.365/L (US\$ 0.89/gal to US\$ 1.38/gal) as the price of corn increased from US\$ 0.071/kg to US\$ 0.125/kg (US\$ 1.80/bu to US\$ 3.20/bu). Another example gave a reduction from 151 to 140 million l/year as the amount of starch in the feed was lowered from 59.5% to 55% (w/w).

Gnansounou et al. (2005) evaluate the sweet sorghum and/or sweet sorghum bagasse-based bioethanol production in North China in a paper with 285 citations. They find that the production of bioethanol from hemicellulose and cellulose in bagasse was more favorable than burning it to make power, but the relative merits of producing bioethanol or sugar from the juice were very sensitive to the price of sugar in China. Thus, they recommend a flexible plant capable of making both sugar and bioethanol fuel from the sorghum juice. Overall, bioethanol production from sorghum bagasse is very favorable.

87.3.3 Energy Assessments of Bioethanol Fuels

The brief information about two most-prolific papers with at least 590 citations each in the field of the other energy assessments for bioethanol fuels is given in Table 87.6. Furthermore, the brief notes about each paper are given.

Macedo et al. (2008) evaluate the energy balance in the production and use of first generation sugarcane-based bioethanol fuels in Brazil in a paper with 601 citations. They find that fossil energy ratio was 9.3 for 2005/2006 and may reach 11.6 in 2020. They find the high impact of sugarcane productivity and bioethanol yield variation on the energy and emission balances and of sugarcane bagasse and bioelectricity surpluses on GHG emissions avoidance based on a sensitivity analysis.

Zhu and Pan (2010) evaluate the energy consumption of woody biomass pretreatment for cellulosic bioethanol production in a paper with 590 citations. They introduce a concept of pretreatment energy efficiency (kg/MJ) based on the total sugar recovery (kg/kg wood) divided by the energy consumption in pretreatment (MJ/kg wood). They then use this concept to evaluate the performances of steam explosion, organosolv, and sulfite pretreatment to overcome lignocellulose recalcitrance (SPORL) for softwood pretreatment. They find that SPORL was the most-efficient process and produced highest sugar yield.

TABLE 87.6
The Energy Assessment of Bioethanol Fuels

No.	Papers	Biomass	Issues	Research Fronts	Cits.
1	Macedo et al. (2008)	Sugarcane	Energy balance in the production and use of bioethanol fuels from sugarcane in Brazil	Emissions, energy assessments	601
2	Zhu and Pan (2010)	Softwood biomass	Energy consumption of woody biomass pretreatment for cellulosic bioethanol production	Energy assessments	590

TABLE 87.7
The Policy and Economic Issues in Bioethanol Fuels

No.	Papers	Biomass	Issues	Research Fronts	Cits.
1	Solomon et al. (2007)	Corn grain biomass, sugarcane, cellulosic biomass	History, economics, and policy of grain and cellulosic biomass-based bioethanol production in Brazil and the USA	Policy, economics	336

87.3.4 Policy and Economics for Bioethanol Fuels

The brief information about a most-prolific paper with 336 citations each in the field of the policy and economics of bioethanol fuels is given in Table 87.7. Furthermore, the brief notes about this paper are given.

Solomon et al. (2007) discusses the history, economics, and policy of grain and cellulosic biomass-based bioethanol production in Brazil and the USA in a paper with 336 citations. National policies have supported the production and use of first generation bioethanol from corn and sugarcane. US support in particular has included exemption from federal gasoline excise taxes, whole or partial exemption from road use (sales) taxes in nine states, a federal production tax credit, and a federal blender's credit. In the last decade, the subsidization of first generation grain-based bioethanol has been increasingly criticized as economically inefficient and of questionable social benefit. In addition, much greater production of corn grain-based bioethanol may conflict with food production needs. The second generation lignocellulosic bioethanol is projected to be much more cost-effective and environmentally beneficial, and have a greater net energy than grain bioethanol. The technology is being developed in North America, Brazil, Japan, and Europe. They note the historical evolution of US federal and state energy policy support for and the currently attractive economics of the production and use of biomass-based bioethanol.

87.4 DISCUSSION

87.4.1 Introduction

Crude oil-based gasoline and petrodiesel fuels have been widely used in the transportation sector. However, there have been great public concerns over the adverse environmental impact of these fuels. Hence, biomass-based bioethanol fuels (Hill et al., 2006, 2009; Konur, 2012, 2015, 2019, 2020a) have increasingly been used in blending gasoline fuels (Hsieh et al., 2002; Najafi et al., 2009) as well as petrodiesel fuels in recent years (Hansen et al., 2005; Li et al., 2005). In this context, there has been a significant focus on the evaluative studies in bioethanol fuels (Farrell et al., 2006; Hamelinck et al., 2005; Hill et al., 2006).

There have been two primary research fronts in this field: environmental impact of bioethanol fuels (100%) and technoeconomics of bioethanol fuels (44%) (Farrell et al., 2006; Hamelinck et al., 2005; Hill et al., 2006; Schmer et al., 2008; Sheehan et al., 2003). There have also been two other research fronts: energy assessments of bioethanol fuels (Macedo et al., 2008; Zhu and Pan, 2010) and policy and economic issues in bioethanol production (Solomon et al., 2007).

Furthermore, there have been four secondary research fronts for the research front of environmental issues: Emissions during bioethanol production (Goldemberg et al., 2008; Hertel et al., 2010; Hill et al., 2006; Macedo et al., 2008; Pimentel, 2003; Schmer et al., 2008; Sheehan et al., 2008), emissions during bioethanol utilization (Al-Hasan, 2003; Hsieh et al., 2002; Kwanchareon et al., 2007; Macedo et al., 2008; Najafi et al., 2009; Sheehan et al., 2003), LCA of bioethanol fuels (Farrell et al., 2006; Hill et al., 2006; Sheehan et al., 2003), and other environmental issues (de Oliveira et al., 2005; Goldemberg et al., 2008; Hertel et al., 2010; Pimentel, 2003).

However, it is essential to develop efficient incentive structures for the primary stakeholders to enhance the research in this field. Although there have been a number of reviews on the evaluative studies in bioethanol fuels, there has been no review of the research of the most-cited 25 articles in the field of the evaluative studies in bioethanol fuels. Hence, this book chapter presents a review of the research of the most-cited 25 articles in this field. Then, it discusses the key findings of these highly influential papers and comments on the future research priorities in this field.

87.4.2 Environmental Impact of Bioethanol Fuels

87.4.2.1 Emissions from Bioethanol Fuels

In this study, the studies on the emissions emanating from the bioethanol production and bioethanol utilization in gasoline and diesel engines are differentiated and given separately below.

87.4.2.1.1 Emissions from Bioethanol Fuel Production

The brief information about eight most-prolific papers with at least 343 citations each in the field of the emissions from bioethanol fuel production is given in Table 87.1. Furthermore, the brief notes about each paper are given.

These studies employ corn grains, switchgrass, corn stover, and sugarcane as a feedstock for the bioethanol production, mostly in the USA, Brazil, and China (de Oliveira et al., 2005; Goldemberg et al., 2008; Hertel et al., 2010; Hill et al., 2006; Macedo et al., 2008; Pimentel, 2003; Schmer et al., 2008). The most-studied emissions are CO_2, NO_x, CH_4, CO, and SO_x.

As expected, the emissions of second generation bioethanol fuels are less than those of the first generation bioethanol fuels and much less than those of gasoline or petrodiesel fuels in general. Furthermore, besides the type of the biomass, the biomass productivity, bioethanol yield, and the efficiency of the conversion technologies strongly impact the emissions of the resulting bioethanol fuels and their blends.

87.4.2.1.2 Emissions from Bioethanol Fuel Utilization in Gasoline and Diesel Engines

The brief information about ten most-prolific papers with at least 285 citations each in the field of the emissions from bioethanol fuel utilization in gasoline or diesel engines is given in Table 87.2. Furthermore, the brief notes about each paper are given.

These studies employ sugarcane, corn stover, and undefined biomass for he bioethanol production (Al-Hasan, 2003; He et al., 2003; Hsieh et al., 2002; Hulwan and Joshi, 2011; Kwanchareon et al., 2007; Li et al., 2005; Macedo et al., 2008; Najafi et al., 2009; Sheehan et al., 2003; Xing-Cai et al., 2004).

These studies evaluate the emissions from the neat bioethanol fuels as well as their bioethanol–gasoline fuel and bioethanol–petrodiesel blends. The most-studied emissions are CO_2, HC, CO, NO_x, SO_x, smoke, acetaldehyde, and CH_4.

As expected, the emissions of bioethanol fuels are less than those of gasoline or petrodiesel fuels in general. Furthermore, the operating conditions of the gasoline or diesel engines as well as the proportions of bioethanol in these blends strongly impact the emissions of these bioethanol fuel blends. With the optimization of operating conditions of the engines as well as the proportions of bioethanol, these studies show that bioethanol blends might have better emission performance than both gasoline and petrodiesel fuels.

The emissions (Lapuerta et al., 2008; Searchinger et al, 2008) of these fuels are also of the most importance in light of the public concerns about climate change (Change, 2007), GHG emissions (Carlson et al., 2017), and global warming (Kerr, 2007). As these concerns have been certainly behind the boom in the research in this field in the last two decades, it is not surprising that the bulk of the majority of the most-cited papers sought to evaluate the environmental impact of bioethanol fuels during its utilization in both gasoline and diesel engines.

The emissions of CO_2 (Ang, 2007; Holtz-Eakin and Selden, 1995) are especially important, and any potential of the reduction of these emissions by the bioethanol blends compared to gasoline or petrodiesel fuels would have significant implications for the environmental protection. The emissions of the PM are also important for the protection of the environment (Abdullahi et al., 2013; Klimont et al., 2017).

87.4.2.2 LCAs of Bioethanol Fuels

The brief information about three most-prolific papers with at least 433 citations each in the field of the LCA of bioethanol fuels is given in Table 87.3. Furthermore, the brief notes about each paper are given.

These studies employ corn grains, corn stover, and undefined cellulosic biomass (Farrell et al., 2006; Hill et al., 2006; Sheehan et al., 2003). The key findings from these studies hint that the methodological issues for the LCA strongly impact the outcome of these assessments (Ayres, 1995; Finnveden et al., 2009).

The key findings from these studies also hint that bioethanol fuels have comparable net energy gain, better emissions, and comparable economic competitiveness following the list of the issues given by Hill et al. (2006). As expected, the emissions, net energy, and economic competitiveness of second generation bioethanol fuels are better than those of the first generation bioethanol fuels and much better than those of gasoline or petrodiesel fuels in general.

87.4.2.3 Other Environmental Issues for Bioethanol Fuels

The brief information about four most-prolific papers with at least 343 citations each in the field of the environmental issues other than emissions for bioethanol fuels is given in Table 87.4. Furthermore, the brief notes about each paper are given.

These studies employ sugarcane and corn grains for the bioethanol production in the USA and Brazil (Goldemberg et al., 2008; Pimentel, 2003; Hertel et al., 2010; de Oliveira et al., 2005).

The studies evaluate the impact of the production and utilization of bioethanol fuels on the biodiversity, deforestation, soil biodegradation, soil decarbonization, water resources contamination and depletion, competition between food and fuels, water pollution, air pollution, land use, global warming, and ecological footprints. It is essential that these issues should be incorporated in the life cycle and technoeconomic assessments of bioethanol fuels.

As expected the impact of second generation bioethanol fuels on these issues are better than those of the first generation bioethanol fuels and much better than those of gasoline or petrodiesel fuels in general.

The change in the land use (Foley et al., 2005; Lambin et al., 2001) due to the production of bioethanol fuels emerges as a critical public policy issue to be considered by the major stakeholders. Such changes in land use potentially undermine the capacity of ecosystems to sustain food production, maintain freshwater and forest resources, regulate climate and air quality, and ameliorate infectious diseases. The global community face the challenge of managing trade-offs between immediate human needs for bioethanol and maintaining the capacity of the biosphere to provide goods and services in the long term (Foley et al., 2005).

Next, with the increasing change in the land use due to the bioethanol production, change in biodiversity (Jenkins, 2003; Pimm et al., 1995), deforestation (Achard et al., 2002; Allen and Barnes, 1985), and water pollution (Moss, 2008; Schwarzenbach et al., 2010) also emerge as a critical public policy issues.

One of the major problems with the use of the first generation food crop-based feedstocks for the bioethanol production is the potential threat to food security (Maxwell, 1996; Pinstrup-Andersen, 2009) for the global society (Ajanovic, 2011; Naylor et al., 2007). For example, growing crops for bioethanol fuels squanders land, water, and energy resources vital for the production of food for human consumption. Using corn for ethanol increases the price of US foods and causes food shortages for the poor of the world (Pimentel et al., 2009).

87.4.3 Technoeconomics of Bioethanol Fuels

The brief information about 11 most-prolific papers with at least 285 citations each in the field of the technoeconomics of bioethanol fuels is given in Table 87.5. Furthermore, brief notes about each paper are given.

These studies employ corn grains, undefined lignocellulosic biomass, switchgrass, corn stover, spruce, salix, sweet sorghum, and sorghum bagasse for the production of bioethanol fuels (Farrell e al., 2006; Gnansounou et al., 2005; Hamelinck et al., 2005; Hill et al., 2006; Kazi et al., 2010; Kwiatkowski et al., 2006; Pimentel, 2003; Sassner et al., 2008; Schmer et al., 2008; Sheehan et al., 2003; Wyman, 1994).

The key findings from these studies hint that as expected second generation bioethanol fuels have better net energy gain and higher economic competitiveness than those of the first generation bioethanol fuels and much better than those of gasoline or petrodiesel fuels in general. These studies also highlight the strong impact of the methodological issues as well as the efficiency of the conversion technologies on the outcome of the technoeconomic assessments as these issues account for the different assessment outcomes.

87.4.4 Energy Assessments of Bioethanol Fuels

The brief information about two most-prolific papers with at least 590 citations each in the field of the energy assessments for bioethanol fuels is given in Table 87.6. Furthermore, the brief notes about each paper are given.

These studies employ sugarcane or softwood biomass for the bioethanol production (Macedo et al., 2008; Zhu and Pan, 2010). These studies focus only on the net energy gain of the bioethanol fuels rather than both the net energy gain and economic competitiveness of these fuels as in the technoeconomic studies.

The key findings from these studies hint that as expected second generation bioethanol fuels have better net energy gain than those of the first generation bioethanol fuels and much better than those of gasoline or petrodiesel fuels in general. These studies also highlight the strong impact of the methodological issues as well as the efficiency of the conversion technologies on the outcome of the net energy gain as these issues account for the different assessment outcomes.

87.4.5 Policy and Economics for Bioethanol Fuels

The brief information about a most-prolific paper with 336 citations each in the field of the policy and economics of bioethanol fuels is given in Table 87.7. Furthermore, the brief notes about this paper are given.

This study employs corn grains, sugarcane, and undefined lignocellulosic biomass (Solomon et al., 2007). This study shows the economic and social policies strongly impact the development of the bioethanol industry in Brazil and the USA.

87.5 CONCLUSION AND FUTURE RESEARCH

This book chapter presented and discussed the key findings from 25 most-cited papers with at least 285 citations each. The brief information about the key research fronts for the field of the evaluative studies in bioethanol fuels and their gasoline or petrodiesel fuel blends is given in Table 87.8.

These papers were grouped under four headings: environmental impact, technoeconomics, and, to a lesser extent, energy assessments and policy and economic issues for bioethanol production (Table 87.8).

The first group of papers were further grouped under the headings of emissions during the production and utilization of bioethanol fuels and, to a lesser extent, LCA of bioethanol fuels, and environmental issues other than emissions for bioethanol fuels.

TABLE 87.8
The Research Fronts in the Evaluation of Bioethanol Fuels

No.	Research Fronts	No. Papers (%)
1.	Environmental impact of bioethanol fuels	100
2.	Emissions during bioethanol production	32
3.	Emissions during bioethanol utilization	40
4.	LCA of bioethanol fuels	12
5.	Other environmental issues	16
6.	Technoeconomics of bioethanol fuels	44
7.	Energy assessments of bioethanol fuels	8
8.	Policy and economic issues in bioethanol production	4

LCA, life cycle assessment.
The total number of papers = 25.

The key findings on these research fronts should be read in the light of the increasing public concerns about climate change, GHG emissions, and global warming as these concerns have been certainly behind the boom in the research in this field in the last two decades. It is therefore not surprising that the bulk of the majority of the most-cited papers sought to evaluate the environmental impact of bioethanol fuels both during its production and utilization in gasoline and diesel engines.

These findings confirm that bioethanol fuels are a viable alternative to crude oil-based gasoline and petrodiesel fuels; have a net energy gain; have environmental benefits impacting favorably global warming, GHG emissions, and climate change; are economically competitive; and are producible in large quantities without reducing food supplies, especially using the second generation bioethanol feedstocks such as lignocellulosic feedstocks using the criteria introduced by Hill et al. (2006).

There is a persistent trend for the bioethanol research to focus more on the second and third generation feedstocks and to move away from the first generation bioethanol feedstocks. These feedstocks would certainly impact the properties, engine performance, emissions as well as the net energy gain, economic competitiveness, and emissions of the resulting bioethanol fuels and their blends. This trend should be followed closely by the major stakeholders in the future. For this reason, the impact of source of the biomass on the properties, engine performance as well as the net energy gain, economic competitiveness, and emissions of bioethanol fuels and their gasoline or petrodiesel blends in the diesel or gasoline engines should be focused on to establish structure–property–processing relationships for these bioethanol fuels in the future studies as it has been done in the other research streams of bioethanol production and utilization.

The emissions of these fuels are also of utmost importance in the light of the public concerns about climate change, GHG emissions, and global warming. As these concerns have been certainly behind the boom in the research in this field in the last two decades, it is not surprising that the bulk of the majority of the most-cited papers sought to evaluate the environmental impact of bioethanol fuels during its production and utilization in both gasoline and diesel engines.

The emissions of CO_2 are especially important, and any potential of the reduction of these emissions by the bioethanol blends compared to gasoline or petrodiesel fuels would have significant implications for the environmental protection. The emissions of the PM are also important for the protection of the environment.

These most-cited studies highlight the determinants of the better emission performance for these blends. The operating conditions of the engines as well as the proportions of bioethanol concentrations in the fuel blends play a key role in determining the emission performance of these blends. With the optimization of operating conditions of the engines as well as proportions of bioethanol, these studies show that bioethanol blends might have better emission performance than both gasoline and petrodiesel fuels.

As expected, the emissions of second generation bioethanol fuels are less than those of the first generation bioethanol fuels and much less than those of gasoline or petrodiesel fuels in general. Furthermore, besides the type of the biomass, biomass productivity, bioethanol yield, and efficiency of the conversion technologies strongly impact the emissions of the resulting bioethanol fuels and their blends.

The key findings from LCA studies hint that the methodological issues for the LCA strongly impact the outcome of these assessments. As expected, the emissions, net energy, and economic competitiveness of second generation bioethanol fuels are better than those of the first generation bioethanol fuels and much better than those of gasoline or petrodiesel fuels in general.

The other environmental studies evaluate the impact of the production and utilization of bioethanol fuels on the biodiversity, deforestation, soil biodegradation, soil decarbonization, water resources contamination and depletion, competition between food and fuels, water pollution, air pollution, land use, global warming, and ecological footprints. It is essential that these issues should be incorporated in the life cycle and technoeconomic assessments of bioethanol fuels. As expected the impact of second generation bioethanol fuels on these issues are better than those of the first generation bioethanol fuels and much better than those of gasoline or petrodiesel fuels in general.

The change in the land use due to the production of bioethanol fuels emerges as a critical public policy issue to be considered by the major stakeholders. Such changes in land use potentially undermine the capacity of ecosystems to sustain food production, maintain freshwater and forest resources, regulate climate and air quality, and ameliorate infectious diseases (Foley et al., 2005). Next, with the increasing change in the land use due to the bioethanol production, change in biodiversity, deforestation, and water pollution also emerge as a critical public policy issues.

One of the major problems with the use of the first generation food crop-based feedstocks for the bioethanol production is the potential threat to the food security for the global society. For example, growing crops for bioethanol fuels squanders land, water, and energy resources vital for the production of food for human consumption. Furthermore, using corn for ethanol increases the price of US foods and causes food shortages for the poor of the world (Pimentel et al., 2009).

The key findings from the technoeconomic studies hint that as expected second generation bioethanol fuels have better net energy gain and higher economic competitiveness than those of the first generation bioethanol fuels and much better than those of gasoline or petrodiesel fuels in general. These studies also highlight the strong impact of the methodological issues as well as the efficiency of the conversion technologies on the outcome of the technoeconomic assessments as these issues account for the different assessment outcomes.

In the end, these most-cited papers in the field of the production and utilization of bioethanol fuels and their blends hint that the benefits sought from these fuels could be maximized using the structure, processing, and property relationships of these fuels as in materials science and engineering (Formela et al., 2016; Konur, 2018a, 2020b, 2021a -d; Konur and Matthews, 1989).

These findings confirm that bioethanol fuels are a viable alternative to crude oil-based gasoline and petrodiesel fuels and have environmental and technoeconomic benefits impacting favorably global warming, GHG emissions, and climate change.

It is recommended that such review studies should be performed for the other research fronts on both the production and utilization of bioethanol fuels complementing the corresponding scientometric studies.

ACKNOWLEDGMENTS

The contribution of the highly cited researchers in the field of the evaluative studies in bioethanol fuels has been gratefully acknowledged.

REFERENCES

Abdullahi, K. L., J. M. Delgado-Saborit and R. M. Harrison. 2013. Emissions and indoor concentrations of particulate matter and its specific chemical components from cooking: A review. *Atmospheric Environment* 71:260–294.

Achard, F., H. D. Eva and H. J. Stibig, et al. 2002. Determination of deforestation rates of the world's humid tropical forests. *Science* 297:999–1002.

Ajanovic, A. 2011. Biofuels versus food production: Does biofuels production increase food prices? *Energy* 36:2070–2076.

Al-Hasan, M. 2003. Effect of ethanol-unleaded gasoline blends on engine performance and exhaust emission. *Energy Conversion and Management* 44:1547–1561.

Allen, J. C. and D. F. Barnes. 1985. The causes of deforestation in developing countries. *Annals of the Association of American Geographers* 75:163–184.

Ang, J. B. 2007. CO_2 emissions, energy consumption, and output in France. *Energy Policy* 35:4772–4778.

Ayres, R. U. 1995. Life cycle analysis: A critique. *Resources, Conservation and Recycling* 14:199–223.

Bosmann, A., L. Datsevich and A. Jess, et al. 2001. Deep desulfurization of diesel fuel by extraction with ionic liquids. *Chemical Communications* 23:2494–2495.

Burnham, J. F. 2006. Scopus database: A review. *Biomedical Digital Libraries* 3:1–8.

Carlson, K. M., J. S. Gerber and N. D. Mueller, et al. 2017. Greenhouse gas emissions intensity of global croplands. *Nature Climate Change* 7:63–68.

Change, C. 2007. Climate change impacts, adaptation and vulnerability. *Science of the Total Environment* 326:95–112.

de Oliveira, M. E. D., B. E. Vaughan and E. J. Rykiel. 2005. Ethanol as fuel: Energy, carbon dioxide balances, and ecological footprint. *BioScience* 55:593–602.

Farrell, A. E., R. J. Plevin and B. T. Turner, et al. 2006. Ethanol can contribute to energy and environmental goals. *Science* 311:506–508.

Finnveden, G., M. Z. Hauschild and T. Ekvall, et al. 2009. Recent developments in life cycle assessment. *Journal of Environmental Management* 91:1–21.

Foley, J. A., R. DeFries and G. P. Asner, et al. 2005. Global consequences of land use. *Science* 309:570–574.

Formela, K., A. Hejna, L. Piszczyk, M. R. Saeb and X. Colom. 2016. Processing and structure-property relationships of natural rubber/wheat bran biocomposites. *Cellulose* 23:3157–3175.

Giakoumis, E. G., C. D. Rakopoulos, A. M. Dimaratos and D. C. Rakopoulos. 2013. Exhaust emissions with ethanol or n-butanol diesel fuel blends during transient operation: A review. *Renewable and Sustainable Energy Reviews* 17:170–190.

Gnansounou, E. and A. Dauriat. 2010. Techno-economic analysis of lignocellulosic ethanol: A review. *Bioresource Technology* 101:4980–4991.

Gnansounou, E., A. Dauriat and C. E. Wyman. 2005. Refining sweet sorghum to ethanol and sugar: Economic trade-offs in the context of North China. *Bioresource Technology* 96:985–1002.

Goldemberg, J., S. T. Coelho and P. Guardabassi. 2008. The sustainability of ethanol production from sugarcane. *Energy Policy* 36:2086–2097.

Hamelinck, C. N., G. van Hooijdonk and A. P. C. Faaij. 2005. Ethanol from lignocellulosic biomass: Techno-economic performance in short-, middle- and long-term. *Biomass and Bioenergy* 28:384–410.

Hammerschlag, R. 2006. Ethanol's energy return on investment: A survey of the literature 1990: Present. *Environmental Science and Technology* 40:1744–1750.

Hansen, A. C, Q. Zhang and P. W. L. Lyne. 2005. Ethanol-diesel fuel blends: A review. *Bioresource Technology* 96:277–285.

He, B. Q., S. J. Shuai, J.X. Wang and H. He. 2003. The effect of ethanol blended diesel fuels on emissions from a diesel engine. *Atmospheric Environment* 37:4965–4971.

Hertel, T. W., A. A. Golub and A. D. Jones, et al. 2010. Effects of US maize ethanol on global land use and greenhouse gas emissions: Estimating market-mediated responses. *BioScience* 60:223–231.

Hill, J., E. Nelson, D. Tilman, S. Polasky and D. Tiffany. 2006. Environmental, economic, and energetic costs and benefits of biodiesel and ethanol biofuels. *Proceedings of the National Academy of Sciences* 103:11206–11210.

Hill, J., S. Polasky and E. Nelson, et al. 2009. Climate change and health costs of air emissions from biofuels and gasoline. *Proceedings of the National Academy of Sciences of the United States of America* 106:2077–2082.

Holtz-Eakin, D. and T. M. Selden. 1995. Stoking the fires? CO_2 emissions and economic growth. *Journal of Public Economics* 57:85–101.

Hsieh, W. D., R. H. Chen, T. L. Wu and T. H. Lin. 2002. Engine performance and pollutant emission of an SI engine using ethanol-gasoline blended fuels. *Atmospheric Environment* 36:403–410.

Hulwan, D. B. and S. V. Joshi. 2011. Performance, emission and combustion characteristic of a multicylinder DI diesel engine running on diesel-ethanol-biodiesel blends of high ethanol content. *Applied Energy* 88:5042–5055.

Jenkins, M. 2003. Prospects for biodiversity. *Science* 302:1175–1177.

Kazi, F. K., J. A. Fortman and R. P. Anex, et al. 2010. Techno-economic comparison of process technologies for biochemical ethanol production from corn stover. *Fuel* 89:S20–S28.

Kerr, R. A. 2007. Global warming is changing the world. *Science* 316:188–190.

Klimont, Z., K. Kupiainen and C. Heyes, et al. 2017. Global anthropogenic emissions of particulate matter including black carbon. *Atmospheric Chemistry and Physics* 17:8681–8723.

Konur, O. 2000. Creating enforceable civil rights for disabled students in higher education: An institutional theory perspective. *Disability & Society* 15:1041–1063.

Konur, O. 2002a. Access to nursing education by disabled students: Rights and duties of nursing programs. *Nurse Education Today* 22:364–374.

Konur, O. 2002b. Assessment of disabled students in higher education: Current public policy issues. *Assessment and Evaluation in Higher Education* 27:131–152.

Konur, O. 2002c. Access to employment by disabled people in the UK: Is the Disability Discrimination Act working? *International Journal of Discrimination and the Law* 5:247–279.

Konur, O. 2006a. Participation of children with dyslexia in compulsory education: Current public policy issues. *Dyslexia* 12:51–67.

Konur, O. 2006b. Teaching disabled students in higher education. *Teaching in Higher Education* 11:351–363.

Konur, O. 2007a. A judicial outcome analysis of the Disability Discrimination Act: A windfall for the employers? *Disability & Society* 22:187–204.

Konur, O. 2007b. Computer-assisted teaching and assessment of disabled students in higher education: The interface between academic standards and disability rights. *Journal of Computer Assisted Learning* 23:207–219.

Konur, O. 2012. The evaluation of the research on the bioethanol: A scientometric approach. *Energy Education Science and Technology Part A: Energy Science and Research* 28:1051–1064.

Konur, O. 2015. Current state of research on algal bioethanol. In *Marine Bioenergy: Trends and Developments*, Eds. S. K. Kim and C. G. Lee, pp. 217–244. Boca Raton, FL: CRC Press.

Konur, O., Ed. 2018. *Bioenergy and Biofuels*. Boca Raton, FL: CRC Press.

Konur, O. 2019. Cyanobacterial bioenergy and biofuels science and technology: A scientometric overview. In *Cyanobacteria: From Basic Science to Applications*, Eds. A. K. Mishra, D. N. Tiwari and A. N. Rai, pp. 419–442. Amsterdam: Elsevier.

Konur, O. 2020a. The scientometric analysis of the research on the bioethanol production from green macroalgae. In *Handbook of Algal Science, Technology and Medicine,* Ed. O. Konur, pp. 385–401. London: Academic Press.

Konur, O., Ed. 2020b. *Handbook of Algal Science, Technology and Medicine.* London: Academic Press.

Konur, O., Ed. 2021a. *Handbook of Biodiesel and Petrodiesel Fuels: Science, Technology, Health, and Environment.* Boca Raton, FL: CRC Press.

Konur, O., Ed. 2021b. *Handbook of Biodiesel and Petrodiesel Fuels: Science, Technology, Health, and Environment. Volume 1. Biodiesel Fuels: Science, Technology, Health, and Environment.* Boca Raton, FL: CRC Press.

Konur, O., Ed. 2021c. *Handbook of Biodiesel and Petrodiesel Fuels: Science, Technology, Health, and Environment. Volume 2. Biodiesel Fuels Based on the Edible and Nonedible Feedstocks, Wastes, and Algae: Science, Technology, Health, and Environment.* Boca Raton, FL: CRC Press.

Konur, O., Ed. 2021d. *Handbook of Biodiesel and Petrodiesel Fuels: Science, Technology, Health, and Environment. Volume 3. Petrodiesel Fuels: Science, Technology, Health, and Environment.* Boca Raton, FL: CRC Press.

Konur, O. 2023. Evaluation of bioethanol fuels: Scientometric study. In *Evaluation and Utilization of Bioethanol Fuels. I.: Evaluation of Bioethanol Fuels, Transport Engines, and Bioethanol Sensors. Handbook of Bioethanol Fuels Volume 5*, Ed. O. Konur, pp. 195–213. Boca Raton, FL: CRC Press.

Konur, O. and F. L. Matthews. 1989. Effect of the properties of the constituents on the fatigue performance of composites: A review. *Composites* 20:317–328.

Kwanchareon, P., A. Luengnaruemitchai and S. Jai-In. 2007. Solubility of a diesel-biodiesel-ethanol blend, its fuel properties, and its emission characteristics from diesel engine. *Fuel* 86:1053–1061.

Kwiatkowski, J. R., A. J. McAloon, F. Taylor and D. B. Johnston. 2006. Modeling the process and costs of fuel ethanol production by the corn dry-grind process. *Industrial Crops and Products* 23:288–296.

Lambin, E. F., B. L. Turner and H. J., Geist, et al. 2001. The causes of land-use and land-cover change: Moving beyond the myths. *Global Environmental Change* 11:261–269.

Lapuerta, M., O. Armas and J. M. Herreros. 2008. Emissions from a diesel-bioethanol blend in an automotive diesel engine. *Fuel* 87:25–31.

Li, D. G., H. Zhen, L. Xingcai, Z. Wu-Gao and Y. Jian-Guang. 2005. Physico-chemical properties of ethanol-diesel blend fuel and its effect on performance and emissions of diesel engines. *Renewable Energy* 30:967–976.

Lynd, L. R. 1996. Overview and evaluation of fuel ethanol from cellulosic biomass: Technology, economics, the environment, and policy. *Annual Review of Energy and the Environment* 21:403–465.

Lynd, L. R., J. H. Cushman, R. J. Nichols and C. E. Wyman. 1991. Fuel ethanol from cellulosic biomass. *Science* 251:1318–1323.

Ma, X., L. Sun and C. Song. 2002. A new approach to deep desulfurization of gasoline, diesel fuel and jet fuel by selective adsorption for ultra-clean fuels and for fuel cell applications. *Catalysis Today* 77:107–116.

Macedo, I. C., J. E. A. Seabra and J. E. A. R. Silva. 2008. Green house gases emissions in the production and use of ethanol from sugarcane in Brazil: The 2005/2006 averages and a prediction for 2020. *Biomass and Bioenergy* 32:582–595.

Maxwell, S. 1996. Food security: A post-modern perspective. *Food Policy* 21:155–170.

Moss, B. 2008. Water pollution by agriculture. *Philosophical Transactions of the Royal Society B: Biological Sciences* 363:659–666.

Najafi, G., B. Ghobadian and T. Tavakoli, et al. 2009. Performance and exhaust emissions of a gasoline engine with ethanol blended gasoline fuels using artificial neural network. *Applied Energy* 86:630–639.

Naylor, R. L., A. J. Liska and M. B. Burke, et al. 2007. The ripple effect: Biofuels, food security, and the environment. *Environment: Science and Policy for Sustainable Development* 49:30–43.

Newman, P. W. G. and J. R. Kenworthy. 1989. Gasoline consumption and cities: A comparison of U.S. cities with a global survey. *Journal of the American Planning Association* 55:24–37.

North, D. C. 1991. Institutions. *Journal of Economic Perspectives* 5:97–112.

Pimentel, D. 2003. Ethanol fuels: Energy balance, economics, and environmental impacts are negative. *Natural Resources Research* 12:127–134.

Pimentel, D., A. Marklein and M. A. Toth, et al. 2009. Food versus biofuels: Environmental and economic costs. *Human Ecology* 37:1–12.

Pimm, S. L., G. J. Russell, J. L. Gittleman and T. M. Brooks. 1995. The future of biodiversity. *Science* 269:347–350.

Pinstrup-Andersen, P. 2009. Food security: Definition and measurement. *Food Security* 1:5–7.

Sassner, P., M. Galbe and G. Zacchi. 2008. Techno-economic evaluation of bioethanol production from three different lignocellulosic materials. *Biomass and Bioenergy* 32:422–430.

Schauer, J. J., M. J. Kleeman, G. R. Cass and B. R. T. Simoneit. 1999. Measurement of emissions from air pollution sources. 2. C1 through C30 organic compounds from medium duty diesel trucks. *Environmental Science & Technology* 33:1578–1587.

Schauer, J. J., M. J. Kleeman, G. R. Cass and B. R. T. Simoneit. 2002. Measurement of emissions from air pollution sources. 5. C1-C32 organic compounds from gasoline-powered motor vehicles. *Environmental Science and Technology* 36:1169–1180.

Schmer, M. R., K. P. Vogel, R. B. Mitchell and P. K. Perrin. 2008. Net energy of cellulosic ethanol from switchgrass. *Proceedings of the National Academy of Sciences of the United States of America* 105:464–469.

Schwarzenbach, R. P., T. Egli, T. B. Hofstetter, U. Von Gunten and B. Wehrli. 2010. Global water pollution and human health. *Annual Review of Environment and Resources* 35:109–136.

Searchinger, T., R. Heimlich and R. A. Houghton, et al. 2008. Use of US croplands for biofuels increases greenhouse gases through emissions from land-use change. *Science* 319:1238–1240.

Sheehan, J., A. Aden and K. Paustian, et al. 2003. Energy and environmental aspects of using corn stover for fuel ethanol. *Journal of Industrial Ecology* 7:117–146.

Solomon, B. D., J. R. Barnes and K. E. Halvorsen. 2007. Grain and cellulosic ethanol: History, economics, and energy policy. *Biomass and Bioenergy* 31:416–425.

Wyman, C. E. 1994. Ethanol from lignocellulosic biomass: Technology, economics, and opportunities. *Bioresource Technology* 50:3–15.

Xing-Cai, L., Y. Jian-Guang, Z. Wu-Gao and H. Zhen. 2004. Effect of cetane number improver on heat release rate and emissions of high speed diesel engine fueled with ethanol-diesel blend fuel. *Fuel* 83:2013–2020.

Zhu, J. Y. and X. J. Pan. 2010. Woody biomass pretreatment for cellulosic ethanol production: Technology and energy consumption evaluation. *Bioresource Technology* 101:4992–45002.

88 Impact of Corn Ethanol Production on Land Use in Brazil

Luís A. B. Cortez
University of Campinas

88.1 INTRODUCTION: A BRIEF HISTORY OF LAND USE AND DEFORESTATION IN BRAZILIAN AMAZON

To understand the logic behind the deforestation process, one should try to see how an observer would see Brazil at the beginning of its colonization by the Europeans, which officially started in the early 16th century. Brazil was considered a vast tropical continent in which the tropical forest was seen as 'green hell' (da Silva Queiroz, 2017).

The history of economic interest in the Amazon Forest began with the rubber boon in the late 19th century, which attracted a significant number of migrants mainly from the Brazilian Northeast (Bueno, 2012). However, the rubber cycle entered decline by the mid-20th century with the rise of petroleum rubber and the rubber tree being planted in Southeast Asia. An emblematic land occupation initiative during that period was the Fordland project for rubber tree (*Hevea brasiliensis*) plantation, which was verified to be unsustainable because in their natural habitat these plants do not grow together.[1]

Later, President Vargas' government, the civilian governments, and then the militaries after 1964 promoted the colonization process in the Amazon region. The implementation of the Trans-Amazonic Road (BR-230) crossing the forest from east to west was considered the most important initiative for the colonization process. The region was considered a national security by the military. The Trans-Amazonic Road was part of the National Integration Plan (NIP) and a vector to promote land occupation, for both relatively small 'agrovilas' and extensive pastureland (Chagas et al., 2017).

Other roads were planned and implemented, particularly between 1955 and 1985, transforming the region's accessibility by road. The Belem–Brasilia (BR-010), Cuiaba–Porto Velho (BR-364), Cuiaba–Santarem (BR-163), Porto Velho–Manaus (BR-319), and Manaus–Boa Vista (BR-174) are among the most important roads. These roads together represent the main venues for the installation of large pastureland areas with correspondent deforestation.[2] The colonization along the roads was of the 'fishbone' type.

Deforestation in the Amazon Forest was used for road construction and opening fields for agriculture and new pastureland.

Cattle expansion in Brazil, which happened mainly in the North and Central-West region, has been one of the causes of Amazon Forest deforestation (Alencar et al., 2016). However, according to Martha et al. (2012), frontier expansion to increase beef production happened only from 1950 to 1975; expansion of extensive pasture occurred due to low opportunity costs and governmental incentives to develop the Cerrado biome and parts of the Amazon Forest, plus the need to secure land property. Since 1975, cattle production has increased due to productivity and genetic improvement, better animal health and forage quality, better animal management practices, and improved nutrition, but without causing pastureland expansion. Currently, Amazon Forest deforestation is

DOI: 10.1201/9781003226567-118

driven by illegal logging, lack of land tenure, deforestation of small area plots, and deforestation due to indigenous and rural settlements, with only a relatively small part due to livestock and agricultural expansion (Brazilian Roundtable on Sustainable Livestock, 2016).

The premise is that cleared land has some value in contrast to land with natural cover, and the tropical rainforest does not have value. However, as stated by Abramovay (2019), Amazon Forest preservation should be more profitable than cutting it down. The problem is complex because a great effort should be made to create new products from the preserved forest and its environment.

In the last decades, the occupation with correspondent deforestation was somehow supported and encouraged by an official government agency,[3] the National Institute for Colonization and Agrarian Reform (INCRA), part of the Ministry of Agriculture (MAPA). The logic behind the NIP was that land titles were given to new settlers, but the property was only consolidated if a certain percentage, usually at least 50%, was cleared within a few years after the colonization started. In practice, the Brazilian Government was supporting the deforestation process in an attempt to maintain control of the land. The problem with the deforestation with the elimination of biodiversity, besides the first negative impact, is the long-term problems such as increased greenhouse gas (GHG) emissions and regional climatic change with evident impact on the continent's environmental stability.

Due to all these issues, several actions are needed and are presented below: The ongoing deforestation should be immediately stopped, to put into practice the new Brazilian Forest Code[4] in which at least 20% of the land should be preserved.

Of course, this will not be sufficient to revert the devastation process that has already occurred. It is estimated that deforestation has already destroyed nearly 20% of the total Amazon Forest biome. Therefore, a large-scale action should be proposed to improve the overall situation. So, here it is proposed to direct the action of the most abundant reason, the extensive beef production.

88.2 PRESENT AND TRENDS OF LAND USE IN BRAZIL

The evolution of land use in Brazil has been dramatic, particularly in the last decades. A glance at that can be seen in Table 88.1. As stated before, the land change was the result of a policy defined by the Brazilian government oriented by the desires of the Brazilian population but also as a reaction to international community pressures.

Brazil Congress approved the new Brazilian Forest Code in 2012.[5] The new Forest Code intends to limit deforestation by establishing a minimum of 20% of land to be protected on every property. After all, it can be considered progress.

TABLE 88.1
Evolution of Land Use in Brazil, 1985–2020

Land Use	1985	2020
Native forest	646	564
Pastureland	110	154
Agriculture	20	56
Silviculture	1.4	7.5
Total land (Mha)	851	

https://mapbiomas.org/infograficos-1?cama_set_language=pt-BR.

The 154 Mha on pastureland is, in fact, a conservative figure. The exact number is difficult to know. Probably, it is more realistic to say that pastureland occupies anything from 150 to 180 Ma. Some references mention 200 Mha.

It is relevant to mention that in the last decades improvements have been made in animal density and, more importantly, in overall productivity, measured by the production of meat per hectare per year. The animal density is, therefore, an indirect measurement.

A remark here is that pasture is the only large 'land frontier' or opportunity to expand Brazilian agriculture, for food and other uses such as biofuel production. It is recognized as 'future trend' to recover pastureland and intensify cattle stocking rate to release area for crop production (Brazilian Roundtable on Sustainable Livestock, 2016).

According to Vale (2014) and Latawiec et al. (2014), pasture intensification in Brazil is a viable way to increase agriculture production and spare land, consequently causing no deforestation or indirect land use change (iLUC). In Brazil, most of the agriculture expansion is already happening in pasturelands.

88.3 BEEF CATTLE PRODUCTION IN BRAZIL

Cattle were introduced in Brazil in the 16th century first to be used for labor and later to produce meat until today. Brazil is the second largest beef producer in the world (USDA, 2016). Currently, according to the Ministry of Agriculture, Livestock and Supply, Brazil has the second largest cattle herd in the world with 213 million heads (BRASIL, 2015) and a cattle stocking rate of around 1.2 heads per hectare. The cattle sector is one of the main pillars of the Brazilian economy and plays an important economic role in exports.

In Brazil, the Nelore[6] breed is mostly used for beef production, and the management system varies among the regions (EMBRAPA, 2005), but it is usually predominantly extensive pasture grazed all year long with a small fraction of beef finished in feedlots (FAO, 2006). Due to extensive management with no advanced technology or proper management, Brazilian beef production has one of the lowest costs, but also the lowest productivity rates per hectare in the world (BOVIPLAN, 2015). Moreover, EMBRAPA (2014) estimates that 50% of pastureland in Brazil is hardly degraded, which means around 85 million hectares need intervention.

The Brazilian Central-West region, located on the South border of the Amazon region, includes Mato Grosso, Mato Grosso do Sul, and Goias State, which have the largest pasture areas, cattle herds, and slaughter rates.

There are three types of beef cattle production systems in Brazil: cow–calf, backgrounding and stocker, and finishing. Usually, calves and heifers are raised in pasture and fattening can occur in pasture or feedlots. In a complete cycle, cows represent 35% of the herd and calves 20% of the herd. The higher the herd productivity, the higher the calf representation will be (EMBRAPA, 2011).

In Brazil, there are different types of climates and soil, and three beef cattle production systems can be managed in different systems: extensive, semi-intensive, and intensive, which can also be intensive-irrigated.

The extensive system uses basically only pastureland at the maximum of the existing natural resources. In the semi-intensive system, the animals are maintained in large pastureland, without supplementary feed, which increases the time to slaughter, which basically leads to a less efficient system. With the increasing number of animals per hectare, there will be a need for confinement, which implies complementary feed and water.

These are the classic cattle management systems, but there are variations according to the context and local conditions.

88.4 PRESENT SITUATION OF LAND USE FOR BEEF PRODUCTION IN BRAZIL

One important point here is to remember that whatever the criteria for beef productivity based on occupied area, by either animal density (animal units/ha) or meat production (kg of meat/ha year), the distribution of number of animals/ha is not homogeneous throughout the country.

Assuming its distribution follows a normal curve, it would be reasonable to consider that the optimization procedure addresses the less productive farms because it is exactly there where the intensification will probably require less economic effort and make more sense. However, more studies are needed to verify the correct distribution and opportunities, particularly in the more critical areas located on the south border of the Amazon region (mainly Acre, Amazon, Rondonia, Para, Tocantins, and Maranhao).

The South border of the Amazon region is also the area where corn ethanol distilleries are being installed in Brazil. They are already about ten in number, with half of them located on the state of Mato Grosso because of the largest availability of low-cost corn.

88.5 WHY NOT SUGARCANE? WHY CORN?

Sugarcane is indeed a very good crop and feedstock for ethanol production. Brazil presently cultivates nearly 10 Mha of sugarcane. With that, the country is the first producer and exporter of sugar[7] and uses sugarcane ethanol as fuel. Ethanol represents around 43% of light vehicles' fuel consumption in Brazil.[8] However, sugarcane has difficulties expanding due to the limits of the sugar market as its coproducts. In the present model, 50% of sucrose goes to sugar and 50% to ethanol, and there are difficulties associated with this interdependence.

According to Souza (2017), there are few successful cases of sugarcane ethanol–beef cattle integration still in operation. Vale do Rosario Mill and Usina Estiva Mill are two real cases of integration still in operation in Sao Paulo State. The Vale do Rosario Mill is located in Morro Agudo, Sao Paulo. The plant has an annual processing capacity of 6.5 million tons of sugarcane (BIOSEV, 2022). Their feedlot functions as a 'hotel' to finish cattle from other properties around the plant, properties that are rented to produce more sugarcane. In 2010, 163 cattle producers finished about 20,000 heads in this feedlot, and in 2008, it had been better: 30,000 heads (PORTALDBO, 2021). The feed using 80% of sugarcane by-products (*in natura* bagasse, hydrolyzed bagasse, molasses, and wet yeast) is about R$100[9] cheaper (per ton) and more efficient than traditional ones (BIOSEV, 2022). According to technical visits to the Vale do Rosario plant, they have been doing this integration for more than 25 years.

Usina Estiva Mill has been integrating sugarcane ethanol with cattle for about 18 years. The plant has an annual processing capacity of 2.8 million tons of sugarcane, and the feedlot created in 1999 has 11,000 heads of static capacity and about 20,000 heads per year. Integration happens using the feed to finish cattle composed of 50% of sugarcane by-products: *in natura* bagasse, hydrolyzed bagasse, molasses, and wet yeast, and using the manure as crop fertilizer. Through this production model, they have better cattle performance (better average daily gain), which is 15% higher than the national average, <20% of the total cost, and avoid expanding 12,000 hectares per year necessary to produce the same amount of cattle without integration. This extra area is used to produce more sugarcane (USINA ESTIVA, 2015).

Although the integration of beef cattle with sugarcane ethanol went relatively well in Brazil, it could not be considered a great success. The reason remains in the fact that technically sugarcane is not exactly a good source of protein and needs to be supplemented with other more costly ingredients.

However, corn is the most important feedstock utilized for ethanol production. It has been extensively used in the USA. According to the U.S. Department of Agriculture (USDA), in 2019 the USA produced nearly 347 M tons of corn[10] using nearly 30 Mha.[11] Of that, about 125 M tons of corn was used to produce nearly 60 billion liters of ethanol allowing the USA to substitute around 10% of gasoline for light vehicles.

However, corn ethanol has received criticism in the last decades due to possible competition with food and poor GHG emission mitigation. In the first years of expansion of corn ethanol production in the USA, there was indeed an impact on corn prices, but in the long run, corn prices went back to normal. Also, it was verified that commodity prices, not only corn, usually follow oil prices and that particular price increase could at least in part be devoted to oil price instabilities during the first decade of this century. Regarding the poor GHG emissions, in the USA natural gas is used as an energy source during processing, which dramatically worsens the overall energy balance of corn ethanol.

It is important to mention that a reasonable strategy would be to combine the positive points of sugarcane and corn. Sugarcane is a well-succeeded raw material for ethanol production in Brazil. It offers high overall productivity and high energy ratio[12] and produces its own fuel, bagasse, which can be stored and used to produce bioelectricity and all energy needs in the mill. However, as mentioned before, corn offers the possibility to produce protein, which allows integration with beef cattle. Besides, corn grains can be stored allowing the mill to operate year-round. Therefore, an interesting opportunity to be better explored would be to combine the advantages of both. For example, a case study could be the State of Sao Paulo with its 150 sugarcane ethanol mills, introducing the 'flex mills' concept,[13] although this is a subject for another discussion.

88.6 THE US CORN ETHANOL PRODUCTION INTEGRATED WITH BEEF PRODUCTION

In the USA, there are two basic processes to produce corn ethanol: the wet and dry methods. In the wet method, besides ethanol, a residue named wet distillers' grain (WDG) is produced. In the dry method, besides ethanol, a residue named dried distillers' grain (DDG) is produced.

Both WDG and DDG are utilized for animal feed in the USA. They are used for beef cattle, dairy cattle, swine, poultry, aquaculture, horse, sheep, goat, and turkey. The most important user of corn ethanol distillers' grains in the USA is beef cattle (US Grains Council, 2015). WDG is considered the best coproduct to feed cattle, and it has the highest energy content among the others. It is mostly used in the United States, because it has better animal performance. However, because it is wet, the WDG is too perishable and hard to handle and store, and it is not economically viable to transport over long distances.

In the USA, the corn ethanol industry is totally integrated with the beef cattle industry. The USA produces around 11 M tons of beef meat, and a significant part of that benefits from the use of corn ethanol, the high-protein WDG or DDG.

In the USA, much of the distillers' grains with solubles are fed wet to finishing cattle, and as a result, they have a higher energy value than DDG. Feeding WDG results in better growth performance compared with feeding DDG to finishing cattle (Erickson et al., 2005).

88.7 CAN THE US CORN–BEEF MODEL BE ADAPTED IN BRAZIL?

Several authors have recently reported the opportunities for corn ethanol production in Brazil (da Silva and Castaneda-Ayarza, 2021; Moreira et al., 2020; Santos and Franco, 2018). However, only a few authors have highlighted the opportunity that corn ethanol production represents for the Brazilian feed industry (Neves, 2020).

In Brazil, a great part of beef cattle is raised extensively using much more land than in the USA. Historically, land was not considered a problem in Brazil and cattle was seen as an important vector for land occupation in different directions, including the so-called colonization of the far-North Amazon region.

Today, around 90% of beef meat is produced extensively. The area occupied by beef cattle reaches 160–180 Mha, around 20% of Brazilian territory, but the distribution of this cattle is unequal. Its density can vary from 0.3 to almost 2 animals/ha, with 1 animal/ha as its average. Also, the national distribution is very unequal, and as the production goes North closer to the Amazon region, the animal density tends to decrease. It can be said that the animal assumes less of the task of producing meat and more of the role of guaranteeing the land property. The land occupation, particularly in the Central and Northern regions (together 65%–70% of Brazilian territory), was encouraged by government policies, particularly during the military regime (1964–1985). This practice implemented by INCRA, a national organization for colonization and agrarian reform, was not forgotten even after the militaries left the government to civilians. The deforestation and introduction of low-density cattle were used as a way to guarantee land property in this vast region.

Therefore, if the process is to be reversed, eliminating deforestation and unsustainable land use practices such as low-grade beef production with soil impoverishment, an anti-cyclic policy, needs to be put in place.

This new strategy needs to recognize that the extensive unsustainable beef cattle production needs to give place to a more rational sustainable beef cattle industry through its gradual intensification.

Corn residues such as distiller's dried grains (DDG) (also named distiller's dried grains with solubles (DDGS)) and wet distillers grains (WDG) (also named wet distillers grains plus solubles (WDGS)) obtained from corn ethanol production are rich in protein and other nutrients and can effectively be used to promote this gradual intensification of the Brazilian beef industry, particularly that located in Central and North Brazil, which creates an environmental threat to the Brazilian Amazon Forest.

Therefore, if the US model is adapted to the conditions prevailing in Brazilian territory, the region could have an important reduction in land used for beef cattle while creating exceptional conditions for land use reform in Brazil.

Of course, there are adjustments to be made, coupling these two industries. The Brazilian beef cattle breeding systems need to be properly adjusted to the different corn ethanol production systems (wet and dry). The resulting residues must be offered according to the needs established by the correspondent client.

88.8 HOW MUCH LAND CAN BE SPARED?

Berndes et al. (2016) state that the released pasture area from intensification could accommodate expansion of crop production and decrease in deforestation. According to the authors, pasture intensification is possible with an improvement in land productivity; the improvement in meat and dairy production is essential to release area for bioenergy crops since most of the available land for bioenergy production is currently used as extensive pasture.

The ultimate objective of this text is to demonstrate with arguments and educated guess estimations that it could be possible to implement a strategy to reduce or even abolish deforestation in the Brazilian Amazon region by expanding corn ethanol production in that area. Naturally, as explained before, the full strategy would need to be complemented with the adoption of necessary policies and economic incentives, such as the implementation of a tropical economy based on the sustainable exploitation of natural resources and adding value to existing and new products. The implementation of this strategy should be mandatory to promote a new land use model for the Brazilian Amazon region. The important question at this point is how much land could be spared.

First, it is important to recognize the complexity of this question. However, the only data for corn ethanol integration with beef cattle come from the USA.

To make a rough estimate of that, initially some numbers are as follows:

Corn in Brazil: Corn productivity in Brazil, as the second crop: 6 tons/ha year.

Basic parameters from corn ethanol in the USA: the amount of the high-protein residue, here generically named DDG, produced per ton of corn processed ~400 kg of WDG[14]/ton of corn (US Grains Council, 2015), and a rough estimation of how much WDG is eaten by an animal per day, for a finishing cattle: ~1% of the animal's weight.

Beef cattle features in Brazil: expected weight gain with the semi-intensive raising method for finishing: 120 kg (from 350 to 470 kg/animal[15]).

Assuming a farm with 100,000 ha of pastureland and 1 animal unit/ha wishes to plant corn for ethanol production, in their plans, 50,000 ha of land would have to maintain 100,000 animals. The remaining 50,000 ha of land would be used for corn, eucalyptus, and other uses.

Assuming each animal averages weighs ~400 kg, 100,000 animals weigh 40,000 tons.

If they eat ~1% of their weight/day, they will eat ~400 tons of WDG/day. In 110 days of finishing,[16] it will result in 44,000 tons of WDG/year.

Considering ~400 kg of WDG/ton of corn, 110,000 tons of corn will be needed for that ethanol plant. Adopting a productivity of 6 tons of corn/ha (as the second crop), ~20,000 ha will have to be planted.

As far as the ethanol plant is considered, ~400 liters of ethanol/ton of corn are expected, with 110,000 tons of corn resulting in 44,000 liters of ethanol/year.

Therefore, the final number could be summarized as follows:

Before integration: system: extensive; animal density: 1 animal: 1 ha; and area occupied by pastureland: 100,000 ha.

After integration: system: semi-intensive; animal density: 2 animals: 1 ha, area occupied by pastureland: 50,000 ha; area occupied by corn: 20,000 ha; ethanol production: 44,000 liters/year; area occupied by eucalyptus (for energy): 3,000 ha; total area (corn + eucalyptus): 23,000 ha; and total spared land: 27,000 ha.

Based on the estimations above, 27,000 ha of land could be spared with 23,000 ha dedicated to corn and eucalyptus and a small corn ethanol distillery. This spared land could be used for food crops or even to replant the previously existing tropical forest.

Naturally, these are estimations based on figures that need to be checked and proven to be feasible, considering the Brazilian conditions.

Among the scientific challenges involved in expanding corn ethanol production integrated into beef cattle in Brazil, we can mention the following:

Cattle:

- What is the capacity of Brazilian cattle to adapt to the new feeding model that involves the use of corn ethanol coproducts, such as WDG/DDG?
- Should Brazil develop specific breeding for semi-intensive beef cattle production eating WDG/DDG as animal feed? In the USA, the majority of breeds integrated with corn ethanol are British breeds (*Bos taurus*).
- The beef cattle intensification process. How will it be done?

Corn:

- Can the double cropping system be fully utilized? Presently, there are about 35 Mha on soybean and half of that on corn as the second crop.
- What if corn is to be produced as the main crop?
- Climate change may be a threat to double cropping.

GHG emission reductions and environmental issues:

- What is the carbon intensity of corn ethanol and the new beef produced in Brazil?
- What are the effective global and local (deforestation) contributions?

Ethanol market and economic impacts:

- How much low-cost ethanol could be produced using the corn ethanol–beef cattle integration?
- Once the market is fully supplied with fuel ethanol, will there be a possibility to export ethanol to their regions in Brazil?
- Is there a tangible socioeconomic benefit in implementing this new strategy?

Policies needed:

- What is the set of policies that need to be implemented?

88.9 LUC AND iLUC CONCEPTS AND THEIR IMPACT ON BIOFUELS

An article published by Searchinger et al. (2008) introduced land use change (LUC) and iLUC concepts. The author explained that when biomass is planted it will ultimately cause two effects: a direct LUC and an iLUC.

The LUC can be defined as the change in land use when another crop is grown. It is a direct effect, and one crop substitutes for the other. For example, LUC happens when a given crop, such as biomass 'pushes' a previous crop or pasture. This is the 'direct' LUC.

The iLUC refers to the second effect, the 'land whose ultimate purpose is essentially changed from its previous use'.[17] It can be referred to as 'indirect' LUC and understood as a kind of 'domino effect'. According to the same source: 'an example would be forest land that was cleared for the cultivation of biofuel crops'.

These two concepts (LUC and iLUC) are considered very important in this text because they impact the life cycle analysis (LCA) for a given biofuel. It can be stated that a given biofuel LCA could be highly influenced by the iLUC if the indirect change is such as displacing a dense forest.

In that paper (Searchinger et al., 2008), it was stated that the negative effects were sufficient to offset the positive effects. So, somehow, all the identified direct and indirect benefits of biofuels were considered important but not sufficient, particularly when reconsidering land use and overall GHG emissions.

Considering these concepts and their impact on biofuel LCA, it can be stated that a given biofuel LCA could highly depend on which crops it displaces, directly and indirectly. Searchinger concluded that the sugarcane ethanol produced in Brazil performed better than the corn ethanol produced in the USA but still resulted in a 'small carbon debt'. This happened because of iLUC in the Brazilian situation because sugarcane planted in Southeast Brazil, ultimately, was displacing tropical rainforest on the South of the Amazon Forest. Directly sugarcane displaced pastureland, but indirectly pastureland was 'reappearing' on the South of the Amazon region as deforestation. Therefore, Searchinger concluded that sugarcane was indirectly causing deforestation and that should be considered by adding more CO_2 equivalent emissions to the sugarcane ethanol produced in Brazil.[18] Figure 88.1 shows the iLUC effect attributed to sugarcane in Brazil.

It is very important to understand that this iLUC logic could be reversed. In this way, Searchinger et al. (2008) presented that the calculated iLUC was adding all the carbon emitted by cutting down the tropical forest in the Amazon Forest because the trees were cut due to the iLUC effect. An illustration of the iLUC effect attributed to sugarcane is presented by Souza (2017) to better visualize iLUC.

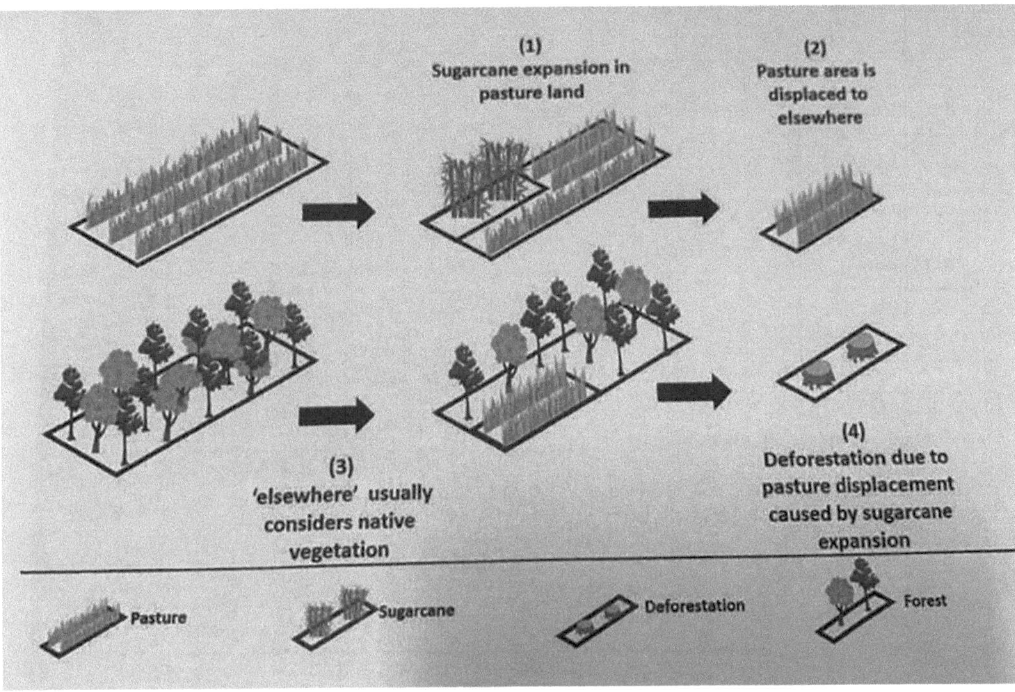

FIGURE 88.1 Illustration representing the iLUC effect attributed to sugarcane. (Souza (2017).)

This logic could be used, therefore, in the opposite direction. If introducing corn ethanol production integrated with beef production, land could be spared freeing land for different uses, including the regrowth of the forest and other carbon-intensive crops in that area; ultimately, the new iLUC derived from corn ethanol would have to be discounted and not added. This would considerably improve the LCA of corn ethanol and the LCA of beef production.

This strategy was suggested by Cortez et al. (2014) in a study of bioethanol and animal breeding integration dynamics in Brazil and exemplified by Figure 88.2. In that study, processed sugarcane bagasse was used as animal feed. Sugarcane bagasse is a low-protein raw material that needs to be supplemented. Corn ethanol DDG is a more complete substrate for beef production under confined conditions.

88.10 WHY THIS STRATEGY IS IMPORTANT FOR BRAZIL?

First, it is important to recognize that Brazil is making good progress in food production and the country became an import food exporter. Also, in the last decades, Brazil has made considerable progress in energy security. Renewables (hydro, wind, solar, and bioenergy) respond for nearly 46% of all energy consumed, making Brazil the best case among the top ten economies in the world.

When GHG emissions are analyzed (Figure 88.3), Brazil emits <1/3 for energy, residues, and industry and nearly two-thirds for agriculture, beef, LUC, and forests. So, if Brazil wants to reduce its GHG emissions in the next decades, clearly it needs to do more work on land-related activities.

The strategy proposed in this text is exactly to make a solid answer to this important question. Contrary to what many specialists believe, land use, beef production, and the Amazon Forest could greatly benefit from the expansion of corn ethanol integrated into beef production. Furthermore, it is important that it happens on the southern border of the Amazon region where extensive beef production is located and where overall agricultural conditions are highly favorable for using second crop corn.

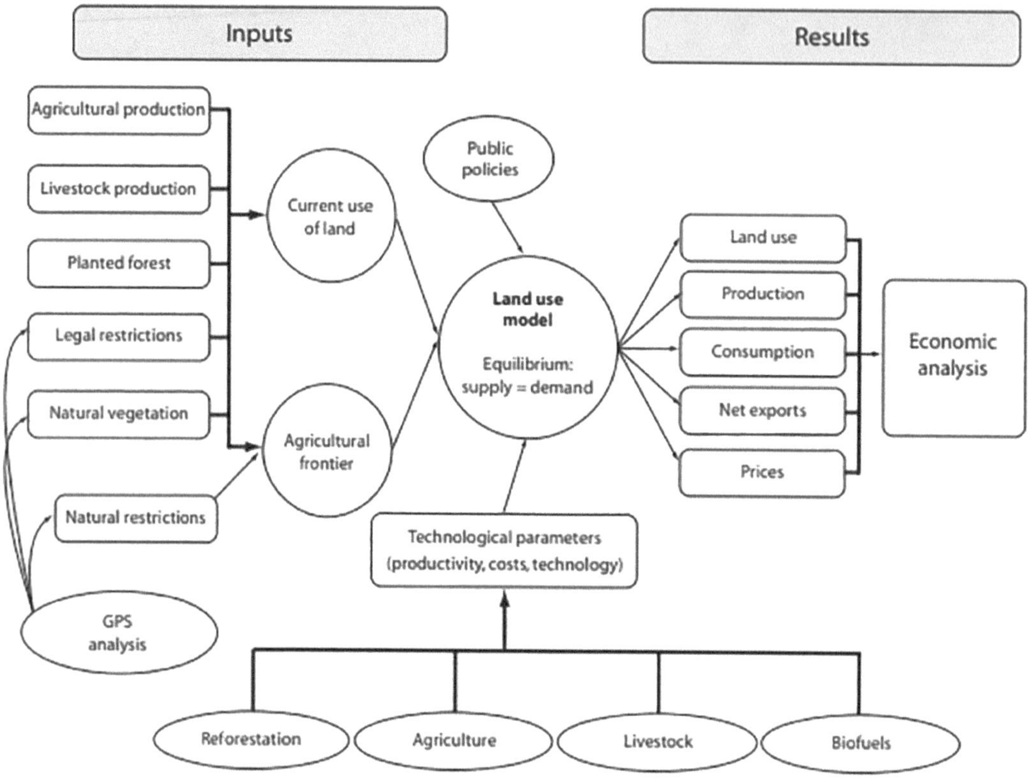

FIGURE 88.2 Methodology diagram of the Brazilian Land Use Model (BLUM). (Cortez et al. (2014).)

88.11 IMPLEMENTING NEW POLICIES FOR LAND USE IN THE AMAZON REGION

As demonstrated above, the high deforestation in the Amazon Forest is a result of several policies driven by several stakeholders, including the Brazilian government, under the perception of a threat to national security.

True or false, the international community perceives Amazon Forest deforestation as the single most important threat to the environment caused by Brazil. The international community also tries, in its own way, using the press, and different political pressure methods to make the Brazilian government act more firmly to reduce deforestation in the region. These pressures also address issues regarding the use of pesticides in crops such as soybeans. Besides that, the critics are also addressed against beef production in the Amazon region. All these combined are understood by the Brazilian Government as foreign intervention, a threat to national security, and end up creating the opposite effect, making the Brazilian Government reluctant to make the necessary changes.

Therefore, what is proposed here is a change in the way both players, the Brazilian Government together with the international community and other important stakeholders, will address the issue and create the necessary environment to make substantial changes in the land use in the Amazon region, while satisfying the preconditions, such as Brazilian sovereignty over the region, preserving

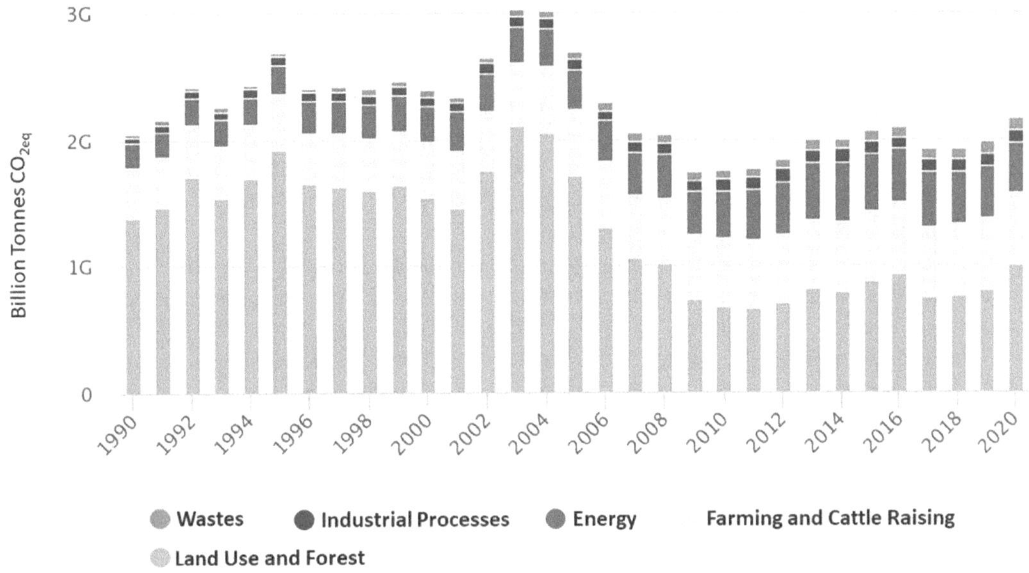

FIGURE 88.3 Brazilian emissions of CO_2 equivalent in tons by sector, 1990–2020, in billion tons of CO_2. (SEEG. https://plataforma.seeg.eco.br/total_emission.)

the Amazon Forest, increasing sustainable food production, and increasing income for the Amazon region population (Cortez and Baldassin, 2016; Cortez et al., 2019).

88.12 BUILDING A TROPICAL BIOECONOMY

An expected, question regarding the sustainability of the Amazon Forest in Brazil is that its economy should guarantee enough income for the local population; otherwise, it would always bring social inequalities. In fact, the population in the Amazon region is not homogeneous. There are large cities such as Manaus and Belem, with more than 2 million, a large number, probably 100, of medium-sized cities, and thousands of small villages. Besides that, there are a great number of small farms and 'agrovilas', and several legal and illegal mining. Finally, a considerable number of indigenous populations usually occupy reserves.

The question is how this population will guarantee a decent social–economic status. The answer to that question is not simple and probably passes by several actions, such as maintaining the free zone (Zona Franca) in Manaus; introducing crops and agricultural activities that will preserve the environment and at the same time guarantee enough revenue; and developing more substantial bioeconomy activities focusing on higher-value products.

An experience of developing a tropical bioeconomy in Brazil was conducted by the Agropolo Campinas – Brasil coordinated by the Agronomic Institute of Campinas (IAC) inspired by the Agropolis Project in Montpellier, France (Carbonell et al., 2021). A roadmap was developed for ten agricultural activities in Sao Paulo State in important research areas such as crops (coffee and sugarcane) and environment (water). The same kind of roadmap could be developed for the Amazon region focusing on local products such as fruits, essential oils, and pharmaceutical products.

88.13 POSSIBLE TRANSFORMATION BY ADOPTING CORN ETHANOL PRODUCTION INTEGRATED WITH BEEF PRODUCTION IN BRAZIL

Brazil still has great potential to improve beef production. Great efforts have already been made by EMBRAPA and can be accomplished by improving the quality of its pastureland, genetic improvements of its herd, and utilization of better feed and managing practices.

This strategy could be at least partially accomplished by adapting the ethanol–livestock integration that happens successfully in the United States.

What is proposed here is a radical change in the way land is being used in the Amazon region through the intensification of beef production and the introduction of corn ethanol production in the region.

Table 88.2 summarizes the present and future situations of the Amazon region, before and after several actions are implemented. As can be seen, the positive impact can be very important, and contrary to belief, large-scale corn ethanol integrated with beef production can greatly benefit the Amazon region. Of course, several other actions will be needed, including the introduction of a tropical bioeconomy in the region, to include more people in this effort. Also, the international community should be prepared to accept and promote these actions validating the entire process.

88.14 CONCLUSIONS

The strategy proposed in the present text is highly controversial. This can be understood for two main reasons.

First, there is a general tendency to believe that to preserve the Amazon Forest and other ecological sanctuaries the best approach is to avoid disturbing the existing environment and rather use conservative protection methods.

Regarding this critic, the argument is that the introduction of corn ethanol production needs to be done on abundant and already disturbed lands, such as pastureland. Pastureland in the Amazon

TABLE 88.2
Present and Future Situations of the Amazon Region, by Sector

Present situation	Policies	Previous policies (INCRA) encouraged colonization and malpractices, including deforestation
	Amazon forest	High deforestation rate
	Beef production	Highly extensive system; high GHG emissions
	Food production	Highly concentrated on soybeans/corn; corn as the second crop
	Energy security	Region with energy insecurity; high imports of ethanol
	Socioeconomic	Low income; economy based on extractivism; small regional GDP
Situation with the implementation of corn ethanol and tropical bioeconomy	Policics	Implement a new set of policies to promote the end of deforestation and more sustainable development in the Amazon region
	Amazon forest	Abolition of deforestation; full implementation of Forest Code
	Beef production	Implement a semi-intensive system; integration of corn ethanol and beef production with DDG
		Integrated corn ethanol–beef production will generate negative iLUC for both productions
	Food production	More diversified crops; more corn as the second crop for ethanol and DDG, and food
	Energy security	Corn ethanol production can eliminate ethanol imports
	Socioeconomic	Improve income; introduction of bioeconomy to add value in the Amazon region

region needs to be reduced and beef cattle needs to stop being used as 'sentinels', or animals, whose main task is to help farmers control the land.

Second, corn ethanol has a relatively bad reputation because of its poor energy ratio and GHG emissions.

The arguments here are that the poor energy ratio presented by corn ethanol is due to the fact that corn does not have its own fuel, as sugarcane does. This also results in bad GHG emissions because more emissions are produced when extra fuel is burned to maintain the mill operation.

However, if the scenario changes and corn ethanol is produced in different circumstances, such as on the South border of the Amazon region, the picture can change dramatically in a positive way.

As explained, first, corn is largely produced in Brazil as the second crop, sharing the land with soybeans. This results in a much better condition because all related inputs are shared. Second, the proposed integration with extensive beef cattle in Brazil will result in a large capacity to save and spare land for other uses. This 'negative iLUC' if added to the corn ethanol LCA will completely change the interpretation.

Therefore, what is proposed here is the use of corn ethanol's 'shrinking potential' over pastureland in Brazil. More studies are naturally needed to evaluate how this adaptation and integration will be conducted.

Finally, it was the intention here to demonstrate that, contrary to what many specialists and the public believe, corn ethanol can help Brazil significantly improve its land use and sustainability indicators and help preserve the Brazilian Amazon Forest. To accomplish these goals, Brazil should implement anticyclical policies to construct a new model of national development, involving the present stakeholders in this challenging process. All these measures together could be used by Brazil, with the help of the international community, to implement a positive agenda for the Amazon Forest.

NOTES

1 https://fee.org/articles/fordlandia-henry-fords-amazon-dystopia/
2 https://www.wwf.org.br/natureza_brasileira/areas_prioritarias/amazonia1/ameacas_riscos_amazonia/infraestrutura_na_amazonia/estradas_na_amazonia/#:~:text=Principais%20estradas%20que%20cortam%20a,Regi%C3%A3o%20da%20rodovia%20BR%2D163.
3 https://imazon.org.br/desmatamento-nos-assentamentos-de-reforma-agraria-na-amazonia/.
4 https://www.embrapa.br/codigo-florestal/entenda-o-codigo-florestal.
5 https://www.embrapa.br/en/codigo-florestal.
6 The Nelore is of the *Bos indicus* species and has the characteristic hump above its shoulders and loose skin. It is usually white in color with black skin, and they have comparatively long legs, which helped them walk in water when grazing. The Nelore has heat and insect resistance due to its loose, thick black skin and covering of white hair, which reflects the sun's rays, and this breed has sweat glands that are twice as big and 30% more numerous compared with European breeds. It has a slow metabolism, which creates less heat and enables it to feed less often adding to its high resistance to bloat (source: https://www.thecattlesite.com/breeds/beef/75/nelore/).
7 https://www.statista.com/statistics/249646/exported-amount-of-sugar-from-brazil/.
8 https://www.epe.gov.br/sites-pt/publicacoes-dados-abertos/publicacoes/PublicacoesArquivos/publicacao-601/topico-588/BEN_S%C3%ADntese_2021_PT.pdf.
9 At that moment, the exchange rate was somewhere between R$ 2.50/US dollar.
10 https://grains.org/corn_report/corn-harvest-quality-report-2019-2020/25/#:~:text=U.S.%20Average%20Production%20and%20Yields,tons%20(13%2C661%20million%20bushels).
11 The average corn productivity in the USA is between 10 and 12 tons of corn/ha.
12 Defined here as the output energy/input energy.
13 More about sugarcane–corn flex mills at Oliverio Dedini's PPT presentation.
14 These estimations used WDG as corn ethanol residue, but the numbers will not change significantly for DDG.
15 The average initial weight is around 350 kg and the final weight is around 470 kg, aging 24–36 months (EMBRAPA, 2011).

16 In feedlots, finishing time varies from 60 to 110 days (240 days in premature system) (EMBRAPA, 2005).
17 https://farm-energy.extension.org/what-is-direct-land-use-or-direct-land-use-change/.
18 Even with that, still sugarcane ethanol produced in Brazil was still considered an 'advanced biofuel' by the US Environmental Protection Agency (EPA) (Cortez et al., 2009; Nassar et al., 2008).

REFERENCES

Abramovay, R. 2019. Preservar a Amazônia é mais lucrativo que desmatar [Preserving the Amazon is more lucrative than destroying]. *Folha de Sao Paulo Newspaper.* https://www1.folha.uol.com.br/ilustrissima/2019/09/preservar-amazonia-e-mais-lucrativo-que-desmatar-diz-economista.shtml.

Alencar, A., C. Pereira and I. Castro, et al. 2016. Desmatamento nos Assentamentos da Amazonia: Historico, Tendencias e Oportunidades. [Deforestation in Amazon Settlements: History, Trends and Opportunities] Brasilia: Amazonia Environmental Research Institute.

Berndes, G., H. Chum, M. R. L. V. Leal, G. Sparovek and A. Walter. 2016. *Bioenergy Feedstock Production on Grasslands and Pastures: Brazilian Experiences and Global Outlook.* IEA. https://www.ieabioenergy.com/wp-content/uploads/2018/01/IEA-Bioenergy-Task-43-TR2016-06.pdf.

BIOSEV. 2022. *Unidades Operacionais da Biosev, 2022 [Biosev Operational Units].* https://www.biosev.com/a-biosev/unidades/, Accessed in April 18th, 2022.

BOVIPLAN. 2015. *Formação e Manejo de Pastagens [Training and Management of Pastagens].* Rio de Janeiro: BOVIPLAN.

BRASIL. 2015. *Dados de Rebanho Bovino e Bubalino no Brasil. [Bovine and Buffalo Herd Cubes in Brazil].* Rio de Janeiro: Ministry of Agriculture, Livestock and Supply.

Brazilian Roundtable on Sustainable Livestock. 2016. *Brazilian Livestock and Its Contribution to Sustainable Development.* Brazilian Roundtable on Sustainable Livestock. https://www.agroicone.com.br/portfolio/gtps-brazilian-livestock-and-its-contribution-to-sustainable-development/ Accessed in April 4th, 2022.

Bueno, R. 2012. *Borracha na Amazonia: as Cicatrizes de um Ciclo Fugaz e o Inicio da Industrializacao* [*Amazon: The Scars of a Fleeting Cycle and the Beginning of Industrialization*]. Porto Alegre: Quattro projetos. https://premiocnh.com.br/livros/livro2012.pdf.

Carbonell, S. A. M., L. A. B. Cortez and L. F. C. Madi, et al. 2021. Bioeconomy in Brazil: Opportunities and guidelines for research and public policy for regional development. *Biofuels, Bioproducts and Biorefining* 15:1675–1695.

Chagas, A. M., P. A. Hecktheuer and F. R. Hecktheuer. 2017. O discurso da internacionalização da Amazonia: do imaginario das narrativas a racionalidade instrumental dos projetos. [The discourse of the internationalization of the Amazon: From the imaginary of the narratives to the instrumental rationality of the projects]. *Novos Estudos Juridicos* 22:3.

Cortez, L. A. B., J. L. Oliverio and J. Goldemberg. 2019. Como reduzir o desmatamento da Amazonia [How to reduce deforestation in the Amazon]. *Valor Econemico Newspaper*, December 12, 2019.

Cortez, L. A. B., M. R. L. V. Leal, A. M. Nassar, M. M. R. Moreira, S. F. Taube-Netto and M. da Silva. 2014. Land requirements for producing ethanol in Brazil. In *Sugarcane Bioethanol: R&D for Productivity and Sustainability*, Ed. L. A. B. Cortez, pp. 301–318. Sao Paulo: Editora Edgard Blucher.

Cortez, L. A. B. and R. Baldassin. 2016. Policies towards bioethanol and their implications: Case Brazil. In *Global Bioethanol: Evolution, Risks, and Uncertainties*, Eds. S. L. M. Salles-Filho and L. A. B. Cortez, pp. 142–162. Cambridge, MA: Academic Press.

Cortez, M. M. 2019. *Uma Proposta de Agronegocio Sustentavel para a Amazonia. [A Sustainable Agribusiness Proposal for the Amazon].* Campinas: State University of Campinas.

da Silva Queiroz, J. F. 2017. Amazania: Inferno Verde ou paraíso perdido? Cenário e território na literatura escrita por Alberto Rangel e Euclides da Cunha [Amazonia: Green Hell or Paradise Lost? Scenario and territory in the literature written by Alberto Rangel and Euclides da Cunha]. *Nova Revista Amazonica* 5:6256.

da Silva, A. L. and J. A. Castaneda-Ayarza. 2021. Macro-environment analysis of the corn ethanol fuel development in Brazil. *Renewable and Sustainable Energy Reviews* 135:110387.

EMBRAPA. 2005. *Sistemas de Producao de Gado de Corte no Brasil: Uma Descricao com Enfase no Regime Alimentar e no Abate* [*Beef Cattle Production Systems in Brazil: A Description with Emphasis on Food and Slaughtering*]. Brasilia: Embrapa. https://www.embrapa.br/en/busca-de-publicacoes/-/publicacao/326307/sistemas-de-producao-de-gado-de-corte-no-brasil-uma-descricao-com-enfase-no-regime-alimentar-e-no-abate, Accessed in April 4th, 2022.

EMBRAPA. 2011. *Gado de Corte. [Beef Cattle].* 2nd ed. Brasília: Embrapa.

EMBRAPA. 2014. *Diagnostico das Pastagens no Brasil* [*Diagnosis of Pastures in Brazil*]. https://www.embrapa.br/en/busca-de-publicacoes/-/publicacao/986147/diagnostico-das-pastagens-no-brasil, Accessed in April 4th, 2022.

Erickson, G. E., T. J. Klopfenstein, D. C. Adams and R. J. Rasby. 2005. *Utilization of Corn Coproducts in the Beef Industry*. Lincoln, NE: Nebraska Corn Board.

FAO. 2006. *Country Pasture/Forage Resource Profiles: Brazil*. Rome: Food and Agriculture Organization.

Latawiec, A. E., B. B. N. Strassburg, J. F. Valentim, F. Ramos and H. N. Alves-Pinto. 2014. Intensification of cattle ranching production systems: Socioeconomic and environmental synergies and risks in Brazil. *Animal* 8:1255–1263.

Martha, G. B., E. Alves and E. Contini. 2012. Land-saving approaches and beef production growth in Brazil, 2012. *Agricultural Systems* 110:173–177.

Moreira, M. M. R., J. E. A. Seabra, L. R. Lynd, S. M. Arantes, M. P. Cunha and J. J. M. Guilhoto. 2020. Social-environmental and land-use impacts of double-cropped maize ethanol in Brazil. *Nature Sustainability* 3:209–216.

Nassar, A. M., L. B. Rudorff and L. B. Antoniazzi, et al. 2008. Prospects of sugarcane expansion in Brazil: Impacts on direct and indirect land use changes. In *Sugarcane Ethanol*, Eds. P. Zuurbier and J. Van de Vooren, pp. 63–93. Wageningen: Wageningen Academic Publishers.

Neves, M. F. 2020. *A Sustentabilidade do Etanol de Milho no Brasil* [*The Sustainability of Corn Ethanol in Brazil*]. Sugar Cane, Corn and Animal Protein Integration. Webmeeting Fermentec.

PORTALDBO. 2021. *Um Quarto de Seculo de Integracao Cana-Pecuaria, [A Quarter of a Century of Sugarcane-Livestock Integration]*. https://www.portaldbo.com.br, Accessed in April 4th, 2022.

Santos, J. C. and N. Franco. 2018. *Potencial Econômico e Produtivo do Etanol de Milho no Brasil [Economic and Productive Potential of Corn Ethanol in Brazil]*. University of Sao Paulo.

Searchinger, T., R. Heimlich and R. A. Houghton, et al. 2008. Use of U.S. croplands for biofuels increases greenhouse gases through emissions from land-use change. *Science* 319:1238–1240.

Souza, N. R. D. 2017. *Techno-Economic and Environmental Evaluation of Beef Pasture Intensification with Sugarcane Ethanol*. M.Sc. thesis, State University of Campinas.

US Grains Council. 2015. *A Guide to Distiller's Dried Grains with Solubles (DDGS)*. https://www.canr.msu.edu/uploads/236/58572/cfans_asset_417244.pdf.

USDA. 2016. *Livestock and Poultry: World Markets and Trade*. https://apps.fas.usda.gov/psdonline/circulars/livestock_poultry.pdf&gt. Accessed in April 4th, 2022.

USINA ESTIVA. 2015. Confinamento [Lockdown]. https://estiva.com.br/pagina.asp?pagina=30&gt, Accessed in April 5th, 2022.

Vale, P. M. 2014. *The Conservation Versus Production Trade-Off: Does Livestock Intensification Increase Deforestation? Evidence from the Brazilian Amazon*. Centre for Climate Change Economics and Policy.

89 Ethanol Fuel from Sugarcane *Present and Future Contribution to Climate Change Mitigation*

Jose R. Moreira and José Goldemberg
University of Sao Paulo

89.1 ETHANOL AS A FUEL FOR OTTO CYCLE MOTORS: ADVANTAGES AND DISADVANTAGES AS COMPARED WITH GASOLINE

Ethanol is being used in more than 50 countries as fuel for road transportation (USEIA, 2021a). The common practice is a blend with gasoline, but, in a few countries, it is used as an almost stand-alone fuel. This is the case in Brazil with 38 million light-duty vehicles (LDVs), of which 87% can use either gasoline or ethanol (flex fuel vehicles) (EPE, 2020a,b), and in the USA where more than 21 million use a blend of gasoline with 85% ethanol, named E85 (USDOE, 2021b). Worldwide ethanol fuel production reached 105 million m^3 in 2020, which represents 6.9% of gasoline consumption by volume, and the USA is the largest producer with 60 mm^3 (USEIA, 2021b). Major motivations for its use are increased concerns with the environment, the risk of full dependence on external oil supply, and as a stop technology able to control world market oil price.

Almost all ethanol fuel is produced from two crops: corn and sugarcane. Global corn production in 2019 was 1.15 Gt in 2019, harvested over an area of 197 Mha. This means an average global yield of 5.825 t/ha. For sugarcane, the production was 1.95 Gt; the yield was 72.8 t/ha; and harvested area was 26.8 Mha (OECD/FAO, 2020). Assuming that around 50% of ethanol fuel comes from each crop and considering its yields of 80 L/t for sugarcane and 400 L/t for corn, production of 105 mm^3 required 33.6% and 11.4% of the global harvested area for sugarcane and corn, respectively.

The major environmental benefit, measured by the amount of greenhouse gas (GHG) emissions avoided, can be estimated based on several authors as 70.9 $MtCO_2e$ average value, ranging from 32.6 to 108.6 $MtCO_2e$ for the ethanol consumed in 2020 (Table 89.1), taking into account mainly the liquid fossil fuel displaced in the road transportation sector. Considering that sugarcane ethanol mills usually generate bioelectricity using cane residues, this figure is enlarged. Only for Brazil, 23 TWh of such bioelectricity has been delivered to third-party users by 2020, displacing bioelectricity from the grid mix. Assuming a grid mix electricity carbon intensity (CI) of 260 gCO_2e/kWh, in Brazil, and 70 gCO_2e/kWh for such bioelectricity (GREET 2019, 2020), 4.37 $MtCO_2e$ has been avoided, enlarging by a few percent the above-quoted world offset GHG capacity of ethanol. Unfortunately, this figure is small when compared with the global commitment required to accomplish the Paris Agreement. In this text, it is our goal to present data about the past use, the potential future of ethanol fuel use in the world, and its impact on climate change.

89.2 THE PRESENT ETHANOL PRODUCTION: 'STATUS', NUMBER OF PLANTS, AND PRODUCTION IN BRAZIL AND OTHER COUNTRIES

Figure 89.1 presents the global evolution of ethanol fuel production up to 2019. A significant increase occurred in the period 2001–2010, in the USA, and in the period 2008–2013, in Brazil. A

DOI: 10.1201/9781003226567-119

TABLE 89.1
GHG Emissions Avoided Due to the Displacement of Gasoline by Ethanol in 2020

Ethanol Feedstock	Average	Max.	Min.	
Corn	14.6	0.7	28.6	$MtCO_2e$
Sugarcane	56.3	31.9	80.0	$MtCO_2e$
Total	70.0	32.6	108.6	$MtCO_2e$

iLUC emissions ignored.
Source: Prepared by the authors with emissions data from Pereira et al. (2019) and Wicke et al. (2012).

total number of installed refineries in the USA reached 210 facilities in 2013 and stabilized since then. From this, 199 were in operation by 2019, with an average capacity of 295 million liters per year (ML/year) (RFA, 2019).

In Brazil, the number of installed mills is 411 units of which 370 are operational with an average installed capacity of 102 ML/year. Stabilization after 2013 in the USA is explained by the 10% 'blend wall' and in Brazil by sectorial economic barriers caused by large investments made in the sector just before the world economic crisis. The 'others' category includes 33 countries with an average production by 2019 of 10,000 bbl/day. There are 55 countries using ethanol fuel blends (REN21, 2020).

Most of the ethanol (94.3%) produced in the USA is from corn starch, followed by refineries handling a mix of corn, sorghum, and cellulosic biomass (2.9%), handling a mix of corn and sorghum biomass, and just 0.5% on cellulosic biomass (RFA, 2019). In Brazil, most of the ethanol comes from sugarcane, and at the end of 2019, some 370 sugar ethanol mills were operating across the country. Brazil also produced around 1.4 billion liters of ethanol from corn (up 75% from 2018), with ten production plants in operation and more corn-based capacity under construction to take advantage of the expected rise in ethanol demand under RenovaBio (REN21, 2020). RenovaBio is a market-driven incentive mechanism, which creates 'carbon credits' per unit of energy efficiency increase in the production of biofuels, certified individually by the producer (Morandi, 2020).

The productivity of corn in the USA and in the world has increased significantly in the last decades as shown in Figure 89.2 (OECD/FAO, 2020). Consequently, in the USA where yield increased at an annual rate of 1.72% and the harvested area has increased modestly (from 23.3 Mha by 1961 to 32.9 Mha by 2019), it becomes possible to increase the amount of available feedstock dedicated to ethanol fuel from 35 million bushels by 1980 to 4,856 million bushels by 2019, without impact on food and feed corn domestic availability (from 4,213 by 1980 to 7,340 million bushels by 2019). The exportation share declined from an average of 28.1% of all production during the 1980s decade to 13.5% during the 2010s decade (USDA, 2021a). At the global level, the yield increased even further (1.91%), as well as in Brazil (2.59%), but starting from a very low number when compared with the USA that implies a present productivity of only 55% of the US figure (OECD/FAO, 2020).

It is noticeable that there is the low energy density per unit of land area used for ethanol fuel production (see Figure 89.2). Taking the USA as a reference and noting that ethanol production was 59 Mm^3 by 2019, consuming corn harvested over 39.4% of the total crop area of 32.95 Mha (USDA, 2021a; USDOE, 2021a), the calculated figure is 4,545 L/ha, which translates into 96.7 GJ/ha. Considering that the present consumption of motor gasoline, at the global level, is 26 million bbl/day (USEIA, 2021a), some 496 Mha of corn-harvested land would be required to displace the fossil fuel. This is a quite huge area when compared with agricultural areas used for some important

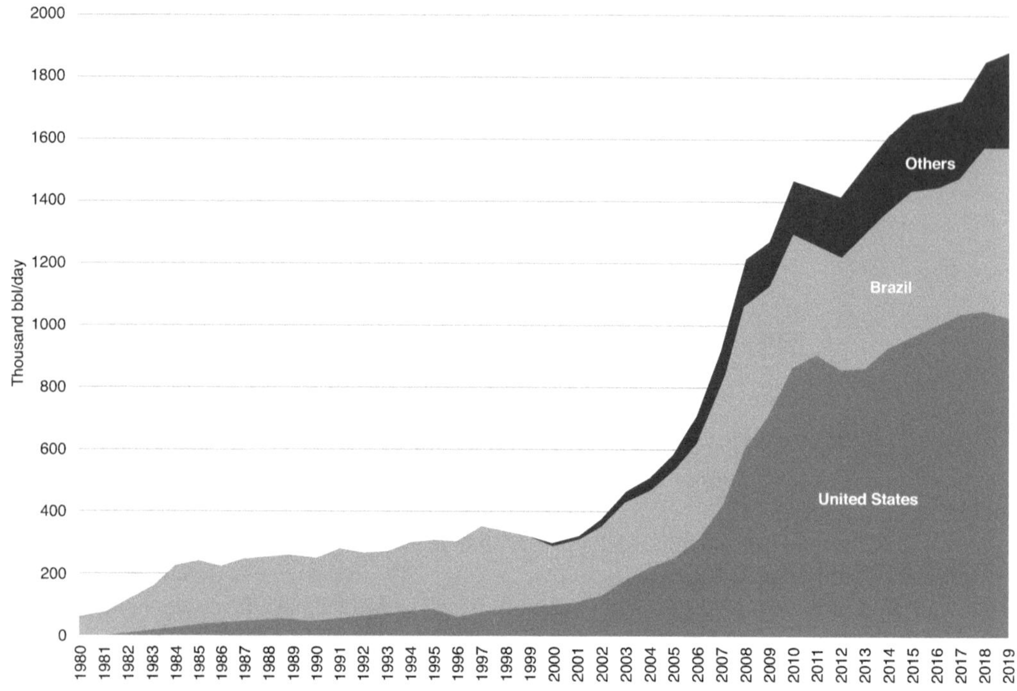

FIGURE 89.1 World ethanol fuel used in the period 1980–2019. (USEIA (2021a).)

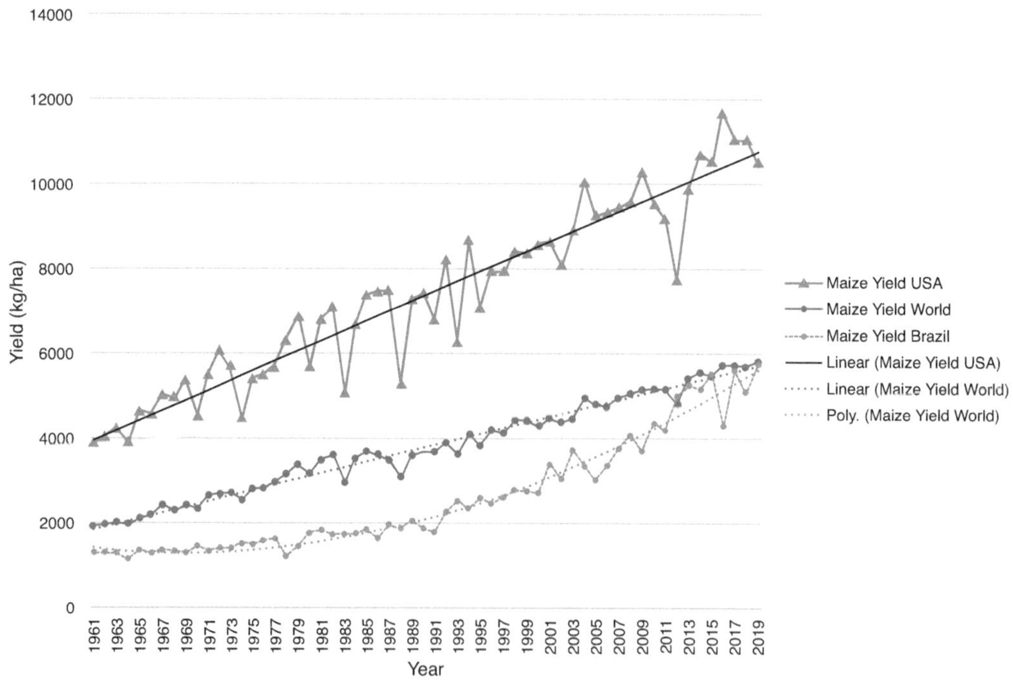

FIGURE 89.2 Maize crop productivity since 1961. Values for the world, USA, and Brazil. (Prepared by the authors based in OECD/FAO (2020).)

food crops such as wheat (216 Mha) and corn (197 Mha) (OECD/FAO, 2020) and has been one of the strongest arguments against enlarging the share of ethanol fuel at a global scale.

The easiest route for ethanol production is using sugarcane, since one step of the process involving starch hydrolysis, required for corn ethanol production, is avoided. It is a tropical culture and is usually selected as the main feedstock by such countries. Sugarcane productivity has increased over time but at more modest rates than corn, as seen in Figure 89.3.

Brazil, as the first country to rely strongly on ethanol fuel, had productivity lower than the rest of the world in 1961 but by 1983 was able to reach the same productivity as the rest of the world. After that, the yield increased continuously up to 2009, with an annual growth rate of 1.40%, and then decreased. The decrease is significant compared with the rest of the world's yield (with an annual increase of only 0.57%, presently, approaching the Brazilian figure). Such a decrease is partially explained by harvesting mechanization that has been initially imposed by environmental legislation (Goldemberg et al., 2008) but is being practiced nowadays due to its better economic performance. Another reason for the yield decline is attributed to poor crop management. Indeed, during the period 2003–2009, there was a large increase in ethanol production through the deployment of Greenfield Mills, which demanded huge investments. By 2010, as a consequence of the global economic recession, the national currency has been devaluated, increasing the loan costs used for the investment, most of which was carried out in hard currency.

The first factor – mechanization – implies further soil compaction due to more frequent displacement of heavy equipment and a decline in the harvested cane quality due to the excess of soil impurities and/or imperfect cutting of the cane stalk. The second factor – costs – puts pressure on new investment in sugarcane reforming, which must be done every 5 or 6 years due to the decline in productivity of this semi-perennial culture. To save money, the reforming period was increased by 1 or 2 years as shown in Figure 89.4. Reforming the plantation after 5 years increases

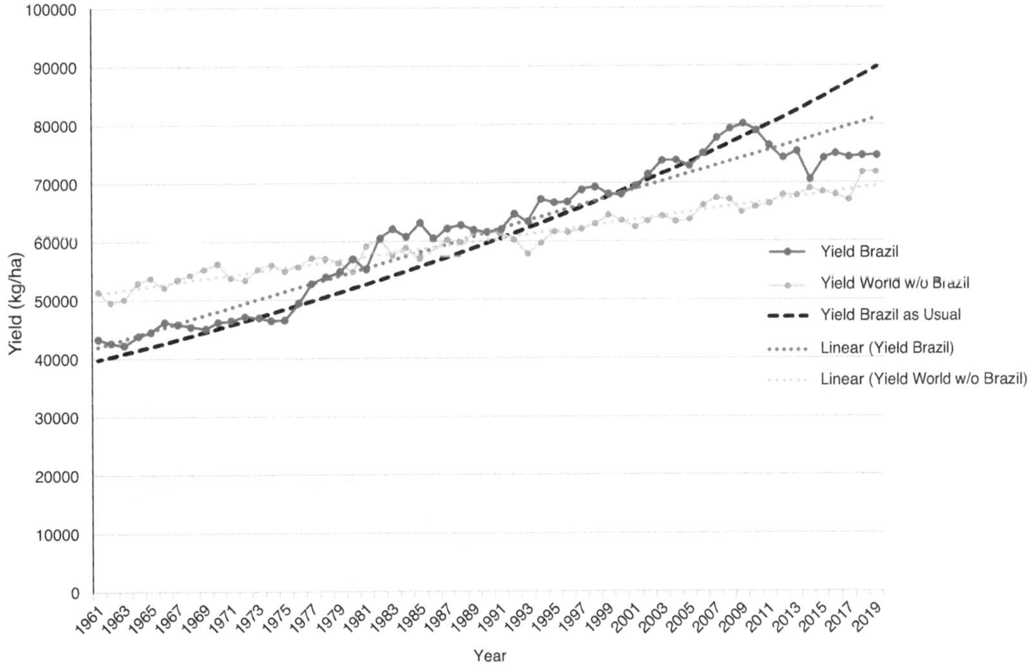

FIGURE 89.3 Sugarcane productivity since 1961. Values for world less Brazil, Brazil, and Brazil if the growth rate up to 2009 was maintained to 2019. (Prepared by the authors using data from OECD/FAO (2020).)

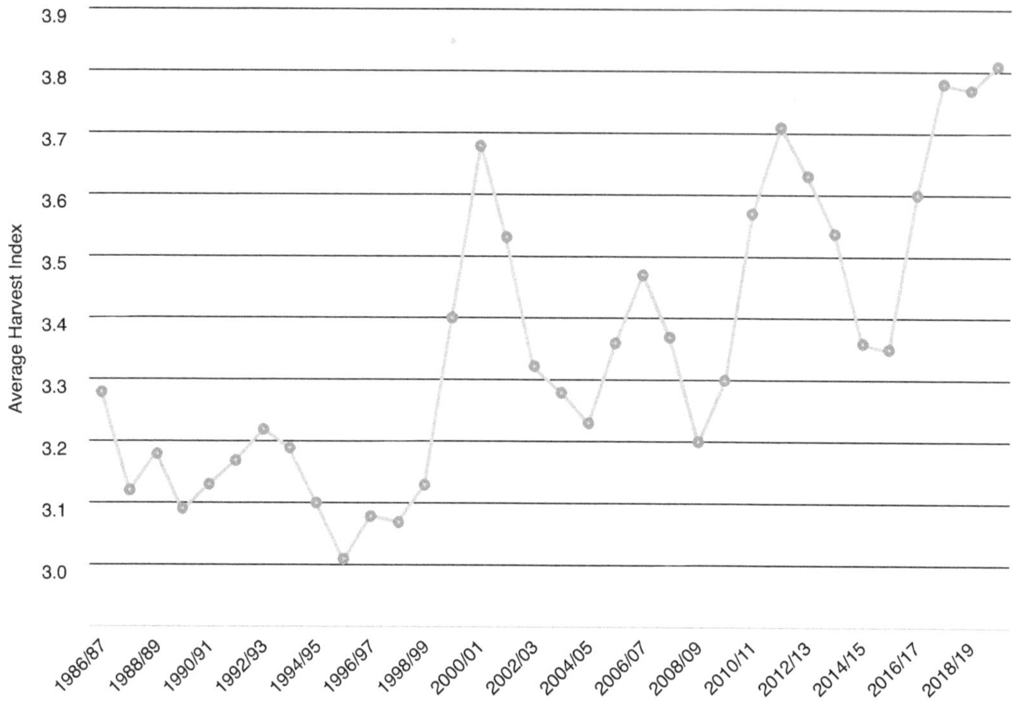

FIGURE 89.4 Historical evolution of Average Harvest Index in the Center–South region of Brazil. (Authors with data from Do Canto Braga Junior et al. (2021).) The values shown on the y-axis are nondimensional figures calculated for semi-perennial crops as weight average of annual harvest productivity divided by the total accumulated productivity in all years. For sugarcane, the weights are, respectively, 1, 2, 3, 4, and 5 for the first five harvests and 6.58 for all harvests after the fifth one Do Canto Braga Junior et al. (2021).

the average yield by 8.9% compared with 6 years of reforming (Do Canto Braga Junior et al., 2021). Thus, instead of 74.7 t/ha for 2019, we could expect a value of 81.4 t/ha. Maintaining the average growth rate of the period 1961–2009, it should be possible to achieve a yield of 89.6 t/ha, and soil compaction was responsible for a decline of 8.2 t/ha.

89.3 PRESENT PRODUCTION AND EVOLUTION UP TO 2020

Sugarcane in 2019 was planted in 103 countries with a total harvest area of 26.8 Mha and a production of 1.94 billion tons occurred in 2019 (OECD/FAO, 2020), most of it as feedstock for sugar production (see Figure 89.5). Global sugarcane used for sugar production in the period 2017–2019 averaged 1.14 billion tons (see Figure 89.5). Considering its modest productivity increase, the production of fuel ethanol in the period 1975–2019 required land area expansion. For Brazil, the area used for sugar and ethanol production expanded from 1.37 Mha by 1961 to 10.1 Mha by 2019 (OECD/FAO, 2020). Such expansion was possible due to the large geographical area of the country and the availability of new agricultural area, most of them in land previously used for pasture or unused due to its poor soil quality (CONAB, 2019). Most of this poor-quality land is in the 'Cerrado region' and was upgraded through proper pH soil correction (Karp et al., 2021). This was performed by sacrificing natural vegetation, which was not dense, but had an environmental cost (Pacca and Moreira, 2009). Nevertheless, with the area expansion and the gain in crop yield it was possible to produce fuel ethanol in an amount sufficient to displace, in the most favorable years as much as 75% of gasoline (Observatorio da Cana, 2021), while significantly increasing domestic sugar supply

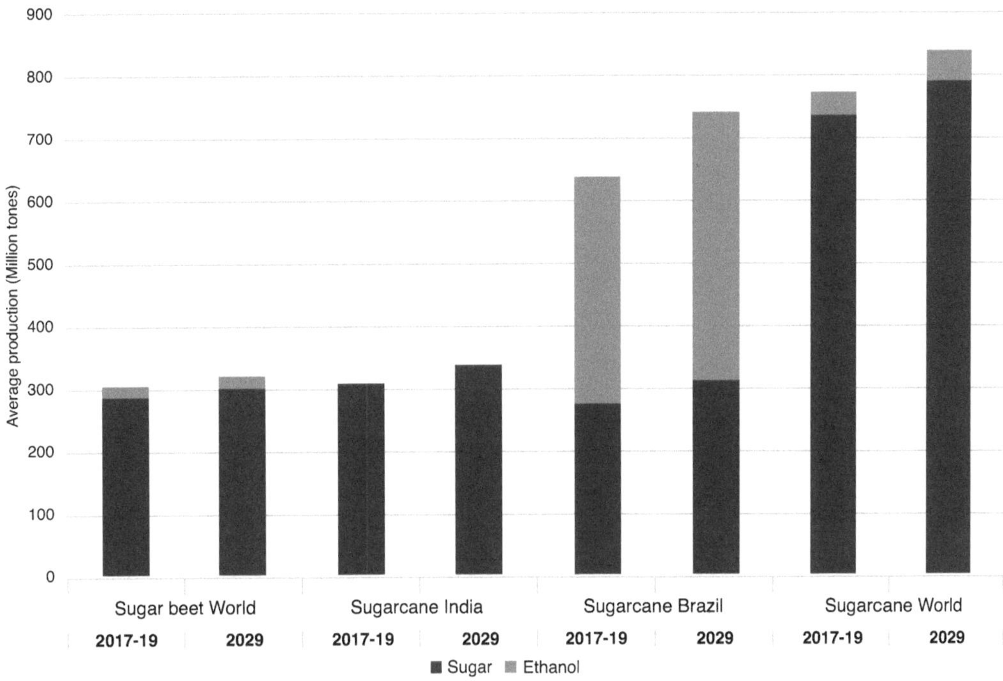

FIGURE 89.5 Average world sugar crop production in the period 2017–2019, and expected values in 2029 for the World, India, and Brazil, the two largest producers. (Authors with data from OECD/FAO (2020).)

and exportation (USDA, 2021b,c). Sugar production increased from 17.1 Mtones in 2000 to 34.2 by 2019. Sugar exports increased, in the same period, from 7.7 to 23.6 Mtones (ISA, 2021).

The quite modest increase in sugarcane productivity deserves to be examined more carefully since Brazilian agriculture has presented particularly good performance in the last three decades. An example of such a statement is shown in Figure 89.3 for corn, but other good results are noticed for soybeans (1.81% annual increase in the period 1961–2019), wheat (2.79%), rice (2.29%), and coffee (2.30%) (OECD/FAO, 2020). Instead of looking for the country's average sugarcane yield, it is worthwhile to examine the results for some mills, as shown in Figure 89.6. Values are presented for a sample of sugar mills located in the Center–South region of Brazil, the main sugarcane production region that shows better performance than the Northeast region. As shown, average productivity, by 2016, was 83 t/ha, instead of the country's average value of 75.2. Another observation is that there are sugar mills with values of 100 t/ha and 140 kg/t of total recovered sugar. With these figures, ethanol yields of over 8,600 L/ha are achieved, much above the current Brazilian average figure of 6,900 L/ha (CONAB, 2019). Since the world's consumption is 26 Mbbl of motor gasoline/day, it would be necessary to harvest 262 Mha of sugarcane all over the globe. As mentioned, when discussing corn, the area extension still sounds too large since it is comparable with the area dedicated to global corn production.

89.3.1 Sugarcane Coproducts

An important aspect of ethanol production is the coproducts generated. When corn is the feedstock, there is the production of distillers' dry grain (DDG) and corn oil. These products contribute economically and sustainably to the ethanol fuel market. They represent 22% and 4%, respectively, of the total sales price of the complex ethanol fuel+coproducts (RFA, 2019). From an environmental perspective, these products offset GHG emissions that should be emitted by their production through conventional routes. As an example, for a factory processing corn through dry milling

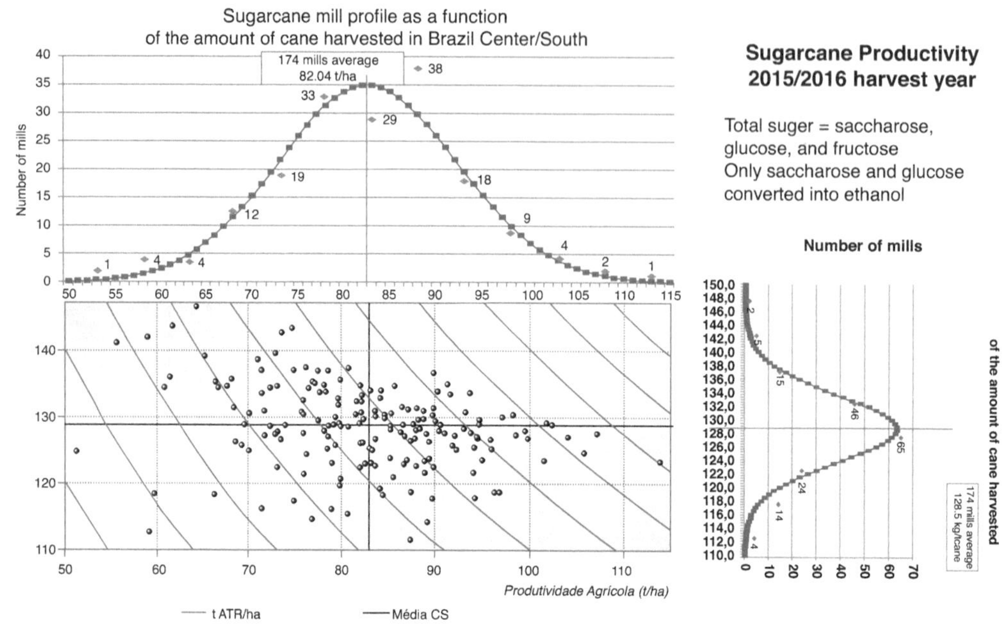

FIGURE 89.6 Profile of agricultural cane and total reduced sugar productivity from sugarcane for several sugar mills located in the Brazil Center–South region for the 2016/2017 harvest season. The numbers shown in the upper figure refer to the number of sugar mills for each interval of five units shown on the x-axis. The same applies to the right figure. (Authors with data from CTC (2016).)

technology without extracting corn oil, the DDG displaces 4.4 pounds of corn, 1.7 of soybeans meal, and 0.13 of urea per gallon of ethanol produced since these products are conventional animal feeds (Arora et al., 2008; GREET2019, 2020).

Figure 89.7 shows the carbon footprint for ethanol derived from corn and sugarcane, and as noted, for corn there is a negative contribution due to the production of DDGs as a coproduct displacing other sources of animal feed (Mekonnen et al., 2018). For sugarcane, according to the same publication, bioelectricity is considered a coproduct of ethanol, but there is no negative value since the authors used the approach developed on the greenhouse gases, regulated emissions, and energy use in transportation (GREET) model (Wang et al., 2019) where such bioelectricity and associated heat are considered energy sources displacing other energy sources that would be used at the fuel preparation stage, transforming the feedstock into ethanol. This approach has a significant impact on GHG emissions calculation, and we will return to this point in Section 89.5.

When sugarcane is the feedstock, bioelectricity, biogas, and fertilizers are coproduced. Bioelectricity is traditionally generated from sugarcane bagasse. A share of this bioelectricity is consumed in the ethanol process, and the surplus is exported to the electric grid. In Brazil, by 2020, 22.6 TWh of bioelectricity has been exported, while 15 TWh has been consumed for ethanol processing (UNICA, 2019, 2021). There are 441 electric plants installed, not necessarily in operation every year, providing all such bioelectricity. Considering the amount of sugarcane processed and that only 90% of such biomass is used by the sugar mills (CONAB, 2019) as much as 40 kWh/tcane, on average, is sold to the grid, providing a revenue of US$ 2.4/tcane.

The bioelectricity production capacity is not explored at its highest level since it became an important market only in this new century, and many old mills did not replace their old heat and electricity-producing facilities. Modern mills are quite productive as is the case of Cerradinho Mills, which generate 96 kWh/tcane (CerradinhoBio, 2020). Such good performance is observed in other countries. In Guatemala, sugar mills generate 2.62 TWh by 2020, and the average efficiency was 103 kWh/tcane, while the most efficient unit generates 125 kWh/tcane and consumes

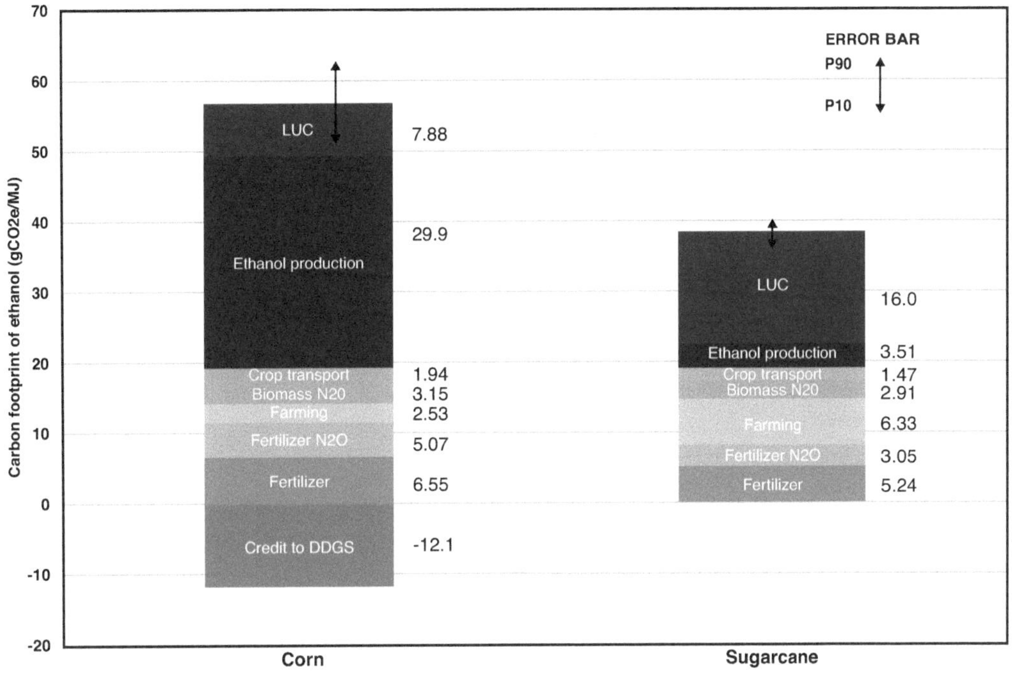

FIGURE 89.7 Carbon footprint of bioethanol from corn in the USA and sugarcane in Brazil. (Mekonnen et al. (2018).

23 kWh/tcane (Cengicana, 2021). With US$ 60/kWh sales price of electricity, the complimentary budget of the sugar mill is increased by US$ 6.1/tcane.

Such bioelectricity has other impacts than just revenue. Focusing on global bioelectricity generation that amounted to 591 TWh by 2019 (REN21, 2020) and considering the global annual cane harvest of 1.9 billion tons by 2019, it should be possible to offer 194 TWh to users if all sugar mills were as efficient as the leading ones. This represents 32.8% of all bioelectricity generation using all types of biomass feedstock and surpasses the amount of bioelectricity (90 TWh) derived from the use of 17 million tons of pellets that are commercialized worldwide (REN21, 2020).

Biogas is produced through anaerobic fermentation of vinasses and filter cakes that are residues of ethanol distillation. There were several attempts to commercially use such biogas as fuel on diesel-type truck engines, as a source of heat to concentrate vinasses, as fuel for electricity production with the use of otto or diesel-type power generators, and as a replacement for natural gas (NG) for delivery through gas ducts (PDE 2030, 2021b). In Brazil, there are commercial plants in operation. One of them with an electricity generation capacity of 15 MW was installed in 2020 using biogas from vinasses (Sebigas, 2021) and designed to use 2.3 billion m^3 of biogas produced in a sugar mill processing 5 Mt of cane/year. Considering that vinasses are produced at a rate of 13 L/L of ethanol, this means that 57,600 MWh are produced from 177 ML ethanol, which is 0.32 kWh/L of ethanol or 28 kWh/tcane (Sebigas, 2021). Thus, US$ 1.68/tcane is added as additional sales for the sugar mill. In conclusion, bioelectricity and biogas revenues can represent 18% of the total sales of the mill. Two other facilities started operation in the last 2 years, with 21 and 10 MW of installed capacity and a 10 MW (Cruz, 2021).

Regarding fertilizers produced, Table 89.2 shows the typical composition of nutrients applied to the soil in a sugarcane plantation and their presence in commercial fertilizers and the vinasses. Thus, by spreading vinasses over the planted area, money spent on commercial fertilizers is saved. Nevertheless, no negative value is shown in Figure 89.7 as this coproduct is used in the ethanol production process.

TABLE 89.2
Quantitative Nutritional Value of Vinasses Compared with Commercial Fertilizers Used in Sugarcane

Commercial Fertilizer	Chemical Element (%wt)	Amount of Fertilizer in Vinasses (kg/tcane)	Amount of Fertilizer Used (kg/tcane)	Vinasses Fertilizer Share (%)
Urea (N)	45.5	0.550	1.015	24.7
Superphosphate triple (P_2O_5)	46.0	0.389	0.360	49.7
Potassium chloride (K_2O)	60	2.19	1.235	106.5

Source: Authors based on data from Germeck and Feigl (1987), p. 44, Pereira et al. (2019), and Simoes et al. (2004). Considering the range of values quoted by Pereira et al. (2019), we used the average value for table preparation.

Another coproduct that can be listed is CO_2 derived from sugar fermentation processing into ethanol. Sucrose ($C1_2H_{22}O_{11}$) is converted to ethanol (C_2H_5OH) and carbon dioxide (CO_2), and the stoichiometric values are 1 kg of ethanol and 0.96 kg of CO_2. This CO_2 is usually vented from the fermentation vessel after being cooled since it carries some valuable amount of ethanol. Being produced by biochemical fermentation, it is essentially free of impurities and classified as proper for human consumption (Sanchez et al., 2018). There are some uses for it, like in the production of refrigeration systems, welding systems, water treatment processes (to stabilize the pH of water), and carbonated beverages (https://sciencing.com/inconel-7687773.htm), as well as a feedstock for the production of oil, and, more recently, being stored underground to reduce climate change impacts (ADM, 2020; Finley, 2014).

The first set of uses is limited due to seasonal availability since ethanol production is carried out during the harvest season and due to the limited market size of the demand (IEA, 2019). Carbon dioxide gas is used in enhanced oil recovery (EOR). The last use mentioned that CO_2 storage has a promising large market due to the growing concern with the environment. Many studies indicate that this practice is the lowest cost possibility for carbon capture and storage (CCS) technology since all that is required is its transportation and storage (Fuss et al., 2016; Sanchez et al., 2018). Capture is cost-free, except for the necessity to remove some small amount of water that also flows out of the fermentation vessel and remains after cooling down to recover ethanol (Sanchez et al., 2018).

89.4 HOW MUCH GASOLINE HAS BEEN REPLACED UP TO 2019 BY ETHANOL FUEL AND HOW LARGE HAS ITS CONTRIBUTION TO THE REDUCTION OF CO_2e EMISSIONS BEEN?

Environmental analysis of the impact of ethanol covers many aspects such as GHG emissions, use of land and water, and local air pollution. Many studies have analyzed these topics (AQEG, 2011; Capodaglio and Bolognesi, 2019; Dominguez-Faus et al., 2009; Giuntoli, 2018; Valin et al., 2015). We will discuss the first issue: GHG emissions. A complete analysis of this issue involves direct emissions associated with the agricultural and industrial activities involved in ethanol production, as well as its combustion in the vehicle engine.

The agricultural stage is quite complex since it involves land use change (LUC), the largest contributor to the ethanol CI index (see Figure 89.7). LUC occurs every time land is an object of human intervention covering from soil preparation up to harvest. These steps are named direct LUC and are part of the GHG accounting process. Another part to be accounted for is the indirect LUC (iLUC) that refers to GHG emissions due to the possibility that, when using the soil for ethanol feedstock, some other type of biomass or pasture has been removed and must be replanted or displaced elsewhere since it was used for food, feed, or fiber by the global population. The replanting will occur in another land area located anywhere in the world (see Woltjer et al., 2017, for a more complete explanation of iLUC).

When ethanol feedstock displaces previous crop used for food or feed, it is quite easy to understand that some other area will be used to fulfill the demand, but not necessarily of the same crop type. A shortage of food or feed supply at the global level can be offset by any crop already used for such purpose. The very comprehensive iLUC-detailed explanation is provided by Woltjer et al. (2017). In the lifecycle results shown in Figure 89.7, iLUC is not considered, but a complete lifecycle evaluation must include its contribution (Bieker, 2021; GREET2019, 2020; USEPA, 2010). Proper accounting of iLUC has been performed using complex models with an ample database covering global land characteristics (Bird et al., 2013; Laborde et al., 2015; Tyner et al., 2016; Wicke et al., 2012; Zhao et al., 2017).

Even considering the effort and value of these calculations, the results are used with some concern and involve a large error bar due to possible uncertainties. On the contrary, for LUC evaluation, it is enough to have full knowledge of the removed vegetation and some of the soil components, which, in principle, can be well determined, but in practice are costly and time-consuming to measure. Fortunately, cataloged values are available and can be used with some accuracy once a few soil parameters are identified. For this lifecycle accounting, it is necessary to compute emissions due to fertilizer manufacture, N_2O emissions due to fertilizer degradation in soil, biomass N_2O in the crop, farming activities that include soil preparation up to harvesting, manufacture of pesticides and herbicides, and crop transportation from the field to the ethanol factory, as shown in Figure 89.7.

At the industrial stage, it is necessary to account for GHG emissions due to the transformation of the feedstock to ethanol. The complete process includes sugarcane unloading from trucks, cane stalk physical preparation through cutters and fiber separation, cane crushing, collection of extracted juice, separation of physical impurities, chemical and biological control to avoid juice infection, fermentation, distillation, and rectification to achieve over 99% ethanol concentration. To carry out all these processes, it is necessary to consider heat and electricity, which are sources of GHG emissions, and products that require energy and proper feedstocks, which are also sources of further emissions.

Once the fuel is ready for delivery, further energy and associated emissions occur during its transportation to the service stations.

GHG emissions due to the combustion of fuel in the vehicle operation must be added. Some small uncertainty is associated with the quality of the fuel and the engine due to the inevitable fraction of unburned fuel and the evaporation of the fuel that occurs due to fueling service, car operation, or even standby.

Based on the available literature, it is possible to calculate GHG emissions from ethanol and compare its value to the traditional fuel used – gasoline, provided a complete emission cycle is also available for this last fuel.

To answer the question about the total amount of displaced gasoline by ethanol up to the present, we need a statistical table covering the world production of this particular biofuel. The Energy Information Administration provides this information covering the period 1980–2019 (USEIA, 2021c). Another factor to consider is the energy content of ethanol and gasoline. The ethanol value is well known and does not change with time since this is a pure chemical compound. The gasoline value, a combination of many hydrocarbons, is time-sensitive since gasoline average composition is not obtained from a unique oil feedstock and the share of different hydrocarbons has changed not only due to the oil feedstock type but also due to legislation demanded mainly by local pollution and environmental concerns. Thus, a precise calculation would require complete knowledge of the share of hydrocarbons, in each particular country and each particular year.

On top of this limitation, another issue is that there is no reason to assume with full confidence that 1 MJ of ethanol displaces 1 MJ of gasoline. This is explained by the specific design of vehicle engines and characteristics other than fuel heat value for which experimental measurements and/or careful analysis are required (GREET2019, 2020). When ethanol is blended with gasoline at a low blending level, there are measurements that confirm such a relation. When ethanol is used in high

TABLE 89.3
Accumulated Amount of Fuel Ethanol and Gasoline Produced and Lifecycle Emissions in the World in the Period 1980–2019

Fuels	Production (Mbbl)	Emissions ($GtCO_2e$)
All ethanol	10,281.4	1.765
Corn ethanol	5,322.0	1.073
Sugarcane ethanol	4,959.4	0.697
Motor gasoline	287,555.2	130.397
Gasoline displaced	6,991.2	3.337
Gasoline–ethanol emissions		1.571

Source: Prepared by the authors using data from USEIA (2021b).

proportion or even as a neat fuel, the blend usually has a higher octane value than gasoline and this allows the use of internal combustion engines (ICEs) with higher compression rates than the ones used for gasoline (Larsen et al., 2009).

Based only on the energy content of each fuel, the relation is 76,330 Btu for 1 gallon of ethanol and 112,114 Btu of gasoline (GREET2019, 2020) – that is, 68.03%. In Brazil, the automotive industry claims that the ratio is higher and suggests the use of a value of 70% (Nigro and Szwarc, 2009). Nevertheless, for this calculation we used the 68.03% figure. Table 89.3 shows the amount of ethanol and gasoline displaced in the period 1980–2019. The figures may sound large, but oil consumption nowadays is near 100 Mbbl/day. Thus, all the effort done so far to reduce GHG pollution from oil represents the displacement of around 70 days of oil consumption.

Another important number to account for is the amount of GHG avoided so far due to the use of ethanol. The lifecycle value of gasoline is known with reasonable accuracy since most of the emissions are due to its combustion. The fraction due to oil exploration and oil refining is available but is country-specific (GREET2019, 2020). Nevertheless, it is reasonable to agree that an average value for 10% ethanol blend of 92 gCO_2e/MJ, by the year 2021, is a good one (Bieker, 2021). As already discussed, the value of ethanol is not well agreed upon either for corn-derived or for sugar-derived, mainly due to the difficulty in iLUC contribution. Nevertheless, there are figures in the literature that we can rely on to make a reasonable evaluation. For corn ethanol, a recent review quotes the CI as 51.4 with an estimated range of 37.6–65.1 gCO_2e/MJ (Scully et al., 2021). For sugarcane ethanol, a study analyzed direct emissions using four different life cycle assessment (LCA) data models and, after data harmonization, concluded that the agreed value is 16–17 gCO_2e/MJ, for the average activity in Brazil (Pereira et al., 2019). To account for LUC, we used the value of 18.1 gCO_2e/MJ (GREET 2019, 2020) and for iLUC the value of 15.8 gCO_2e/MJ (Woltjer et al., 2017).

Considering corn ethanol CI of 51.4, sugarcane ethanol of 49.8, and gasoline of 96 gCO_2e/MJ and considering that 51.8% of the world ethanol was derived from corn and the remaining from sugarcane, the total amount of GHG emissions avoided in the period 1980–2019 is 1.572 $GtCO_2e$ (see Table 89.3). This figure gives us an idea of how large efforts must be to mitigate climate change, comparing all the accumulated mitigation obtained from the production and use of ethanol fuel with 2019 global energy-related emissions of 31.5 $GtCO_2$ (USEIA, 2021b).

89.5 FUTURE PERSPECTIVES – MORE ETHANOL USE WITH MODEST LAND AREA REQUIREMENT

What area is needed to supply the world's automobile fleet with ethanol fuel? To answer this question, we can rely on two very detailed studies: one carried out by the Argonne National Laboratory in 2016 (Elgowainy et al., 2016), focusing on the US market, and the other prepared by International

Council on Clean Transportation (ICCT) in 2021 (Bieker, 2021), covering Europe, US, China, and India markets. These studies' main goal is to evaluate the amount of GHG emissions from vehicles used for personal transportation, in the year they were published and for 2030, but they provided the necessary information to address our question. Unfortunately, both studies ignored the relevance of sugarcane ethanol fuels used in plug-in hybrid electric vehicles (PHEVs). The Argonne report only presents results for corn ethanol and the ICCT, based on an European Union (EU) directive that limits the use of ethanol produced from food and feed crops to only 2% of the fuel blend. Thus, both ignored the fuel alternative based on large-scale sugarcane ethanol blends.

Results from both studies are reasonably in agreement concluding that ICE powered by gasoline blended with a small amount of ethanol has emissions of around 460 gCO_2/mile in 2015, 395 gCO_2e/mile in 2021, and 240 gCO_2e/mile in 2030, as a consequence of efficiency improvements in the fuel and vehicle technology. For PHEV powered with the same fuel, the emissions are lower than for ICE and are around 325 in 2015, 290 in 2021, and 270 to 240 gCO_2e/mile in 2030. The best performance occurs for battery electric vehicles (BEVs) with around 300 in 2015, 130 in 2021, and 95–240 gCO_2e/mile. For this last vehicle type, the agreement between both studies is not good as a consequence of the significant improvements in batteries in the last few years (Bieker, 2021).

To overcome the unavailability of results for PHEV using pure ethanol, we replicate the analysis carried out by Argonne using the same software. One of the most used models appropriate for such evaluation is known as GREET and was developed by the National Argonne Laboratory in the USA. The purpose of such evaluation is to address the question – how much GHG emissions are avoided when cars, specifically LDVs that are used for personal transportation, are fed with ethanol instead of gasoline.

To provide a figure, it is necessary to completely describe the type of fuel in use. Considering that there is a trend to blend ethanol with gasoline in some countries and that the blend level is 10% in the USA, 27% in Brazil, and 7% in EU countries, we assumed for ethanol a value of 10% blending for our calculation. Regarding ethanol, since some countries are concerned with the potential use of neat ethanol fuel as an alcoholic beverage, we assumed that such fuel is blended with 2% gasoline as a denaturant. The next issue to consider deals with the types of LDVs being evaluated.

Traditional ICE vehicles using gasoline (ICEG100) are still the most common ones. Since the 1980s, there are LDVs running with 100% ethanol (ICEE100) in Brazil and, since the 2000s, with 85% ethanol in the USA (ICEE85). At the beginning of this century, electric vehicles using batteries (BEV) and hybrid electric vehicles using a combination of batteries and gasoline (HEV) are found in the market, and in the last decade, plug-in electric vehicles (PHEVs) are being commercialized. Furthermore, such PHEVs, which have the capacity of storing electricity from the grid in batteries, can use gasoline (PHEVG100) or ethanol (PHEVE100). PHEV's main merit is the possibility of traveling a limited distance using electricity from the grid and, consequently, avoiding fuel emissions, while using fuel for long-distance travel or when the batteries are discharged. They are quite appreciated since most city driving involves short distance and then zero local emissions. Accounting for the present trend of the market, we selected for our evaluation ICEG100, ICEE100, BEV, PHEVE100, and PHEVG100.

At first sight, considering that ethanol and gasoline are burned in the vehicle engine, BEV can be understood as the lowest emission one, providing electricity from the grid is environmentally clean. In principle, this is the case for electricity generated by combustion-free processes, such as hydroelectricity, wind, and solar sources. Nevertheless, even considering that these renewable sources imply some GHG emissions and that ethanol is produced using biomass – also a renewable feedstock – that grows using solar energy, the 'a priory' conclusion deserves to be confirmed by careful calculation. Another evident insight is that since ethanol, either produced by corn or sugarcane, requires large land areas, the use of hybrid vehicles that share the propulsion energy between fuel and electricity is less land area-intensive.

CALCULATION ASSUMPTIONS BOX

The calculation is performed for vehicles in operation in Brazil, using gasoline, ethanol from sugarcane, and electricity derived from the grid mix and NG sources. The reason to include NG sources as the displaced feedstock for electricity generation is that NG is considered the most resilient fossil fuel to be used in environmentally sound scenarios due to its lower fossil emission and the necessity to guarantee supply as intermittent renewables increase their share.

When using electricity, it is important to account for the bioelectricity produced as a coproduct of sugarcane-based ethanol. As pointed out in Section 89.2, optimized mills can generate 96 kWh/tcane. Such a mill uses all the available sugarcane bagasse and a small fraction (10%) of sugarcane leaves (straw) to generate heat and power, mainly through cogeneration. The amount of straw produced by green cane harvest is significant (140 dry kg/tcane), and it is possible to collect and use up to 50% of it while preserving the soil quality (SUCRE, 2020). Under this scenario, 161 kWh/tcane is generated and 121 kWh/tcane is exported as a coproduct. Thus, 121 kWh is available for third-party users, for every 86 L of ethanol produced, since we assume that each ton of cane produces 86 L of ethanol. This coproduct has an impact on our calculation when considering ICEE100 and PHEVE100 vehicles. For the ICEE100, this bioelectricity displaces either grid mix or NG-based electricity. For PHEVE100, this bioelectricity is directly used in the vehicle. For ICEE100, bioelectricity becomes available for all other uses that rely on electricity from the grid and consequently displaces generation either by the grid mix or by NG-based plants. For ICEG100 and BEV bioelectricity, it is not available since these cars do not require ethanol. Thus, it is necessary to determine the lifecycle GHG emissions for bagasse + straw bioelectricity generation. The value can be taken from the GREET code, assuming values listed for forest residues as a proxy. A correction is applied since in the agricultural phase bagasse and straw feedstock emission is already accounted for in the ethanol lifecycle. Only emissions associated with straw transportation are considered since bagasse is already available at the mill (USEPA, 2010).

Results are presented for the 2030 technology (ethanol production and vehicle technology). These results can be calculated, provided the grid electricity mix is defined. For 2030, we use a grid emission factor of 197 gCO_2e/kWh, considering the share of primary energy sources forecasted by the Brazilian Planning Agency (PDE 2030, 2021a), the C intensity of fossil sources provided by the GREET code, and for renewables the value provided by the Intergovernmental Panel on Climate Change (IPCC) (IPCC, 2011a) that amounts to 40 gCO_2e/kWh for wind and solar and 150 gCO_2e/kWh for hydro.

For LUC, we use the GREET code default values of 18.06 gCO_2e/MJ for cane ethanol, while for iLUC we used a range of values from 15.75 to 47.25 gCO_2e/MJ, as suggested by Woltjer et al (2017), but the most probable one is near the lower range value (Valin et al., 2015). Furthermore, we assumed that CCS is performed in sugar mills, capturing CO_2 from ethanol fermentation. This technology is already commercial (ADM, 2020). CO_2 is released during the biochemical conversion of sucrose to ethanol at a stoichiometric rate of 0.768 kg/L of ethanol, yielding a capture of 35.56 gCO_2/MJ of ethanol.

Results from our evaluation are shown in Figure 89.8, for ICEE100, and in Figure 89.9, for PHEVE100, both compared with ICEG100 and BEV cars. We evaluated results for several values of iLUC GHG emission contribution since this is a much-debated contribution that cannot be measured and is only calculated through models that assume a large amount of agronomic and economic data for the globe. Thus, we include four possible iLUC values for ethanol derived from

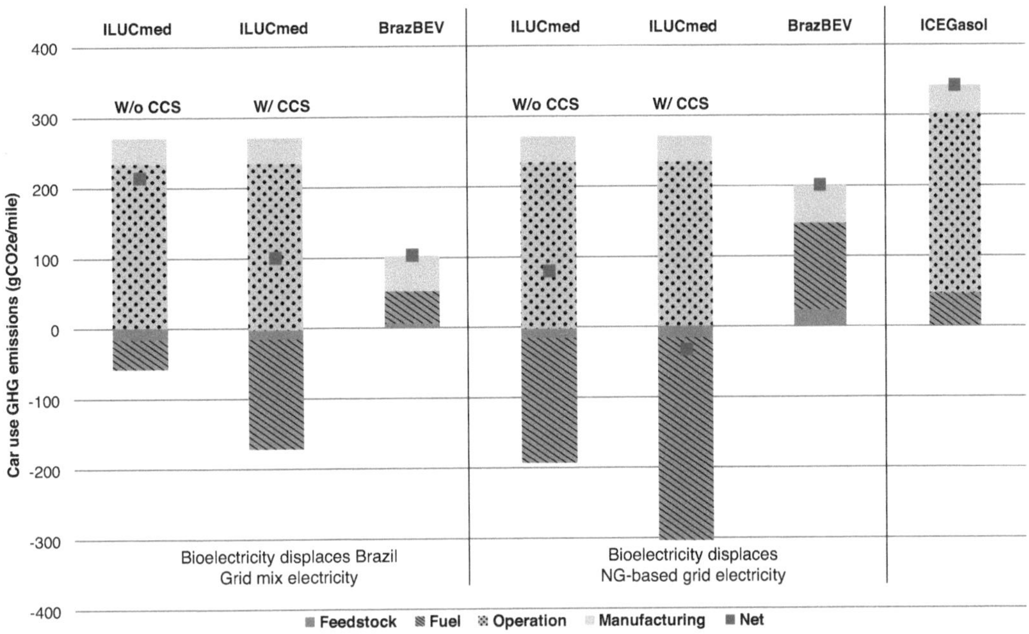

FIGURE 89.8 Well-to-wheel lifecycle emissions due to the use of ICE ethanol fuel-powered cars and BEV by 2030 as a function of the type of grid electricity displaced by sugarcane generated bioelectricity and the use of CCS technology on the ethanol fermentation process.

sugarcane, ranging from 0 to 47.3 gCO_2e/MJ (Woltjer et al., 2017). Considering the LUC value of 18.08 gCO_2e/MJ and adding the direct emission figure of 18.0 gCO_2e/MJ (Chum, 2018), well-to-wheel GHG emissions range from 36.1 to 83.4 gCO_2e/MJ. Considering that for gasoline such emissions are 92 gCO_2e/MJ and that ethanol is considered an environmentally sound replacement for gasoline, as is the case for the USA that the Environmental Protection Agency (EPA) agency has rewarded sugarcane as the most efficient feedstock considering it advanced rather than conventional (Nassar and Moreira, 2013; USEPA, 2010), we understand that the iLUCMax figure (47.3) is over-evaluated (see Bieker, 2021; USEPA, 2021) and we will consider as the most probable figure the one for iLUCmed (31.5).

We also present results for ethanol being produced in sugar mills that capture CO_2 from the biochemical sugar fermentation process. This technology has already been commercialized (ADM, 2020) and has some potential to be used by 2030.

Another issue considered in the calculation is the amount of bioelectricity exported by the sugar mills. As we pointed out in the calculation assumptions box, we assumed a figure of 121 kWh/tcane derived from the use of sugarcane residues and included another 57 kWh/tcane generated using eucalyptus. The reason for this extra source of bioelectricity is twofold. The first one is that sugar mills operate only part of the year and consequently export bioelectricity mainly during the harvest season. High-efficiency biomass-based electricity generation unit is costlier when compared with traditional power plants in use in most of the mills. One way to justify higher investments in efficient power plants is to operate them most of the year. This can be achieved, with low GHG emission impact, using supplementary biomass, such as eucalyptus, planted at a reasonable distance from the mills (<200 km) (Gabrielli et al., 2013). This approach is being commercially used by sugar mills in Brazil that produce ethanol from sugarcane and corn. The second point is that PHEVE100 consumes more electricity than can be generated by sugarcane bagasse and straw and to guarantee that the vehicle is fed fully by the sugar mills we need extra biomass electricity generation at the ethanol factory.

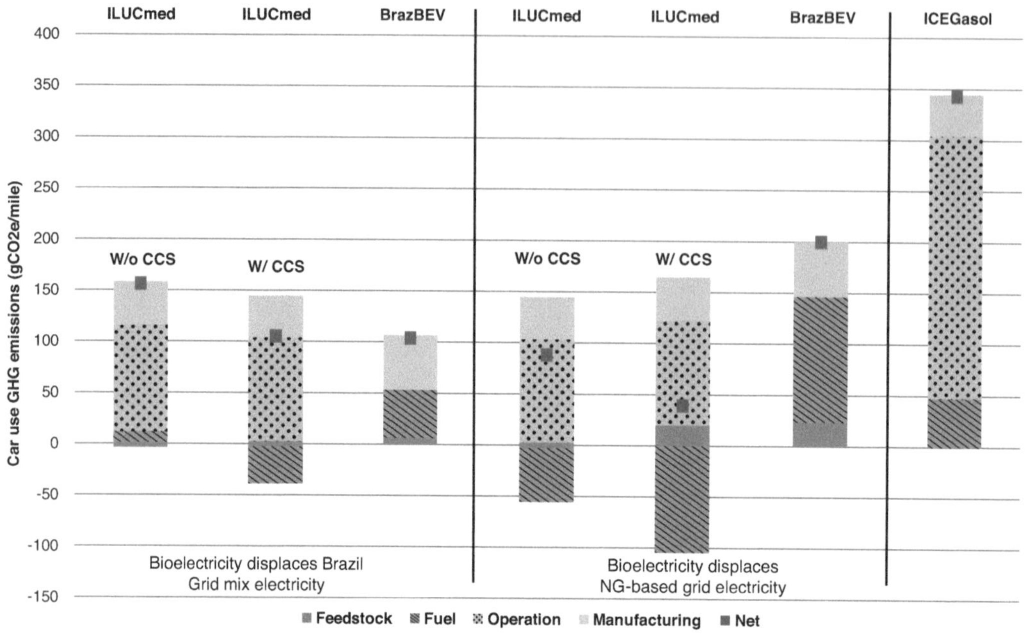

FIGURE 89.9 Well-to-wheel lifecycle emissions due to the use of PHEV ethanol fuel-powered cars and BEV by 2030 as a function of the type of grid electricity displaced by sugarcane generated bioelectricity and the use of CCS technology on the ethanol fermentation process. (Authors using GREET2019 code.)

In Figures 89.8 and 89.9, it is possible to see the complete lifecycle GHG emissions due to the feedstock, fuel preparation, vehicle operation, and vehicle manufacturing, as well as the net value of all these contributions, when bioelectricity exported from sugar mills displaces the electricity grid mix and electricity generated only by NG.

In Figure 89.8, we observe the relevance of the bioelectricity coproduct of ethanol production in the change in values of the net emissions as a function of the type of grid electricity displaced. When bioelectricity displaces electricity generated by NG turbines, the ICEE100 GHG emissions decrease to –32 gCO_2e/mile from the 81 gCO_2e/mile for electricity generated by the grid mix. Also, it is possible to note the impact of CCS, since with its use in the sugar mills the ICEE100 emissions decline to 102 and –32 gCO_2e/mile, instead of 215 and 81 gCO_2e/mile when compared with results without CCS, respectively, for each one of the assumed grid electricity scenarios. The negative figure above must be interpreted as CO_2 removal from the atmosphere when constructing and operating this type of vehicle, and it is not a surprise (IPCC, 2011b; Muratori et al., 2017).

For BEV, the result is independent of CCS since if this is the dominant car technology by 2030 there is no motivation for ethanol production. Nevertheless, the emission result is sensitive to the primary source of electricity accounted for in the grid supply. Thus, we note from Figure 89.8 that BEV emissions values are 105 and 201 gCO_2e/mile.

Finally, we add the result for ICEG100, which is not sensitive to the grid mix or CCS since there will be no ethanol production under this scenario. Its GHG emission is 360 GCO_2e/mile, in good accordance with the results presented by the Argonne Lab (Elgowainy et al., 2016).

Figure 89.9 shows the same results, this time for PHEVE100 car in comparison with BEV and ICEG100. Again, it is easy to see the impact of the electricity grid displaced by bioelectricity, as well as the contribution of CCS technology, significantly reducing car emissions.

An analysis of the presented results allows the extraction of some conclusions for 2030. The ICEE100 car, even with CCS technology, shows similar emission with BEV (102 and 105 gCO_2e/mile, respectively) when displacing grid mix electricity (Figure 89.8). This result is a consequence of the relatively clean grid mix electricity in the country (197 gCO_2e/kWh), considering that the average C intensity of the global electric grid by 2020 is above 400 gCO_2e/kWh, and the value for 2030 would be 400 gCO_2e/kWh, and the result will be different than the one for Brazil. Under such a situation, the ICEE100 emission decreases since bioelectricity with an emission of 70 gCO_2e/kWh displaces the 400 gCO_2e/kWh grid emission.

With an ethanol consumption of 0.1541 L/mile, the coproduced bioelectricity is 0.3190 kWh/mile, implying an emission reduction of (0.3190×(400 – 70) 105 gCO_2e/mile instead of (0.3190×(197 – 70) 41 gCO_2e/mile calculated for Brazil. Under this same global C intensity of electricity, the BEV, which consumes 0.2820 kWh/mile, emissions increase by 93 gCO_2e/mile, instead of 36 gCO_2e/mile for Brazil. The final result is that ICEE100 average emission for the globe would be 102 – (105 – 41) = 41 gCO_2e/km, while the BEV emissions would be 105 + (93 – 36) = 162 gCO_2e/km, quite different from what is observed from 89.80. Thus, even considering that bioelectricity displaces the electricity grid mix, ICEE100 cars would be more environmentally sound than BEV cars by a difference of 121 gCO_2e/mile. This difference would be higher for the scenario where bioelectricity displaces NG-based electricity from the grid, since, as shown in Figure 89.8, ICEE100 is a better environmental choice than BEV for Brazil.

For PHEVE100, the results (see Figure 89.9), without CCS when displacing grid mix electricity, are 105 gCO_2e/mile, comparable with BEV (also 105 gCO_2e/kWh). When NG-based electricity is displaced by bioelectricity, its environmental performance improves (89 and 39 gCO_2e/mile, respectively) and is better than the BEV, in both electricity grid scenarios. This ICEE100's better performance for Brazil would improve even further for the average globe scenario, as discussed in the previous paragraph.

The central point that deserves comment deals with the extension of land area required for the operation of the vehicles considered. This can be calculated based on the ICEE100 performance of 0.154 L/mile and the PHEVE100 of 0.0681 L/mile. For an ethanol production of 7,740 L (86 L/tcane and 90 tcane/ha), and assuming an annual vehicle displacement of 10,000 miles, each one of these cars requires the harvest of 0.195 and 0.0880 ha of sugarcane, respectively. When supplying fuel for the 2020 global car fleet (1 billion units), it is necessary to harvest 195 and 88 Mha of land distributed over the potential sugarcane-producing countries. Obviously, the larger value sounds high, but the lower one is more reasonable when compared with the present world sugarcane harvest area – 27 Mha by 2019 (OECD/FAO, 2020). Furthermore, according to the IPCC statement that all technologies are necessary to mitigate climate change, sugarcane ethanol can be used on a share of the global fleet, let us say 50%. Under such a scenario, a harvest area of 44 Mha could be used, which requires the duplication of the present global land area dedicated to the sugarcane crop. It is useful to highlight that we used a modest sugarcane yield for such conclusions and assumed no efficiency improvement in ethanol production in the period 2020–2030. This sounds like a very conservative evaluation, since, taking corn ethanol as a reference, the relevant improvement in its environmental merit in less than a decade is impressive (RFA, 2019).

89.6 CONCLUSION

Accelerated transformation of vehicle propulsion technology to reduce CO_2e emissions must be done in a balanced way to avoid 'putting all eggs in the same basket'. Most emphasis has been put recently on electrification. The existence of other short-term possibilities must be considered, and in this text, we try to demonstrate that ethanol from sugarcane is one of these possibilities. In particular, for the case of PHEV, it uses bioelectricity that is a coproduct of sugarcane ethanol, and this implies zero LUC contribution, as well as fitting in the category of production of low-risk iLUC biofuels, as is the use of residues and by-products (Woltjer et al., 2017).

Economic considerations are not analyzed in this study, but the current vehicles' market cost of straight electrification demonstrates that they are costlier than conventional ones and are sold under the label of an environmentally friendly alternative. More realistic expectations for the future rely on a much cleaner electric grid and the continuous improvement of battery technology.

Such an approach ignores other solutions and reduces the chances of their progress. In the text, we have described the modest improvement in fuel ethanol technology in the last decades and pointed out the existence of improvements for the next decade. This is clearly the case for biofuels, mainly one that provides the same feedstock of liquid fuel and bioelectricity. The complementarity of these energy sources when both are used in road vehicles must be considered when addressing environmental policies worldwide.

REFERENCES

ADM (Archer Daniels Midland). 2020. *Watch for ADM to Pioneer Biofuels, More Carbon Capture Projects.* https://www.greenbiz.com/article/watch-adm-pioneer-biofuels-more-carbon-capture-projects.

AQEG (Air Quality Expert Group). 2011. *Road Transport Biofuels: Impact on UK Air Quality.* London: Department for Environment, Food and Rural Affairs.

Arora, S., M. Wu and M. Wang. 2008. *Update of Distillers Grains Displacement Ratios for Corn Ethanol Life-Cycle Analysis.* Argonne, IL: Argonne National Laboratory. https://greet.es.anl.gov/publication-3bi0z09m.

Bieker, G. 2021. *A Global Comparison of the Life-Cycle Greenhouse Gas Emissions of Combustion Engine and Electric Passenger Cars.* International Council on Clean Transportation. https://theicct.org/sites/default/files/publications/Global-LCA-passenger-cars-jul2021_0.pdf.

Bird, D. N., G. Zanchi and N. Pena. 2013. A method for estimating the indirect land use change from bioenergy activities based on the supply and demand of agricultural-based energy. *Biomass & Bioenergy* 59:3–15.

Capodaglio, A. G. and S. Bolognesi. 2019. Ecofuel feedstocks and their prospects. In *Advances in Eco-Fuels for a Sustainable Environment*, Ed. K. Azad, pp. 15–51. Cambridge: Woodhead Publishing.

Cengicana. 2021. *Boletín Estadistico, Generacion ee Energia* [*Statistical Bulletin, Power Generation*]. Guatemala: Guatemalan Center for Sugar Cane Research and Training. https://cengicana.org/files/20201204105723342.pdf.

CerradinhoBio. 2020. *Usina Cerradinho Bio Production.* https://www.cerradinhobio.com.br/conteudo_pti.asp?idioma=0&conta=45&tipo=51240.

Chum, H. L. 2018. *Biomass Pyrolysis to Hydrocarbon Fuels in the Petroleum Refining Context: Cooperative Research and Development Final Report.* Golden, CO: National Renewable Energy Laboratory.

CONAB. 2019. *Perfil do Setor do Acucare do Etanol no Brasil; Edição para a Safra 2015/16,* [*Profile of the Ethanol Sugar Setor in Brazil; Edition for the Season 2015/16*] Brasilia: Compania Nacional de Abastecimento. https://www.conab.gov.br/info-agro/safras/cana.

Cruz, D. M. 2021. *Biogas de Cana Chega de Vez ao Mercado* [*Sugarcane Biogas Hits the Market for Good*]. Energia que fala com voce [Energy that speaks to you]. https://www.energiaquefalacomvoce.com.br/2020/08/20/biogas-de-cana-chega-de-vez-ao-mercado/.

CTC. 2016. *Centro de Tecnologia Canavieira* [*Canavieira Technology Center*] https://www.ctcanavieira.com.br/. Accessed in 19 June 2021.

Do Canto Braga Junior, R. L., M. G. de Andrade Landell and D. N. da Silva, et al. 2021. Censo varietal IAC de cana de acucar no Brasil [Varietal census of sugar cane in Brazil]. Boletin Technico IAC 225. Instituto Agronomico de Campinas [Agronomic Institute of Campinas]. https://www.iac.sp.gov.br/publicacoes/arquivos/iacbt225.pdf.

Dominguez-Faus, R., S. E. Powers, J. G. Burken and P. J. Alvarez. 2009. The water footprint of biofuels: A drink or drive issue? The water consumption and agrochemical use during biofuel production could adversely impact both availability and quality of a precious resource. *Environmental Science and Technology* 43:3005–3010.

Elgowainy, A., J. Han and J. Ward, et al. 2016. *Cradle-to-Grave Lifecycle Analysis of U.S. Light-Duty Vehicles-Fuel Pathways: A Greenhouse and Economic Assessment of Current (2015) and Future (2025-2030) Technologies.* Argonne, IL: Argonne National laboratory.

EPE. 2020a. *Anuario Estatisticode Energia Eletrica 2020* [*Electric Energy Statistical Yearbook 2020*]. Rio de Janeiro: Empresa de Planejamento Energetico [Energy Planning Company].

EPE. 2020b. *Demanda de Energia dos Veículos Leves do Ciclo Otto: 2021-2030* [*Energy Demand of Otto Cycle Light Vehicles: 2021–2030*]. Rio de Janeiro: Empresa de Planejamento Energetico [Energy Planning Company].

Finley R. J. 2014. An overview of the Illinois basin-Decatur project. *Greenhouse Gases: Science and Technology* 4:571–579.
Fuss, S., C. D. Jones and F. Kraxner, et al. 2016. Research priorities for negative emissions. *Environmental Research Letters* 11:115007.
Gabrielli, B., N. N. The, P. Maupu and E. Vial. 2013. Life cycle assessment of eucalyptus short rotation coppices for bioenergy production in Southern France. *GCB Bioenergy* 5:30–42.
Germeck, H. A. and G. F. Feigl. 1987. Processo de reducao de vinhaca [Vinasse reduction process]. *STAB* 6:42–50.
Giuntoli, J. 2018. *Final Recast Renewable Energy Directive for 2021-2030 in the European Union*. International Council on Clean Transportation. https://theicct.org/sites/default/files/publications/EU_Fuels_Policy_Update_20180719.pdf.
Goldemberg, J., S. Coelho and P. Guardabassi. 2008. The sustainability of ethanol production from sugarcane. *Energy Policy* 36:2086–2097.
GREET2019. 2020. *Argonne GREET Model*. Argonne, IL: Argonne National Laboratory.
IEA. 2019. *Putting CO_2 to Use: Creating Value from Emissions*. International Energy Agency. https://iea.blob.core.windows.net/assets/50652405-26db-4c41-82dc-c23657893059/Putting_CO2_to_Use.pdf.
IPCC. 2011a. Technical Summary, Chapters 3, 4 and 5. In *IPCC Special Report on Renewable Energy Sources and Climate Change Mitigation*, Eds. O. Edenhofer, R. Pichs-Madruga and Y. Sokona, et al, pp. 7–19. Cambridge: Cambridge University Press.
IPCC. 2011b. Technical summary, Chapter 9. In *IPCC Special Report on Renewable Energy Sources and Climate Change Mitigation*, Eds. O. Edenhofer, R. Pichs-Madruga and Y. Sokona, et al., p. 26. Cambridge: Cambridge University Press.
ISA. 2021. International Sugar Association. https://www.isosugar.org/
Karp, S. G., J. D. C. Medina and L. A. J. Letti, et al. 2021. Bioeconomy and biofuels: The case of sugarcane ethanol in Brazil. *Biofuels, Bioproducts and Biorefining* 15:899–912.
Laborde, D., M. Padella, R. Edwards and L. Marelli. 2015. *Progress in Estimates of ILUC with MIRAGE Model*. European Commission.
Larsen, U., T. Johansen and J. Schramm. 2009. *Ethanol as a Fuel for Road Transportation-Main Report*. Technical University of Denmark.
Mekonnen, M. M., T. L. Romanelli and C. Ray, et al. 2018. Water, energy, and carbon footprints of bioethanol from the US and Brazil. *Environmental Science & Technology* 52:14508–14518.
Morandi, M. A. B. 2020. *The Science behind Brazilian Biofuels Policy*. Brasilia: Brazilian Agricultural Research Company. https://www.embrapa.br/busca-de-noticias/-/noticia/54067756/article-the-science-behind-brazilian-biofuels-policy--renovabio, Accessed in July 29, 2021.
Muratori, M., H. Kheshgi and B. Mignone, et al. 2017. Carbon capture and storage across fuels and sectors in energy system transformation pathways. *International Journal of Greenhouse Gas Control* 57:34–41.
Nassar, A. M. and M. Moreira. 2013. *Evidences on Sugarcane Expansion and Agricultural Land Use Changes in Brazil Report*. ICONE-Institute for International Trade Negotiation. https://www.iconebrasil.com.br/datafiles/publicacoes/estudos/2013/evidences_on_sugarcane_expansion_and_agricultural_land_use_changes_in_brazil_1206_2.pdf.
Nigro, F. and A. Szwarc. 2009. Ethanol as a fuel. In *Ethanol and Bioelectricity: Sugarcane in the Future of the Energy Matrix*, Eds. E. L. L. de Souza and I. C. Macedo, pp. 156–189. Sao Paulo: UNICA.
Observatorio da Cana. 2021. *Fuel Consumption*. https://observatoriodacana.com.br/listagem.php?idMn=11&idioma=2.
OECD/FAO. 2020. *OECD-FAO Agricultural Outlook*. OECD. https://www.agri-outlook.org/.
Pacca, S. and J. R. Moreira. 2009. Historical carbon budget of the Brazilian ethanol program. *Energy Policy* 37:4863–4873.
PDE 2030. 2021a. *Plano Decenal de Expansao de Energia 2030* [*Ten-Year Energy Expansion Plan 2030*]. Brasilia: Empresa de Pesquisa Energetica.
PDE 2030. 2021b. *Plano Decenal de Expansao de Energia 2030-Item 8.5.1* [*Ten-Year Energy Expansion Plan 2030-Item 8.5.1*]. Brasilia: Empresa de Pesquisa Energetica.
Pereira, L. G., O. Cavaletta and A. Bonomia, et al. 2019. Comparison of biofuel life-cycle GHG emissions assessment tools: The case studies of ethanol produced from sugarcane, corn, and wheat. *Renewable and Sustainable Energy Reviews* 110:1–12.
REN21 (Renewables Now). 2020. *Renewables 2020 Global Status Report: A Comprehensive Annual Overview of the State of Renewable Energy*. REN21-Renewables Now. https://www.ren21.net/gsr-2020/.
RFA. 2019. *Powered with Renewed Energy*. Renewable Fuel Association.

Sanchez, D. L., N. Johnson, S. T. McCoy, P. A. Turner and K. J. Mach, 2018. Near-term deployment of carbon capture and sequestration from biorefineries in the United States, *Proceedings of the National Academy of Sciences* 115:4875–4880.

Scully, M. J., G. A. Norris, T. M. A. Falconi and D. L. MacIntosh. 2021. Carbon intensity of corn ethanol in the United States: State of the science. *Environmental Research Letters* 16:043001.

Sebigas Cotica. 2021. *Biogas de Vinhaaa: Uma realidade.* [*Vinhaca Biogas: A reality*]. Sebigas Cotica. https://sebigascotica.com.br/artigo/biogas-de-vinhaca-uma-realidade.html.

Simoes, C. L. N., E. R. de Sena and R. Campos, 2004. Estudo da viabilidade economica da concentracao de vinhoto atraves de osmose inversa. [Study of the economic feasibility of the concentration of wine through reverse osmose]. In *XXIV Encontro Nacional de Engenharia de Producao.* Florianpolis, SC, Brazil, November 03-05, 2004. https://www.abepro.org.br/biblioteca/ENEGEP2004_Enegep1004_1360.pdf.

SUCRE. 2020. *Sugarcane Renewable Electricity (SUCRE) Project.* Eds. M. R. L. Leal and T. A. D. Hernandes. Brazilian Biorenewables National Laboratory.

Tyner, W. E., F. Taheripour and K. Hoekman, et al. 2016. *Follow-on Study of Transportation Fuel Life Cycle Analysis: Review of Current Carb and EPA Estimates of Land Use Change (LUC) Impacts.* Sacramento, CA: Sierra Research.

UNICA. 2019. *A Bioeletricidade da Cana* [*The Bioelectricity of Sugarcane*]. Union of the Sugar Cane Industry. https://www.unica.com.br/wp-content/uploads/2019/07/UNICA-Bioeletricidade-julho2019-1.pdf.

UNICA. 2021. *Bioeletricidade* [*Bioelectricity*]. Union of the Sugar Cane Industry https://unica.com.br/setor-sucroenergetico/bioeletricidade/

USDA. 2021a. *National Agricultural Statistics Service, Charts and Maps,* https://www.nass.usda.gov/Charts_and_Maps/Field_Crops/index.php.

USDA. 2021b. *Sugar: World Markets and Trade.* United States Department of Agriculture. https://apps.fas.usda.gov/psdonline/circulars/sugar.pdf.

USDA. 2021c. *Economic Research Service.* USDA. https://data.ers.usda.gov/FEED-GRAINS-custom-query.aspx#ResultsPanel.

USDOE. 2021a. *Alternative Production and Distribution, Maps & Data.* United States Department of Energy. https://afdc.energy.gov/fuels/ethanol_production.html.

USDOE. 2021b. *Flex Fuel Vehicles*, Alternative Fuel Data Center, United States Department of Energy. https://afdc.energy.gov/vehicles/flexible_fuel.html.

USEIA. 2021a. *Biofuels.* U. S. Energy Information Administration. https://www.iea.org.

USEIA. 2021b. *Global Energy Review 2021: CO_2 Emissions.* U. S. Energy Information Administration. https://www.iea.org/reports/global-energy-review-2021/co2-emissions.

USEIA. 2021c. *Petroleum and Other Liquids.* U. S. Energy Information Administration https://www.iea.org.

USEPA. 2010. *Renewable Fuel Standard Program (RFS2) Regulatory Impact Analysis.* Washington, DC: Environmental Protection Agency.

USEPA. 2021. *Lifecycle Analysis of Greenhouse Gas Emissions under the Renewable Fuel Standard.* https://www.epa.gov/fuels-registration-reporting-and-compliance-help/lifecycle-greenhouse-gas-results.

Valin, H., D. Peters and M. van den Berg, et al. 2015. *The Land Use Change Impact of Biofuels Consumed in the EU: Quantification of Area and Greenhouse Gas Impacts.* Utrecht, ECOFYS Netherlands B.V.

Wang, Z., F. P. Kamali, P. Osseweijer and J. A. Posada. 2019. Socioeconomic effects of aviation biofuel production in Brazil: A scenarios-based input-output analysis. *Journal of Cleaner Production* 230:1036–1050.

Wicke, B., P. Verweij and H. Meijl, et al. 2012. Indirect land use change: Review of existing models and strategies for mitigation. *Biofuels* 3:87–100.

Woltjer, G, V. Daioglou and B. Elbersen, et al. 2017. *Reporting Requirements on Biofuels and Bioliquids Stemming from the Directive (EU) 2015/1513.* PBL Netherlands Environmental Assessment Agency.

Zhao, X., D. Y. van der Mensbrugghe and W. E. Tyner. 2017. Modeling land physically in CGE models: New insights on intensive and extensive margins. *Presented at the 20th Annual Conference on Global Economic Analysis*, West Lafayette, IN. https://www.gtap.agecon.purdue.edu/resources/res_display.asp?RecordID=5291.

Part 27

Utilization of Bioethanol Fuels

90 Utilization of Bioethanol Fuels
Scientometric Study

Ozcan Konur
(Formerly) Ankara Yildirim Beyazit University

90.1 INTRODUCTION

Crude oil-based gasoline fuels (Ma et al., 2002; Newman and Kenworthy, 1989) have been widely used in transportation sector since the 1920s. However, there have been great public concerns over adverse environmental and human impact of these fuels (Hill et al., 2006, 2009). Hence, biomass-based bioethanol fuels (Hill et al., 2006; Konur, 2012e, 2015, 2019, 2020a) have increasingly been used in blending gasoline fuels (Hsieh et al., 2002; Najafi et al., 2009) and in fuel cells (Antolini, 2007, 2009). Additionally, bioethanol fuels have been used to produce valuable biochemicals (Liu and Hensen, 2013; Wang et al., 2004) in a biorefinery (Maity, 2015a-b) context.

The primary focus of research in this area has been utilization of bioethanol fuels in fuel cells (Antolini, 2007; Liang et al., 2009), utilization of bioethanol fuels in gasoline and diesel engines (Hansen et al., 2005; Kohse-Hoinghaus et al., 2010), production of biohydrogen fuels from bioethanol fuels for fuel cells (Haryanto et al., 2005; Ni et al., 2007), and, to a lesser extent, development and utilization of bioethanol sensors (Liu et al., 2005; Wan et al., 2004), and production of biochemicals from bioethanol fuels (Liu and Hensen, 2013; Wang et al., 2004). Additionally, research on gasoline fuels (Khalili et al., 1995; Song, 2003) and applications of nanotechnology (Murdoch et al., 2011; Wan et al., 2004) in this field has also been closely related to this field.

However, it is essential to develop efficient incentive structures (North, 1991) for the primary stakeholders to enhance research in this field (Konur, 2000, 2002a–c, 2006a,b, 2007a,b). Scientometric analysis has been used in this context to inform the primary stakeholders about the current state of research in a selected research field (Garfield, 1955; Konur, 2011, 2012a–i, 2015, 2018b, 2019, 2020a).

As there have been no scientometric studies on utilization of bioethanol fuels, this book chapter presents a scientometric study of research in utilization of bioethanol fuels. It examines scientometric characteristics of both the sample and population data presenting scientometric characteristics of these both datasets in the order of documents, authors, publication years, institutions, funding bodies, source titles, countries, Scopus subject categories, keywords, and research fronts.

90.2 MATERIALS AND METHODS

Search for this study was carried out using Scopus database (Burnham, 2006) in November 2022.

As the first step for search of relevant literature, keywords were selected using the first 200 most-cited papers for each substituent research front. The selected keyword list was optimized to obtain a representative sample of papers for each research field, and they are integrated to form the keyword list for this study. This keyword list was provided in appendices of Konur (2023c–g) for future replication studies.

As the second step, two sets of data were used for this study. First, a population sample of over 14,000 papers was used to examine scientometric characteristics of the population data. Secondly, a sample of 280 most-cited papers, corresponding to 2% of the population papers, was used to examine scientometric characteristics of these citation classics.

DOI: 10.1201/9781003226567-121

Scientometric characteristics of these both sample and population datasets were presented in the order of documents, authors, publication years, institutions, funding bodies, source titles, countries, Scopus subject categories, keywords, and research fronts.

Lastly, the key scientometric findings for both datasets were discussed to highlight research landscape for utilization of bioethanol fuels. Additionally, a number of brief conclusions were drawn and a number of relevant recommendations were made to enhance future research landscape.

90.3 RESULTS

90.3.1 The Most Prolific Documents in Utilization of Bioethanol Fuels

Information on the types of documents for both datasets is given in Table 90.1. Articles, review papers, and conference papers dominate both sample and population papers. Further, review papers and conference papers gave a surplus and deficit, respectively.

It is also interesting to note that all of the papers in the sample dataset were published in journals, while only 96% of the papers were published in journals for the population dataset. Furthermore, 2.4% and 1.1% of the papers were published in the book series and books, respectively.

90.3.2 The Most Prolific Authors in Utilization of Bioethanol Fuels

Information about the 52 most prolific authors with at least two sample papers and ten population papers each is given in Table 90.2.

The most prolific authors are Claude Lamy and Gongquan Sun with eight sample papers each, followed by Panagiotis Tsiakaras and Qing Qin with seven sample papers each. Other prolific authors are Shuqin Song, Christophe Coutanceau, Peikang Shen, Miguel Laborde, Stefano Cavallaro, Shi-Jin Shuai, Jian-Xin Wang, Weijiang Zhou, Zhenhua Zhou, El Mustapha Belgsir, and Changwei Xu with five to six sample papers each.

The most prolific institution for the sample dataset is the University of Poitiers with six authors, followed by the Chinese Academy of Sciences and Italian National Research Council (CNR) with five and four sample papers, respectively. Other prolific institutions are Ciudad University, National Technical University of Athens, National Technological Institute, Southern University of Science and Technology, University of Sao Paulo, and Xiamen University with two or three sample papers each. In total, 31 institutions house these prolific authors.

The most prolific country for the sample dataset is China with 20 authors, followed by France with 6 authors. Other prolific countries are Argentina, Brazil, Germany, Greece, and Italy with two to four authors each. In total, 13 countries house these authors.

TABLE 90.1
Documents in Utilization of Bioethanol Fuels

Documents	Sample Dataset (%)	Population Dataset (%)	Surplus (%)
Article	87.9	87.4	0.5
Review	7.1	1.5	5.6
Conference paper	4.3	9.0	–4.7
Note	0.4	0.3	0.1
Short survey	0.4	0.1	0.3
Book chapter	0.0	1.5	–1.5
Letter	0.0	0.1	–0.1
Editorial	0.0	0.0	0.0
Book	0.0	0.0	0.0
Sample size	280	14,009	

TABLE 90.2
Most-Prolific Authors in Utilization of Bioethanol Fuels

No.	Author Name	Author Code	Sample Papers (%)	Population Papers (%)	Institution	Country	Res. Front
1	Sun, Gongquan	7402760735	3.2	0.1	Chinese Acad. Sci.	China	F
2	Lamy, Claude	7007017658	3.2	0.0	Univ. Poitiers	France	F
3	Tsiakaras, Panagiotis	7003948427	2.8	0.1	Univ. Thessalia	Greece	F
4	Qin, Qing	57206386894	2.8	0.0	Xiamen Univ.	China	H
5	Song, Shuqin	7403349881	2.4	0.0	Chinese Acad. Sci.	China	F
6	Coutanceau, Christophe	8714035200	2.4	0.0	Univ. Poitiers	France	F
7	Shen, Peikang	7201767641	2.0	0.1	Guangxi Univ.	China	F
8	Laborde, Miguel	7003342826	2.0	0.0	Ciudad Univ.	Argentina	H
9	Cavallaro, Stefano	7006497260	2.0	0.0	Univ. Messina	Italy	H
10	Shuai, Shi-Jin	6603356005	2.0	0.0	Tsinghua Univ.	China	E
11	Wang, Jian-Xin	35254238800	2.0	0.0	Tsinghua Univ.	China	e
12	Zhou, Weijiang	56003143300	2.0	0.0	Nanyang Technol. Univ.	Singapore	F
13	Zhou, Zhenhua	7406094713	2.0	0.0	Headwaters Technol.	USA	F
14	Belgsir, El Mustapha	6701740559	2.0	0.0	Univ. Poitiers	France	F
15	Xu, Changwei	9248835900	2.0	0.0	Jinan Univ.	China	F
16	Gonzalez, Ernesto R.	57199756260 35596037000	1.6	0.1	Univ. Sao Paulo	Brazil	F
17	Zhao, Tianshou	13004121800	1.6	0.1	Southern Univ. Sci. Technol.	China	F
18	Freni, Salvatore	7004444222	1.6	0.0	CNR	Italy	H
19	Jiang, Luhua	57209054216	1.6	0.0	Qingdao Univ. Sci. Technol.	China	F
20	Leger, Jean M	7201980020	1.6	0.0	Univ. Poitiers	France	F
21	Frusteri, Francesco	7003418964	1.6	0.0	CNR	Italy	H
22	Rakopoulos, Constantine D.	6506189320	1.6	0.0	Natl. Tech. Univ. Athens	Greece	E
23	Marino, Fernando	7101879626	1.6	0.0	Ciudad Univ.	Argentina	H
24	Rakopoulos, Dimitrios C.	6603012578	1.6	0.0	Natl. Tech. Univ. Athens	Greece	E
25	Rousseau, Severine*	8714035100	1.6	0.0	Univ. Poitiers	France	F
26	Su, Dong	55154198900	1.6	0.0	Chinese Acad. Sci.	China	F
27	Llorca, Jordi	26039349400	1.2	0.1	Tech. Univ. Catalonia	Spain	H
28	Noronha, Fabio V.	7004514118	1.2	0.1	Natl. Technol. Inst.	Brazil	H
29	Mattos, Lisiane V.*	7005199691	1.2	0.1	Natl. Technol. Inst.	Brazil	H
30	Idriss, Hicham	7006768868	1.2	0.0	Karlsruhe Inst. Technol.	Germany	H
31	Lee, Jong-Heun	26643283000	1.2	0.0	Korea Univ.	S. Korea	S
32	Jacobs, Gary	7203024020	1.2	0.0	Univ. Texas. S. A.	USA	H
33	Ticianelli, Edson A.	35444263400	1.2	0.0	Univ. Sao Paulo	Brazil	H
34	He, Hong	55229497700	1.2	0.0	Chinese Acad. Sci.	China	E
35	Masjuki, Haji H.	57175108000	1.2	0.0	Univ. Malaya	Malaysia	E
36	Davis, Burtron H.	36043712200	1.2	0.0	Univ. Kentucky	USA	H
37	Li, Yinshi	15762935100	1.2	0.0	Xi'an Jiaotong Univ.	China	F

(Continued)

TABLE 90.2 (*Continued*)
Most-Prolific Authors in Utilization of Bioethanol Fuels

No.	Author Name	Author Code	Sample Papers (%)	Population Papers (%)	Institution	Country	Res. Front
38	Wang, Taihong	35241217600	1.2	0.0	Southern Univ. Sci. Technol.	China	S
39	Verykios, Xenophon E.	35551305100	1.2	0.0	Univ. Patras	Greece	H
40	Adzic, Radoslav R	7006804065	1.2	0.0	Brookhaven Natl. Lab.	USA	F
41	Behm, Rolf J.	36885065400	1.2	0.0	Univ. Ulm	Germany	F
42	Chiodo, Vitaliano	6603101616	1.2	0.0	CNR	Italy	H
43	Guo, Shaojun	16202647200	1.2	0.0	Peking Univ.	China	F
44	Li, Ya-Dong	57192004602	1.2	0.0	Tsinghua Univ.	China	S
45	Sayin, Cenk	7801370427	1.2	0.0	Marmara Univ.	Turkey	E
46	Spadaro, Lorenzo	6602740818	1.2	0.0	CNR	Italy	H
47	Tian, Na	57203214786	1.2	0.0	Xiamen Univ.	China	F
48	Tong, Ye-Xiang	55596802500	1.2	0.0	Sun Yat-Sen Univ.	China	F
49	Vigier, Fabrice	6507964980	1.2	0.0	Univ. Poitiers	France	F
50	Zhou, Zhi-You	7406098551	1.2	0.0	Xiamen Univ.	China	F
51	Chen, Yujin	7601437135	1.2	0.0	Chinese Acad. Sci.	China	S
52	Li, Wenzhen	8922678800	1.2	0.0	Iowa State Univ.	China	F

*, Female researchers; Author code: Unique code given by Scopus to authors. Sample papers: Number of papers authored in the sample dataset. Population papers: Number of papers authored in the population dataset.
E, Utilization of bioethanol fuels in gasoline or diesel engines; F, Utilization of bioethanol fuels in fuel cells; H, Production of biohydrogen fuels from bioethanol fuels; S, Bioethanol sensors.

The most prolific research front is fuel cells with 25 authors, followed by biohydrogen fuels produced from bioethanol fuels with 16 authors. Other prolific research fronts are utilization of bioethanol fuels in gasoline and diesel engines and bioethanol sensors with seven and four sample papers, respectively. It is notable that there are no authors on biochemicals produced from bioethanol fuels.

On the other hand, there is a significant gender deficit (Beaudry and Lariviere, 2016) for the sample dataset as surprisingly only two of these top researchers are female with a 4% representation rate.

On the other hand, a large number of authors with relative low citations have contributed immensely to research in this field: Almir O. Neto, Estevam V. Spinace, Vincenzo Palma, Yaping Du, Marcelo Linardi, Guido Busca, Germano Tremiliosi-Filho, Ilenia Rossetti, Suttichai Assabumrungrat, Wen-Bin Cai, Shash Zhao, Keqiang Ding, Mauro C. de Santos, Jose R. Sodre, Lucia G. Appel, Pavlo E. Kyriienko, Concetta Ruocko, Sergiy O. Soloviev, Wen Zeng, Ermete Antolini, Jose L. G. Fierro, Peter G. Pickup, Julio C. M. da Silva, Bengt Johansson, Bunjerd Jongsomjit, Gianguido Ramis, In-Kyu Song, Qin Xin, Olga V. Larina, Timothy H. Lee, Bianca M. Vaglieco, Han Xu, Elsabete M. Assaf, Angelo B. Basile, Narcis Homs, Chia-Fon Lee, Sunghoon Park, Jayati Datta, Juan M. Feliu, Chang Sik Lee, Rodrigo F. B. de Souza, Yadollah Mortazavi, Giancarlo R. Salazar-Banda, Wenjie Shen, Tong Zhang, Mingyuan Zheng, Christina Abello, Gabriella Garbarino, Jinlong Gong, Georgios Karavalakis, Chongmu Lee, Elena Pastor, and Xuesog Wu.

90.3.3 The Most Prolific Research Output by Years in Utilization of Bioethanol Fuels

Information about papers published between 1970 and 2022 is given in Figure 90.1. This figure clearly shows that the bulk of research papers in the population dataset were published primarily in the 2010s, early 2020, and 2010s with 57%, 20%, and 16% of the population dataset, respectively.

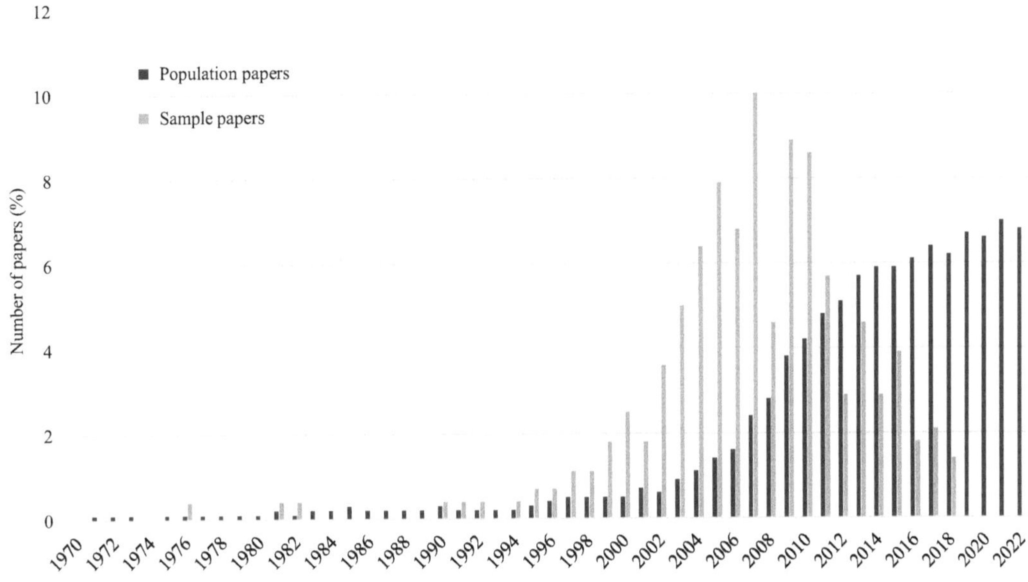

FIGURE 90.1 Research output by years regarding utilization of bioethanol fuels.

The publication rates for the 1990s, 1980s, 1970s, and pre-1970s were 3%, 2%, 1%, and 1% respectively. The number of population papers rose between 2003 and 2017. However, it steadied after 2017 around 6% of the population papers for each year, losing its momentum.

Similarly, the bulk of research papers in the sample dataset were published in the 2000s and 2010s with 58% and 25% of the sample dataset, respectively. Publication rates for the 1990s, 1980s, and 1970s were 7%, 1%, and 0.4% of the sample papers, respectively.

The most prolific publication years for the population dataset were between 2019 and 2022 with at least 6.6% each year. Over 86% of the population papers were published between 2007 and 2022. Similarly, 92% of the sample papers were published between 1999 and 2017. The most prolific publication years were in 2007, 2009, and 2010 with 10%, 8.9%, and 8.6% of the sample papers each.

90.3.4 The Most Prolific Institutions in Utilization of Bioethanol Fuels

Information about the 46 most prolific institutions publishing papers on utilization of bioethanol fuels with at least three sample papers each is given in Table 90.3.

The most prolific institutions are Chinese Academy of Sciences and University of Poitiers with 5.7% of the sample papers, followed by Tsinghua University with 4.3% of the sample papers. University of Poitiers, Sun Yat-Sen University, University of Sao Paulo, Brookhaven National Laboratory, University of Thessaly, and University of Patras are other prolific institutions with 2.1%–2.9% of the sample papers each.

The top country for these most prolific institutions is China with 12 institutions, followed by the USA with 5 institutions. Other prolific countries are Brazil, France, Greece, Italy, S. Korea, Spain, and Turkey with two or three institutions each. In total, 20 countries house these top institutions.

On the other hand, institutions with the most citation impact are Tsinghua University and Chinese Academy of Sciences with 3.2% and 2.9% surplus, respectively. Other prolific institutions are Sun Yat-Sen University, University of Poitiers, University of Thessaly, and Brookhaven National University with 2.1%–2.5% surplus each.

On the other hand, there are a large number of institutions with relatively low citation impact contributing immensely to the research in this field: Jilin University, University of Rio de Janeiro, Institute of Energy and Nuclear Research, National Council for Scientific and Technical Research

TABLE 90.3
The Most Prolific Institutions in Utilization of Bioethanol Fuels

No.	Institutions	Country	Sample Papers (%)	Population Papers (%)	Surplus (%)
1	Chinese Acad. Sci.	China	5.7	2.8	2.9
2	Tsinghua Univ.	China	4.3	1.1	3.2
3	Univ. Poitiers	France	2.9	0.5	2.4
4	Sun Yat-Sen Univ.	China	2.9	0.4	2.5
5	Univ. Sao Paulo	Brazil	2.5	2.0	0.5
6	Brookhaven Natl. Lab.	USA	2.5	0.4	2.1
7	Univ. Thessaly	Greece	2.5	0.3	2.2
8	Univ. Patras	Greece	2.1	0.3	1.8
9	Natl. Res. Counc. (CNR)	Italy	1.8	1.1	0.7
10	CSIC	Spain	1.8	0.6	1.2
11	Natl. Cheng Kung Univ.	Taiwan	1.8	0.5	1.3
12	Lund Univ.	Sweden	1.8	0.4	1.4
13	Hong Kong Univ. Sci. Technol.	China	1.8	0.3	1.5
14	Univ. Messina	Italy	1.8	0.3	1.5
15	Tianjin Univ.	China	1.4	1.8	–0.4
16	CNRS	France	1.4	0.7	0.7
17	Xiamen Univ.	China	1.4	0.7	0.7
18	Shanghai Jiao Tong Univ.	China	1.4	0.7	0.7
19	Univ. Buenos Aires	Argentina	1.4	0.3	1.1
20	Univ. Lyon	France	1.4	0.3	1.1
21	Natl. Tech Univ. Athens	Greece	1.4	0.2	1.2
22	Soochow Univ.	China	1.1	0.7	0.4
23	Chulalongkorn Univ.	Thailand	1.1	0.6	0.5
24	Natl. Inst. Technol.	Brazil	1.1	0.6	0.5
25	Beijing Univ. Chem. Technol.	China	1.1	0.5	0.6
26	Univ. Castilla la Mancha	Spain	1.1	0.4	0.7
27	Nanyang Technological Univ.	S. Korea	1.1	0.4	0.7
28	Natl. Inst. Adv. Ind. Sci. Technol.	Japan	1.1	0.4	0.7
29	Univ. Malaya	Malaysia	1.1	0.3	0.8
30	Pacific NW Natl. Lab.	USA	1.1	0.3	0.8
31	Korea Univ.	S. Korea	1.1	0.3	0.8
32	Univ. Porto	Portugal	1.1	0.3	0.8
33	Univ. Wisconsin-Madison	USA	1.1	0.3	0.8
34	Fudan Univ.	China	1.1	0.2	0.9
35	Hong Kong Polytech. Univ.	China	1.1	0.2	0.9
36	Korea Adv. Inst. Sci. technol.	S. Korea	1.1	0.2	0.9
37	Nanyang Technol. Univ.	Singapore	1.1	0.2	0.9
38	Univ. Ulm	Germany	1.1	0.2	0.9
39	Gazi Univ.	Turkey	1.1	0.2	0.9
40	Jinan Univ.	China	1.1	0.2	0.9
41	Leiden Univ.	Netherlands	1.1	0.2	0.9
42	Univ. Kentucky	USA	1.1	0.2	0.9
43	Case Western Reserve Univ.	USA	1.1	0.1	1.0
44	Karabuk Univ.	Turkey	1.1	0.1	1.0
45	Marmara Univ.	Turkey	1.1	0.1	1.0
46	Univ. Auckland	N. Zealand	1.1	0.1	1.0

(CONICET), Xi'an Jiaotong University, University of Science and Technology of China, Russian Academy of Sciences, State University of Campinas, Zhejiang University, National Commission of Nuclear Energy Rio de Janeiro, National Academy of Sciences in Ukraine, State University of Paulista, Beijing Institute of Technology, Taiyuan University of Technology, Jiangsu University, Hunan University, University of Barcelona, Chongqing University, Argonne National Laboratory, Federal University of Rio Grande do Sul, Seoul National University, University of Genoa, federal University of Sao Carlos, Indian Institute of Technology Delhi, Dalian University of Technology, Huazhong University of Science and Technology, State Key Laboratory of Physical Chemistry of Solid Surfaces, Polytechnic University of Catalonia, Wuhan University of Technology, Shandong University, Key Laboratory for Green Chemical Technology of Ministry of Education, University of Michigan Ann Arbor, King Abdullah University of Science and Technology, and Ford Motor Company.

90.3.5 The Most Prolific Funding Bodies in Utilization of Bioethanol Fuels

Information about the 20 most prolific funding bodies funding at least 1.1% of the sample papers each is given in Table 90.4.

The most prolific funding body is National Natural Science Foundation of China funding 11.4% of the sample papers. Other prolific funding bodies are US Department of Energy, European Commission, Chinese Academy of Sciences, Sao Paulo State Research Foundation, National Science Foundation, and Basic Energy Sciences with 1.8%–2.9% of the sample papers each.

It is notable that 46% and 49% of the sample and population papers are funded, respectively.

TABLE 90.4
The Most Prolific Funding Bodies in Utilization of Bioethanol Fuels

No.	Funding Bodies	Country	Sample Paper No. (%)	Population Paper No. (%)	Surplus (%)
1	National Natural Science Foundation of China	China	11.4	14.5	–3.1
2	US Dept. Energy	USA	2.9	1.7	1.2
3	European Commission	Europe	2.1	0.7	1.4
4	Chinese Acad. Sci.	China	2.1	0.6	1.5
5	Sao Paulo State Res. Found	Brazil	1.8	2.1	–0.3
6	National Science Foundation	USA	1.8	1.5	0.3
7	Basic Ener. Sci.	USA	1.8	0.6	1.2
8	Eng. Phys. Sci. Res. Counc.	UK	1.4	0.7	0.7
9	Ministry Educ.	China	1.4	0.6	0.8
10	Univ. Buenos Aires	Argentina	1.4	0.1	1.3
11	German Res. Found	Germany	1.4	0.1	1.3
12	Natl. Counc. Sci. Technol. Dev.	Brazil	1.1	3.4	–2.3
13	Natl. Key Res. Devnt. Pr.	China	1.1	1.4	–0.3
14	Eur. Reg. Devnt. Fund	EU	1.1	1.0	0.1
15	Pr. Acad. Prg. Devnt. Jiangsu HEIs	China	1.1	0.7	0.4
16	Natl. Sci. Found. Guangdong Pr.	China	1.1	0.3	0.8
17	Govnt. Catalunya	Spain	1.1	0.1	1.0
18	Army Res. Off.	USA	1.1	0.1	1.0
19	Natl. Inst. Sci. Technol. Excitotoc. Neur.	Portugal	1.1	0.1	1.0
20	French Env. Ener. Mng. Agcy.	France	1.1	0.1	1.0

The most prolific countries for these top funding bodies are China, the USA, the EU, and Brazil with six, four, two, and two funding bodies, respectively. In total, eight countries and the European Union house these top funding bodies.

Funding bodies with the most citation impact are Chinese Academy of Sciences, European Commission German Research Foundation, and University of Buenos Aires with 1.3%–1.5% surplus each. Similarly, funding bodies with the least citation impact are National Natural Science Foundation of China and National Council for Science and Technology Development with 3.1% and 2.3% deficit, respectively.

On the other hand, there are a large number of funding bodies with relatively low citation impact in this field: Fundamental Research Funds for the Central Universities, Higher Education Personnel Improvement Coordination, National Research Foundation of Korea, China Postdoctoral Science Foundation, Natural Sciences and Engineering Research Council of Canada, Japan Society for the Promotion of Science, China Scholarship Council, Ministry of Economy and Competitiveness, Ministry of Science and Technology India, Office of Science, Thailand Research Fund, National Basic Research Program of China (973 Program), Ministry of Education, National Agency for Scientific and Technological Promotion, Natural Science Foundation of Shandong Province, National Council for Science and Technology, Ministry of Science and Technology Taiwan, Ministry of Science and Innovation, Ministry of Education, Science and Technology, Program for New Century Excellent Talents in University, Russian Foundation for Basic Research, Ministry of Science and Technology of China, Ministry of Education, Culture, Sports, Science and Technology, Council of Scientific and Industrial Research India, and National Science Council.

90.3.6 The Most Prolific Source Titles in Utilization of Bioethanol Fuels

Information about the 26 most prolific source titles publishing at least 1.1% of the sample papers each in utilization of bioethanol fuels is given in Table 90.5.

The most prolific source titles are Journal of Power Sources with 8.9% of the sample papers, followed by Applied Catalysis B Environmental, International Journal of Hydrogen Energy, Fuel, and Journal of Catalysis with 7.5%, 7.1%, 7.1%, and 5.0% of the sample papers, respectively. Other prolific titles are Electrochimica Acta, Renewable Energy, Journal of the American Chemical Society, Atmospheric Environment, and Renewable and Sustainable Energy Reviews with 2.5%–3.6% of the sample papers each.

On the other hand, source titles with the most citation impact are Journal of Power Sources and Applied Catalysis B Environmental with 6.7% and 6.0% surplus, respectively. Other influential titles are Journal of Catalysis, Fuel, Renewable Energy, Renewable and Sustainable Energy Reviews, Atmospheric Environment, Journal of the American Chemical Society, and Advanced Materials with 2%–4% surplus each.

On the other hand, a large number of journals with a relative low citations have contributed immensely to the research in this field: SAE Technical Papers, Electrochimica Acta, Applied Surface Science, RSC Advances, Journal of Alloys and Compounds, Catalysis Letters, Journal of the Electrochemical Society, Chemical Engineering Journal, Industrial and Engineering Chemistry Research, Journal of Physical Chemistry C, Materials Letters, International Journal of Electrochemical Science, Catalysts, Advanced Materials Research, Applied Thermal Engineering, Fuel Processing Technology, Journal of Materials Science Materials in Electronics, SAE International Journal of Fuels and Lubricants, Journal of Colloid and Interface Science, Electrochemistry Communications, Journal of Materials Chemistry A, Ceramics International, Proceedings of the Combustion Institute, Energies, Electroanalysis, Materials Chemistry and Physics, Nanotechnology, Topics in Catalysis, ACS Sustainable Chemistry and Engineering, Catalysis Science and Technology, Materials Today Proceedings, and Physical Chemistry Chemical Physics.

TABLE 90.5
The Most Prolific Source Titles in Utilization of Bioethanol Fuels

No.	Source Titles	Sample Papers (%)	Population Papers (%)	Surplus (%)
1	Journal of Power Sources	8.9	2.2	6.7
2	Applied Catalysis B Environmental	7.5	1.5	6.0
3	International Journal of Hydrogen Energy	7.1	5.2	1.9
4	Fuel	7.1	3.8	3.3
5	Journal of Catalysis	5.0	1.0	4.0
6	Electrochimica Acta	3.6	1.8	1.8
7	Renewable Energy	3.2	0.5	2.7
8	Journal of the American Chemical Society	2.5	0.3	2.2
9	Atmospheric Environment	2.5	0.3	2.2
10	Renewable and Sustainable Energy Reviews	2.5	0.2	2.3
11	Sensors and Actuators B Chemical	2.1	3.5	−1.4
12	Advanced Materials	2.1	0.1	2.0
13	Applied Catalysis A General	1.8	1.3	0.5
14	Energy	1.8	0.8	1.0
15	Journal of Electroanalytical Chemistry	1.8	0.8	1.0
16	Catalysis Communications	1.8	0.6	1.2
17	Angewandte Chemie International Edition	1.8	0.2	1.6
18	Catalysis Today	1.4	1.3	0.1
19	Applied Energy	1.4	0.7	0.7
20	Energy Conversion and Management	1.4	0.6	0.8
21	Journal of Physical Chemistry B	1.4	0.1	1.3
22	Applied Physics Letters	1.4	0.1	1.3
23	Energy and Fuels	1.1	1.1	0.0
24	Combustion and Flame	1.1	0.8	0.3
25	ACS Applied Materials and Interfaces	1.1	0.7	0.4
26	ACS Catalysis	1.1	0.6	0.5

90.3.7 The Most Prolific Countries in Utilization of Bioethanol Fuels

Information about the 21 most prolific countries publishing at least 1.4% of the sample paper each in utilization of bioethanol fuels is given in Table 90.6.

The most prolific countries are China and the USA with 27% and 20% of the sample papers, respectively. India, France, Greece, Brazil, Spain, Turkey, and Italy are other prolific countries with 3.6%–6.0% of the sample papers each. Further, nine European countries listed in Table 90.6 produce 36% and 22% of the sample and population papers, respectively.

On the other hand, countries with the most citation impact are the USA and Greece with 7% and 5% surplus, respectively. France, Turkey, Argentina, and Netherlands follow these countries with 1.2%–3.6% surplus each. Similarly, countries with the least citation impact are India and Brazil with 2.3% deficit each. Japan, Canada, S. Korea, Italy, and Thailand follow these countries with 0.4%–1.3% deficit each.

On the other hand, a large number of countries with relatively low citation rates have contributed immensely to the research in this field. Iran, Russia, Poland, Saudi Arabia, Mexico, Egypt, Czech Republic, Singapore, Ukraine, South Africa, Portugal, Vietnam, Colombia, Finland, Indonesia, Romania, Switzerland, Belgium, Hungary, and Denmark.

90.3.8 The Most Prolific Scopus Subject Categories in Utilization of Bioethanol Fuels

Information about the nine most prolific Scopus subject categories indexing at least 2.5% of the sample papers each is given in Table 90.7.

TABLE 90.6
The Most Prolific Countries in Utilization of Bioethanol Fuels

No.	Countries	Sample Papers (%)	Population Papers (%)	Surplus (%)
1	China	27.1	27.1	0.0
2	USA	20.4	13.1	7.3
3	India	6.4	8.7	−2.3
4	France	6.4	2.8	3.6
5	Greece	6.4	1.0	5.4
6	Brazil	5.7	8.0	−2.3
7	Spain	4.3	4.3	0.0
8	Turkey	4.3	1.7	2.6
9	Italy	3.6	4.2	−0.6
10	Japan	3.2	4.5	−1.3
11	UK	3.2	3.2	0.0
12	Germany	3.2	2.9	0.3
13	Argentina	2.9	1.3	1.6
14	S. Korea	2.5	3.3	−0.8
15	Sweden	2.1	1.1	1.0
16	Netherlands	2.1	0.9	1.2
17	Canada	1.8	2.7	−0.9
18	Taiwan	1.8	1.8	0.0
19	Australia	1.8	1.5	0.3
20	Malaysia	1.8	1.5	0.3
21	Thailand	1.4	1.8	−0.4

TABLE 90.7
The Most Prolific Scopus Subject Categories in Utilization of Bioethanol Fuels

No.	Scopus Subject Categories	Sample Papers (%)	Population Papers (%)	Surplus (%)
1	Chemistry	52.5	43.6	8.9
2	Chemical Engineering	46.4	35.4	11.0
3	Energy	42.1	28.0	14.1
4	Engineering	24.3	32.8	−8.5
5	Environmental Science	18.9	15.3	3.6
6	Physics and Astronomy	16.4	23.8	−7.4
7	Materials Science	15.0	25.4	−10.4
8	Biochemistry, Genetics and Molecular Biology	3.2	2.9	0.3
9	Earth and Planetary Sciences	2.5	1.5	1.0

The most prolific Scopus subject categories in utilization of bioethanol fuels are chemistry, chemical engineering, and energy with 53%, 46%, and 42% of the sample papers, respectively. Engineering, environmental science, physics and astronomy, and materials science follow these top categories with 15%–24% of sample papers each.

On the other hand, Scopus subject categories with the most citation impact are the energy and chemical engineering with 14% and 11% surplus, respectively, followed by chemistry and environmental science with 9% and 4% surplus, respectively. Similarly, Scopus subject categories with the least citation impact are materials science, engineering, and physics and astronomy with 7%–10% deficit each.

90.3.9 The Most Prolific Keywords in Utilization of Bioethanol Fuels

Information about keywords used in at least 3.6% of the sample or population papers each is given in Table 90.8. For this purpose, keywords related to keyword set, given in appendices of Konur (2023c–g), are selected from a list of the most prolific keyword set provided by Scopus database.

TABLE 90.8
The Most Prolific Keywords in Utilization of Bioethanol Fuels

No.	Keywords	Sample Papers (%)	Population Papers (%)	Surplus (%)
1.			**Bioethanol Fuels**	
	Ethanol	85.4	81.0	4.4
	Direct ethanol fuel cells	13.6	7.0	6.6
	Ethanol oxidation	11.8	7.8	4.0
	Ethanol fuels	10.7	13.2	–2.5
	Ethanol electro-oxidation	6.4	5.5	0.9
	Bioethanol	5.4	5.4	0.0
	Ethanol steam reforming	4.6	4.8	–0.2
	Ethanol oxidation reaction	3.2	5.4	–2.2
2.			**Catalysis**	
	Catalysts	25.7	11.0	14.7
	Catalysis	15.0	5.5	9.5
	Catalyst activity	12.1	10.5	1.6
	Electrocatalysts	11.8	7.5	4.3
	Electrocatalysis	5.0	3.7	1.3
	Electrocatalytic activity	3.6	3.7	–0.1
	Catalyst selectivity	2.9	3.8	–0.9
3.			**Fuel Cells**	
	Fuel cells	26.4	9.0	17.4
	Oxidation	26.1	12.4	13.7
	Platinum	12.9	7.6	5.3
	Electrooxidation	10.4	7.4	3.0
	Palladium	8.6	4.7	3.9
	Electrochemistry	5.7	1.6	4.1
	Nanoparticles	5.0	6.5	–1.5
	Catalytic oxidation	3.6	3.7	–0.1
	Platinum alloys	3.6	2.6	1.0
4.			**Engines**	
	Diesel engines	19.6	9.3	10.3
	Gasoline	11.8	10.3	1.5
	Carbon monoxide	11.8	4.4	7.4
	Exhaust emission	10.4	1.6	8.8
	Emissions	10.0	3.3	6.7
	Carbon dioxide	9.3	6.0	3.3
	Diesel fuels	9.3	2.6	6.7
	Combustion	8.9	8.2	0.7

(Continued)

TABLE 90.8 (*Continued*)
The Most Prolific Keywords in Utilization of Bioethanol Fuels

No.	Keywords	Sample Papers (%)	Population Papers (%)	Surplus (%)
	Engines	7.1	4.9	2.2
	Engine cylinders	7.1	3.4	3.7
	Nitrogen oxides	6.4	3.7	2.7
	Engine performance	6.4	1.8	4.6
	Biofuels	6.1	4.0	2.1
	Biodiesel	6.1	3.5	2.6
	Particulate emissions	6.1	1.9	4.2
	Ignition	5.0	4.8	0.2
	Fuel consumption	5.0		5.0
	Gas emissions	3.6	1.9	1.7
	Internal combustion engines	3.6	1.7	1.9
	Direct injection	2.9	3.6	–0.7
5.			**Biochemicals**	
	Platinum	12.9	7.6	5.3
	Carbon monoxide	11.8	4.4	7.4
	Methanol	10.4	5.9	4.5
	Carbon dioxide	9.3	6.0	3.3
	Acetaldehyde	5.4	4.2	1.2
	Aldehydes	4.3	1.6	2.7
	Acetic acid	3.6	1.9	1.7
	Ethylene	2.1	4.3	–2.2
6			**Biohydrogen Fuels**	
	Hydrogen	12.9	6.2	6.7
	Steam reforming	12.1	7.7	4.4
	Hydrogen production	9.6	7.9	1.7
	Reforming reactions	8.6	1.6	7.0
	Nickel	6.8	4.6	2.2
	Steam	5.7	2.9	2.8
	Alumina	4.6	1.9	2.7
	Cerium compounds	4.3		4.3
	Rhodium	3.9		3.9
7.			**Sensors**	
	Chemical sensors	7.5	5.5	2.0
	Zinc oxide	5.7	4.0	1.7
	Gas detectors	2.1	5.7	–3.6

These keywords are grouped under seven headings: bioethanol fuels, catalysis, utilization of bioethanol fuels in fuel cells and transport engines, production of biohydrogen fuels and biochemicals, and development and utilization of bioethanol sensors.

The most prolific keyword used related to bioethanol fuels is ethanol with 85% of the sample papers. Other prolific keywords are direct ethanol fuel cells, ethanol oxidation, and ethanol fuels with 11%–14% of the sample papers each. It is notable that bioethanol keyword appears in the sample and population paper keyword list with around 5% of both papers each.

TABLE 90.9
The Most Prolific Research Fronts in Utilization of Bioethanol Fuels

No.	Research Fronts	Papers (%)
1	Utilization of bioethanol fuels in fuel cells	34
2	Utilization of bioethanol fuels in transport engines	30
3	Production of biohydrogen fuels from bioethanol fuels	22
4	Bioethanol sensors	11
5	Production of biochemical from bioethanol fuels	8

Papers (%): Sample of the 280 most-cited papers.

The prolific keywords related to the catalysis are catalysts, catalysis, catalyst activity, electrocatalysts, and electrooxidation with 10%–26% of the sample papers each, while prolific keywords related to fuel cells are fuel cells, oxidation, platinum, and electrooxidation with 10%–26% of the sample papers each.

Prolific keywords related to utilization of bioethanol fuels in gasoline and diesel engines are diesel engines, gasoline, carbon monoxide, exhaust emission, and emissions with 10%–20% of the sample papers each, while prolific keywords related to biochemicals are platinum, carbon monoxide, methanol, and carbon dioxide with 9%–13% of the sample papers each.

Prolific keywords related to production of biohydrogen fuels from bioethanol fuels are hydrogen, steam reforming, and hydrogen production with 10%–13% of the sample papers each, while prolific keywords related to bioethanol sensors are chemical sensors, zinc oxide, and gas detectors with 2%–8% of the sample papers each.

On the other hand, the most prolific keywords across all research fronts are ethanol, fuel cells, oxidation, catalysts, diesel engines, catalysis, direct ethanol fuel cells, platinum, hydrogen, catalyst activity, steam reforming, gasoline, ethanol oxidation, electrocatalysts, carbon monoxide, ethanol fuels, electrooxidation, methanol, exhaust emission, and emissions with 10%–85% of the sample papers each. Similarly, the most influential keywords across all research fronts are fuel cells, catalysts, oxidation, diesel engines, and catalysis with 10%–17% surplus each.

90.3.10 The Most Prolific Research Fronts in Utilization of Bioethanol Fuels

Information about the most prolific research fronts for the sample papers in bioethanol fuel evaluation and utilization is given in Table 90.9.

As Table 90.9 shows there are five primary research fronts for these highly cited papers (HCPs): utilization of bioethanol fuels in fuel cells, utilization of bioethanol fuels in gasoline and diesel engines, production of biohydrogen fuels from bioethanol fuels for fuel cells, development and utilization of bioethanol sensors, and production of biochemicals from bioethanol fuels with 34%, 30%, 22%, 11%, and 8% of the sample papers each.

90.4 DISCUSSION

90.4.1 Introduction

Crude oil-based gasoline fuels have been widely used in transportation sector since the 1920s. However, there have been great public concerns over adverse environmental and human impact of these fuels. Hence, biomass-based bioethanol fuels have increasingly been used in blending gasoline fuels. Additionally, bioethanol fuels have been used to produce valuable biochemicals in a biorefinery context.

The primary focus of research in this area has been utilization of bioethanol fuels in fuel cells, utilization of bioethanol fuels in gasoline and diesel engines, production of biohydrogen fuels from bioethanol fuels for fuel cells, and, to a lesser extent, development and utilization of bioethanol sensors and production of biochemicals from bioethanol fuels. Additionally, research on gasoline fuels (Konur, 2023a) and applications of nanotechnology (Konur, 2023b) in this field has also been related to this field.

However, it is essential to develop efficient incentive structures for the primary stakeholders to enhance research in this field. Scientometric analysis has been used in this context to inform the primary stakeholders about the current state of research in a selected research field.

As the first step for search of relevant literature, keywords were selected using the first 200 most-cited papers for each substituent research front. The selected keyword list was optimized to obtain a representative sample of papers for each research field and they are integrated to form the keyword list for this study. This keyword list was provided in appendices of Konur (2023c–g) for future replication studies.

As the second step, two sets of data were used for this study. First, a population sample of over 14,000 papers was used to examine scientometric characteristics of the population data. Secondly, a sample of 280 most-cited papers, corresponding to 2% of the population papers, was used to examine scientometric characteristics of these citation classics.

The scientometric characteristics of these both sample and population datasets were presented in the order of documents, authors, publication years, institutions, funding bodies, source titles, countries, Scopus subject categories, keywords, and research fronts.

Lastly, the key scientometric findings for both datasets were discussed to highlight research landscape for utilization of bioethanol fuels. Additionally, a number of brief conclusions were drawn and a number of relevant recommendations were made to enhance future research landscape.

90.4.2 The Most Prolific Documents in Utilization of Bioethanol Fuels

Articles and review papers dominate both the sample and population datasets, while review papers and conference papers have a surplus and deficit, respectively.

Scopus differs from Web of Science database in differentiating and showing articles and conference papers published in journals separately. Similarly, Scopus differs from Web of science database in introducing short surveys. Hence, the total number of articles and review papers in the sample dataset are 92% and 8%, respectively.

It is observed during the search process that there has been inconsistency in classification of documents in Scopus as well as in other databases such as Web of Science. This is especially relevant for classification of papers as reviews or articles as the papers not involving a literature review may be erroneously classified as a review paper. There is also a case of review papers being classified as articles. For example, although there are 21 review papers and short surveys (7.5%) as classified by Scopus database, 25 of the sample papers (8.9%) are review papers based on literature reviews.

In this context, it would be helpful to provide a classification note for the published papers in books and journals at the first instance. It would also be helpful to use document types listed in Table 90.1 for this purpose. Book chapters may also be classified as articles or reviews as an additional classification to differentiate review chapters from experimental chapters as it is done by Web of Science. It would be further helpful to additionally classify conference papers as articles or review papers as well as it is done in Web of Science database.

90.4.3 The Most Prolific Authors in Utilization of Bioethanol Fuels

There have been 52 most-prolific authors with at least three sample papers and ten population papers each as given in Table 90.2. These authors have shaped development of research in this field.

The most prolific authors are Claude Lamy, Gongquan Sun, Panagiotis Tsiakaras, Qing Qin, and, to a lesser extent, Shuqin Song, Christophe Coutanceau, Peikang Shen, Miguel Laborde, Stefano Cavallaro, Shi-Jin Shuai, Jian-Xin Wang, Weijiang Zhou, Zhenhua Zhou, El Mustapha Belgsir, and Changwei Xu.

It is important to note inconsistencies in indexing of author names in Scopus and other databases. It is especially an issue for names with more than two components such as 'Judge Alex de Camp Lamy'. The probable outcomes are 'Lamy, J.A.D.C.', 'de Camp Lamy, J.A.', or 'Camp Lamy, J.A.D.'. The first choice is the gold standard of publishing sector as the last word in the name is taken as the last name. In most of the academic databases such as PubMed, EBSCO databases, this version is used predominantly. The second choice is a strong alternative, while the last choice is an undesired outcome as two last words are taken as the last name. It is good practice to combine the words of the last name with a hyphen: 'Camp-Lamy, J.A.D.'. It is notable that inconsistent indexing of author names may cause substantial inefficiencies in the search process of papers as well as allocating credit to the authors as there are different author entries for each outcome in the databases.

There is also a case of shortening Chinese names. For example, 'Yuoyang Zhang is often shortened as 'Zhang, Y.', 'Zhang, Y.-Y.', and 'Zhang Y.Y.' as it is done in Web of Science database as well. However, the gold standard in this case is 'Zhang Y' where the last word is taken as the last name and the first word is taken as a single forename. In most of the academic databases such as PubMed and EBSCO, this first version is used predominantly. However, it makes sense to use the third option to differentiate Chinese names efficiently: 'Zhang Y.Y.'. Therefore, there have been difficulties to locate papers for the Chinese authors. In such cases, use of unique author codes provided for each author by Scopus database has been helpful.

There is also a difficulty in allowing credit for authors, especially for authors with common names such as 'Wang, Y', or 'Huang, Y' or 'Zhu, Y.' in conducting scientometric studies. These difficulties strongly influence efficiency of scientometric studies as well as allocating credit to authors as there are the same author entries for different authors with the same name, e.g., 'Wang Y.' in the databases.

In this context, coding of authors in Scopus database is a welcome innovation compared to other databases such as Web of Science. In this process, Scopus allocates a unique number to each author in the database (Aman, 2018). However, there might still be substantial inefficiencies in this coding system especially for common names. For example, some of the papers for a certain author maybe allocated to another researcher with a different author code. It is possible that Scopus uses a number of software programs to differentiate author names, and the program may not be false-proof (Kim, 2018).

In this context, it does not help that author names are not given in full in some journals and books. This makes difficult to differentiate authors with common names and makes the scientometric studies further difficult in the author domain. Therefore, author names should be given in all books and journals at the first instance. There is also a cultural issue where some authors do not use their full names in their papers. Instead, they use initials for their forenames: 'Aden, A.P.' or just 'Aden' instead of 'Aden, Alas Padras'.

There are also inconsistencies in naming of authors with more than two components by authors themselves in journal papers and book chapters. For example, 'Aden, APC' might be given as 'Lamy, A', 'Lamy, P', 'Lamy, P.C.', or 'Lamy, C' in journals and books. This also makes scientometric studies difficult in the author domain. Hence, contributing authors should use their name consistently in their publications.

Another critical issue regarding author names is spelling of author names in national spellings (e.g., Üzümcü, Çiğdemsüt) rather than in English spellings (e.g., Uzumcu, Cigdemsut) in Scopus database. Scopus differs from Web of science database and many other databases in this respect where author names are given only in English spellings. It is observed that national spellings of author names do not help in conducting scientometric studies as well as in allocating credits to authors as sometimes there are different author entries for English and National spellings in Scopus database.

The most prolific institutions for sample dataset are University of Poitiers, Chinese Academy of Sciences, CNR, and, to a lesser extent, Ciudad University, National Technical University of Athens, National Technological Institute, Southern University of. Science and Technology, University of Sao Paulo, and Xiamen University. The most prolific countries for sample dataset are China, and, to a lesser extent, France, Argentina, Brazil, Germany, Greece, and Italy.

The most prolific research front is utilization of bioethanol fuel in fuel cells, biohydrogen fuels produced from bioethanol fuels, and, to a lesser extent, utilization of bioethanol fuels in gasoline and diesel engines and bioethanol sensors. It is notable that there are no authors for biochemicals produced from bioethanol fuels.

It is also notable that there is a significant gender deficit for sample dataset as surprisingly nearly all of these top researchers are male. This finding is the most thought-provoking with strong public policy implications. Hence, institutions, funding bodies, and policy-makers should take efficient measures to reduce the gender deficit in this field as well as other scientific fields with strong gender deficit. In this context, it is worth to note the level of representation of researchers from minority groups in science on the basis of race, sexuality, age, and disability, besides gender (Blankenship, 1993; Dirth and Branscombe, 2017; Konur, 2000, 2002a–c, 2006a,b, 2007a,b).

90.4.4 The Most Prolific Research Output by Years in Utilization of Bioethanol Fuels

Research output observed between 1970 and 2022 is illustrated in Figure 90.1. This figure clearly shows that the bulk of research papers in population dataset was published primarily in the 2010s, early 2020, and 2010s. Similarly, the bulk of research papers in sample dataset was published in the 2000s and 2010s.

The bulk of population papers were published between 2019 and 2022, while the bulk of sample papers were published between 1999 and 2007. The number of population papers rose between 2003 and 2017. However, it steadied after 2017 around 6% of the population papers for each year, losing its momentum.

These data suggest that the most-cited sample and population papers were primarily published in the 2000s, 2010s, and the early 2020s. These are thought-provoking findings as there has been no significant research in this field prior to 2005 for the population papers, but there has been significant research boom after that with a rising trend. However, the number of publications for population papers steadied after 2017. In this context, the increasing public concerns about climate change (Change, 2007), greenhouse gas emissions (Carlson et al., 2017), and global warming (Kerr, 2007) have been certainly behind the boom in the research in this field in the last two decades.

Based on these findings, the size of population papers is likely to more than double in the current decade, provided that public concerns about climate change, greenhouse gas emissions, and global warming are translated efficiently to research funding in this field.

90.4.5 The Most Prolific Institutions in Utilization of Bioethanol Fuels

The 46 most prolific institutions publishing papers on utilization of bioethanol fuels with at least three sample papers each given in Table 90.3 have shaped the development of research in this field.

The most prolific institutions are Chinese Academy of Sciences, University of Poitiers, and, to a lesser extent, Tsinghua University, University of Poitiers, Sun Yat-Sen University, University of Sao Paulo, Brookhaven National Laboratory, University of Thessaly, and University of Patras. Further, the top countries for these most prolific institutions are China and, to a lesser extent, the USA, Brazil, France, Greece, Italy, S. Korea, Spain, and Turkey. In total, 20 countries house these top institutions.

On the other hand, institutions with the most citation impact are Tsinghua University, Chinese Academy of Sciences, and, to a lesser extent, Sun Yat-Sen University, University of Poitiers, University of Thessaly, and Brookhaven National University.

Thus, it appears that institutions from China and, to a lesser extent, the USA, Brazil, S. Korea, and Europe are the major contributors to research in this field as they are also the major producers and consumers of bioethanol fuels.

90.4.6 The Most Prolific Funding Bodies in Utilization of Bioethanol Fuels

The 20 most prolific funding bodies funding at least 1.1% of the sample papers each is given in Table 90.4. It is notable that 46% and 49% of the sample and population papers are funded, respectively.

The most prolific funding bodies are National Natural Science Foundation of China and, to a lesser extent, US Department of Energy, European Commission, Chinese Academy of Sciences, Sao Paulo State Research Foundation, National Science Foundation, and Basic Energy Sciences.

The most prolific countries for these top funding bodies are China, the USA, the EU, and Brazil. In total, eight countries and the EU house these top funding bodies.

The heavy funding by Chinese, and to a lesser extent, the US, European, and Brazilian funding bodies are notable as these countries are major producers of research in this field. They are also the major producers and consumers of bioethanol fuels. However, it is notable that six Chinese funding bodies fund over 18% of both the sample and population papers each showing awareness of China of funding-scientific output relationships. On the contrary, five US funding bodies fund only 9% and 5% of the sample and population papers, respectively, falling well behind China in funding the research on utilization of bioethanol fuels. Similarly, seven funding bodies from Europe fund only 9% and 3% of the sample and population papers, respectively, falling well behind of China in funding the research on utilization of bioethanol fuels. Another major competitor of China, Brazil, funds only 3% and 5% of the sample and population papers, respectively.

Funding bodies with the most citation impact are the Chinese Academy of Sciences, European Commission German Research Foundation, and University of Buenos Aires. Similarly, funding bodies with the least citation impact are the National Natural Science Foundation of China and National Council for Science and Technology Development.

These findings on funding of research in this field suggest that the level of the funding, mostly in the last two decades, is highly intensive, and it has been largely instrumental in enhancing research in this field (Ebadi and Schiffauerova, 2016) in the light of North's institutional framework (North, 1991).

90.4.7 The Most Prolific Source Titles in Utilization of Bioethanol Fuels

The 26 most prolific source titles publishing at least 1.1% of the sample papers each in utilization of bioethanol fuels have shaped the development of research in this field (Table 90.5).

The most prolific source titles are Journal of Power Sources, Applied Catalysis B Environmental, International Journal of Hydrogen Energy, Fuel, Journal of Catalysis, and, to a lesser extent, Electrochimica Acta, Renewable Energy, Journal of the American Chemical Society, Atmospheric Environment, and Renewable and Sustainable Energy Reviews.

On the other hand, source titles with the most citation impact are the Journal of Power Sources, Applied Catalysis B Environmental, and, to a lesser extent, Journal of Catalysis, Fuel, Renewable Energy, Renewable and Sustainable Energy Reviews, Atmospheric Environment, Journal of the American Chemical Society, and Advanced Materials.

It is notable that these top source titles are primarily related to energy and catalysis. This finding suggests that journals in these fields have significantly shaped the development of research in this field as they focus on utilization of bioethanol fuels where focus is on the development of catalyst systems.

90.4.8 The Most Prolific Countries in Utilization of Bioethanol Fuels

The 21 most prolific countries publishing at least 1.4% of the sample papers each have significantly shaped the development of research in this field (Table 90.6).

The most prolific countries are the USA, China, and, to a lesser extent, India, France, Greece, Brazil, Spain, Turkey, and Italy. Further, nine European countries listed in Table 90.6 published 36% and 22% of the sample and population papers, respectively, with 14% surplus.

On the other hand, countries with the most citation impact are the USA, Greece, and to a lesser extent, France, Turkey, Argentina, and Netherlands. Similarly, countries with the least citation impact are India, Brazil, and, to a lesser extent, Japan, Canada, S. Korea, Italy, and Thailand.

A close examination of these findings suggests that Europe, China, and the USA are the major producers of research in this field. Other prolific countries are India, Brazil, Japan, Argentina, S. Korea, Canada, Taiwan, Malaysia, and Australia. It is a fact that the USA has been a major player in science (Leydesdorff and Wagner, 2009; Leydesdorff et al., 2014). The USA has further developed a strong research infrastructure to support its corn and grass-based bioethanol industry (Vadas et al., 2008).

However, China has been a rising star in scientific research in competition with the USA and Europe (Leydesdorff and Zhou, 2005). China is also a major player in this field as a major producer of bioethanol (Li and Chan-Halbrendt, 2009).

Next, Europe has been a persistent player in scientific research in competition with both the USA and China (Leydesdorff, 2000). Europe has also been a persistent producer of bioethanol along with the USA and Europe (Gnansounou, 2010).

90.4.9 The Most Prolific Scopus Subject Categories in Utilization of Bioethanol Fuels

The nine most prolific Scopus subject categories indexing at least 2.5% of the sample papers each, given in Table 90.7 have shaped the development of research in this field.

The most prolific Scopus subject categories in utilization of bioethanol fuels are chemistry, chemical engineering, energy, and, to a lesser extent, engineering, environmental science, physics and astronomy, and materials science.

On the other hand, Scopus subject categories with the most citation impact are energy, chemical engineering, and, to a lesser extent, chemistry and environmental science, while Scopus subject categories with the least citation impact are materials science, engineering, and physics and astronomy.

These findings are thought-provoking suggesting that the primary subject categories are related to chemistry, chemical engineering, and energy. It is also notable that 42% and 28% of the sample and population papers are indexed by energy subject category with 14% surplus, showing the impact of this category. It also shows that most of the papers are indexed by other subject categories.

Another key finding is that social sciences are not well represented in both the sample and population papers with 0% and 1.5% of the sample and population papers, respectively. These findings are not surprising as the key research fronts in this field are technical studies on utilization of bioethanol fuels.

90.4.10 The Most Prolific Keywords in Utilization of Bioethanol Fuels

A limited number of keywords have shaped the development of research in this field as shown in Table 90.8 and appendices of Konur (2023c–g).

These keywords are grouped under seven headings: bioethanol fuels, catalysis, utilization of bioethanol fuels in fuel cells and transport engines, production of biohydrogen fuels and biochemicals, and development and utilization of bioethanol sensors. These prolific keywords highlight the key research fronts in this field and reflect well the keywords used in sample and population papers. It also indicates the relative research in intensity of these research fronts for both sample and population papers.

On the other hand, the most prolific keywords across all research fronts are ethanol, fuel cells, oxidation, catalysts, diesel engines, catalysis, direct ethanol fuel cells, platinum, hydrogen, catalyst activity, steam reforming, gasoline, ethanol oxidation, electrocatalysts, carbon monoxide, ethanol fuels, electrooxidation, methanol, exhaust emission, and emissions. Similarly, the most influential keywords across all research fronts are fuel cells, catalysts, oxidation, diesel engines, and catalysis.

90.4.11 The Most Prolific Research Fronts in Utilization of Bioethanol Fuels

As Table 90.9 shows there are five primary research fronts for these 280 most-cited papers: utilization of bioethanol fuels in fuel cells, utilization of bioethanol fuels in gasoline and diesel engines, production of biohydrogen fuels from bioethanol fuels for fuel cells, development and utilization of bioethanol sensors, and production of biochemicals from bioethanol fuels.

Bioethanol fuels are directly used in fuel cells to produce bioelectricity (Antolini, 2007; Lamy et al., 2004). Catalysts play a crucial role in oxidation of bioethanol fuels (Antolini, 2007; Liang et al., 2009). These findings are thought-provoking in seeking ways to optimize utilization of bioethanol fuels in fuel cells. These studies focused mostly on electrooxidation of bioethanol fuels (Dong et al., 2010; Kowal et al., 2009). Activity and mechanisms of electrooxidation of bioethanol fuels were extensively evaluated in these studies. Catalysts, catalyst supports, mechanisms and activity of bioethanol electrooxidation, and applications of nanotechnology in catalysts emerge as the key issues for this field. It is expected that the nanotechnology applications would increase to improve the utility of bioethanol fuel utilization in fuel cells in this field in future. It is also expected that bioethanol fuels would increasingly be used in fuel cells to produce bioelectricity at a global scale considering recent supply shocks caused by COVID-19 pandemic and Ukrainian war (Aday and Aday, 2020; Chenarides et al., 2021; Jagtap et al., 2022).

As bioethanol fuels are one of the most-used biofuels in gasoline and diesel engines to blend gasoline and petrodiesel fuels, it is not surprising that research in this field is substantial. A wide range of issues are covered in these papers: properties, combustion, performance, and greenhouse gas emissions such as CO, CO_2, HC, and NO_x emissions. The focus is more on gasoline and bioethanol fuel blends (Al-Hasan, 2003; Hsieh et al., 2002) compared to petrodiesel and bioethanol fuel blends (Hansen et al., 2005; Kwanchareon et al., 2007). These findings are thought-provoking in seeking ways to optimize utilization of bioethanol fuels in gasoline and diesel engines to improve combustion, performance, and GHG emissions of petrodiesel and gasoline fuels in these engines. It is expected that the nanotechnology applications (Geim, 2009; Geim and Novoselov, 2007) would increase to optimize utilization of bioethanol fuels in gasoline and diesel engines to improve combustion, performance, and GHG emissions of petrodiesel and gasoline fuels in these engines in future. It is also expected that bioethanol fuels would increasingly be used in these engines at a global scale considering recent supply shocks caused by COVID-19 pandemic and Ukrainian war (Aday and Aday, 2020; Chenarides et al., 2021; Jagtap et al., 2022).

Fuels cells are used to produce bioelectricity from biohydrogen fuels, produced from bioethanol fuels (Haryanto et al., 2005; Ni et al., 2007). Catalysts play a crucial role in oxidation of bioethanol fuels (Murdoch et al., 2011). These studies focus mostly on steam reforming of bioethanol fuels (Haryanto et al., 2005; Ni et al., 2007), photocatalytic hydrogen production (Bamwenda et al., 1995; Murdoch et al., 2011), and autothermal reforming of bioethanol fuels (Deluga et al. 2004). These findings are thought-provoking in seeking ways to optimize utilization of bioethanol fuels for biohydrogen production for fuel cells. Catalyst types, catalyst support types, nanostructured catalyst systems, steam reforming, autothermal reforming of bioethanol fuels, photocatalytic biohydrogen production, catalyst loading, preparation methods, bioethanol conversion, biohydrogen selectivities, and catalytic activity emerge as the key issues for this field. It is expected that nanotechnology applications would increase to improve utility of bioethanol fuel utilization to produce biohydrogen fuels for fuel cells in this field in future. It is also expected that bioethanol fuels would increasingly be used to produce biohydrogen fuels for fuel cells at a global scale considering recent supply shocks caused by COVID-19 pandemic and Ukrainian war (Aday and Aday, 2020; Chenarides et al., 2021; Jagtap et al., 2022).

Production of biochemicals from bioethanol fuels in a biorefinery context is a useful strategy to reduce production cost of bioethanol fuels. Catalysts play a crucial role in production and utilization of biochemicals from bioethanol fuels as a green alternative to crude oil-based chemicals (Angelici et al., 2013; Sun and Wang, 2014). Issues studied in these studies are bioethanol photooxidation, visible and UV irradiation, bioethanol electrooxidation mechanisms, catalyst types, and electrocatalytic activity. Biochemicals produced in these studies include CO_2, acetaldehyde, formic acid, adsorbed CH_3CO, acetic acid, butanol, methanol, and CO. These findings are thought-provoking in seeking ways to optimize development and production of biochemical from bioethanol fuels though oxidation of bioethanol fuels. It is expected that nanotechnology applications would increase to optimize utilization of bioethanol fuels to produce biochemicals in the future. It is also expected that bioethanol fuels would increasingly be used in biochemical production in future considering recent supply shocks caused by COVID-19 pandemic and Ukrainian war (Aday and Aday, 2020; Chenarides et al., 2021; Jagtap et al., 2022).

Another research front is related to bioethanol sensors with a focus on development of effective materials for high bioethanol selectivity (Liu et al., 2005; Wan et al., 2004). Issues covered in these studies include nanosensors, biosensors, fabrication of sensors, MEMS, bioethanol gas sensitivity, hydrothermal method, and low-temperature bioethanol gas sensitivity. These findings are thought-provoking in seeking ways to optimize development and utilization of bioethanol sensors using nanomaterials. It is expected that nanotechnology applications would increase to improve bioethanol gas sensitivity of bioethanol sensors in future using a wide range of one- and two-dimensional nanomaterials. It is also expected that research in this area would intensify in future due to expected rise in utilization of bioethanol fuels in fuels cells, gasoline and diesel engines, biochemical and biohydrogen fuel production considering recent supply shocks caused by COVID-19 and Ukrainian war (Aday and Aday, 2020; Chenarides et al., 2021; Jagtap et al., 2022).

In the end, these most-cited papers in this field hint that efficiency of utilization of bioethanol fuels could be optimized using structure, processing, and property relationships of bioethanol fuels and their derivatives (Formela et al., 2016; Konur, 2018a, 2020b, 2021a–d; Konur and Matthews, 1989).

90.5 CONCLUSION AND FUTURE RESEARCH

Research on utilization of bioethanol fuels has been mapped through a scientometric study of both sample ($N=280$) and population ($N=14,009$) datasets.

The critical issue in this study has been to obtain a representative sample of research as in any other scientometric study. Therefore, the keyword set has been carefully devised and optimized after a number of runs in Scopus database. It is a representative sample of the wider population studies which include large numbers of countries in this field.

Another issue has been the selection of a multidisciplinary database to carry out scientometric study of the research in this field. For this purpose, Scopus database has been selected. Journal coverage of this database has been wider than that of Web of Science.

The key scientometric properties of the research in this field have been determined and discussed in this book chapter. It is evident that a limited number of documents, authors, institutions, publication periods, institutions, funding bodies, source titles, countries, Scopus subject categories, keywords, and research fronts have shaped the development of research in this field.

There is ample scope to increase the efficiency of scientometric studies in this field in the author and document domains by developing consistent policies and practices in both domains across all the academic databases. In this respect, authors, journals, and academic databases have a lot to do. Furthermore, the significant gender deficit as in most scientific fields emerges as a public policy issue. Potential deficits on the basis of age, race, disability, and sexuality need also to be explored in this field as in other scientific fields.

Research in this field has boomed in the 2000s, 2010s, and the early 2020s possibly promoted by public concerns on global warming, greenhouse gas emissions, and climate change. Institutions from China and, to a lesser extent, the USA, Brazil, S. Korea, and Europe have mostly shaped the research in this field.

The relatively low funding rate of 46% and 49% for sample and population papers, respectively, suggests that funding in this field significantly enhanced the research in this field primarily in the 2010s and the early 2020s, possibly more than doubling in the current decade. However, there is ample room for more funding.

The most prolific journals are primarily related to energy and catalysis. This finding suggests that these journals have significantly shaped the development of research in this field as they focus on utilization of bioethanol fuels where the focus is on development of catalyst systems for efficient utilization of bioethanol fuels.

Europe, China, the USA, and, to a lesser extent India, Brazil, Japan, Argentina, S. Korea, Canada, Taiwan, Malaysia, and Australia have been the major producers of research in this field as the major producers and users of bioethanol fuels from different types of biomass materials such as corn, sugarcane, and grass as well as other types of biomass materials. These countries have well-developed research infrastructure in bioethanol fuels.

The primary subject categories have been chemistry, chemical engineering, energy, and to a lesser extent, engineering, environmental science, physics and astronomy, and materials science as the focus of the sample papers have been on utilization of bioethanol fuels. It is notable that social sciences and humanities are not well represented as it would be expected. Thus, there is ample room for interdisciplinary research related to social sciences and humanities in this field.

Ethanol is more popular than bioethanol as a keyword with strong implications for search strategy. In other words, the search strategy using only bioethanol keyword would not be much helpful. It is also recommended that following the increasing trend in this field, the term 'bioethanol fuels' is used rather than 'ethanol', 'bio-ethanol', 'bioethanol', or 'fuel ethanol' in titles of papers by their authors.

These keywords are grouped under seven headings: bioethanol fuels, catalysis, utilization of bioethanol fuels in fuel cells and transport engines, production of biohydrogen fuels and biochemicals, and development and utilization of bioethanol sensors. It is important to note that these groups of keywords highlight the potential primary research fronts for these fields for both sample and population papers.

As Table 90.9 shows there are five primary research fronts for these 280 most-cited papers: utilization of bioethanol fuels in fuel cells, utilization of bioethanol fuels in gasoline and diesel engines, production of biohydrogen fuels from bioethanol fuels for fuel cells, development and utilization of bioethanol sensors, and production of biochemicals from bioethanol fuels.

It is very important to produce biochemicals and biohydrogen fuels from bioethanol fuels, and bioelectricity in fuel cells fueled with bioethanol fuels or bioethanol fuel-based biohydrogen fuels within biorefinery context to reduce production costs of bioethanol fuels as well as to develop human- and environment-friendly biochemicals, biohydrogen fuels, and bioelectricity as a sustainable alternative to crude oil-based chemicals, water-based hydrogen fuels, and natural gas-, coal-, and syngas-based power generation. Therefore, it is important to devise proper incentive structures for efficient development of bioethanol industry and research in these fields in the light of North's institutional framework (North, 1991). This is especially important in the face of recent supply shocks caused by COVID-19 pandemic and the Ukrainian war.

It is expected the nanotechnology would play more vital role in development of catalyst systems for oxidation of bioethanol fuels for fuel cells to produce bioelectricity, production of biohydrogen fuels and biochemical from bioethanol fuels, and development of bioethanol sensors as well as better combustion of bioethanol-gasoline and bioethanol-petrodiesel blends in gasoline and petrodiesel engines, respectively.

In the end, these most-cited papers in this field hint that efficiency of bioethanol-based biohydrogen fuel production, bioelectricity production in fuel cells fueled with bioethanol fuels or bioethanol-based biohydrogen fuels, and bioethanol fuel-based biochemical production as well as utilization of bioethanol fuels in gasoline and diesel engines and development and utilization of bioethanol sensors could be optimized using structure, processing, and property relationships of bioethanol fuels and their derivatives.

It is important to provide efficient incentive structures for the development of bioethanol industry and research in this field at a global scale in the light of North's institutional framework (North, 1991). In this context, the major producers and users of bioethanol fuels such as Brazil, the USA, China, and Europe have developed strong incentive structures for the effective development of bioethanol industry and research in these fields. It is thus expected that research on utilization of bioethanol fuels would intensify in the future with the efficient incentive structures for the major stakeholders.

Thus, scientometric analysis has a great potential to gain valuable insights into evolution of research in this field as in other scientific fields.

It is recommended that further scientometric studies are carried out about the primary research fronts of utilization of bioethanol fuels. It is further recommended that reviews of the most-cited papers are carried out for each research front to complement these scientometric studies. Next, the scientometric studies of hot papers in these primary fields are carried out.

ACKNOWLEDGMENTS

Contribution of the highly cited researchers in the field of utilization of bioethanol fuels has been gratefully acknowledged.

REFERENCES

Aday, S. and M. S. Aday. 2020. Impact of COVID-19 on the food supply chain. *Food Quality and Safety* 4:167–180.

Al-Hasan, M. 2003. Effect of ethanol-unleaded gasoline blends on engine performance and exhaust emission. *Energy Conversion and Management* 44:1547–1561.

Aman, V. 2018. Does the Scopus author ID suffice to track scientific international mobility? A case study based on Leibniz laureates. *Scientometrics* 117:705–720.

Angelici, C., B. M. Weckhuysen and P. C. A. Bruijnincx. 2013. Chemocatalytic conversion of ethanol into butadiene and other bulk chemicals. *ChemSusChem* 6:1595–1614.

Antolini, E. 2007. Catalysts for direct ethanol fuel cells. *Journal of Power Sources* 170:1–12.

Antolini, E. 2009. Palladium in fuel cell catalysis. *Energy and Environmental Science* 2:915–931.

Bamwenda, G. R., S. Tsubota, T. Nakamura and M. Haruta. 1995. Photoassisted hydrogen production from a water-ethanol solution: A comparison of activities of Au/TiO_2 and Pt/TiO_2. *Journal of Photochemistry and Photobiology, A: Chemistry* 89:177–189.

Beaudry, C. and V. Lariviere. 2016. Which gender gap? Factors affecting researchers' scientific impact in science and medicine. *Research Policy* 45:1790–1817.

Blankenship, K. M. 1993. Bringing gender and race in: US employment discrimination policy. *Gender & Society* 7:204–226.

Burnham, J. F. 2006. Scopus database: A review. *Biomedical Digital Libraries* 3:1–8.

Carlson, K. M., J. S. Gerber and N. D. Mueller, et al. 2017. Greenhouse gas emissions intensity of global croplands. *Nature Climate Change* 7:63–68.

Change, C. 2007. Climate change impacts, adaptation and vulnerability. *Science of the Total Environment* 326:95–112.

Chenarides, L., M. Manfredo and T. J. Richards. 2021. COVID-19 and food supply chains. *Applied Economic Perspectives and Policy* 43:270–279.

Deluga, G. A., J. R. Salge, L. D. Schmidt and X. E. Verykios. 2004. Renewable hydrogen from ethanol by autothermal reforming. *Science* 303:993–997.

Dirth, T. P. and N. R. Branscombe. 2017. Disability models affect disability policy support through awareness of structural discrimination. *Journal of Social Issues* 73:413–442.

Dong, L., R. R. S. Gari, Z. Li, M. M. Craig and S. Hou. 2010. Graphene-supported platinum and platinum-ruthenium nanoparticles with high electrocatalytic activity for methanol and ethanol oxidation. *Carbon* 48:781–787.

Ebadi, A. and A. Schiffauerova. 2016. How to boost scientific production? A statistical analysis of research funding and other influencing factors. *Scientometrics* 106:1093–1116.

Formela, K., A. Hejna, L. Piszczyk, M. R. Saeb and X. Colom. 2016. Processing and structure-property relationships of natural rubber/wheat bran biocomposites. *Cellulose* 23:3157–3175.

Garfield, E. 1955. Citation indexes for science. *Science* 122:108–111.

Geim, A. K. 2009. Graphene: Status and prospects. *Science* 324:1530–1534.

Geim, A. K. and K. S. Novoselov. 2007. The rise of graphene. *Nature Materials* 6:183–191.

Gnansounou, E. 2010. Production and use of lignocellulosic bioethanol in Europe: Current situation and perspectives. *Bioresource Technology* 101:4842–4850.

Hansen, A. C., Q. Zhang and P. W. L. Lyne. 2005. Ethanol-diesel fuel blends: A review. *Bioresource Technology* 96:277–285.

Haryanto, A., S. Fernando, N. Murali and S. Adhikari. 2005. Current status of hydrogen production techniques by steam reforming of ethanol: A review. *Energy and Fuels* 19:2098–2106.

Hill, J., E. Nelson, D. Tilman, S. Polasky and D. Tiffany. 2006. Environmental, economic, and energetic costs and benefits of biodiesel and ethanol biofuels. *Proceedings of the National Academy of Sciences of the United States of America* 103:11206–11210.

Hill, J., S. Polasky and E. Nelson, et al. 2009. Climate change and health costs of air emissions from biofuels and gasoline. *Proceedings of the National Academy of Sciences of the United States of America* 106:2077–2082.

Hsieh, W. D., R. H. Chen, T. L. Wu and T. H. Lin. 2002. Engine performance and pollutant emission of an SI engine using ethanol-gasoline blended fuels. *Atmospheric Environment* 36:403–410.

Jagtap, S., H. Trollman and F. Trollman et al. 2022. The Russia-Ukraine conflict: Its implications for the global food supply chains. *Foods* 11:2098.

Kerr, R. A. 2007. Global warming is changing the world. *Science* 316:188–190.

Khalili, N. R., P. A. Scheff and T. M. Holsen. 1995. PAH source fingerprints for coke ovens, diesel and, gasoline engines, highway tunnels, and wood combustion emissions. *Atmospheric Environment* 29:533–542.

Kim, J. 2018. Evaluating author name disambiguation for digital libraries: A case of DBLP. *Scientometrics* 116:1867–1886.

Kohse-Hoinghaus, K., P. Osswald and T. A. Cool, et al. 2010. Biofuel combustion chemistry: From ethanol to biodiesel. *Angewandte Chemie: International Edition* 49:3572–3597.

Konur, O. 2000. Creating enforceable civil rights for disabled students in higher education: An institutional theory perspective. *Disability & Society* 15:1041–1063.

Konur, O. 2002a. Access to nursing education by disabled students: Rights and duties of nursing programs. *Nurse Education Today* 22:364–374.

Konur, O. 2002b. Assessment of disabled students in higher education: Current public policy issues. *Assessment and Evaluation in Higher Education* 27:131–52.

Konur, O. 2002c. Access to employment by disabled people in the UK: Is the Disability Discrimination Act working? *International Journal of Discrimination and the Law* 5:247–279.

Konur, O. 2006a. Participation of children with dyslexia in compulsory education: Current public policy issues. *Dyslexia* 12:51–67.

Konur, O. 2006b. Teaching disabled students in higher education. *Teaching in Higher Education* 11:351–363.

Konur, O. 2007a. A judicial outcome analysis of the Disability Discrimination Act: A windfall for the employers? *Disability & Society* 22:187–204.

Konur, O. 2007b. Computer-assisted teaching and assessment of disabled students in higher education: The interface between academic standards and disability rights. *Journal of Computer Assisted Learning* 23:207–219.

Konur, O. 2011. The scientometric evaluation of the research on the algae and bio-energy. *Applied Energy* 88:3532–3540.

Konur, O. 2012a. Prof. Dr. Ayhan Demirbas' scientometric biography. *Energy Education Science and Technology Part A: Energy Science and Research* 28:727–738.

Konur, O. 2012b. The evaluation of the biogas research: A scientometric approach. *Energy Education Science and Technology Part A: Energy Science and Research* 29:1277–1292.

Konur, O. 2012c. The evaluation of the global energy and fuels research: A scientometric approach. *Energy Education Science and Technology Part A: Energy Science and Research* 30:613–628.

Konur, O. 2012d. The evaluation of the research on the biodiesel: A scientometric approach. *Energy Education Science and Technology Part A: Energy Science and Research* 28:1003–1014.

Konur, O. 2012e. The evaluation of the research on the bioethanol: A scientometric approach. *Energy Education Science and Technology Part A: Energy Science and Research* 28:1051–1064.

Konur, O. 2012f. The evaluation of the research on the biofuels: A scientometric approach. *Energy Education Science and Technology Part A: Energy Science and Research* 28:903–916.

Konur, O. 2012g. The evaluation of the research on the biohydrogen: A scientometric approach. *Energy Education Science and Technology Part A: Energy Science and Research* 29:323–338.

Konur, O. 2012h. The evaluation of the research on the microbial fuel cells: A scientometric approach. *Energy Education Science and Technology Part A: Energy Science and Research* 29:309–322.

Konur, O. 2012i. The scientometric evaluation of the research on the production of bioenergy from biomass. *Biomass and Bioenergy* 47:504–515.

Konur, O. 2015. Current state of research on algal bioethanol. In *Marine Bioenergy: Trends and Developments*, Eds. S. K. Kim and C. G. Lee, pp. 217–244. Boca Raton, FL: CRC Press.

Konur, O., Ed. 2018a. *Bioenergy and Biofuels*. Boca Raton, FL: CRC Press.

Konur, O. 2018b. Bioenergy and biofuels science and technology: Scientometric overview and citation classics. In *Bioenergy and Biofuels*, Ed. O. Konur, pp. 3–63. Boca Raton: CRC Press.

Konur, O. 2019. Cyanobacterial bioenergy and biofuels science and technology: A scientometric overview. In *Cyanobacteria: From Basic Science to Applications*, Eds. A. K. Mishra, D. N. Tiwari and A. N. Rai, pp. 419–442. Amsterdam: Elsevier.

Konur, O. 2020a. The scientometric analysis of the research on the bioethanol production from green macroalgae. In *Handbook of Algal Science, Technology and Medicine*, Ed. O. Konur, pp. 385–401. London: Academic Press.

Konur, O., Ed. 2020b. *Handbook of Algal Science, Technology and Medicine*. London: Academic Press.

Konur, O., Ed. 2021a. *Handbook of Biodiesel and Petrodiesel Fuels: Science, Technology, Health, and Environment*. Boca Raton, FL: CRC Press.

Konur, O., Ed. 2021b. *Handbook of Biodiesel and Petrodiesel Fuels: Science, Technology, Health, and Environment. Volume 1. Biodiesel Fuels: Science, Technology, Health, and Environment*. Boca Raton, FL: CRC Press.

Konur, O., Ed. 2021c. *Handbook of Biodiesel and Petrodiesel Fuels: Science, Technology, Health, and Environment. Volume 2. Biodiesel Fuels Based on the Edible and Nonedible Feedstocks, Wastes, and Algae: Science, Technology, Health, and Environment*. Boca Raton, FL: CRC Press.

Konur, O., Ed. 2021d. *Handbook of Biodiesel and Petrodiesel Fuels: Science, Technology, Health, and Environment. Volume 3. Petrodiesel Fuels: Science, Technology, Health, and Environment*. Boca Raton, FL: CRC Press.

Konur, O. 2023a. Gasoline fuels: Scientometric study. In *Evaluation and Utilization of Bioethanol Fuels. I.: Evaluation of Bioethanol Fuels, Transport Engines, and Bioethanol Sensors. Handbook of Bioethanol Fuels Volume 5*, Ed. O. Konur, pp. 87–106. Boca Raton, FL: CRC Press.

Konur, O. 2023b. Nanotechnology applications in bioethanol fuels: Scientometric study. In *EvaluatiEvaluation and Utilization of Bioethanol Fuels. I.: Evaluation of Bioethanol Fuels, Transport Engines, and Bioethanol Sensors. Handbook of Bioethanol Fuels Volume 5*, Ed. O. Konur, pp. 120–139. Boca Raton, FL: CRC Press.

Konur, O. 2023c. Utilization of bioethanol fuels in the transport engines: Scientometric study. In *Evaluation and Utilization of Bioethanol Fuels. I.: Evaluation of Bioethanol Fuels, Transport Engines, and Bioethanol Sensors. Handbook of Bioethanol Fuels Volume 5*, Ed. O. Konur, pp. 157–175. Boca Raton, FL: CRC Press.

Konur, O. 2023d. Bioethanol fuel sensors: Scientometric study. In *Evaluation and Utilization of Bioethanol Fuels. I.: Evaluation of Bioethanol Fuels, Transport Engines, and Bioethanol Sensors. Handbook of Bioethanol Fuels Volume 5*, Ed. O. Konur, pp. 317–334. Boca Raton, FL: CRC Press.

Konur, O. 2023e. Bioethanol fuel-based biohydrogen fuels: Scientometric study. In *Evaluation and Utilization of Bioethanol Fuels. II.: Biohydrogen Fuels, Fuel Cells, Biochemicals, and Country Experiences. Handbook of Bioethanol Fuels Volume 6*, Ed. O. Konur, pp. 215–236. Boca Raton, FL: CRC Press.

Konur, O. 2023f. Bioethanol fuel cells: Scientometric study. In *Evaluation and Utilization of Bioethanol Fuels. II.: Biohydrogen Fuels, Fuel Cells, Biochemicals, and Country Experiences. Handbook of Bioethanol Fuels Volume 6*, Ed. O. Konur, pp. 277–297. Boca Raton, FL: CRC Press.

Konur, O. 2023g. Bioethanol fuel-based biochemical production: Scientometric study. In *Evaluation and Utilization of Bioethanol Fuels. II.: Biohydrogen Fuels, Fuel Cells, Biochemicals, and Country Experiences. Handbook of Bioethanol Fuels Volume 6*, Ed. O. Konur, pp. 317–337. Boca Raton, FL: CRC Press.

Konur, O. and F. L. Matthews. 1989. Effect of the properties of the constituents on the fatigue performance of composites: A review. *Composites* 20:317–328.

Kowal, A., M. Li and M. Shao, et al. 2009. Ternary Pt/Rh/SnO_2 electrocatalysts for oxidizing ethanol to CO_2. *Nature Materials* 8:325–330.

Kwanchareon, P., A. Luengnaruemitchai and S. Jai-In. 2007. Solubility of a diesel-biodiesel-ethanol blend, its fuel properties, and its emission characteristics from diesel engine. *Fuel* 86:1053–1061.

Lamy, C., S. Rousseau, E. M. Belgsir, C. Coutanceau and J. M. Leger. 2004. Recent progress in the direct ethanol fuel cell: Development of new platinum-tin electrocatalysts. *Electrochimica Acta* 49:3901–3908.

Leydesdorff, L. 2000. Is the European Union becoming a single publication system? *Scientometrics* 47:265–280.

Leydesdorff, L. and C. Wagner. 2009. Is the United States losing ground in science? A global perspective on the world science system. *Scientometrics* 78:23–36.

Leydesdorff, L., C. S. Wagner and L. Bornmann. 2014. The European Union, China, and the United States in the top-1% and top-10% layers of most-frequently cited publications: Competition and collaborations. *Journal of Informetrics* 8:606–617.

Leydesdorff, L. and P. Zhou. 2005. Are the contributions of China and Korea upsetting the world system of science? *Scientometrics* 63:617–630.

Li, S. Z. and C. Chan-Halbrendt. 2009. Ethanol production in (the) People's Republic of China: Potential and technologies. *Applied Energy* 86:S162–S169.

Liang, Z. X., T. S. Zhao, J. B. Xu and L. D. Zhu. 2009. Mechanism study of the ethanol oxidation reaction on palladium in alkaline media. *Electrochimica Acta* 54:2203–2208.

Liu, J., X. Wang, Q. Peng and Y. Li. 2005. Vanadium pentoxide nanobelts: Highly selective and stable ethanol sensor materials. *Advanced Materials* 17:764–767.

Liu, P. and E. J. M. Hensen. 2013. Highly efficient and robust Au/$MgCuCr_2O_4$ catalyst for gas-phase oxidation of ethanol to acetaldehyde. *Journal of the American Chemical Society* 135:14032–14035.

Ma, X., L. Sun and C. Song. 2002. A new approach to deep desulfurization of gasoline, diesel fuel and jet fuel by selective adsorption for ultra-clean fuels and for fuel cell applications. *Catalysis Today* 77:107–116.

Maity, S. K. 2015a. Opportunities, recent trends and challenges of integrated biorefinery: Part I. *Renewable and Sustainable Energy Reviews* 43:1427–1445.

Maity, S. K. 2015b. Opportunities, recent trends and challenges of integrated biorefinery: Part II. *Renewable and Sustainable Energy Reviews* 43:1446–1466.

Murdoch, M., G. I. N. Waterhouse and M. A. Nadeem, et al. 2011. The effect of gold loading and particle size on photocatalytic hydrogen production from ethanol over Au/TiO_2 nanoparticles. *Nature Chemistry* 3:489–492.

Najafi, G., B. Ghobadian and T. Tavakoli, et al. 2009. Performance and exhaust emissions of a gasoline engine with ethanol blended gasoline fuels using artificial neural network. *Applied Energy* 86:630–639.

Newman, P. W. G. and J. R. Kenworthy. 1989. Gasoline consumption and cities: A comparison of U.S. cities with a global survey. *Journal of the American Planning Association* 55:24–37.

Ni, M., D. Y. C. Leung and M. K. H. Leung. 2007. A review on reforming bio-ethanol for hydrogen production. *International Journal of Hydrogen Energy* 32:3238–3247.

North, D. C. 1991. Institutions. *Journal of Economic Perspectives* 5:97–112.

Song, C. 2003. An overview of new approaches to deep desulfurization for ultra-clean gasoline, diesel fuel and jet fuel. *Catalysis Today* 86:211–263.

Sun, J. and Y. Wang. 2014. Recent advances in catalytic conversion of ethanol to chemicals. *ACS Catalysis* 4:1078–1090.

Vadas, P. A., K. H. Barnett and D. J. Undersander. 2008. Economics and energy of ethanol production from alfalfa, corn, and switchgrass in the Upper Midwest, USA. *Bioenergy Research* 1:44–55.

Wan, Q., Q. H. Li and Y. J. Chen, et al. 2004. Fabrication and ethanol sensing characteristics of ZnO nanowire gas sensors. *Applied Physics Letters* 84:3654–3656.

Wang, H., Z. Jusys and R. J. Behm. 2004. Ethanol electrooxidation on a carbon-supported Pt catalyst: Reaction kinetics and product yields. *Journal of Physical Chemistry B* 108:19413–19424.

91 Utilization of Bioethanol Fuels
Review

Ozcan Konur
(Formerly) Ankara Yildirim Beyazit University

91.1 INTRODUCTION

Crude oil-based gasoline fuels (Ma et al., 2002; Newman and Kenworthy, 1989) have been widely used in the transportation sector since the 1920s. However, there have been great public concerns over the adverse environmental and human impact of these fuels (Hill et al., 2006, 2009). Hence, biomass-based bioethanol fuels (Hill et al., 2006; Konur, 2012, 2015, 2019, 2020a) have increasingly been used in blending gasoline fuels (Hsieh et al., 2002; Najafi et al., 2009), in fuel cells (Antolini, 2007, 2009), and in biochemical production (Angelici et al., 2013; Morschbacker, 2009) in a biorefinery context (Fernando et al., 2006; Huang et al., 2008).

The primary focus of the research in this area has been the utilization of bioethanol fuels in fuel cells (Antolini, 2007; Liang et al., 2009), utilization of bioethanol fuels in gasoline and diesel engines (Hansen et al., 2005; Kohse-Hoinghaus et al., 2010), production of biohydrogen fuels from bioethanol fuels (Haryanto et al., 2005; Ni et al., 2007), development and utilization of bioethanol sensors (Liu et al., 2005; Wan et al., 2004), and production of biochemicals from bioethanol fuels (Liu and Hensen, 2013; Wang et al., 2004). Additionally, the research on gasoline fuels (Khalili et al., 1995; Konur, 2023f; Song, 2003) and applications of nanotechnology (Konur, 2023g; Murdoch et al., 2011; Wan et al., 2004) in this field has also been closely related to this field.

However, it is essential to develop efficient incentive structures (North, 1991) for the primary stakeholders to enhance the research in this field (Konur, 2000, 2002a–c, 2006a,b, 2007a,b). Although there have been several review papers on the evaluation and utilization of bioethanol fuels (Antolini, 2007; Hansen et al., 2005; Haryanto et al., 2005; Kohse-Hoinghaus et al., 2010), there has been no review of the 25 most-cited articles in this field.

Thus, this book chapter presents a review of the 25 most-cited articles in these fields. Then, it discusses the key findings of these highly influential papers and comments on future research priorities in this field.

91.2 MATERIALS AND METHODS

The search for this study was carried out using the Scopus database (Burnham, 2006) in November 2022.

As a first step for the search of the relevant literature, the keywords were selected using the first 200 most-cited papers for each research front. The selected keyword list was optimized to obtain a representative sample of papers for the searched research field (Konur, 2023a–e). This keyword list for each research front was collected to form a combined keyword set for all seven research fronts, and this combined set was provided in the appendix of the constituent studies for future replication studies (Konur, 2023a–e).

As a second step, a sample dataset was used for this study. The first 25 articles with at least 414 citations each were selected for the review study. Key findings from each paper were taken from the abstracts of these papers and were discussed. Additionally, several brief conclusions were drawn and many relevant recommendations were made to enhance the future research landscape.

 DOI: 10.1201/9781003226567-122

91.3 RESULTS

Brief information about the 25 most-cited papers with at least 414 citations each on the utilization of bioethanol fuels is given as follows.

91.3.1 Utilization of Bioethanol Fuels in Fuel Cells

Brief information about 12 prolific studies with at least 414 citations each on the utilization of bioethanol fuels in fuel cells is given in Table 91.1.

Liang et al. (2009) evaluated bioethanol oxidation reaction (BOR) mechanisms on palladium (Pd) electrode in alkaline media using the cyclic voltammetry method in a paper with 720 citations. They found that the dissociative adsorption of bioethanol proceeded rather quickly and the rate-determining step was the removal of the adsorbed ethoxy by the adsorbed hydroxyl on the Pd electrode. The adsorption of OH^- ions followed the Temkin-type isotherm on the Pd electrode. Finally, at higher potentials, the reaction kinetics was affected by the adsorption of the OH^- ions and by the formation of the inactive oxide layer on the Pd electrode.

Marinov (1999) developed a chemical kinetic model for high-temperature bioethanol oxidation in a paper with 701 citations. They used laminar flame speed data, ignition delay data, and bioethanol oxidation product profiles from a jet-stirred and turbulent flow reactor. They found good agreement in the modeling of the datasets obtained from the five different experimental systems. They observed

TABLE 91.1
Utilization of Bioethanol Fuels in Fuel Cells

No.	Papers	Catalysts	Parameters	Country	Cits.
1	Liang et al. (2009)	Pd	Bioethanol oxidation reaction (BOR) mechanisms, electrocatalytic activity	China	720
2	Marinov (1999)	Na	High-temperature bioethanol oxidation, kinetic models	USA	701
3	Kowal et al. (2009)	C/Pt-Rh-SnO_2	Bioethanol oxidation, electrooxidation, electrocatalytic activity, catalyst types	USA	666
4	Zhou et al. (2003)	C/Pt-Sn, C/Pt-Ru, C/Pt-W, C/Pt-Pd	Bioethanol electrooxidation, additives, DEFC performance, temperature effect, catalyst types	China	660
5	Lamy et al. (2004)	Pt-Sn	Bioethanol electrooxidation, electrocatalytic activity	France	572
6	Dong et al. (2010)	Gr/Pt NP, Gr/Pt-Ru NP, C/Pt NP, C/Pt-Ru NP	Bioethanol electrooxidation, electrocatalytic activity, catalyst support types, nanocatalysts, catalyst types	USA	561
7	Klosek and Raftery (2002)*	V-TiO_2	Bioethanol photooxidation, CO_2, acetaldehyde, formic acid, CO, visible and UV irradiation	USA	556
8	Vigier et al. (2004)*	Pt, Pt-Sn	Bioethanol electrooxidation mechanisms, catalyst types, electrocatalytic activity	France	489
9	Xu et al. (2007a)	C/Pt, C/Pd	Bioethanol electrooxidation, catalyst types, C black, C microsphere supports	China	481
10	Xu et al. (2007b)	Pd nanowire arrays	Bioethanol electrooxidation, electrocatalytic activity, nanocatalysts	China	467
11	Rousseau et al. (2006)*	Pt/C, Pt-Sn/C, Pt-Sn-Ru/C	Bioethanol electrooxidation, electrocatalytic activity, catalyst types	France	457
12	Tian et al. (2010)	Pd NCs, Pd	Bioethanol electrooxidation, index facets, catalytic activity, surface atomic step density	China	414

* Listed also in Table 91.4.
Na, nonavailable.

that the high-temperature bioethanol oxidation exhibited strong sensitivity to the fall-off kinetics of bioethanol decomposition, branching ratio selection for $C_2H_5OH+OH \leftrightarrow$ products, and reactions involving the hydroperoxyl radical.

Kowal et al. (2009) evaluated the C/platinum (Pt)–rhodium (Rh)–tin dioxide (SnO_2) electrocatalysts for bioethanol electrooxidation in a paper with 666 citations. They found that this electrocatalyst effectively splits the C–C bond in bioethanol at room temperature in acid solutions, facilitating its oxidation at low potentials to CO_2. They reasoned that this superb electrocatalytic activity was due to the specific property of each of its constituents, induced by their interactions.

Zhou et al. (2003) evaluated the C/Pt-Sn, C/Pt-Ru, C/Pt-W, and C/Pt-Pd catalyst systems for bioethanol electrooxidation in a paper with 660 citations. They found that the presence of tin (Sn), ruthenium (Ru), and tungsten (W) enhanced the activity of Pt toward bioethanol electrooxidation: $Pt_1Sn_1/C > Pt_1Ru_1/C > Pt_1W_1/C > Pt_1Pd_1/C > Pt/C$. Moreover, Pt-Ru/C, further modified by W and molybdenum (Mo), showed improved bioethanol electrooxidation activity, but its direct ethanol fuel cell (DEFC) performance was lower than that measured for Pt-Sn/C. Furthermore, the single DEFC with Pt_1Sn_1/C or Pt_3Sn_2/C or Pt_2Sn_1/C showed better performances than those with Pt_3Sn_1/C or Pt_4Sn_1/C. The latter two DEFCs exhibited higher performances than the single DEFC using Pt_1Ru_1/C. They reasoned that this distinct difference in DEFC performance between these catalysts was due to the bifunctional mechanism and the electronic interaction between Pt and additives. Finally, at lower temperatures or at low current density regions, Pt_3Sn_2/C was more suitable for the DEFC. At 75C, the single DEFC with Pt_3Sn_2/C showed comparable performance to that with Pt_2Sn_1/C, but at a higher temperature of 90C, the latter was a much better performance.

Lamy et al. (2004) evaluated the Pt-Sn electrocatalysts for bioethanol electrooxidation in a paper with 572 citations. They found that the overall electrocatalytic activity was greatly enhanced at low potentials. The optimum composition in Sn was in the range of 10–20 at.%. With this composition, they observed that poisoning by adsorbed CO was greatly reduced leading to a significant enhancement of the electrode activity. However, the oxidation of ethanol was not complete, leading to the formation of C_2 products.

Dong et al. (2010) evaluated graphene (Gr)-supported Pt and Pt-Ru nanoparticles (NPs) with high electrocatalytic activity of bioethanol electrooxidation in a paper with 561 citations. They found that these nanocatalysts enhanced efficiency for ethanol electrooxidation with regard to diffusion efficiency, oxidation potential, forward oxidation peak current density, and the ratio of the forward peak current density to the reverse peak current density, compared with the C black catalyst supports. Furthermore, the forward peak current density of bioethanol electrooxidation for Gr- and C black-supported Pt NPs was 16.2 and 13.8 mA/cm^2, respectively; the ratios were 3.66 and 0.90, respectively.

Klosek and Raftery (2002) evaluated the visible light-driven vanadium (V)-doped titanium dioxide (TiO_2) photocatalyst for the photooxidation of bioethanol fuels in a paper with 556 citations. They found that this catalyst photooxidized bioethanol fuels to produce mostly CO_2 with small amounts of acetaldehyde, formic acid, and CO under visible irradiation. Under UV irradiation, this catalyst had comparable activity and product distribution as a similarly prepared TiO_2 thin-film monolayer catalyst.

Vigier et al. (2004) evaluated the mechanisms of bioethanol electrooxidation on Pt and Pt-Sn catalysts in a paper with 489 citations. First, they established the beneficial effect of Sn for ethanol electrooxidation where the Pt-Sn catalyst activity was almost double that of Pt. They identified adsorbed CO, adsorbed CH_3CO, acetaldehyde (CH_3CHO), acetic acid (CH_3COOH), and CO_2 as reaction products and intermediates. They established that two effects were involved in ethanol electrooxidation on Pt-Sn: the bifunctional mechanism and the ligand effect. The presence of Sn allowed bioethanol to adsorb dissociatively and then to break the C–C bond, at lower potentials and with a higher selectivity than on pure Pt. It then allowed the formation of acetic acid at lower potentials than on Pt alone.

Xu et al. (2007a) evaluated bioethanol electrooxidation on Pt and Pd electrocatalysts with C microsphere and C black supports in alkaline media in a paper with 481 citations. They found that these electrocatalyst systems gave better performance than those with C black supports. Pd showed excellently higher activity and better steady-state electrolysis than Pt for bioethanol electrooxidation in alkaline media. Finally, there was a synergistic effect by the interaction between Pd and C microspheres, while the Pd supported on C microspheres had excellent electrocatalytic properties.

Xu et al. (2007b) evaluated Pd nanowire arrays (NWAs) as electrocatalysts for bioethanol electrooxidation in DEFCs in a paper with 467 citations. They observed that Pd nanowires (NWs) were highly ordered with uniform diameter and length. Further, the NWs were uniform, well isolated, parallel to one another, and standing vertically to the electrode substrate surface, while the Pd NWAs exhibited a face-centered cubic (FCC) lattice structure. Finally, they observed that the high electrocatalytic activity of the Pd NWA electrode resulted in its superior performance for the electrooxidation reaction of bioethanol fuels in DEFC.

Rousseau et al. (2006) studied ethanol electrooxidation at different Pt-based electrodes in a single DEFC in terms of reaction product distribution depending on the anode catalyst in a paper with 457 citations. In DEFC experiments, they detected only three reaction products: acetaldehyde (AAL), acetic acid (AA), and CO_2. The addition of tin (Sn) to Pt increased the activity of the catalyst by several orders of magnitude and enhanced the electrical performance of the DEFC greatly from a few mW/cm^2 to 30 mW/cm^2 at 80°C, with Pt/C and Pt-Sn/C catalysts, respectively. Furthermore, at Pt-Sn/C and Pt-Sn-Ru/C the formation of CO_2 and AAL was lowered, whereas the formation of AA was increased compared with a Pt/C catalyst. Thus, the addition of Ru to Pt-Sn only enhanced the electrical performance of the DEFC, i.e., the activity of the catalyst, but did not modify the product distribution. They observed very good stability in the open circuit voltage of the DEFC (close to 0.75 V) over a period of 2 weeks at 90°C, the cell undergoing start–run–stop cycles each day. They also observed good stability under operating conditions at a given current density over 6 h.

Tian et al. (2010) produced the direct electrodeposition of tetrahexahedron Pd nanocrystals (THH Pd NCs) with {730} high-index facets and high catalytic activity for ethanol electrooxidation in a paper with 414 citations. They produced these NCs on a glassy carbon substrate in a dilute $PdCl_2$ solution by a newly developed programmed electrodeposition method. They observed that these NCs had four to six times higher catalytic activity than commercial Pd black catalyst toward ethanol electrooxidation in alkaline solutions due to their high density of surface atomic steps.

91.3.2 Utilization of Bioethanol Fuels for Biohydrogen Fuel Production for Fuel Cells

Brief information about seven prolific studies with at least 442 citations each on the utilization of bioethanol fuels for biohydrogen production for fuel cells is given in Table 91.2.

Murdoch et al. (2011) evaluated the effect of gold (Au) loading and NP size on photocatalytic hydrogen production from bioethanol fuels over Au/TiO_2 NPs in a paper with 1,013 citations. They showed that Au NPs in the size range of 3–30 nm on TiO_2 were very active in biohydrogen production. Au NPs of similar size on anatase NPs resulted in a rate of two orders of photoreaction magnitude higher than that for Au on rutile NPs. Surprisingly, Au NP size did not affect the photoreaction rate over the 3–12 nm range. They obtained a high biohydrogen yield with these nanocatalysts.

Deluga et al. (2004) evaluated biohydrogen production from bioethanol fuels by autothermal reforming in a paper with 918 citations. They converted the bioethanol and bioethanol–water mixtures into biohydrogen with nearly 100% selectivity and at least 95% conversion by catalytic partial oxidation on Rh–ceria (CeO_2) catalysts. They carried out rapid vaporization and mixing with air with an automotive fuel injector at temperatures sufficiently low and times sufficiently fast that homogeneous reactions producing biochemical such as bioethylene could be minimized.

Liguras et al. (2003) evaluated biohydrogen production by steam reforming of bioethanol over noble metal catalysts such as Rh, Ru, Pt, and Pd with catalyst supports such as Al_2O_3, MgO, and TiO_2 in a paper with 619 citations. They used a temperature range of 600°C–850°C and a metal loading of 0–5 wt%. They found that for low-loaded catalysts, Rh was significantly more

TABLE 91.2
Utilization of Bioethanol Fuels for Biohydrogen Production for Fuel Cells

No.	Papers	Catalysts	Parameters	Country	Cits.
1	Murdoch et al. (2011)	Au/TiO_2 NPs	Photocatalytic hydrogen production, catalyst loading, NP size, nanocatalysts	UK	1013
2	Deluga et al. (2004)	Rh/CeO_2	Biohydrogen production by autothermal reforming	Greece	918
3	Liguras et al. (2003)	Rh, Ru, Pt, Pd/Al_2O_3, MgO, TiO_2	Biohydrogen production by steam reforming, catalyst loading, bioethanol conversions and biohydrogen selectivities, catalyst types	Greece	619
4	Bamwenda et al. (1995)	Au/TiO_2, Pt/TiO_2	Photocatalytic biohydrogen fuel production, photocatalytic activity, catalyst types, preparation method	Japan	556
5	Fatsikostas and Verykios (2004)	Ni/ γ-Al_2O_3, Ni/La_2O_3, Ni/La_2O_3-γ-Al_2O	Steam reforming of bioethanol fuels, catalyst support types	Greece	543
6	Llorca et al. (2002)	Co/MgO, Co/γ-Co/Al_2O_3, Co/SiO_2, Co/TiO_2, Co/ V_2O_5, Co/ZnO, Co/ La_2O_3, Co/CeO_2	Steam reforming of bioethanol fuels, catalyst support types, catalytic performance, bioethanol selectivity	Spain	504
7	Fatsikostas et al. (2002)	Ni/La_2O_3, Ni/Al_2O_3, NiYSZ, Ni/MgO	Steam reforming of bioethanol fuels, catalyst support types, catalytic performance, bioethanol selectivity, stability, reaction temperature, water-to-ethanol ratio, and space velocity	Greece	442

Cits.: the number of citations received by each paper.

active and selective toward biohydrogen formation compared with Ru, Pt, and Pd. The catalytic performance of Rh and, particularly, Ru was significantly improved with increasing metal loading, leading to higher bioethanol conversion and biohydrogen selectivities at given reaction temperatures. The catalytic activity and selectivity of high-loaded Ru catalysts were comparable to those of Rh. They observed that, under certain reaction conditions, the 5% Ru/Al_2O_3 catalyst system completely converted bioethanol with selectivities toward biohydrogen above 95% with methane coproduct.

Bamwenda et al. (1995) evaluated the photocatalytic biohydrogen fuel production from a water–ethanol solution using Au/TiO_2 and Pt/TiO_2 catalyst systems in a paper with 555 citations. They deposited Au and Pt using TiO_2 powders in aqueous suspensions by deposition–precipitation (DP), impregnation (IMP), photodeposition (FD), and, in the case of Au, mixing TiO_2 with colloidal Au suspensions (MIX). The main reaction products were hydrogen, methane, carbon dioxide (CO_2), and acetaldehyde. They found that the overall activity of Au samples was generally about 30% lower than that of Pt samples. The activity of Au samples strongly depended on the method of preparation and decreased in the order Au/TiO_2-FD > Au/TiO_2-DP > Au/TiO_2-IMP > Au/TiO_2-MIX. The activities of the Pt samples were less sensitive to the preparation method and decreased in the order Pt/TiO_2-FD > Pt/TiO_2-DP ≈ Pt/TiO_2-IMP. Au and Pt precursors calcined in air at 573 K showed the highest activity toward H_2 generation, followed by a decline in activity with increasing calcination temperature. The H_2 yield was dependent on the metal content of TiO_2 and showed a maximum in the ranges 0.3–1 wt% Pt and 1–2 wt% Au. Finally, the rate of H_2 production was strongly dependent on the initial pH of the suspension as pH values in the range of 4–7 gave better yields, whereas highly acidic and basic suspensions resulted in a considerable decrease in the H_2 yield.

Fatsikostas and Verykios (2004) evaluated the reaction network of steam reforming of bioethanol fuels over Ni-based catalysts in a paper with 543 citations. They used alumina (γ-Al_2O_3), lanthana (La_2O_3), and La_2O_3/γ-Al_2O_3 as catalyst supports. They found that bioethanol fuels interacted strongly with Al_2O_3 and less strongly with La_2O_3. In the presence of Ni, catalytic activity was shifted toward lower temperatures. In addition to the above reactions, reforming, water–gas shift, methanation, and carbon deposition contributed significantly to product distribution. They then found that the rate of carbon deposition was a strong function of the carrier, the steam-to-bioethanol ratio, and reaction temperature as the presence of La_2O_3 on the catalyst, high steam-to-bioethanol ratio, and high temperature offer enhanced resistance toward carbon deposition.

Llorca et al. (2002) evaluated the steam reforming of bioethanol fuels over supported cobalt (Co) catalysts to produce biohydrogen fuels in a paper with 491 citations. They used a temperature range of 573–723 K at atmospheric pressure. They prepared catalysts (1%) by IMP of $Co_2(CO)_8$ on MgO, γ-Al_2O_3, SiO_2, TiO_2, V_2O_5, ZnO, La_2O_3, CeO_2, and Sm_2O_3. They found that bioethanol steam reforming took place largely over ZnO-, La_2O_3-, Sm_2O_3-, and CeO_2-supported catalysts where CO-free biohydrogen was produced. Depending on the catalyst support, they identified different Co-based phases: metallic Co particles, Co_2C, CoO, and La_2CoO_4. Furthermore, the extent and nature of carbon deposition depended on the sample and on the reaction temperature. Finally, ZnO-supported samples showed the best catalytic performances, while under 100% bioethanol conversion, they obtained selectivity up to 73.8% to H_2 and 24.2% to CO_2.

Fatsikostas et al. (2002) produced biohydrogen fuels for fuel cells by steam-reforming bioethanol fuels in a paper with 442 citations. They performed steam reforming of ethanol over Ni catalysts supported on La_2O_3, Al_2O_3, yttria-stabilized zirconia (YSZ), and magnesia (MgO). They examined the effect of reaction temperature, water-to-ethanol ratio, and space velocity on catalytic activity and selectivity. They observed that the Ni/La_2O_3 catalyst had high activity and selectivity toward hydrogen production and long-term stability for steam reforming of ethanol due to scavenging of coke deposition on the Ni surface by La_2O_3 species, which exist on top of the Ni particles under reaction conditions.

91.3.3 Utilization of Bioethanol Fuels in the Transport Engines

Brief information about four prolific studies with at least 433 citations each on the utilization of bioethanol fuels in transportation is given in Table 91.3.

Macedo et al. (2008) evaluated the net energy and greenhouse gas (GHG) emission balance in the production and utilization of bioethanol fuels from sugarcane and sugarcane bagasse in Brazil in a paper with 645 citations. They used two scenarios: 2005/2006 for a sample of mills processing up to 100 million tons of sugarcane per year and 2020. They found that the net energy balance was 9.3 for 2005/2006 and might reach 11.6 in 2020. The total GHG emissions were 436 kg carbon dioxide (CO_2) eq/m^3 bioethanol for 2005/2006, decreasing to 345 kg CO_2 eq/m^3 in the 2020 scenario. However, the avoided GHG emissions depended on the final use. For E100 use in Brazil, they were 2,181 kg CO_2 eq/m^3 bioethanol in 2005/2006, and for E25, they were 2,323 kg CO_2 eq/m^3 bioethanol. Both values would increase about 26% for 2020 mostly due to the large increase in sales of power surpluses. Finally, there was a high impact of sugarcane productivity and bioethanol yield variation on these balances and of sugarcane bagasse and power surpluses on GHG emission avoidance.

Hsieh et al. (2002) evaluated the properties, engine performance, and gas emissions of a spark ignition (SI) engine using bioethanol–gasoline fuel blends as a function of bioethanol content for E0, E5, E10, E20, and E30 in a paper with 525 citations. They found that by increasing the bioethanol content, the heating value of the blended fuels decreased, while the octane number of the blended fuels increased. With increasing the bioethanol content, the Reid vapor pressure of the blended fuels initially increased to a maximum of 10% bioethanol addition and then decreased. Furthermore, using these blends, torque output and fuel consumption of the engine slightly increased, while CO and unburned hydrocarbon (HC) emissions decreased dramatically as a result of the ethanol

TABLE 91.3
Utilization of Bioethanol Fuels in the Transport Engines

No.	Papers	Blends	Parameters	Country	Cits.
1	Macedo et al. (2008)	Gasoline–ethanol blends	Net energy balance, GHG emissions, CO_2 emissions, E25, E100	Brazil	645
2	Hsieh et al. (2002)	Bioethanol–gasoline fuel blends	Properties, performance, gas emissions, heating values, octane numbers, Reid vapor pressure, torque output, fuel consumption, CO, CO_2, HC, NO_x emissions, bioethanol content, E0, E5, E10, E20, E30 blends	Taiwan	525
3	Kwanchareon et al. (2007)	Diesel–biodiesel–ethanol blends	Solubility, fuel properties, and emission characteristics, ethanol purity, temperature, density, heat of combustion, cetane number, flash point and pour point, CO, NO_x, HC	Thailand	441
4	Al-Hasan (2003)	Unleaded gasoline–bioethanol blends	SI engine performance, exhaust emissions, equivalence air–fuel ratio, fuel consumption, volumetric efficiency, brake thermal efficiency, brake power, engine torque and brake-specific fuel consumption, CO, CO_2, HC, engine speed	Jordan	433

Cits.: the number of citations received by each paper.

addition; CO_2 emission increased due to improved combustion. Finally, nitrogen oxide (NO_x) emission depended on the engine operating conditions rather than the bioethanol content.

Kwanchareon et al. (2007) studied the solubility, fuel properties, and emission characteristics of diesel–biodiesel–ethanol blends at different purities of ethanol and different temperatures in a paper with 441 citations. They examined the density, heat of combustion, cetane number, flash point, and pour point of the selected blends compared with base petrodiesel fuels. They found that the fuel properties were close to the standard limit for diesel fuel. However, the flash point of blends containing ethanol was quite different from that of conventional diesel, while the high cetane number of biodiesel could compensate for the decrease in the cetane number of the blends caused by the presence of ethanol. The heating value of the blends containing lower than 10% ethanol was not significantly different from that of diesel. CO and HC reduced significantly at high engine load, whereas NO_x increased, compared with those of diesel. They recommended a blend of 80% diesel, 15% biodiesel, and 5% ethanol (D80/B15/E5) for diesohol production because of the acceptable fuel properties (except flash point) and the reduction in emissions.

Al-Hasan (2003) studied the effect of ethanol-unleaded gasoline blends on SI engine performance and exhaust emission in a paper with 433 citations. He used a four-stroke, four-cylinder SI engine and performed performance tests for equivalence air–fuel ratio, fuel consumption, volumetric efficiency, brake thermal efficiency (BTE), brake power, engine torque, and brake-specific fuel consumption (BSFC), while he analyzed exhaust emissions for CO, CO_2, and HC. They used these blends with different percentages of fuel at the three-fourth throttle opening position and variable engine speed ranging from 1,000 to 4,000 rpm. They observed that blending unleaded gasoline with ethanol increased brake power, torque, volumetric and BTEs, and fuel consumption, while it decreased the BSFC and equivalence air–fuel ratio. Furthermore, the CO and HC emission concentrations in the engine exhaust decreased, while the CO_2 concentration increased. Finally, he found that the 20 vol.% ethanol (E20) in the fuel blend gave the best results for all measured parameters at all engine speeds.

91.3.4 Utilization of Bioethanol Fuels for Biochemical Production

Brief information about three prolific studies with at least 457 citations each on the utilization of bioethanol fuels for biochemical production is given in Table 91.4. Further information about these studies is given in Section 91.3.1.

TABLE 91.4
Utilization of Bioethanol Fuels for Biochemical Production

No.	Papers	Catalysts	Parameters	Country	Cits.
7	Klosek and Raftery (2002)*	V-TiO_2	Bioethanol photooxidation, CO_2, acetaldehyde, formic acid, CO, visible and UV irradiation	USA	556
8	Vigier et al. (2004)*	Pt, Pt-Sn	Bioethanol electrooxidation mechanisms, catalyst types, electrocatalytic activity	France	489
11	Rousseau et al. (2006)*	Pt/C, Pt-Sn/C, Pt-Sn-Ru/C	Bioethanol electrooxidation, electrocatalytic activity, catalyst types	France	457

* Listed also in Table 91.1.
Cits.: the number of citations received by each paper.

TABLE 91.5
Bioethanol Sensors

No.	Papers	Materials	Parameters	Country	Cits.
1	Wan et al. (2004)	ZnO nanowires	Nanosensors, biosensor fabrication, MEMS, bioethanol gas sensitivity	China	1908
2	Liu et al. (2005)	V_2O_5 nanobelts	Nanosensors, biosensor fabrication, hydrothermal method, low-temperature bioethanol gas sensitivity	China	526

Cits.: the number of citations received by each paper.

91.3.5 Bioethanol Sensors

Brief information about two prolific studies with at least 526 citations each on the ethanol sensors is given in Table 91.5.

Wan et al. (2004) evaluated zinc oxide (ZnO) nanowire-based bioethanol sensors in a paper with 1908 citations. They used microelectromechanical system (MEMS) technology to fabricate these sensors. They found that these highly sensitive nanosensors exhibited a very high sensitivity to bioethanol gas with a fast response time at 300°C.

Liu et al. (2005) evaluated single-crystalline divanadium pentoxide (V_2O_5) nanobelts as bioethanol sensor materials in a paper with 508 citations. They produced these nanobelts by a simple mild hydrothermal method with a high yield. They found that the sensors fabricated using these nanobelts showed great potential for the detection of bioethanol molecules at a relatively low temperature. Finally, there were no problems of interference with bioethanol due to relative humidity and other gases.

91.4 DISCUSSION

91.4.1 Introduction

Crude oil-based gasoline fuels have been widely used in the transportation sector since the 1920s. However, there have been great public concerns over the adverse environmental and human impact of these fuels. Hence, biomass-based bioethanol fuels have increasingly been used in blending gasoline and petrodiesel fuels, in fuel cells, and in biochemical production in a biorefinery context.

The research in the field of utilization of bioethanol fuels has also intensified in recent years. The primary focus of the research in this area has been the utilization of bioethanol fuels in fuel cells, utilization of bioethanol fuels in gasoline and diesel engines, production of biohydrogen fuels and biochemicals from bioethanol fuels, and development and utilization of bioethanol sensors. Additionally, the research on gasoline fuels and applications of nanotechnology in this field has also been related to this field.

However, it is essential to develop efficient incentive structures for the primary stakeholders to enhance the research in this field. Although there have been several review papers on the utilization of bioethanol fuels, there has been no review of the 25 most-cited articles in this field.

Thus, this book chapter presents a review of the 25 most-cited articles in these fields. Then, it discusses the key findings of these highly influential papers and comments on future research priorities in this field.

As a first step for the search of the relevant literature, the keywords were selected using the first 200 most-cited papers for each research front. The selected keyword list was optimized to obtain a representative sample of papers for the searched research field (Konur, 2023a–e). This keyword list for each research front was collected to form a combined keyword set for all seven research fronts, and this combined set was provided in the appendix of the constituent studies for future replication studies (Konur, 2023a–e).

As a second step, a sample dataset was used for this study. The first 25 articles with at least 414 citations each were selected for the review study. Key findings from each paper were taken from the abstracts of these papers and were discussed. Additionally, several brief conclusions were drawn and many relevant recommendations were made to enhance the future research landscape.

As Table 91.6 shows, there are five primary research fronts for these 25 most-cited papers: utilization of bioethanol fuels in fuel cells, utilization of bioethanol fuels for biohydrogen production for fuel cells, utilization of bioethanol fuels in transportation such as gasoline and diesel engines, and to a lesser extent utilization of bioethanol fuels for biochemical production, and bioethanol sensors with 48%, 28%, 16%, 12%, and 8% of the reviewed papers, respectively.

Table 91.6 also shows that the first research front is overrepresented in the reviewed papers compared with a wider sample of 280 highly cited papers by 14%. On the contrary, the third research front is underrepresented in the reviewed papers by 14%. The difference between these two samples is relatively small compared with the other research fronts.

91.4.2 Utilization of Bioethanol Fuels in Fuel Cells

Over 48% and 34% of the reviewed papers and 280 highly cited papers, respectively, were related to the utilization of bioethanol fuels in fuel cells (Table 91.6). These studies were carried out in the USA, China, and France with four, five, and three papers, respectively (Table 91.1).

A number of catalysts were used in these studies: Pt (Dong et al., 2010; Kowal et al., 2009; Rousseau et al., 2006; Lamy et al., 2004; Vigier et al., 2004; Xu et al., 2007a; Zhou et al., 2003), Pd (Liang et al., 2009; Tian et al., 2010; Xu et al., 2007a,b; Zhou et al., 2003), Ru (Dong et al., 2010; Rousseau et al., 2006; Zhou et al., 2003), Sn (Kowal et al., 2009; Lamy et al., 2004; Rousseau et al., 2006; Vigier et al., 2004; Zhou et al., 2003), Rh (Kowal et al., 2009), W (Zhou et al., 2003), and V (Klosek and Raftery, 2002). Only Dong et al. (2010), Tian et al. (2010), and Xu et al. (2007b) considered the applications of nanotechnology in these catalysts.

Five studies focused on the C as catalyst supports (Dong et al., 2010; Kowal et al., 2009; Rousseau et al., 2006; Xu et al., 2007a; Zhou et al., 2003). Additionally, Dong et al. (2010) used graphene as a catalyst support compared with C catalyst supports.

These studies focused mostly on the electrooxidation of bioethanol fuels (Dong et al., 2010; Kowal et al., 2009; Lamy et al., 2004; Liang et al., 2009; Rousseau et al., 2006; Tian et al., 2010; Vigier et al., 2004; Xu et al., 2007a,b, Zhou et al., 2003). Additionally, Marinov (1999) and Klosek and Raftery (2002) studied the oxidation and photooxidation of bioethanol fuels, respectively. The activity and mechanisms of electrooxidation of bioethanol fuels were extensively evaluated in these studies.

TABLE 91.6
Utilization of Bioethanol Fuels: Research Fronts

No.	Research Fronts	Reviewed Papers (%)	Sample Papers (%)	Surplus (%)
1	Utilization of bioethanol fuels in fuel cells	48	34	14
2	Utilization of bioethanol fuels for biohydrogen production for fuel cells	28	22	6
3	Utilization of bioethanol fuels in transportation such as gasoline engines	16	30	−14
4	Utilization of bioethanol fuels for biochemical production	12	8	4
5	Bioethanol sensors	8	11	−3
	Sample size	25	280	

Reviewed Papers: 25 papers reviewed in this study. Sample Papers: the sample of the most-cited 280 papers.

These findings are thought-provoking in seeking ways to optimize the utilization of bioethanol fuels in fuel cells. The catalysts, catalyst supports, mechanisms and activity of bioethanol electro-oxidation, and applications of nanotechnology in catalysts emerge as the key issues for this field. It is expected that nanotechnology applications would increase to improve the utility of bioethanol fuel utilization in fuel cells in this field in the future.

91.4.3 Utilization of Bioethanol Fuels for Biohydrogen Production for Fuel Cells

Around 28% and 22% of the reviewed papers and 280 highly cited papers, respectively, were related to the utilization of bioethanol fuels for biohydrogen production for the fuel cells (Table 91.6). These studies were carried out in Greece, Japan, Spain, and the UK with four, one, one, and one papers, respectively (Table 91.2).

A number of catalysts were used in these studies: Au (Bamwenda et al., 1995; Murdoch et al., 2011), Rh (Deluga et al., 2004; Liguras et al., 2003), Ru (Liguras et al., 2003), Pt (Bamwenda et al., 1995; Liguras et al., 2003), Pd (Liguras et al., 2003), Ni (Fatsikostas and Verykios, 2004; Fatsikostas et al., 2002), and Co (Llorca et al., 2002).

Similarly, several catalyst supports were used in these studies: TiO_2 (Bamwenda et al., 1995; Liguras et al., 2003; Llorca et al., 2002; Murdoch et al., 2011), CeO_2 (Deluga et al., 2004; Llorca et al. (2002), Al_2O_3 (Liguras et al., 2003; Llorca et al., 2002; Fatsikostas and Verykios, 2004; Fatsikostas et al., 2002), MgO (Fatsikostas et al., 2002; Liguras et al., 2003; Llorca et al., 2002), γ-Al_2O_3 (Fatsikostas and Verykios, 2004), La_2O_3 (Fatsikostas and Verykios, 2004; Fatsikostas et al., 2002; Llorca et al., 2002), La_2O_3-γ-Al_2O (Fatsikostas and Verykios, 2004), SiO_2 (Llorca et al., 2002), V_2O_5 (Llorca et al., 2002), YZS (Fatsikostas et al., 2002), and ZnO (Llorca et al., 2002). Only Murdoch et al. (2011) considered nanotechnology applications.

These studies focused mostly on the steam reforming of bioethanol fuels (Fatsikostas and Verykios, 2004; Fatsikostas et al., 2002; Liguras et al., 2003; Llorca et al., 2002), photocatalytic hydrogen production (Bamwenda et al., 1995; Murdoch et al., 2011), and autothermal reforming of bioethanol fuels (Deluga et al. 2004). The key issues considered were the catalyst loading (Liguras et al., 2003; Murdoch et al., 2011), catalyst types (Bamwenda et al., 1995; Liguras et al., 2003), catalyst support types (Fatsikostas and Verykios, 2004; Fatsikostas et al., 2002; Liguras et al., 2003; Llorca et al., 2002), nanostructured catalyst systems (Murdoch et al., 2011), preparation methods (Bamwenda et al., 1995), bioethanol conversion and biohydrogen selectivities (Fatsikostas et al., 2002; Liguras et al., 2003; Llorca et al., 2002), and catalytic activity (Bamwenda et al., 1005; Llorca et al., 2002).

These findings are thought-provoking in seeking ways to optimize the utilization of bioethanol fuels for biohydrogen production for fuel cells. The catalyst types, catalyst support types, nanostructured catalyst systems, steam and autothermal reforming of bioethanol fuels, photocatalytic biohydrogen production, catalyst loading, preparation methods, bioethanol conversion, biohydrogen selectivities, and catalytic activity emerge as the key issues for this field. It is expected that nanotechnology applications would increase to improve the utility of bioethanol fuel utilization to produce biohydrogen fuels for fuel cells in this field in the future.

91.4.4 Utilization of Bioethanol Fuels in the Transport Engines

Around, 16% and 30% of the reviewed papers and 280 highly cited papers, respectively, were related to the utilization of bioethanol fuels in transportation such as gasoline or diesel engines (Table 91.6). These studies were carried out in Brazil, Jordan, Thailand, and Taiwan (Table 91.3).

A wide range of issues were covered in these papers: properties, performance, gas emissions, heating values, octane numbers, Reid vapor pressure, torque output, fuel consumption, CO, CO_2, HC, NO_x emissions, bioethanol content, E0, E5, E10, E20, E25, E30, and E85 blends.

The focus was more on the gasoline–bioethanol fuel blends (Al-Hasan, 2003; Hsieh et al., 2002; Macedo et al., 2008) compared with petrodiesel–bioethanol fuel blends (Kwanchareon et al., 2007).

These findings are thought-provoking in seeking ways to optimize the utilization of bioethanol fuels in gasoline and diesel engines to improve combustion, performance, and GHG emissions of the petrodiesel and gasoline fuels in these engines.

It is expected that nanotechnology applications would increase to optimize the utilization of bioethanol fuels in gasoline and diesel engines to improve combustion, performance, and GHG emissions of the petrodiesel and gasoline fuels in these engines in the future. It is also expected that bioethanol fuels would increasingly be used in these engines on a global scale.

91.4.5 Utilization of Bioethanol Fuels for Biochemical Production

Around 12% and 8% of the reviewed papers and 280 highly cited papers, respectively, were related to the utilization of bioethanol fuels for biochemical production (Table 91.6). Both studies were carried out in China (Table 91.4).

A number of catalysts were used in these studies: V (Klosek and Raftery, 2002), Pt (Rousseau et al., 2006; Vigier et al., 2004), Pt-Sn (Rousseau et al., 2006; Vigier et al., 2004), and Pt-Sn-Ru (Rousseau et al., 2006). C and TiO_2 were used as catalyst supports (Klosek and Raftery, 2002).

The issues studied in these studies were bioethanol photooxidation, visible and UV irradiation, bioethanol electrooxidation mechanisms, catalyst types, and electrocatalytic activity. The biochemicals produced in these studies included CO_2, acetaldehyde, formic acid, adsorbed CH_3CO, acetic acid, and CO.

These findings are thought-provoking in seeking ways to optimize the development and production of biochemical from bioethanol fuels through the oxidation of bioethanol fuels. It is expected that nanotechnology applications would increase to optimize the utilization of bioethanol fuels to produce biochemicals in the future. It is also expected that bioethanol fuels would increasingly be used in biochemical production in the future.

91.4.6 Bioethanol Sensors

Around 8% and 11% of the reviewed papers and 280 highly cited papers, respectively, were related to bioethanol sensors (Table 91.6). These studies were carried out in China (Table 91.5).

Two catalysts were used in these studies: ZnO nanowires (Wan et al., 2004) and V_2O_5 nanobelts (Liu et al., 2005). The issues covered in these studies were nanosensors, biosensors, fabrication of sensors, MEMS, bioethanol gas sensitivity, hydrothermal method, and low-temperature bioethanol gas sensitivity.

These findings are thought-provoking in seeking ways to optimize the development and utilization of bioethanol sensors using nanomaterials. It is expected that nanotechnology applications would increase to improve the bioethanol gas sensitivity of bioethanol sensors in the future using a wide range of none and two-dimensional nanomaterials. It is also expected that the research in this area would intensify in the future.

91.4.7 The Overall Remarks

Although there are five major research fronts in the field of utilization of bioethanol fuels, their share among the reviewed papers differs substantially (Table 91.6). It is important to note that the field of the utilization of bioethanol fuels in fuel cells is overrepresented in the reviewed papers by 14%, while the field of the utilization of bioethanol fuels in transportation such as gasoline engines is underrepresented by 14% in this sample. There are also minor fluctuations for the other research fronts.

Considering that biohydrogen fuels are used ultimately in fuel cells for electric vehicles (Clement-Nyns et al., 2010; Lu et al., 2013), the share of the papers related to fuel cells rises to 72% of the reviewed paper sample, compared with 56% of the 280 highly cited paper samples. In light of North's institutional framework (North, 1991), it could be said that these two major research fronts have greater societal importance compared with the other three research fields. The utilization of bioethanol fuels and biohydrogen fuels produced from bioethanol fuels to produce bioelectricity (Kim et al., 2005; Larsson et al., 2014) in fuel cells would improve the GHG emissions of the fuel cells (Gong et al., 2009; Steele and Heinzel, 2001), compared with the gasoline- and diesel-fueled engines and natural gas-based (Zhu et al., 2016) and coal-based (Buhre et al., 2005) electricity generation (Gao et al., 2018; Raghuvanshi et al., 2006).

Thus, findings related to the direct utilization of bioethanol fuels in fuel cells are thought-provoking in seeking ways to optimize the utilization of bioethanol fuels in fuel cells. The catalyst types (de Boer et al., 1965), catalyst support types (Arai and Machida, 1996), mechanisms and activity of bioethanol electrooxidation (Bagotzki and Vassilyev, 1967), and applications of nanotechnology in catalysts (Fihri et al., 2011) emerge as the key issues for this field.

The findings related to the utilization of bioethanol fuels for biohydrogen fuel production for fuel cells are also thought-provoking in seeking ways to optimize the utilization of bioethanol fuels for biohydrogen production for fuel cells. The catalyst types (de Boer et al., 1965), catalyst support types (Arai and Machida, 1996), nanostructured catalyst systems (Fihri et al., 2011), steam (Palo et al., 2007) and autothermal reforming (Ayabe et al., 2003) of bioethanol fuels, photocatalytic biohydrogen production (Tong et al., 2012), catalyst loading (Jusys et al., 2003), bioethanol conversion (Maurya et al., 2015), and catalytic activity (Thiele, 1939) emerge as the key issues for this field.

A wide range of issues were covered in the papers related to the utilization of bioethanol fuels in gasoline and petrodiesel engines: fuel properties (Dooley et al., 2010), fuel performance (Crookes, 2006), heating values (Nhuchhen and Salam, 2012), octane numbers (Albahri, 2003), Reid vapor pressure (Cooper et al., 1995), torque output (Ahmadi et al., 2015), fuel consumption (Ramos-Paja et al., 2008), CO emissions (Muller and Stavrakou, 2005), CO_2 emissions, (Davis et al., 2011), unburned HC emissions (Stanglmaier et al., 1999), NO_x emissions (Muller and Stavrakou, 2005), and bioethanol content (Yunoki and Saito, 2009): E0, E5, E10, E20, E25, E30, and E85 blends.

Similarly, the focus of the influential papers related to bioethanol sensors is on the development of one- and two-dimensional nanomaterials for bioethanol sensors. The research on chemical sensors and biosensors has been very intense (Chaubey and Malhotra, 2002). In this context, the research on ZnO-based sensors has also been very intense (Kumar et al., 2006). Nanomaterials have been widely used in the development of sensors (Howes et al., 2014).

It emerges that the field of the utilization of bioethanol fuel for biochemical production (Angelici et al., 2013; Liu and Hensen, 2013; Wang et al., 2004) is widely related to the field of the utilization of bioethanol fuels in fuel cells. It is very important to produce biochemicals within the biorefinery context to reduce the production costs of bioethanol fuels and to develop human- and environment-friendly biochemicals as an alternative to crude oil-based chemicals.

Since the studies in each of the five research fields take gasoline fuels as a base case, the review of gasoline fuels as a whole was provided in Konur (2023g). The key research fronts for gasoline fuels are desulfurization of gasoline fuels (Ma et al., 2002; Song, 2003), properties (Chica and Corma, 1999; Morgan et al., 2010), combustion (Battin-Leclerc, 2008; Mehl et al., 2011), performance (Dec et al., 2011; Sayin et al., 2007), and emissions of gasoline fuels and their blends (Khalili et al., 1995), gasoline economics (Borenstein et al., 1997; Newman and Kenworthy, 1989), gasoline fuel production (Bjorgen et al., 2008; Meisel et al., 1976), utilization of gasoline fuels in fuel cells (Ogden et al., 1999; Thomas et al., 2000), and adverse environmental and health impact of gasoline fuels (Hill et al., 2009; Jacobson, 2007; Moldovan et al., 2002). Gasoline fuels are also blended with other biofuels: butanol (Gu et al., 2012), methanol (Liu et al., 2007), hydrogen (D'Andrea et al., 2004), and diesel fuels (Ma et al., 2013) among other fuels.

It is important to note that Hill et al. (2006) laid down the rules for the environment- and human-friendly production of bioethanol fuels: They should have a net energy balance, have environmental benefits, have economic competitiveness, have large-scale production, and have no competition to the food production. It appears that these rules are also relevant to the derivatives of bioethanol fuels. The driving force for the development of the research and industry in this field has been the public concerns about climate change (Change, 2007), GHG emissions (Carlson et al., 2017), and global warming (Kerr, 2007) caused by the petrodiesel and gasoline fuels.

It is expected that nanotechnology (Geim, 2009; Geim and Novoselov, 2007) would play a more vital role in the development of catalyst systems for the oxidation of bioethanol fuels, production of biohydrogen fuels and biochemicals from bioethanol fuels, and development of bioethanol sensors and the utilization of bioethanol fuels in gasoline and diesel engines (Konur, 2023f).

In the end, these most-cited papers in this field hint that the efficiency of the biohydrogen and biochemical production from bioethanol fuels, utilization of bioethanol fuels in gasoline and diesel engines and fuel cells, and development and utilization of bioethanol sensors could be optimized using the structure, processing, and property relationships of bioethanol fuels and their derivatives (Formela et al., 2016; Konur, 2018, 2020b, 2021a–d; Konur and Matthews, 1989).

These reviewed studies also show the importance of incentive structures for the development of bioethanol industry and research in these research fronts at a global scale in light of North's institutional framework (North, 1991). In this context, the major producers and users of bioethanol fuels and their derivatives such as Brazil, the USA, China, and Europe had developed strong incentive structures for the effective development of bioethanol industry and research in these fields.

91.5 CONCLUSION AND FUTURE RESEARCH

Brief information about the key research fronts covered by the 25 most-cited papers with at least 414 citations each is given under five headings: utilization of bioethanol fuels in fuel cells, utilization of bioethanol fuels for biohydrogen production for fuel cells, bioethanol sensors, utilization of bioethanol fuels in transportation such as gasoline and diesel engines, and utilization of bioethanol fuels for biochemical production (Table 91.6). In light of North's institutional framework (North, 1991), it could be said that these major research fronts have great societal importance.

The key findings on these research fronts should be read in light of the increasing public concerns about climate change, GHG emissions, and global warming as these concerns have been certainly behind the boom in the research in this field in the last two decades.

The findings related to the direct utilization of bioethanol fuels in fuel cells are thought-provoking in seeking ways to optimize the utilization of bioethanol fuels in fuel cells to produce bioelectricity. The catalysts, catalyst supports, mechanisms and activity of bioethanol electrooxidation, and applications of nanotechnology in catalysts emerge as the key issues for this field. There is ample room for the utilization of one- and two-dimensional nanomaterials in this field.

The findings related to the utilization of bioethanol fuels for biohydrogen fuel production for fuel cells to produce bioelectricity are thought-provoking in seeking ways to optimize the utilization of bioethanol fuels for biohydrogen production for fuel cells. The catalyst types, catalyst support types, nanostructured catalyst systems, steam and autothermal reforming of bioethanol fuels, photocatalytic biohydrogen production, catalyst loading, preparation methods, bioethanol conversion, biohydrogen selectivities, and catalytic activity emerge as the key issues for this field. There is ample room for the utilization of one- and two-dimensional nanomaterials in this field.

The findings related to the utilization of bioethanol fuels for blending with gasoline and petrodiesel fuels in gasoline or diesel engines are thought-provoking in seeking ways to optimize the utilization of bioethanol fuels in these engines. The combustion, performance, and emissions of these bioethanol fuels in these engines emerge as the key issues for this field. There is ample room for the utilization of one- and two-dimensional nanomaterials in this field.

The findings related to the development and utilization of bioethanol sensors are thought-provoking in seeking ways to optimize the bioethanol sensitivity of bioethanol sensors. There is ample room for the utilization of one- and two-dimensional nanomaterials in this field.

The findings related to the utilization of bioethanol for biochemical production are thought-provoking in seeking ways to optimize biochemical production. There is ample room for the utilization of one- and two-dimensional nanomaterials in this field.

It is very important to produce biochemicals and biohydrogen fuels from bioethanol fuels and bioelectricity in fuel cells fueled with bioethanol fuels or bioethanol fuel-based biohydrogen fuels within the biorefinery context to reduce the production costs of bioethanol fuels and to develop human- and environment-friendly biochemicals, biohydrogen fuels, and bioelectricity as a sustainable alternative to crude oil-based chemicals, water-based hydrogen fuels, and natural gas-, coal-, and syngas-based power generation. Therefore, it is important to devise proper incentive structures for the efficient development of bioethanol industry and research in these fields in light of North's institutional framework (North, 1991). This is especially important in the face of recent supply shocks caused by the coronavirus disease 2019 (COVID-19) pandemic and the Ukrainian war.

It is expected that nanotechnology would play a more vital role in the development of catalyst systems for the oxidation of bioethanol fuels for the fuel cells to produce bioelectricity, production of biohydrogen fuels from bioethanol fuels, and development of bioethanol sensors and better combustion of bioethanol–gasoline and bioethanol–petrodiesel blends in the gasoline and petrodiesel engines, respectively.

In the end, these most-cited papers in this field hint that the efficiency of bioethanol-based biohydrogen fuel production, bioelectricity production in fuel cells fueled with bioethanol fuels or bioethanol-based biohydrogen fuels, and bioethanol fuel-based biochemical production and the utilization of bioethanol fuels in gasoline and diesel engines and development and utilization of bioethanol sensors could be optimized using the structure, processing, and property relationships of bioethanol fuels and their derivatives.

These reviewed studies also show the importance of incentive structures for the development of bioethanol industry and research at a global scale in light of North's institutional framework (North, 1991). In this context, the major producers and users of bioethanol fuels such as Brazil, the USA, China, and Europe had developed strong incentive structures for the effective development of bioethanol industry and research in these fields. It is thus expected that the research on the utilization of bioethanol fuels would intensify in the future with efficient incentive structures for the major stakeholders.

It is recommended that such review studies should be performed for the individual research fronts on both the production and utilization of bioethanol fuels complementing the scientometric studies for the evaluation and utilization of bioethanol fuels.

ACKNOWLEDGMENTS

The contribution of highly cited researchers in the field of the utilization of bioethanol fuels has been gratefully acknowledged.

REFERENCES

Ahmadi, M. H., M. A. Ahmadi, S. A. Sadatsakkak and M. Feidt. 2015. Connectionist intelligent model estimates output power and torque of stirling engine. *Renewable and Sustainable Energy Reviews* 50:871–883.

Albahri, T. A. 2003. Structural group contribution method for predicting the octane number of pure hydrocarbon liquids. *Industrial & Engineering Chemistry Research* 42:657–662.

Al-Hasan, M. 2003. Effect of ethanol-unleaded gasoline blends on engine performance and exhaust emission. *Energy Conversion and Management* 44:1547–1561.

Angelici, C., B. M. Weckhuysen and P. C. A. Bruijnincx. 2013. Chemocatalytic conversion of ethanol into butadiene and other bulk chemicals. *ChemSusChem* 6:1595–1614.

Antolini, E. 2007. Catalysts for direct ethanol fuel cells. *Journal of Power Sources* 170:1–12.

Antolini, E. 2009. Palladium in fuel cell catalysis. *Energy and Environmental Science* 2:915–931.

Arai, H. and M. Machida. 1996. Thermal stabilization of catalyst supports and their application to high-temperature catalytic combustion. *Applied Catalysis A: General* 138:161–176.

Ayabe, S., H. Omoto and T. Utaka, et al. 2003. Catalytic autothermal reforming of methane and propane over supported metal catalysts. *Applied Catalysis A: General* 241:261–269.

Bagotzky, V. S. and Y. B. Vassilyev. 1967. Mechanism of electro-oxidation of methanol on the platinum electrode. *Electrochimica Acta* 12:1323–1343.

Bamwenda, G. R., S. Tsubota, T. Nakamura and M. Haruta. 1995. Photoassisted hydrogen production from a water-ethanol solution: a comparison of activities of Au/TiO_2 and Pt/TiO_2. *Journal of Photochemistry and Photobiology, A: Chemistry* 89:177–189.

Battin-Leclerc, F. 2008. Detailed chemical kinetic models for the low-temperature combustion of hydrocarbons with application to gasoline and diesel fuel surrogates. *Progress in Energy and Combustion Science* 34:440–498.

Bjorgen, M., F. Joensen and S. M. Holm, et al. 2008. Methanol to gasoline over zeolite H-ZSM-5: Improved catalyst performance by treatment with NaOH. *Applied Catalysis A: General* 345:43–50.

Borenstein, S., A. C. Cameron and R. Gilbert. 1997. Do gasoline prices respond asymmetrically to crude oil price changes? *Quarterly Journal of Economics* 112:304–339.

Buhre, B. J. P., L. K. Elliott, C. D. Sheng, R. P. Gupta and T. F. Wall. 2005. Oxy-fuel combustion technology for coal-fired power generation. *Progress in Energy and Combustion Science* 31:283–307.

Burnham, J. F. 2006. Scopus database: A review. *Biomedical Digital Libraries* 3:1–8.

Carlson, K. M., J.S. Gerber and N. D. Mueller, et al. 2017. Greenhouse gas emissions intensity of global croplands. *Nature Climate Change* 7:63–68.

Change, C. 2007. Climate change impacts, adaptation and vulnerability. *Science of the Total Environment* 326:95–112.

Chaubey, A. and B. Malhotra. 2002. Mediated biosensors. *Biosensors and Bioelectronics* 17:441–456.

Chica, A. and A. Corma. 1999. Hydroisomerization of pentane, hexane, and heptane for improving the octane number of gasoline. *Journal of Catalysis* 187:167–176.

Clement-Nyns, K., E. Haesen and J. Driesen. 2010. The impact of charging plug-in hybrid electric vehicles on a residential distribution grid. *IEEE Transactions on Power Systems* 25:5356176.

Cooper, J. B., K. L. Wise, J. Groves and W. T. Welch. 1995. Determination of octane numbers and Reid vapor pressure of commercial petroleum fuels using FT-Raman spectroscopy and partial least-squares regression analysis. *Analytical Chemistry* 67:4096–4100.

Crookes, R. J. 2006. Comparative bio-fuel performance in internal combustion engines. *Biomass and Bioenergy* 30:461–468.

D'Andrea, T., P. F. Henshaw and D. S. K. Ting. 2004. The addition of hydrogen to a gasoline-fuelled SI engine. *International Journal of Hydrogen Energy* 29:1541–1552.

Davis, S. J., G. P. Peters and K. Caldeira. 2011. The supply chain of CO_2 emissions. *Proceedings of the National Academy of Sciences* 108:18554–18559.

de Boer, J. H., B. G. Linsen and T. J. Osinga. 1965. Studies on pore systems in catalysts: VI. The universal t curve. *Journal of Catalysis* 4:643–648.

Dec, J. E., Y. Yang and N. Dronniou. 2011. Boosted HCCI: Controlling pressure-rise rates for performance improvements using partial fuel stratification with conventional gasoline. *SAE International Journal of Engines* 4:1169–1189.

Deluga, G. A., J. R. Salge, L. D. Schmidt and X. E. Verykios. 2004. Renewable hydrogen from ethanol by autothermal reforming. *Science* 303:993–997.

Dong, L., R. R. S. Gari, Z. Li, M. M. Craig and S. Hou. 2010. Graphene-supported platinum and platinum-ruthenium nanoparticles with high electrocatalytic activity for methanol and ethanol oxidation. *Carbon* 48:781–787.

Dooley, S., S. H. Won and M. Chaos, et al. 2010. A jet fuel surrogate formulated by real fuel properties. *Combustion and Flame* 157:2333–2339.
Fatsikostas, A. N. and X. E. Verykios. 2004. Reaction network of steam reforming of ethanol over Ni-based catalysts. *Journal of Catalysis* 225:439–452.
Fatsikostas, A. N., D. L. Kondarides and X. E. Verykios. 2002. Production of hydrogen for fuel cells by reformation of biomass-derived ethanol. *Catalysis Today* 75:145–155.
Fernando, S., S. Adhikari, C. Chandrapal and M. Murali. 2006. Biorefineries: Current status, challenges, and future direction. *Energy & Fuels* 20:1727–1737.
Fihri, A., M. Bouhrara, B. Nekoueishahraki, J. M. Basset and V. Polshettiwar. 2011. Nanocatalysts for Suzuki cross-coupling reactions. *Chemical Society Reviews* 40:5181–5203.
Formela, K., A. Hejna, L. Piszczyk, M. R. Saeb and X. Colom. 2016. Processing and structure-property relationships of natural rubber/wheat bran biocomposites. *Cellulose* 23:3157–3175.
Gao, M., G. Beig and S. Song, et al. 2018. The impact of power generation emissions on ambient PM2.5 pollution and human health in China and India. *Environment International* 121:250–259.
Geim, A. K. 2009. Graphene: Status and prospects. *Science* 324:1530–1534.
Geim, A. K. and K. S. Novoselov. 2007. The rise of graphene. *Nature Materials* 6:183–191.
Gong, K., F. Du, Z. Xia, M. Durstock and L. Dai. 2009. Nitrogen-doped carbon nanotube arrays with high electrocatalytic activity for oxygen reduction. *Science* 323:760–764.
Gu, X., Z. Huang and J. Cai, et al. 2012. Emission characteristics of a spark-ignition engine fuelled with gasoline-n-butanol blends in combination with EGR. *Fuel* 93:611–617.
Hansen, A. C., Q. Zhang and P. W. L. Lyne. 2005. Ethanol-diesel fuel blends: A review. *Bioresource Technology* 96:277–285.
Haryanto, A., S. Fernando, N. Murali and S. Adhikari. 2005. Current status of hydrogen production techniques by steam reforming of ethanol: A review. *Energy and Fuels* 19:2098–2106.
Hill, J., E. Nelson, D. Tilman, S. Polasky and D. Tiffany. 2006. Environmental, economic, and energetic costs and benefits of biodiesel and ethanol biofuels. *Proceedings of the National Academy of Sciences of the United States of America* 103:11206–11210.
Hill, J., S. Polasky and E. Nelson, et al. 2009. Climate change and health costs of air emissions from biofuels and gasoline. *Proceedings of the National Academy of Sciences of the United States of America* 106:2077–2082.
Howes, P. D., R. Chandrawati and M. M. Stevens. 2014. Colloidal nanoparticles as advanced biological sensors. *Science* 346:1247390.
Hsieh, W. D., R. H. Chen, T. L. Wu and T. H. Lin. 2002. Engine performance and pollutant emission of an SI engine using ethanol-gasoline blended fuels. *Atmospheric Environment* 36:403–410.
Huang, H. J., S. Ramaswamy, U. W. Tschirner and B. V. Ramarao. 2008. A review of separation technologies in current and future biorefineries. *Separation and Purification Technology* 62:1–21.
Jacobson, M. Z. 2007. Effects of ethanol (E85) versus gasoline vehicles on cancer and mortality in the United States. *Environmental Science and Technology* 41:4150–4157.
Jusys, Z., J. Kaiser, J. and R. J. Behm. 2003. Methanol electrooxidation over Pt/C fuel cell catalysts: Dependence of product yields on catalyst loading. *Langmuir* 19:6759–6769.
Kerr, R. A. 2007. Global warming is changing the world. *Science* 316:188–190.
Khalili, N. R., P. A. Scheff and T. M. Holsen. 1995. PAH source fingerprints for coke ovens, diesel and, gasoline engines, highway tunnels, and wood combustion emissions. *Atmospheric Environment* 29:533–542.
Kim, J. R., B. Min and B. E. Logan. 2005. Evaluation of procedures to acclimate a microbial fuel cell for electricity production. *Applied Microbiology and Biotechnology* 68:23–30.
Klosek, S. and D. Raftery. 2002. Visible light driven V-doped TiO_2 photocatalyst and its photooxidation of ethanol. *Journal of Physical Chemistry B* 105:2815–2819.
Kohse-Hoinghaus, K., P. Osswald and T. A. Cool, et al. 2010. Biofuel combustion chemistry: From ethanol to biodiesel. *Angewandte Chemie: International Edition* 49:3572–3597.
Konur, O. 2000. Creating enforceable civil rights for disabled students in higher education: An institutional theory perspective. *Disability & Society* 15:1041–1063.
Konur, O. 2002a. Access to nursing education by disabled students: Rights and duties of nursing programs. *Nurse Education Today* 22:364–374.
Konur, O. 2002b. Assessment of disabled students in higher education: Current public policy issues. *Assessment and Evaluation in Higher Education* 27:131–152.
Konur, O. 2002c. Access to employment by disabled people in the UK: Is the Disability Discrimination Act working? *International Journal of Discrimination and the Law* 5:247–279.

Konur, O. 2006a. Participation of children with dyslexia in compulsory education: Current public policy issues. *Dyslexia* 12:51–67.

Konur, O. 2006b. Teaching disabled students in higher education. *Teaching in Higher Education* 11:351–363.

Konur, O. 2007a. A judicial outcome analysis of the Disability Discrimination Act: A windfall for the employers? *Disability & Society* 22:187–204.

Konur, O. 2007b. Computer-assisted teaching and assessment of disabled students in higher education: The interface between academic standards and disability rights. *Journal of Computer Assisted Learning* 23:207–219.

Konur, O. 2012. The evaluation of the research on the bioethanol: A scientometric approach. *Energy Education Science and Technology Part A: Energy Science and Research* 28:1051–1064.

Konur, O. 2015. Current state of research on algal bioethanol. In *Marine Bioenergy: Trends and Developments*, Eds. S. K. Kim and C. G. Lee, pp. 217–244. Boca Raton, FL: CRC Press.

Konur, O., Ed. 2018. *Bioenergy and Biofuels.* Boca Raton, FL: CRC Press.

Konur, O. 2019. Cyanobacterial bioenergy and biofuels science and technology: A scientometric overview. In *Cyanobacteria: From Basic Science to Applications*, Eds. A. K. Mishra, D. N. Tiwari and A. N. Rai, pp. 419–442. Amsterdam: Elsevier.

Konur, O. 2020a. The scientometric analysis of the research on the bioethanol production from green macroalgae. In *Handbook of Algal Science, Technology and Medicine,* Ed. O. Konur, pp. 385–401. London: Academic Press.

Konur, O., Ed. 2020b. *Handbook of Algal Science, Technology and Medicine.* London: Academic Press.

Konur, O., Ed. 2021a. *Handbook of Biodiesel and Petrodiesel Fuels: Science, Technology, Health, and Environment.* Boca Raton, FL: CRC Press.

Konur, O., Ed. 2021b. *Handbook of Biodiesel and Petrodiesel Fuels: Science, Technology, Health, and Environment. Volume 1. Biodiesel Fuels: Science, Technology, Health, and Environment.* Boca Raton, FL: CRC Press.

Konur, O., Ed. 2021c. *Handbook of Biodiesel and Petrodiesel Fuels: Science, Technology, Health, and Environment. Volume 2. Biodiesel Fuels Based on the Edible and Nonedible Feedstocks, Wastes, and Algae: Science, Technology, Health, and Environment.* Boca Raton, FL: CRC Press.

Konur, O., Ed. 2021d. *Handbook of Biodiesel and Petrodiesel Fuels: Science, Technology, Health, and Environment. Volume 3. Petrodiesel Fuels: Science, Technology, Health, and Environment.* Boca Raton, FL: CRC Press.

Konur, O. 2023a. Utilization of bioethanol fuels in the transport engines: Scientometric study. In *Evaluation and Utilization of Bioethanol Fuels. I.: Evaluation of Bioethanol Fuels, Transport Engines, and Bioethanol Sensors. Handbook of Bioethanol Fuels Volume 5*, Ed. O. Konur, pp. 157–175. Boca Raton, FL: CRC Press.

Konur, O. 2023b. Bioethanol fuel sensors: Scientometric study. In *Evaluation and Utilization of Bioethanol Fuels. I.: Evaluation of Bioethanol Fuels, Transport Engines, and Bioethanol Sensors. Handbook of Bioethanol Fuels Volume 5*, Ed. O. Konur, pp. 317–334. Boca Raton, FL: CRC Press.

Konur, O. 2023c. Bioethanol fuel-based biohydrogen fuels: Scientometric study. In *Evaluation and Utilization of Bioethanol Fuels. II.: Biohydrogen Fuels, Fuel Cells, Biochemicals, and Country Experiences. Handbook of Bioethanol Fuels Volume 6*, Ed. O. Konur, pp. 215–236. Boca Raton, FL: CRC Press.

Konur, O. 2023d. Bioethanol fuel cells: Scientometric study. In *Evaluation and Utilization of Bioethanol Fuels. II.: Biohydrogen Fuels, Fuel Cells, Biochemicals, and Country Experiences. Handbook of Bioethanol Fuels Volume 6*, Ed. O. Konur, pp. 277–297. Boca Raton, FL: CRC Press.

Konur, O. 2023e. Bioethanol fuel-based biochemical production: Scientometric study. In *Evaluation and Utilization of Bioethanol Fuels. II.: Biohydrogen Fuels, Fuel Cells, Biochemicals, and Country Experiences. Handbook of Bioethanol Fuels Volume 6*, Ed. O. Konur, pp. 317–337. Boca Raton, FL: CRC Press.

Konur, O. 2023f. Nanotechnology applications in bioethanol fuels: Scientometric study. In *Evaluation and Utilization of Bioethanol Fuels. I.: Evaluation of Bioethanol Fuels, Transport Engines, and Bioethanol Sensors. Handbook of Bioethanol Fuels Volume 5*, Ed. O. Konur, pp. 120–139. Boca Raton, FL: CRC Press.

Konur, O. 2023g. Gasoline fuels: Scientometric study. In *Evaluation and Utilization of Bioethanol Fuels. I.: Evaluation of Bioethanol Fuels, Transport Engines, and Bioethanol Sensors. Handbook of Bioethanol Fuels Volume 5*, Ed. O. Konur, pp. 87–106. Boca Raton, FL: CRC Press.

Konur, O. and F. L. Matthews. 1989. Effect of the properties of the constituents on the fatigue performance of composites: A review. *Composites* 20:317–328.

Kowal, A., M. Li and M. Shao, et al. 2009. Ternary Pt/Rh/SnO_2 electrocatalysts for oxidizing ethanol to CO_2. *Nature Materials* 8:325–330.

Kumar, N., A. Dorfman and J. I. Hahm. 2006. Ultrasensitive DNA sequence detection using nanoscale ZnO sensor arrays. *Nanotechnology* 17:2875.

Kwanchareon, P., A. Luengnaruemitchai and S. Jai-In. 2007. Solubility of a diesel-biodiesel-ethanol blend, its fuel properties, and its emission characteristics from diesel engine. *Fuel* 86:1053–1061.

Lamy, C., S. Rousseau, E. M. Belgsir, C. Coutanceau and J. M. Leger. 2004. Recent progress in the direct ethanol fuel cell: Development of new platinum-tin electrocatalysts. *Electrochimica Acta* 49:3901–3908.

Larsson, S., D. Fantazzini, S. Davidsson, S. Kullander and M. Hook. 2014. Reviewing electricity production cost assessments. *Renewable and Sustainable Energy Reviews* 30:170–183.

Liang, Z. X., T. S. Zhao, J. B. Xu and L. D. Zhu. 2009. Mechanism study of the ethanol oxidation reaction on palladium in alkaline media. *Electrochimica Acta* 54:2203–2208.

Liguras, D. K., D. L. Kondarides and X. E. Verykios. 2003. Production of hydrogen for fuel cells by steam reforming of ethanol over supported noble metal catalysts. *Applied Catalysis B: Environmental* 43:345–354.

Liu, J., X. Wang, Q. Peng and Y. Li. 2005. Vanadium pentoxide nanobelts: Highly selective and stable ethanol sensor materials. *Advanced Materials* 17:764–767.

Liu, P. and E. J. M. Hensen. 2013. Highly efficient and robust $Au/MgCuCr_2O_4$ catalyst for gas-phase oxidation of ethanol to acetaldehyde. *Journal of the American Chemical Society* 135:14032–14035.

Liu, S., E. R. C. Clemente, T. Hu and Y. Wei. 2007. Study of spark ignition engine fueled with methanol/gasoline fuel blends. *Applied Thermal Engineering* 27:1904–1910.

Llorca, J., N. Homs, J. Sales and P. R. de la Piscina. 2002. Efficient production of hydrogen over supported cobalt catalysts from ethanol steam reforming. *Journal of Catalysis* 209:306–317.

Lu, L., X. Han, J. Li, J. Hua and M. Ouyang. 2013. A review on the key issues for lithium-ion battery management in electric vehicles. *Journal of Power Sources* 226:272–288.

Ma, S., Z. Zheng, H. Liu, Q. Zhang and M. Yao. 2013. Experimental investigation of the effects of diesel injection strategy on gasoline/diesel dual-fuel combustion. *Applied Energy* 109:202–212.

Ma, X., L. Sun and C. Song. 2002. A new approach to deep desulfurization of gasoline, diesel fuel and jet fuel by selective adsorption for ultra-clean fuels and for fuel cell applications. *Catalysis Today* 77:107–116.

Macedo, I. C., J. E. A. Seabra and J. E. A. R. Silva. 2008. Green house gases emissions in the production and use of ethanol from sugarcane in Brazil: The 2005/2006 averages and a prediction for 2020. *Biomass and Bioenergy* 32:582–595.

Marinov, N. M. 1999. A detailed chemical kinetic model for high temperature ethanol oxidation. *International Journal of Chemical Kinetics* 31:183–220.

Maurya, D. P., A. Singla and S. Negi. 2015. An overview of key pretreatment processes for biological conversion of lignocellulosic biomass to bioethanol. *3 Biotech* 5:597–609.

Mehl, M., W. J., C. K. Westbrook and H. J. Curran. 2011. Kinetic modeling of gasoline surrogate components and mixtures under engine conditions. *Proceedings of the Combustion Institute* 33:193–200.

Meisel, S. L., J. P. McCullough, C. H. Lechthaler and P. B. Weisz. 1976. Gasoline from methanol in one step. *Chemische Technik* 6:86–89.

Moldovan, M., M. A. Palacios and M. M. Gomez, et al. 2002. Environmental risk of particulate and soluble platinum group elements released from gasoline and diesel engine catalytic converters. *Science of the Total Environment* 296:199–208.

Morgan, N., A. Smallbone and A. Bhave, et al. 2010. Mapping surrogate gasoline compositions into RON/MON space. *Combustion and Flame* 157:1122–1131.

Morschbacker, A. 2009. Bio-ethanol based ethylene. *Polymer Reviews* 49:79–84.

Muller, J. F. and T. Stavrakou. 2005. Inversion of CO and NO_x emissions using the adjoint of the IMAGES model. *Atmospheric Chemistry and Physics* 5:1157–1186.

Murdoch, M., G. I. N. Waterhouse and M. A. Nadeem, et al. 2011. The effect of gold loading and particle size on photocatalytic hydrogen production from ethanol over Au/TiO_2 nanoparticles. *Nature Chemistry* 3:489–492.

Najafi, G., B. Ghobadian and T. Tavakoli, et al. 2009. Performance and exhaust emissions of a gasoline engine with ethanol blended gasoline fuels using artificial neural network. *Applied Energy* 86:630–639.

Newman, P. W. G. and J. R. Kenworthy. 1989. Gasoline consumption and cities: A comparison of U.S. cities with a global survey. *Journal of the American Planning Association* 55:24–37.

Nhuchhen, D. R. and P. A. Salam. 2012. Estimation of higher heating value of biomass from proximate analysis: A new approach. *Fuel* 99:55–63.

Ni, M., D. Y. C. Leung and M. K. H. Leung. 2007. A review on reforming bio-ethanol for hydrogen production. *International Journal of Hydrogen Energy* 32:3238–3247.

North, D. C. 1991. Institutions. *Journal of Economic Perspectives* 5:97–112.

Ogden, J. M., M. M. Steinbugler and T. G. Kreutz. 1999. Comparison of hydrogen, methanol and gasoline as fuels for fuel cell vehicles: implications for vehicle design and infrastructure development. *Journal of Power Sources* 79:143–168.

Palo, D. R., R. A. Dagle and J. D. Holladay. 2007. Methanol steam reforming for hydrogen production. *Chemical Reviews* 107:3992–4021.

Raghuvanshi, S. P., A. Chandra and A. K. Raghav. 2006. Carbon dioxide emissions from coal based power generation in India. *Energy Conversion and Management* 47:427–441.

Ramos-Paja, C. A., C. Bordons, C., A. Romero, R. Giral and L. Martínez-Salamero. 2008. Minimum fuel consumption strategy for PEM fuel cells. *IEEE Transactions on Industrial Electronics* 56:685–696.

Rousseau, S., C. Coutanceau, C., Lamy and J. M. Leger. 2006. Direct ethanol fuel cell (DEFC): Electrical performances and reaction products distribution under operating conditions with different platinum-based anodes. *Journal of Power Sources* 158:18–24.

Sayin, C., H. M. Ertunc, M. Hosoz, I. Kilicaslan and M. Canakci. 2007. Performance and exhaust emissions of a gasoline engine using artificial neural network. *Applied Thermal Engineering* 27:46–54.

Song, C. 2003. An overview of new approaches to deep desulfurization for ultra-clean gasoline, diesel fuel and jet fuel. *Catalysis Today* 86:211–263.

Stanglmaier, R. H., J. Li and R. D. Matthews. 1999. The effect of in-cylinder wall wetting location on the HC emissions from SI engines. SAE Transactions 1999:01-0502.

Steele, B. C. H. and A. Heinzel. 2001. Materials for fuel-cell technologies. *Nature* 414:345–352.

Thiele, E. W. 1939. Relation between catalytic activity and size of particle. *Industrial & Engineering Chemistry* 31:916–920.

Thomas, C. E., B. D. James, F. D. Lomax and I. F. Kuhn. 2000. Fuel options for the fuel cell vehicle: Hydrogen, methanol or gasoline? *International Journal of Hydrogen Energy* 25:551–567.

Tian, N., Z. Y. Zhou, N. F. Yu, L. Y. Wang and S. G. Sun. 2010. Direct electrodeposition of tetrahexahedral Pd nanocrystals with high-index facets and high catalytic activity for ethanol electrooxidation. *Journal of the American Chemical Society* 132:7580–7581.

Tong, H., S. Ouyang and Y. Bi, et al. 2012. Nano-photocatalytic materials: Possibilities and challenges. *Advanced Materials* 24:229–251.

Vigier, F., C. Coutanceau, F. Hahn, E. M. Belgsir and C. Lamy. 2004. On the mechanism of ethanol electrooxidation on Pt and PtSn catalysts: Electrochemical and *in situ* IR reflectance spectroscopy studies. *Journal of Electroanalytical Chemistry* 563:81–89.

Wan, Q., Q. H. Li and Y. J. Chen, et al. 2004. Fabrication and ethanol sensing characteristics of ZnO nanowire gas sensors. *Applied Physics Letters* 84:3654–3656.

Wang, H., Z. Jusys and R. J. Behm. 2004. Ethanol electrooxidation on a carbon-supported Pt catalyst: Reaction kinetics and product yields. *Journal of Physical Chemistry B* 108:19413–19424.

Xu, C., H. Wang, P. K. Shen and S. P. Jiang. 2007b. Highly ordered Pd nanowire arrays as effective electrocatalysts for ethanol oxidation in direct alcohol fuel cells. *Advanced Materials* 19:4256–4259.

Xu, C., L. Cheng, P. Shen and Y. Liu. 2007a. Methanol and ethanol electrooxidation on Pt and Pd supported on carbon microspheres in alkaline media. *Electrochemistry Communications* 9:997–1001.

Yunoki, S. and M. Saito. 2009. A simple method to determine bioethanol content in gasoline using two-step extraction and liquid scintillation counting. *Bioresource Technology* 100:6125–6128.

Zhou, W., Z. Zhou and S. Song, et al. 2003. Pt based anode catalysts for direct ethanol fuel cells. *Applied Catalysis B: Environmental* 46:273–285.

Zhu, N., Q. Zhao, L. Tian and Q. Zhang. 2016. Cost analysis and development strategies for China' natural gas power generation industry under the situation of energy price's reformation. *Energy Procedia* 104:203–208.

Part 28

Bioethanol Fuel Sensors

92 Bioethanol Fuel Sensors
Scientometric Study

Ozcan Konur
(Formerly) Ankara Yildirim Beyazit University

92.1 INTRODUCTION

Crude oil-based gasoline fuels (Ma et al., 2002; Newman and Kenworthy, 1989) and petrodiesel fuels (Bosmann et al., 2001; Ma et al., 1994) have been widely used in the transportation sector since the 1920s. However, there have been great public concerns over the adverse environmental impact and sustainability of these fuels (Hill et al., 2009; Schauer et al., 2002).

Hence, biomass-based bioethanol fuels (Konur, 2012e, 2105, 2020a) have increasingly been used in blending gasoline fuels (Hsieh et al., 2002; Najafi et al., 2009) and petrodiesel fuels in recent years (Hansen et al., 2005; Li et al., 2005). In this context, there has been a significant focus on bioethanol fuel sensors (Murdoch et al., 2011; Wan et al., 2004). There are three primary research fronts in this field: zinc oxide (ZnO)-based bioethanol fuel sensors (Wan et al., 2004; Wang et al., 2012), tin oxide (SnO_2)-based bioethanol fuel sensors (Chen et al., 2005, 2006), and other material-based bioethanol fuel sensors (Choi, et al., 2010; Choudhury, 2009). Furthermore, there are further research fronts within each of these research fronts: nanomaterial-based sensors and conventional material-based sensors (Hellegouarc'h et al., 2001; Paraguay et al., 2000).

In the meantime, the research in nanomaterials and nanotechnology has intensified in recent years to become a major research field in scientific research with over one and a half million published papers (Geim, 2009; Geim and Novoselov, 2007). In this context, a large number of nanomaterials have been developed nearly for every research field. These materials offer an innovative way to increase the efficiency in the production and utilization of bioethanol fuels as in the other scientific fields (Konur, 2016a–f, 2017a–e, 2019, 2021a,b). Similarly, there has been intense research in the field of sensors in recent years (Brolo, 2012; Chaubey and Malhotra, 2002; Liu and Zeng, 2003).

However, it is essential to develop efficient incentive structures (North, 1991) for the primary stakeholders to enhance the research in this field (Konur, 2000, 2002a–c, 2006a,b, 2007a,b). The scientometric analysis has been used in this context to inform the primary stakeholders about the current state of the research in a selected research field (Garfield, 1955; Konur, 2011, 2012a–i, 2015, 2016a–f, 2017a–f, 2018b, 2019, 2020a).

As there has not been any scientometric study on bioethanol fuel sensors, this book chapter presents a scientometric study of the research in bioethanol fuel sensors. It examines the scientometric characteristics of both the sample and population data presenting the scientometric characteristics of these both datasets in the order of documents, authors, publication years, institutions, funding bodies, source titles, countries, Scopus subject categories, keywords, and research fronts.

92.2 MATERIALS AND METHODS

The search for this study was carried out using the Scopus database (Burnham, 2006) in August 2021.

As a first step for the search of the relevant literature, the keywords were selected using the first 200 most-cited papers. The selected keyword list was optimized to obtain a representative sample of papers for the searched research field. This keyword list was provided in the appendix for future

DOI: 10.1201/9781003226567-124

replication studies. Additionally, information about the most-used keywords was given in Section 92.3.9 to highlight the key research fronts in Section 92.3.10.

As a second step, two sets of data were used for this study. First, a population sample of over 1,800 papers was used to examine the scientometric characteristics of the population data. Second, a sample of 100 most-cited papers was used to examine the scientometric characteristics of these citation classics with over 95 citations each.

The scientometric characteristics of these both sample and population datasets were presented in the order of documents, authors, publication years, institutions, funding bodies, source titles, countries, Scopus subject categories, keywords, and research fronts.

Lastly, the key scientometric findings for both datasets were discussed to highlight the research landscape for the bioethanol fuel sensors. Additionally, several brief conclusions were drawn and many relevant recommendations were made to enhance the future research landscape.

92.3 RESULTS

92.3.1 The Most Prolific Documents in Bioethanol Fuel Sensors

The information on the types of documents for both datasets is given in Table 92.1. The articles dominate both the sample and population datasets. The articles and conference papers have a slight surplus and deficit, respectively.

It is also interesting to note that all of the papers in the sample dataset were published in journals, while only 97.4% of the papers were published in journals for the population dataset. Furthermore, 2.5% and 0.1% of the population papers were published in books and book series, respectively.

92.3.2 The Most Prolific Authors in Bioethanol Fuel Sensors

The information about the 30 most prolific authors with at least two sample papers and five population papers each is given in Table 92.2.

The most prolific authors are Yujin Chen, Jong-Heun Lee, and Taihong Wang with at least eight sample papers each. Chunling Zhu, X.Y. Xue, Supab Choopun, Il-Doo Kim, and Y.G. Wang follow these top authors with at least four sample papers each.

The most prolific institution for the sample dataset is Jilin University with five authors. Chiang Mai University, Chinese Academy of Sciences, and Harbin Engineering University follow this top institution with three authors each. In total, 18 institutions house these authors.

TABLE 92.1
Documents in Bioethanol Fuel Sensors

Documents	Sample Dataset (%)	Population Dataset (%)	Surplus (%)
Article	95	94.5	0.5
Conference paper	4	4.7	−0.7
Book chapter	0	0.2	−0.2
Review	1	0.3	0.7
Note	0	0.2	−0.2
Letter	0	0.0	0
Editorial	0	0.0	0
Book	0	0.0	0
Short survey	0	0.1	−0.1
Sample size	100	1853	

TABLE 92.2
Most Prolific Authors in Bioethanol Fuel Sensors

No.	Author Name	Author Code	Sample Papers	Population Papers	Institution	Country
1	Chen, Yujin	7601437135	11	12	Harbin Inst. Technol.	China
2	Lee, Jong-Heun	26643283000	8	18	Korea Univ.	S. Korea
3	Wang, Taihong	35241217600	8	12	Chinese Acad. Sci.	China
4	Zhu, Chunling	7403439505	5	11	Harbin Eng. Univ.	China
5	Xue, X.Y.	13611568700	5	5	Harbin Eng. Univ.	China
6	Choopun, Supab	6701740305	4	15	Chiang Mai Univ.	Thailand
7	Kim, Il-Doo	36037925300	4	6	Korea Adv. Inst. Sci. Technol.	S. Korea
8	Wang, Y.G.	55954567600	4	5	Harbin Eng. Univ.	China
9	Li, Jianping	55142372000	3	17	Chinese Acad. Sci.	China
10	Rahman, Muhammad M	56397398200	3	8	Najran Univ.	S. Arabia
11	Faisal, Muhammad	35617425400	3	7	Najran Univ.	S. Arabia
12	Hongsith, Niyom	8653765100	3	6	Univ. Phayao	Thailand
13	Khan, Sher Bahadar	36059229400	3	6	King Abdulaziz Univ.	S. Arabia
14	Mangkorntong, Nikorn	6701740305	3	5	Chiang Mai Univ.	Thailand
15	Mangkorntong, Pongsri	6602631737	3	5	Chiang Mai Univ.	Thailand
16	He, Xiuli	56843987000	3	5	Chinese Acad. Sci.	China
17	Mortazavi, Yadollah	35563647800	2	21	Univ. Tehran	Iran
18	Liu, Jiangyang	56381725600	2	18	Jilin Univ.	China
19	Lu, Geyu	7403460117	2	14	Jilin Univ.	China
20	Song, Peng	57188675724	2	11	Univ. Jinan	China
21	Hsueh, Ting-Jen	9336934600	2	10	Natl. Cheng Kung Univ.	Taiwan
22	Wang, Lili	57037822000	2	10	Jilin Univ.	China
23	Sun, Peng	35254303000	2	10	Jilin Univ.	China
24	Lou, Zheng	36739423500	2	9	Jilin Univ.	China
25	Chang, Shoou-Jinn	57221300233	2	9	Natl. Cheng Kung Univ.	Taiwan
26	Neri, Giovanni	23068135700	2	8	Univ. Messina	Italy
27	Van Hieu, Nguyen	37063946700	2	8	Hanoi Univ. Sci. Technol.	Vietnam
28	Hsu, Cheng-Liang	7404946445	2	6	Natl. Univ. Tainan	Taiwan
29	Llobet, Eduard	35518506500	2	6	Univ. Rovira	Spain
30	Miura, Norio	7401868774	2	5	Kyushu Univ.	Japan

Author Code: the unique code given by Scopus to the authors. Sample Papers: the number of papers authored in the sample dataset. Population Papers: the number of papers authored in the population dataset.

The most prolific country for the sample dataset is China with 13 authors. Thailand, Saudi Arabia, Taiwan, and South Korea (S. Korea) follow this top country with at least two authors each. In total, 10 countries house these authors.

However, there is a significant gender deficit (Beaudry and Lariviere, 2016) for the sample dataset as surprisingly nearly all of these top researchers are male.

92.3.3 The Most Prolific Research Output by Years in Bioethanol Fuel Sensors

Information about papers published between 1970 and 2021 is given in Figure 92.1. This figure clearly shows that the bulk of the research papers in the population dataset were published primarily in the 2010s with 65.5% of the population datasets. This was followed by the early 2020s and 2000s

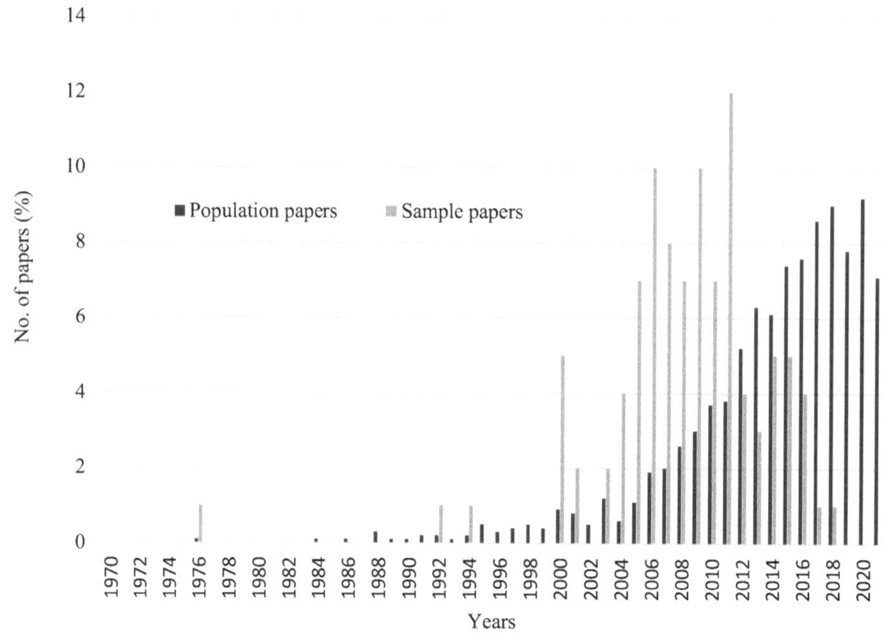

FIGURE 92.1 Research output by years regarding bioethanol fuel sensors.

with 16.3% and 14.6% of the population papers, respectively. The publication rates for the pre-2000s were negligible.

Similarly, the bulk of the research papers in the sample dataset were published in the 2000s and 2010s with 55% and 42% of the sample datasets, respectively. The publication rates for the pre-2000s were negligible.

The most prolific publication years for the population dataset were after 2014 with at least 7.1% each of the dataset. Similarly, 61% of sample papers were published between 2005 and 2011.

92.3.4 The Most Prolific Institutions in Bioethanol Fuel Sensors

Information about the 22 most prolific institutions publishing papers on bioethanol fuel sensors with at least two sample papers and 0.3% of the population papers each is given in Table 92.3.

The most prolific institutions are the Chinese Academy of Sciences and Korea University with 14 and 9 papers, respectively. Jilin University, Harbin Engineering University, and Chiang Mai University follow these top institutions with six sample papers each.

The top country for these most prolific institutions is China with nine institutions. Next, S. Korea and Taiwan follow China with three and two institutions, respectively. In total, 11 countries house these top institutions.

However, the institutions with the most citation impact are the Chinese Academy of Sciences and Korea University with 9.7% and 7.7% surplus, respectively. The Harbin Engineering University and Chiang Mai University follow these top institutions with at least 4.6% surplus each. Similarly, the institutions with the least impact are Chongqing University, Inha University, and University of Tehran with at least 0.5% surplus each.

92.3.5 The Most Prolific Funding Bodies in Bioethanol Fuel Sensors

Information about the seven most prolific funding bodies funding at least two sample papers and 0.9% of the population papers each is given in Table 92.4.

TABLE 92.3
Most Prolific Institutions in Bioethanol Fuel Sensors

No.	Institutions	Country	Sample Papers (%)	Population Papers (%)	Surplus (%)
1	Chinese Acad. Sci.	China	14	4.3	9.7
2	Korea Univ.	S. Korea	9	1.3	7.7
3	Jilin Univ.	China	6	4.2	1.8
4	Harbin Eng. Univ.	China	6	1.4	4.6
5	Chiang Mai Univ.	China	6	1.2	4.8
6	Univ. Tehran	Iran	2	1.5	0.5
7	Chongqing Univ.	China	2	1.5	0.5
8	Inha Univ.	S. Korea	2	1.4	0.6
9	Northwest Normal Univ.	China	2	1.1	0.9
10	Najran Univ.	S. Arabia	2	1.1	0.9
11	Hanoi Univ. Sci. Technol.	Vietnam	2	0.8	1.2
12	Shenzhen Univ.	China	2	0.6	1.4
13	Shandong Univ.	China	2	0.6	1.4
14	Nankai Univ.	China	2	0.6	1.4
15	Univ. Messina	Italy	2	0.5	1.5
16	Kyushu Univ.	Japan	2	0.4	1.6
17	Natl. Univ. Tainan	Taiwan	2	0.4	1.6
18	Nanyang Technol. Univ.	Singapore	2	0.4	1.6
19	Univ. Washington	USA	2	0.3	1.7
20	Univ. Rovira	Spain	2	0.3	1.7
21	Ind. Technol. Res. Inst.	Taiwan	2	0.3	1.7
22	Korea Inst. Sci. Technol.	S. Korea	2	0.3	1.7

TABLE 92.4
Most Prolific Funding Bodies in Bioethanol Fuel Sensors

No.	Funding Bodies	Country	Sample Paper No. (%)	Population Paper No. (%)	Surplus (%)
1	National Natural Science Foundation of China	China	19	23.4	−4.4
2	Ministry of Education	China	7	5.1	1.9
3	Ministry of Science and Technology	China	3	2.4	0.6
4	China Postdoctoral Science Foundation	China	3	1.5	1.5
5	Chinese Academy of Sciences	China	3	0.9	2.1
6	European Commission	EU	2	1.1	0.9
7	National Science Foundation	USA	2	0.9	1.1

The most prolific funding body is the National Natural Science Foundation of China with 19 sample papers. The Ministry of Education follows this top funding body with seven sample papers. It is notable that 42% and 53.3% of the sample and population papers are funded, respectively.

The most prolific country for these top funding bodies is China with five funding bodies. In total, China, the USA, and European Union house these top funding bodies. The citation impact of these funding bodies is also shown in this table.

92.3.6 The Most Prolific Source Titles in Bioethanol Fuel Sensors

Information about the 12 most prolific source titles publishing at least two sample papers and 0.5% of the population papers each in bioethanol fuel sensors is given in Table 92.5.

TABLE 92.5
Most Prolific Source Titles in Bioethanol Fuel Sensors

No.	Source Titles	Sample Papers (%)	Population Papers (%)	Surplus (%)
1	Sensors and Actuators B Chemical	39	19.9	19.1
2	Nanotechnology	7	1.8	5.2
3	ACS Applied Materials and Interfaces	7	1.2	5.8
4	Applied Physics Letters	7	0.7	6.3
5	Chemical Communications	4	0.6	3.4
6	Journal of Physical Chemistry C	4	0.6	3.4
7	Ceramics International	3	2.0	1.0
8	Analytical Chemistry	3	0.6	2.4
9	Applied Surface Science	2	2.9	–0.9
10	Materials Chemistry and Physics	2	1.0	1.0
11	Thin Solid Films	2	0.7	1.3
12	Chemical Physics Letters	2	0.5	1.5

TABLE 92.6
Most Prolific Countries in Bioethanol Fuel Sensors

No.	Countries	Sample Papers (%)	Population Papers (%)	Surplus (%)
1	China	47	42.8	4.2
2	S. Korea	11	6.2	4.8
3	USA	10	5.2	4.8
4	India	5	14.8	–9.8
5	Japan	5	3.6	1.4
6	Italy	4	3.3	0.7
7	Thailand	4	1.8	2.2
8	Iran	3	5.2	–2.2
9	Taiwan	3	3.5	–0.5
10	Australia	3	1.3	1.7
11	Vietnam	3	1.2	1.8
12	S. Arabia	2	2.1	–0.1
13	France	2	1.6	0.4
14	Spain	2	1.5	0.5

The most prolific source title is 'Sensors and Actuators B Chemical' with 39 sample papers. The 'Nanotechnology', 'ACS Applied Materials and Interfaces', and 'Applied Physics Letters' follow this top journal with seven papers each.

However, the source title with the most citation impact is 'Sensors and Actuators B Chemical' with 19.1% surplus. The 'Applied Physics Letters', 'ACS Applied Materials and Interfaces', and 'Nanotechnology' follow this top journal with at least 5.2% surplus each. Similarly, the source titles with the least impact are 'Applied Surface Science', 'Ceramics International', and 'Materials Chemistry and Physics' with at least 1% deficit each. It is notable that seven of these top source titles are related to materials science.

92.3.7 The Most Prolific Countries in Bioethanol Fuel Sensors

Information about the 14 most prolific countries publishing at least two sample papers and 1.5% of the population papers each in bioethanol fuel sensors is given in Table 92.6.

The most prolific country is China with 47 sample papers, and the USA and S. Korea follow China with 11 and 10 sample papers, respectively. However, the countries with the most citation impact are S. Korea and the USA with 4.8% publication surplus each. Similarly, the country with the least citation impact is India with 9.8% citation deficit. Iran, Taiwan, and Saudi Arabia are the other countries with the least citation impact.

92.3.8 The Most Prolific Scopus Subject Categories in Bioethanol Fuel Sensors

Information about the six most prolific Scopus subject categories indexing at least 5% and 2.3% of the sample and population papers, respectively, is given in Table 92.7.

The most prolific Scopus subject category in bioethanol fuel sensors is 'Materials Science' with 82 sample papers. This top category is followed by 'Physics and Astronomy' and 'Engineering' with 63 and 51 sample papers, respectively.

However, the Scopus subject category with the most citation impact is 'Materials Science'. Similarly, the Scopus subject categories with the least citation impact are 'Chemistry' and 'Engineering' with at least 2.5% deficit each.

92.3.9 The Most Prolific Keywords in Bioethanol Fuel Sensors

Information about the keywords used in at least 5% of the sample and population papers each is given in Table 92.8. For this purpose, keywords related to the keyword set given in the appendix are selected from a list of the most prolific keyword set provided by the Scopus database.

These keywords are grouped under four headings: bioethanol fuels, nanomaterials, sensors, and materials.

There are six keywords used related to bioethanol fuels. The prolific keywords are ethanol, ethanol sensors, and ethanol sensing. It is notable that the bioethanol keyword does not appear in these lists.

The prolific keywords related to nanomaterials are nanostructured materials, nanorods, and nanowires. Prolific keywords related to the sensors are chemical sensors, sensors, gas sensors, and gas detectors. The keywords related to the materials are zinc oxide, ZnO, tin dioxides, and tin compounds.

However, the keywords with the most citation impact are nanostructured materials, sensors, and chemical sensors. Similarly, the keywords with the least citation impact are chemical detection, gas sensing electrodes, and gas detectors.

92.3.10 The Most Prolific Research Fronts in Bioethanol Fuel Sensors

Information about the most prolific research fronts for sample papers in bioethanol fuel sensors is given in Table 92.9.

TABLE 92.7
Most Prolific Scopus Subject Categories in Bioethanol Fuel Sensors

No.	Scopus Subject Categories	Sample Papers (%)	Population Papers (%)	Surplus (%)
1	Materials Science	82	67.9	14.1
2	Physics and Astronomy	63	57.8	5.2
3	Engineering	51	53.5	–2.5
4	Chemistry	27	31.1	–4.1
5	Chemical Engineering	17	14.6	2.4
6	Energy	5	2.3	2.7

TABLE 92.8
Most Prolific Keywords in Bioethanol Fuel Sensors

No.	Keywords	Sample Papers (%)	Population Papers (%)	Surplus (%)
1.	**Bioethanol Fuels**			
	Ethanol	84	84.5	−0.5
	Ethanol sensors	25	19.1	5.9
	Ethanol sensing	15	14.7	0.3
	Ethanol vapor	6	0.0	6
	Ethanol concentrations	0	8.0	−8
	Ethanol gas sensors	0	6.0	−6
2.	**Nanomaterials**			
	Nanostructured materials	21	0.0	21
	Nanorods	13	7.7	5.3
	Nanowires	11	5.0	6
	Nanostructures	7	6.4	0.6
	Nanoparticles	6	10.4	−4.4
	Electrospinning	6	0.0	6
	Single crystals	5	0.0	5
	Carbon nanotubes	5	0.0	5
	Nanomaterials	5	0.0	5
	Nanocomposites	0	5.5	−5.5
3.	**Sensors**			
	Chemical sensors	48	29.2	18.8
	Sensors	35	14.9	20.1
	Gas sensors	26	21.0	5
	Gas detectors	20	29.8	−9.8
	Chemical detection	8	19.3	−11.3
	Gas sensing	7	7.0	0
	Gas sensing properties	6	12.0	−6
	Sensing property	5	0.0	5
	Gas sensing electrodes	0	10.8	−10.8
	Sensing performance	0	6.4	−6.4
4.	**Materials**			
	Zinc oxide	26	16.0	10
	ZnO	15	6.6	8.4
	Tin dioxides	14	0.0	14
	Tin compounds	13	0.0	13
	Titanium dioxide	8	0.0	8
	Additives	6	0.0	6
	Tin	5	0.0	5

There are three primary research fronts in this field: ZnO-based bioethanol fuel sensors (Wan et al., 2004; Wang et al., 2012), SnO_2-based bioethanol fuel sensors (Chen et al., 2005, 2006), and other material-based bioethanol fuel sensors (Choi, et al., 2010; Choudhury, 2009) with 40, 28, and 47 sample papers, respectively. Furthermore, there are further research fronts within each of these research fronts: nanomaterial-based sensors and conventional material-based sensors

TABLE 92.9
Most Prolific Research Fronts in Bioethanol Fuel Sensors

No.	Research Fronts	Sample Papers (%)
1	ZnO-based bioethanol fuel sensors	40
	Nanomaterial-based ZnO	35
	Conventional material-based ZnO	5
2	SnO_2-based bioethanol fuel sensors	28
	Nanomaterial-based SnO_2	24
	Conventional material-based SnO_2	4
3	Other material-based bioethanol fuel sensors	47
	Nanomaterial-based sensors	39
	Conventional material-based sensors	8

The total number of reviewed papers = *25. Sample Papers:* the sample of the 100 most-cited papers.

(Hellegouarc'h et al., 2001; Paraguay et al., 2000). In total, there are 98 and 17 papers for the nanomaterial-based sensors and conventional material-based sensors, respectively.

The focus of these 100 most-cited papers is the development of nanomaterials for the most efficient processes for each of these research fronts.

92.4 DISCUSSION

92.4.1 Introduction

Crude oil-based gasoline and petrodiesel fuels have been widely used in the transportation sector. However, there have been great public concerns over the adverse environmental impact of these fuels. Hence, biomass-based bioethanol fuels have increasingly been used in blending gasoline fuels and petrodiesel fuels in recent years. There has been a significant focus on bioethanol fuel sensors. However, it is essential to develop efficient incentive structures for the primary stakeholders to enhance the research in this field.

The scientometric analysis has been traditionally used to inform the primary stakeholders about the current state of the research in a selected research field. Although there have been several scientometric studies on bioethanol fuels, there has been no scientometric study of the research in bioethanol fuel sensors. This book chapter presents a scientometric study of the research in this field. It examines the scientometric characteristics of the sample and population data presenting the scientometric characteristics of these both datasets in the order of documents, authors, publication years, institutions, funding bodies, source titles, countries, Scopus subject categories, keywords, and research fronts.

The search for this study was carried out using the Scopus database in August 2021. As a first step for the search of the relevant literature, the keywords were selected using the first 200 most-cited papers. The selected keyword list was optimized to obtain a representative sample of papers for the searched research field. This keyword list was provided in the appendix for future replication studies. Additionally, information about the most-used keywords was given in Section 92.3.9 to highlight the key research fronts in Section 92.3.10.

As a second step, two sets of data were used for this study. First, a population sample of over 1,800 papers was used to examine the scientometric characteristics of the population data. Second, a sample of 100 most-cited papers was used to examine the scientometric characteristics of these citation classics with over 95 citations each.

92.4.2 The Most Prolific Documents in Bioethanol Fuel Sensors

The articles dominate both the sample and population datasets. The articles and conference papers have a slight surplus and deficit, respectively. Scopus differs from the Web of Science database in differentiating and showing articles and conference papers published in journals separately.

It is observed during the search process that there has been inconsistency in the classification of the documents in Scopus and in other databases such as Web of Science. This is especially relevant for the classification of papers as reviews or articles as papers not involving a literature review may be erroneously classified as a review paper. There is also a case of review papers being classified as articles. In this context, it would be helpful to provide a classification note for the published papers in the books and journals at the first instance. It would also be helpful to use the document types listed in Table 92.1 for this purpose. Book chapters may also be classified as articles or reviews as an additional classification to differentiate review chapters from experimental chapters as it is done by the Web of Science. It would be further helpful to additionally classify the conference papers as articles or review papers and it is done in the Web of Science database.

92.4.3 The Most Prolific Authors in Bioethanol Fuel Sensors

There have been 30 most prolific authors with at least two sample papers and five population papers each as given in Table 92.2. These authors have shaped the development of the research in this field.

The most prolific authors are Yujin Chen, Jong-Heun Lee, Taihong Wang, Chunling Zhu, X.Y. Xue, Supab Choopun, Il-Doo Kim, and Y.G. Wang (Table 92.2).

It is important to note the inconsistencies in the indexing of the author names in Scopus and other databases. It is especially an issue for names with more than two components such as 'Judge Alex de Camp Sirous'. The probable outcomes are 'Sirous J.A.D.C.', 'de Camp Sirous, J.A.', or 'Camp Sirous, J.A.D.'. The first choice is the gold standard of the publishing sector as the last word in the name is taken as the last name. The second choice is a strong alternative, while the last choice is an undesired outcome as two last words are taken as the last name. It is notable that inconsistent indexing of the author names may cause substantial inefficiencies in the search process for the papers and allocating credit to the authors as there are different author entries for each outcome in the databases.

There is also a case of shortening Chinese names. For example, 'Yuoyang Wang' is often shortened as 'Wang, Y.', 'Wang, Y.-Y.', and 'Wang Y.Y.' as it is done in the Web of Science database as well. However, the gold stand in this case is 'Wang Y' where the last word is taken as the last name and the first word is taken as a single forename. Therefore, there have been difficulties to locate papers, for example, by Yujin Chen, Chunling Zhu, and Taihong Wang where there is inconsistency in the shortening of these names. In such cases, the use of the unique author codes provided for each author by the Scopus database has been helpful. However, it makes sense to use the third option to differentiate Chinese names efficiently: 'Wang Y.Y.'.

There is also a difficulty in allowing credit for the authors, especially for the authors with common names such as 'Wang, Y.', or 'Huang, Y.' or 'Zhu, Y.' in conducting scientometric studies. These difficulties strongly influence the efficiency of scientometric studies and allocating credit to the authors as there are the same author entries for different authors with the same name, e.g., 'Wang Y.' in the databases.

In this context, the coding of authors in the Scopus database is a welcome innovation compared with other databases such as Web of Science. In this process, Scopus allocates a unique number to each author in the database. However, there might still be substantial inefficiencies in this coding system, especially for common names. It is possible that Scopus uses many software programs to differentiate the author names (Shin et al., 2014).

In this context, it does not help that author names are not given in full in some journals. This makes it difficult to differentiate authors with common names and makes scientometric studies further difficult in the author domain. Therefore, the author names should be given in all books and journals at the first instance.

There are also inconsistencies in naming of the authors with more than two components by themselves in journal papers and book chapters. For example, 'Faaij, A.P.C', 'Wyman C.E.', and 'Shuai, S.J.' might be given as 'Faaij, A.', 'Wyman C.', or 'Shuai S.' in the journals and books. This also makes scientometric studies difficult in the author domain. Hence, contributing authors should use their name consistently in their publications.

The other critical issue regarding the author names is the spelling of the author names in the national spellings (e.g., Gonçalves, Übeiro) rather than in the English spellings (e.g., Goncalves, Ubeiro) in the Scopus database. Scopus differs from the Web of Science database in this respect where the author names are given only in English spellings. It is observed that the national spellings of the author names do not help in conducting scientometric studies and in allocating credits to the authors as sometimes there are different author entries for the English and national spellings in the Scopus database.

The most prolific institutions for the sample dataset are Jilin University, Chiang Mai University, the Chinese Academy of Sciences, and Harbin Engineering University. Similarly, the most prolific countries for the sample dataset are China, Thailand, Saudi Arabia, Taiwan, and S. Korea.

It is also notable that there is a significant gender deficit for the sample dataset as surprisingly nearly all of these top researchers are male. This finding is the most thought-provoking with strong public policy implications. Hence, institutions, funding bodies, and policymakers should take efficient measures to reduce the gender deficit in this field and other scientific fields with strong gender deficit.

92.4.4 The Most Prolific Research Output by Years in Bioethanol Fuel Sensors

The research output observed between 1970 and 2021 is illustrated in Figure 92.1. This figure clearly shows that the bulk of the research papers in the population dataset were published primarily in the 2010s. This was followed by the early 2020s and 2000s. The publication rates for the pre-2000s were negligible.

Similarly, the bulk of the research papers in the sample dataset were published in the 2000s and 2010s. The publication rates for the pre-2000s were negligible.

These data suggest that the most-cited sample papers were primarily published in the 2000s and 2010s, while the population papers were primarily published in the 2010s. These are thought-provoking findings as there has been no significant research in this field in the pre-2000s, but there has been a significant research boom in the last two decades. In this context, the increasing public concerns about climate change (Change, 2007), greenhouse gas emissions (Carlson et al., 2017), and global warming (Kerr, 2007) have been certainly behind the boom in the research in this field in the last two decades.

The data in Figure 92.1 also suggest that the research in this field has boomed in the last two decades and the size of the population papers is likely to more than double in the current decade, provided that the public concerns about climate change, greenhouse gas emissions, and global warming are translated efficiently to the research funding in this field.

92.4.5 The Most Prolific Institutions in Bioethanol Fuel Sensors

The 22 most prolific institutions publishing papers on bioethanol fuel sensors with at least two sample papers and 0.3% of the population papers each given in Table 92.3 have shaped the development of the research in this field.

The most prolific institutions are the Chinese Academy of Sciences, Korea University, Jilin University, Harbin Engineering University, and Chiang Mai University. The top countries for these most prolific institutions are China, S. Korea, and Taiwan.

However, thc institutions with the most citation impact are the Chinese Academy of Sciences, Korea University, Harbin Engineering University, and Chiang Mai University. Similarly, the institutions with the least impact are Chongqing University, Inha University, and University of Tehran.

92.4.6 The Most Prolific Funding Bodies in Bioethanol Fuel Sensors

The seven most prolific funding bodies funding at least two sample papers and 0.9% of the population papers each are given in Table 92.4.

The most prolific funding bodies are the National Natural Science Foundation of China and the Ministry of Education of China. It is notable that 42% and 53% of the sample and population papers are funded, respectively. The most prolific country for these top funding bodies is China. In total, China, the USA, and European Union house these top funding bodies.

These findings on the funding of the research in this field suggest that the level of funding, mostly in the 2010s, is modest, and they has been largely instrumental in enhancing the research in this field (Ebadi and Schiffauerova, 2016) in light of North's institutional framework (North, 1991).

92.4.7 The Most Prolific Source Titles in Bioethanol Fuel Sensors

The 12 most prolific source titles publishing at least two sample papers and 0.5% of the population papers each in bioethanol fuel sensors have shaped the development of the research in this field (Table 92.5). The most prolific source titles are 'Sensors and Actuators B Chemical', 'Nanotechnology', 'ACS Applied Materials and Interfaces', and 'Applied Physics Letters'.

However, the source titles with the most citation impact are 'Sensors and Actuators B Chemical', 'Applied Physics Letters', 'ACS Applied Materials and Interfaces', and 'Nanotechnology'. Similarly, the source titles with the least impact are 'Applied Surface Science', 'Ceramics International', and 'Materials Chemistry and Physics'.

It is notable that seven of these top source titles are related to materials science. This finding suggests that the journals in this field have significantly shaped the development of the research in this field as they focus on the development of nanomaterials and conventional materials for efficient bioethanol fuel sensors.

92.4.8 The Most Prolific Countries in Bioethanol Fuel Sensors

The 14 most prolific countries publishing at least two papers and 1.5% of the population papers each have significantly shaped the development of the research in this field (Table 92.6).

The most prolific countries are China, the USA, and S. Korea. However, the countries with the most citation impact are China and the USA. Similarly, the countries with the least citation impact are Iran, Taiwan, and Saudi Arabia.

A close examination of these findings suggests that the USA, China, S. Korea, Europe, and to a lesser extent India and Japan are the major producers of the research in this field.

It is a fact that the USA has been a major player in science (Leydesdorff and Wagner, 2009; Leydesdorff et al., 2014). The USA has further developed a strong research infrastructure to support its corn- and grass-based bioethanol industry (Vadas et al., 2008). The USA has also been very active in nanotechnology research (Dong et al., 2016) and the sensor research (Sharma et al., 2015).

However, China has been a rising star in scientific research in competition with the USA and Europe (Leydesdorff and Zhou, 2005). China is also a major player in this field as a major producer of bioethanol (Li and Chan-Halbrendt, 2009). China has also been very active in nanotechnology research (Dong et al., 2016) and the sensor research (Gao et al., 2015).

Next, Europe has been a persistent player in scientific research in competition with both the USA and China (Leydesdorff, 2000). Europe has also been a persistent producer of bioethanol along with the USA and Brazil (Gnansounou, 2010). The European Union has also been very active in nanotechnology research (Baraton, 2003) and the sensor research (Campbell et al., 2013).

Additionally, S. Korea has also been a persistent player in scientific research at a moderate level (Leydesdorff and Zhou, 2005). S. Korea, producing 11.4% of the population papers, has also developed a strong research infrastructure to support its biomass-based bioethanol industry (Lim et al., 2017). S. Korea has also been very active in nanotechnology research (Kostoff et al., 2008) and the sensor research (Lim et al., 2019).

92.4.9 The Most Prolific Scopus Subject Categories in Bioethanol Fuel Sensors

The six most prolific Scopus subject categories indexing at least 5% and 2.3% of the sample and population papers, respectively, given in Table 92.7 have shaped the development of the research in this field.

The most prolific Scopus subject categories in bioethanol fuel sensors are 'Materials Science', 'Physics and Astronomy', and 'Engineering'.

However, the Scopus subject category with the most citation impact is 'Materials Science'. Similarly, the Scopus subject categories with the least citation impact are 'Chemistry' and 'Engineering'.

These findings are thought-provoking, suggesting that the primary subject categories are engineering, physics, and materials science. The other key finding is that social sciences are not well represented in both the sample and population papers, unlike the field of evaluative studies in bioethanol fuels.

These findings are not surprising as the key research fronts in this field are the performance and development of innovative nanomaterials and conventional materials for each research front in this field. All these research fronts are related to the hard sciences and engineering.

92.4.10 The Most Prolific Keywords in Bioethanol Fuel Sensors

A limited number of keywords have shaped the development of the research in this field as shown in Table 92.8 and the appendix.

These keywords are grouped under four headings: bioethanol fuels, nanomaterials, sensors, and materials. There are six keywords used related to bioethanol fuels. The prolific keywords are ethanol, ethanol sensors, and ethanol sensing. It is notable that the bioethanol keyword does not appear in these lists.

The prolific keywords related to nanomaterials are nanostructured materials, nanorods, and nanowires. Prolific keywords related to the sensors are chemical sensors, sensors, gas sensors, and gas detectors. The keywords related to the materials are zinc oxide, ZnO, tin dioxides, and tin compounds.

However, the keywords with the most citation impact are nanostructured materials, sensors, and chemical sensors. Similarly, the keywords with the least citation impact are chemical detection, gas sensing electrodes, and gas detectors.

These prolific keywords highlight the key research fronts in this field and reflect well the keywords used in sample papers.

92.4.11 The Most Prolific Research Fronts in Bioethanol Fuel Sensors

There are three primary research fronts in this field: ZnO-based bioethanol fuel sensors (Wan et al., 2004; Wang et al., 2012), SnO_2-based bioethanol fuel sensors (Chen et al., 2005, 2006), and other material-based bioethanol fuel sensors (Choi, et al., 2010; Choudhury, 2009) with 40, 28, and 47 sample papers, respectively. Furthermore, there are further research fronts within each of these research fronts: nanomaterial-based sensors and conventional material-based sensors (Hellegouarc'h et al.,

2001; Paraguay et al., 2000). In total, there are 98 and 17 papers for the nanomaterial-based sensors and conventional material-based sensors, respectively.

The focus of these 100 most-cited papers is the development of nanomaterials for the most efficient processes for each of these research fronts.

There are strong structure–processing–property relationships for all of nanomaterials and the conventional materials such as thin films or alcohol oxidases used for these research fronts. In the end, these most-cited papers in this field hint that the efficiency of nanotechnology applications could be maximized using the structure, processing, and property relationships of the nanomaterials used in these applications (Formela et al., 2016; Konur, 2018a, 2020b, 2021a–d; Konur and Matthews, 1989).

92.5 CONCLUSION AND FUTURE RESEARCH

The research on bioethanol fuel sensors has been mapped through a scientometric study of both sample and population datasets.

The critical issue in this study has been to obtain a representative sample of the research as in any other scientometric study. Therefore, the keyword set has been carefully devised and optimized after several runs in the Scopus database.

The other issue has been the selection of a multidisciplinary database to carry out the scientometric study of the research in this field. For this purpose, the Scopus database has been selected. The journal coverage of this database has been wider than that of the Web of Science.

The key scientometric properties of the research in this field have been determined and discussed in this book chapter. It is evident that a limited number of documents, authors, institutions, publication periods, institutions, funding bodies, source titles, countries, Scopus subject categories, keywords, and research fronts have shaped the development of the research in this field.

There is ample scope to increase the efficiency of scientometric studies in this field in the author and document domains by developing consistent policies and practices in both domains. In this respect, authors, journals, and academic databases have a lot to do. Furthermore, the significant gender deficit as in most scientific fields emerges as a public policy issue.

The research in this field has boomed in the late 2010s possibly promoted by the public concerns on global warming, greenhouse gas emissions, and climate change. Institutions from the USA, China, Europe, and S. Korea have mostly shaped the research in this field.

The modest funding rate of 42% and 53% for sample and population papers, respectively, suggests that this funding rate significantly enhanced the research in this field primarily in the 2010s, possibly more than doubling in the current decade. However, there is ample room for more funding.

The most prolific journals have been mostly indexed by the subject categories of materials science, engineering, and physics (materials physics).

The USA, China, S. Korea, Europe, and to a lesser extent India and Japan have been the major producers of the research in this field as the major producers of bioethanol fuels from different types of biomass such as corn, sugarcane, and grass and other types of biomass. China and India have also contributed largely to the population papers. These countries have well-developed research infrastructure in bioethanol fuels, sensors, nanomaterials, and conventional materials.

The primary subject categories have been 'Materials Science', 'Engineering', and 'Physics and Astronomy'. Due to the technological emphasis of this field, social sciences have not been fairly represented in both the sample and population papers, unlike the evaluative studies in bioethanol fuels.

Ethanol is more popular than bioethanol as a keyword with strong implications for the search strategy. In other words, the search strategy using only the bioethanol keyword would not be much helpful.

These keywords are grouped under four headings: bioethanol fuels, nanomaterials, sensors, and materials. These groups of keywords highlight the potential primary research fronts for these fields.

There are three primary research fronts in this field: ZnO-based bioethanol fuel sensors, SnO_2-based bioethanol fuel sensors, and other material-based bioethanol fuel sensors with 40, 28, and 47 sample papers, respectively. Furthermore, there are further research fronts within each of these research fronts: nanomaterial-based sensors and conventional material-based sensors. In total, there are 98 and 17 papers for the nanomaterial-based sensors and conventional material-based sensors, respectively, for sample papers.

These findings are thought-provoking. The focus of these 100 most-cited papers is the development of nanomaterials and conventional materials for the most efficient processes for each of these research fronts. There are strong structure–processing–property relationships for all of the materials used for these research fronts. In the end, these most-cited papers in this field hint that the efficiency of nanotechnology and materials science applications could be maximized using the structure, processing, and property relationships of the materials used in these applications.

Thus, the scientometric analysis has a great potential to gain valuable insights into the evolution of the research in this field as in other scientific fields.

It is recommended that further scientometric studies are carried out about the other aspects of both production and utilization of bioethanol fuels. It is further recommended that reviews of the most-cited papers are carried out for each research front to complement these scientometric studies. Next, scientometric studies of the hot papers in these primary fields are carried out.

ACKNOWLEDGMENTS

The contribution of highly cited researchers in the field of bioethanol fuel sensors has been gratefully acknowledged.

APPENDIX: THE KEYWORD SET FOR BIOETHANOL FUEL SENSORS

(((TITLE (ethanol* OR bioethanol* OR c2h5oh) OR SRCTITLE (ethanol* OR bioethanol* OR c2h5oh)) AND (TITLE (*sensor* OR *sensing OR biosensor* OR biosensing OR chemosens* OR nanosens* OR detection OR ((zno OR sno2) AND sensit*)) OR SRCTITLE (sensor* OR sensing OR biosensor* OR biosensing OR chemosens* OR nanosens* OR detection)))) AND NOT ((SRCTITLE (remote) OR TITLE ({remote sens*} OR *sensory OR ethanolamine OR {biofuel cell*} OR sers OR rhodamine OR fe3+ OR fabrics OR bisulfite OR {temperature sensor*}) OR SUBJAREA (phar OR dent OR vete OR heal OR nurs OR psyc OR immu OR neur OR medi OR bioc))) AND (LIMIT-TO (DOCTYPE, "ar") OR LIMIT-TO (DOCTYPE, "cp") OR LIMIT-TO (DOCTYPE, "re") OR LIMIT-TO (DOCTYPE, "no") OR LIMIT-TO (DOCTYPE, "ch") OR LIMIT-TO (DOCTYPE, "sh")) AND (LIMIT-TO (LANGUAGE, "English")) AND (LIMIT-TO (SRCTYPE, "j") OR LIMIT-TO (SRCTYPE, "k") OR LIMIT-TO (SRCTYPE, "b")).

REFERENCES

Baraton, M. I. 2003. Nanowatch Europe: Surface science in nanotechnology. *Electrochemical Society Interface* 12:14.

Beaudry, C. and V. Lariviere. 2016. Which gender gap? Factors affecting researchers' scientific impact in science and medicine. *Research Policy* 45:1790–1817.

Bosmann, A., L. Datsevich and A. Jess, et al. 2001. Deep desulfurization of diesel fuel by extraction with ionic liquids. *Chemical Communications* 23:2494–2495.

Brolo, A. G. 2012. Plasmonics for future biosensors. *Nature Photonics* 6:709–713.

Burnham, J. F. 2006. Scopus database: A review. *Biomedical Digital Libraries* 3:1–8.

Campbell, K., P. Barnes and S. A. Haughey, et al. 2013. Development and single laboratory validation of an optical biosensor assay for tetrodotoxin detection as a tool to combat emerging risks in European seafood. *Analytical and Bioanalytical Chemistry* 405:7753–7763.

Carlson, K. M., J. S. Gerber and D. Mueller, et al. 2017. Greenhouse gas emissions intensity of global croplands. *Nature Climate Change* 7:63–68.
Change, C. 2007. Climate change impacts, adaptation and vulnerability. *Science of the Total Environment* 326:95–112.
Chaubey, A. and B. Malhotra. 2002. Mediated biosensors. *Biosensors and Bioelectronics* 17:441–456.
Chen, Y. J., L. Nie, X. Y. Xue, Y. G. Wang and T. H. Wang. 2006. Linear ethanol sensing of SnO_2 nanorods with extremely high sensitivity. *Applied Physics Letters* 88:083105.
Chen, Y. J., X. Y. Xue, Y. G. Wang and T. H. Wang. 2005. Synthesis and ethanol sensing characteristics of single crystalline SnO_2 nanorods. *Applied Physics Letters* 87:1–3.
Choi, K. I., H. R. Kim and K. M. Kim, et al. 2010. C_2H_5OH sensing characteristics of various Co_3O_4 nanostructures prepared by solvothermal reaction. *Sensors and Actuators, B: Chemical* 146:183–189.
Choudhury, A. 2009. Polyaniline/silver nanocomposites: Dielectric properties and ethanol vapour sensitivity. *Sensors and Actuators, B: Chemical* 138:318–325.
Dong, H., Y. Gao and P. J. Sinko, et al. 2016. The nanotechnology race between China and the United States. *Nano Today* 11:7–12.
Ebadi, A. and A. Schiffauerova. 2016. How to boost scientific production? A statistical analysis of research funding and other influencing factors. *Scientometrics* 106:1093–1116.
Formela, K., A. Hejna, L. Piszczyk, M. R. Saeb and X. Colom. 2016. Processing and structure-property relationships of natural rubber/wheat bran biocomposites. *Cellulose* 23:3157–3175.
Gao, M., J. Cao and E. Seto. 2015. A distributed network of low-cost continuous reading sensors to measure spatiotemporal variations of PM2.5 in Xi'an, China. *Environmental Pollution* 199:56–65.
Garfield, E. 1955. Citation indexes for science. *Science* 122:108–111.
Geim, A. K. 2009. Graphene: Status and prospects. *Science* 324:1530–1534.
Geim, A. K. and K. S. Novoselov. 2007. The rise of graphene. *Nature Materials* 6:183–191.
Gnansounou, E. 2010. Production and use of lignocellulosic bioethanol in Europe: Current situation and perspectives. *Bioresource Technology* 101:4842–4850.
Hansen, A. C., Q. Zhang and P. W. L. Lyne. 2005. Ethanol-diesel fuel blends: A review. *Bioresource Technology* 96:277–285.
Hellegouarc'h, F., F. Arefi-Khonsari, R. Planade and J. Amouroux. 2001. PECVD prepared SnO_2 thin films for ethanol sensors. *Sensors and Actuators, B: Chemical* 73:27–34.
Hill, J., S. Polasky and E. Nelson, et al. 2009. Climate change and health costs of air emissions from biofuels and gasoline. *Proceedings of the National Academy of Sciences of the United States of America* 106:2077–2082.
Hsieh, W. D., R. H. Chen, T. L. Wu and T. H. Lin. 2002. Engine performance and pollutant emission of an SI engine using ethanol-gasoline blended fuels. *Atmospheric Environment* 36:403–410.
Kerr, R. A. 2007. Global warming is changing the world. *Science* 316:188–190.
Konur, O. 2000. Creating enforceable civil rights for disabled students in higher education: An institutional theory perspective. *Disability & Society* 15:1041–1063.
Konur, O. 2002a. Access to nursing education by disabled students: Rights and duties of nursing programs. *Nurse Education Today* 22:364–374.
Konur, O. 2002b. Assessment of disabled students in higher education: Current public policy issues. *Assessment and Evaluation in Higher Education* 27:131–52.
Konur, O. 2002c. Access to employment by disabled people in the UK: Is the Disability Discrimination Act working? *International Journal of Discrimination and the Law* 5:247–279.
Konur, O. 2006a. Participation of children with dyslexia in compulsory education: Current public policy issues. *Dyslexia* 12:51–67.
Konur, O. 2006b. Teaching disabled students in higher education. *Teaching in Higher Education* 11:351–363.
Konur, O. 2007a. A judicial outcome analysis of the Disability Discrimination Act: A windfall for the employers? *Disability & Society* 22:187–204.
Konur, O. 2007b. Computer-assisted teaching and assessment of disabled students in higher education: The interface between academic standards and disability rights. *Journal of Computer Assisted Learning* 23:207–219.
Konur, O. 2011. The scientometric evaluation of the research on the algae and bio-energy. *Applied Energy* 88:3532–3540.
Konur, O. 2012a. Prof. Dr. Ayhan Demirbas' scientometric biography. *Energy Education Science and Technology Part A: Energy Science and Research* 28:727–738.
Konur, O. 2012b. The evaluation of the biogas research: A scientometric approach. *Energy Education Science and Technology Part A: Energy Science and Research* 29:1277–1292.

Konur, O. 2012c. The evaluation of the global energy and fuels research: A scientometric approach. *Energy Education Science and Technology Part A: Energy Science and Research* 30:613–628.

Konur, O. 2012d. The evaluation of the research on the biodiesel: A scientometric approach. *Energy Education Science and Technology Part A: Energy Science and Research* 28:1003–1014.

Konur, O. 2012e. The evaluation of the research on the bioethanol: A scientometric approach. *Energy Education Science and Technology Part A: Energy Science and Research* 28:1051–1064.

Konur, O. 2012f. The evaluation of the research on the biofuels: A scientometric approach. *Energy Education Science and Technology Part A: Energy Science and Research* 28:903–916.

Konur, O. 2012g. The evaluation of the research on the biohydrogen: A scientometric approach. *Energy Education Science and Technology Part A: Energy Science and Research* 29:323–338.

Konur, O. 2012h. The evaluation of the research on the microbial fuel cells: A scientometric approach. *Energy Education Science and Technology Part A: Energy Science and Research* 29:309–322.

Konur, O. 2012i. The scientometric evaluation of the research on the production of bioenergy from biomass. *Biomass and Bioenergy* 47:504–515.

Konur, O. 2015. Current state of research on algal bioethanol. In *Marine Bioenergy: Trends and Developments*, Eds. S. K. Kim and C. G. Lee, pp. 217–244. Boca Raton, FL: CRC Press.

Konur, O. 2016a. Scientometric overview in nanobiodrugs. In *Nanoarchitectonics for Smart Delivery and Drug Targeting*, Eds. A. M. Holban and A. M. Grumezescu, pp. 405–428. Amsterdam: Elsevier.

Konur, O. 2016b. Scientometric overview regarding nanoemulsions used in the food industry. In *Emulsions: Nanotechnology in the Agri-Food Industry*, Ed. A. M. Grumezescu, pp. 689–711. Amsterdam: Elsevier.

Konur, O. 2016c. Scientometric overview regarding the nanobiomaterials in antimicrobial therapy. In *Nanobiomaterials in Antimicrobial Therapy*, Ed. A. M. Grumezescu, pp. 511–535. Amsterdam: Elsevier.

Konur, O. 2016d. Scientometric overview regarding the nanobiomaterials in dentistry. In *Nanobiomaterials in Dentistry*, Ed. A. M. Grumezescu, pp. 425–453. Amsterdam: Elsevier.

Konur, O. 2016e. Scientometric overview regarding the surface chemistry of nanobiomaterials. In *Surface Chemistry of Nanobiomaterials*, Ed. A. M. Grumezescu, pp. 463–486. Amsterdam: Elsevier.

Konur, O. 2016f. The scientometric overview in cancer targeting. In *Nanoarchitectonics for Smart Delivery and Drug Targeting*, Eds. A. M. Holban and A. Grumezescu, pp. 871–895. Amsterdam: Elsevier.

Konur, O. 2017a. Recent citation classics in antimicrobial nanobiomaterials. In *Nanostructures for Antimicrobial Therapy*, Eds. A. Ficai and A. M. Grumezescu, pp. 669–685. Amsterdam: Elsevier.

Konur, O. 2017b. Scientometric overview in nanopesticides. In *New Pesticides and Soil Sensors*, Ed. A. M. Grumezescu, pp. 719–744. Amsterdam: Elsevier.

Konur, O. 2017c. Scientometric overview regarding oral cancer nanomedicine. In *Nanostructures for Oral Medicine*, Eds. E. Andronescu, A. M. Grumezescu, pp. 939–962. Amsterdam: Elsevier;

Konur, O. 2017d. Scientometric overview regarding water nanopurification. In *Water Purification,* Ed. A. M. Grumezescu, pp. 693–716. Amsterdam: Elsevier.

Konur, O. 2017e. Scientometric overview in food nanopreservation. In *Food Preservation,* Ed. A. M. Grumezescu, pp. 703–729. Amsterdam: Elsevier.

Konur, O. 2017f. The top citation classics in alginates for biomedicine. In *Seaweed Polysaccharides: Isolation, Biological and Biomedical Applications*, Eds. J. Venkatesan, S. Anil, S. K. Kim, pp. 223–249. Amsterdam: Elsevier.

Konur, O., Ed. 2018a. *Bioenergy and Biofuels*. Boca Raton, FL: CRC Press.

Konur, O. 2018b. Bioenergy and biofuels science and technology: Scientometric overview and citation classics. In *Bioenergy and Biofuels*, Ed. O. Konur, pp. 3–63. Boca Raton, FL: CRC Press.

Konur, O. 2019. Cyanobacterial bioenergy and biofuels science and technology: A scientometric overview. In *Cyanobacteria: From Basic Science to Applications*, Eds. A. K. Mishra, D. N. Tiwari and A. N. Rai, pp. 419–442. Amsterdam: Elsevier.

Konur, O. 2020a. The scientometric analysis of the research on the bioethanol production from green macroalgae. In *Handbook of Algal Science, Technology and Medicine,* Ed. O. Konur, pp. 385–401. London: Academic Press.

Konur, O., Ed. 2020b. *Handbook of Algal Science, Technology and Medicine.* London: Academic Press.

Konur, O., Ed. 2021a. *Handbook of Biodiesel and Petrodiesel Fuels: Science, Technology, Health, and Environment*. Boca Raton, FL: CRC Press.

Konur, O., Ed. 2021b. *Handbook of Biodiesel and Petrodiesel Fuels: Science, Technology, Health, and Environment. Volume 1. Biodiesel Fuels: Science, Technology, Health, and Environment.* Boca Raton, FL: CRC Press.

Konur, O., Ed. 2021c. *Handbook of Biodiesel and Petrodiesel Fuels: Science, Technology, Health, and Environment. Volume 2. Biodiesel Fuels Based on the Edible and Nonedible Feedstocks, Wastes, and Algae: Science, Technology, Health, and Environment.* Boca Raton, FL: CRC Press.

Konur, O., Ed. 2021d. *Handbook of Biodiesel and Petrodiesel Fuels: Science, Technology, Health, and Environment. Volume 3. Petrodiesel Fuels: Science, Technology, Health, and Environment.* Boca Raton, FL: CRC Press.

Konur, O. and F. L. Matthews. 1989. Effect of the properties of the constituents on the fatigue performance of composites: A review. *Composites* 20:317–328.

Kostoff, R., R. Barth and C. Lau. 2008. Relation of seminal nanotechnology document production to total nanotechnology document production-South Korea. *Scientometrics* 76:43–67.

Leydesdorff, L. 2000. Is the European Union becoming a single publication system? *Scientometrics* 47:265–280.

Leydesdorff, L. and C. Wagner. 2009. Is the United States losing ground in science? A global perspective on the world science system. *Scientometrics* 78:23–36.

Leydesdorff, L. and P. Zhou. 2005. Are the contributions of China and Korea upsetting the world system of science? *Scientometrics* 63:617–630.

Leydesdorff, L., C. S. Wagner and L. Bornmann. 2014. The European Union, China, and the United States in the top-1% and top-10% layers of most-frequently cited publications: Competition and collaborations. *Journal of Informetrics* 8:606–617.

Li, D., Z. Huang, X. C. Lu, W. Zhang and J. Yang. 2005. Physico-chemical properties of ethanol-diesel blend fuel and its effect on performance and emissions of diesel engines. *Renewable Energy* 30:967–976.

Li, S. Z. and C. Chan-Halbrendt. 2009. Ethanol production in (the) People's Republic of China: Potential and technologies. *Applied Energy* 86:S162–S169.

Lim, C. C., H. Kim and M. R. Vilcassim, et al. 2019. Mapping urban air quality using mobile sampling with low-cost sensors and machine learning in Seoul, South Korea. *Environment International* 131:105022.

Lim, S. Y., H. J. Kim and S. H. Yoo. 2017. Public's willingness to pay a premium for bioethanol in Korea: A contingent valuation study. *Energy Policy* 101:20–27.

Liu, B. and H. C. Zeng. 2003. Hydrothermal synthesis of ZnO nanorods in the diameter regime of 50 nm. *Journal of the American Chemical Society* 125:4430–4431.

Ma, X., L. Sun and C. Song. 2002. A new approach to deep desulfurization of gasoline, diesel fuel and jet fuel by selective adsorption for ultra-clean fuels and for fuel cell applications. *Catalysis Today* 77:107–116.

Ma, X. L., K. Y. Sakanish and I. Mochida. 1994. Hydrodesulfurization reactivities of various sulfur-compounds in diesel fuel. *Industrial & Engineering Chemistry Research* 33:218–222.

Murdoch, M., G. I. N. Waterhouse and M. A. Nadeem, et al. 2011. The effect of gold loading and particle size on photocatalytic hydrogen production from ethanol over Au/TiO_2 nanoparticles. *Nature Chemistry* 3:489–492.

Najafi, G., B. Ghobadian and T. Tavakoli, et al. 2009. Performance and exhaust emissions of a gasoline engine with ethanol blended gasoline fuels using artificial neural network. *Applied Energy* 86:630–639.

Newman, P. W. G. and J. R. Kenworthy. 1989. Gasoline consumption and cities: A comparison of U.S. cities with a global survey. *Journal of the American Planning Association* 55:24–37.

North, D. C. 1991. Institutions. *Journal of Economic Perspectives* 5:97–112.

Paraguay, D., F., M. Miki-Yoshida, J. Morales, J. Solis and L. W. Estrada. 2000. Influence of Al, In, Cu, Fe and Sn dopants on the response of thin film ZnO gas sensor to ethanol vapour. *Thin Solid Films* 373:137–140.

Schauer, J. J., M. J. Kleeman, G. R. Cass and B. R. T. Simoneit. 2002. Measurement of emissions from air pollution sources. 5. C_1-C_{32} organic compounds from gasoline-powered motor vehicles. *Environmental Science and Technology* 36:1169–1180.

Sharma, L. K., H. Bu, A. Denton and D. W. Franzen. 2015. Active-optical sensors using red NDVI compared to red edge NDVI for prediction of corn grain yield in North Dakota, USA. *Sensors* 15:27832–27853.

Shin, D., T. Kim, J. Choi and J. Kim. 2014. Author name disambiguation using a graph model with node splitting and merging based on bibliographic information. *Scientometrics* 100:15–50.

Vadas, P. A., K. H. Barnett and D. J. Undersander. 2008. Economics and energy of ethanol production from alfalfa, corn, and switchgrass in the Upper Midwest, USA. *Bioenergy Research* 1:44–55.

Wan, Q., Q. H. Li and Y. J. Chen, et al. 2004. Fabrication and ethanol sensing characteristics of ZnO nanowire gas sensors. *Applied Physics Letters* 84:3654–3656.

Wang, L., Y. Kang and X. Liu, et al. 2012. ZnO nanorod gas sensor for ethanol detection. *Sensors and Actuators, B: Chemical* 162:237–243.

93 Bioethanol Fuel Sensors
Review

Ozcan Konur
(Formerly) Ankara Yildirim Beyazit University

93.1 INTRODUCTION

Crude oil-based gasoline fuels (Ma et al., 2002; Newman and Kenworthy, 1989) have been widely used in the transportation sector since the 1920s. However, there have been great public concerns over the adverse environmental impact of these fuels (Hill et al., 2006, 2009; Schauer et al., 1999, 2002). Hence, biomass-based bioethanol fuels (Konur, 2012, 2015, 2019, 2020a) have been increasingly used in blending gasoline fuels (Al-Hasan, 2003; Hsieh et al., 2002; Najafi et al., 2009).

In this context, there has been a significant focus on bioethanol fuel sensors (Murdoch et al., 2011; Wan et al., 2004). There are three primary research fronts in this field: ZnO-based bioethanol fuel sensors (Wan et al., 2004; Wang et al., 2012), SnO_2-based bioethanol fuel sensors (Chen et al., 2005a, 2006), and other material-based bioethanol fuel sensors (Choudhury, 2009). Furthermore, there are further research fronts within each of these research fronts: nanomaterial-based sensors and conventional material-based sensors (Hellegouarc'h et al., 2001; Paraguay et al., 2000).

In the meantime, research on nanomaterials and nanotechnology has intensified in recent years to become a major research field in scientific research with over one and a half million published papers (Geim, 2009; Geim and Novoselov, 2007). In this context, a large number of nanomaterials have been developed nearly for every research field. These materials offer an innovative way to increase the efficiency of the production and utilization of bioethanol fuels as in other scientific fields (Konur, 2016a–f, 2017a–e, 2021e,f). Similarly, there has been intense research in the field of sensors in recent years (Brolo, 2012; Chaubey and Malhotra, 2002).

However, it is essential to develop efficient incentive structures (North, 1991) for primary stakeholders to enhance research in this field (Konur, 2000, 2002a–c, 2006a,b, 2007a,b).

Although there have been a limited number of review papers on bioethanol fuel sensors (Azevedo et al., 2005; Bhardwaj and Sharma, 2020; Sangeetha et al., 2020), there has been no review of the most-cited 25 articles in this field.

This book chapter presents a review of the most-cited 25 articles in bioethanol fuel sensors. Then, it discusses the key findings of these highly influential papers and comments on future research priorities in this field.

93.2 MATERIALS AND METHODS

The search for this study was carried out using the Scopus database (Burnham, 2006) in August 2021.

As a first step to the search of the relevant literature, keywords were selected using the first most-cited 200 papers. The selected keyword list was optimized to obtain a representative sample of papers for the searched research field. This keyword list is provided in the appendix of Konur (2023) for future replication studies.

As a second step, a sample dataset was used for this study. The first 25 articles in the sample of 100 most-cited papers with at least 181 citations each were selected for the review study. Key findings from each paper were taken from the abstracts of these papers and were discussed.

DOI: 10.1201/9781003226567-125

Additionally, a number of brief conclusions were drawn, and a number of relevant recommendations are made to enhance the future research landscape.

93.3 RESULTS

The brief information about 25 most-cited papers with at least 181 citations each on the bioethanol sensing properties of sensors is given under three headings: zinc oxide (ZnO), tin dioxide (SnO_2), and other material-based bioethanol fuel sensors with 12, 6, and 10 papers, respectively.

93.3.1 Zinc Oxide-based Bioethanol Fuel Sensors

Brief information about 12 prolific studies with at least 181 citations each on bioethanol sensing properties of ZnO-based sensors is given in Table 93.1. Furthermore, brief notes on the contents of these studies are also given.

Wan et al. (2004) fabricate ZnO nanowire gas sensors using the microelectromechanical system technology and study their ethanol sensing characteristics in a paper with 1,834 citations. They find that these sensors exhibited a very high sensitivity to ethanol gas and fast response time at 300°C.

Wang et al. (2012) evaluate ZnO nanorod gas sensors for ethanol detection in a paper with 361 citations. They fabricate these nanorods by a simple low-temperature hydrothermal process in high yield (about 85%), starting with @equ_0001.eps@ aqueous solution. They observe that this gas sensor exhibited a high, reversible, and fast response to bioethanol.

Bie et al. (2007) evaluate ZnO nanorod gas sensors for sensing bioethanol in a paper with 296 citations. They fabricate these aligned nanorods via a two-step solution approach on an Al_2O_3 tube. They observe that these nanorods were uniform with diameters of 10–30 nm and a length of about

TABLE 93.1
ZnO-based Bioethanol Fuel Sensors

No.	Papers	Nanomaterials	Issues	Cits.
1	Wan et al. (2004)	ZnO nanowires	ZnO nanowire-based gas sensors	1,834
2	Wang et al. (2012)	ZnO nanorods	ZnO nanorod-based gas sensors	361
3	Bie et al. (2007)	ZnO nanorods	ZnO nanorod-based gas sensors	296
4	Hsueh et al. (2007a)	ZnO nanowires	ZnO nanowire-based gas sensors	293
5	Paraguay et al. (2000)	ZnO thin films	ZnO thin film-based gas sensors. Effect of dopants	282
6	Rout et al. (2006)	ZnO nanorods, nanowires, and nanotubes	ZnO nanorods, nanowires, and nanotube-based gas sensors	247
7	Na et al. (2011)	Co_3O_4-decorated ZnO nanowires	Co_3O_4-decorated ZnO nanowire-based gas sensors	239
8	Zhu et al. (2018)	ZnO nanoparticles, nanoplates, and nanoflowers	ZnO nanoparticles, nanoplates, and nanoflower-based gas sensors: Effect of nanostructures	206
9	Kim et al. (2007)	SnO_2-ZnO composite thin films	SnO_2-ZnO composite thin film-based gas sensors	201
10	Hsueh et al. (2007b)	ZnO nanowires	ZnO nanowire-based gas sensors	198
11	Rao (2000)	ZnO	ZnO-based gas sensors: Effect of La_2O_3 and Pd	195
12	Hongsith et al. (2008)	ZnO nanowires	ZnO nanowire-based gas sensors: Effect of Au doping	181

Cits.: Number of the citations received by each paper.

1.4 μm. The response of the aligned ZnO nanorod sensor reached 18.29 ppm ethanol, which was a two-fold increase compared with that reported in the literature.

Hsueh et al. (2007a) evaluate the laterally grown ZnO nanowire ethanol gas sensors in a paper with 293 citations. They fabricate these nanowires on ZnO:Ga/glass templates. They observe that growth direction of nanowires depends strongly on growth parameters. Resistivity of the fabricated sensor decreased upon ethanol gas injection. By introducing 1,500 ppm ethanol gas, they find that the device response was around 20%, 35%, 58%, and 61% when the gas sensor was operated at 180°C, 230°C, 260°C, and 300°C, respectively. The device response at 300°C was around 18%, 26%, 43%, 55%, and 61% when the concentration of injected ethanol gas was 50, 100, 500, 1,000, and 1,500 ppm, respectively.

Paraguay et al. (2000) study the effect of dopants on the response of thin film ZnO gas sensors to ethanol vapor in a paper with 282 citations. They obtain ZnO:dopant films doped with different elements of Al, In, Cu, Fe, and Sn by a spray pyrolysis technique. They note that the amount as well as the type of dopant modifies the microstructure and surface morphology of these films. They note the non-oriented growth, (002) oriented growth, and non-oriented growth and poor crystallinity at 0%, 1%, and 3% doping levels, respectively. They observe a better sensitivity for Sn- and Al-doped films, with a dopant/Zn ratio of 0.4 and 1.8 at.%, respectively.

Rout et al. (2006) evaluate bioethanol sensors based on ZnO nanorods, nanowires, and nanotubes in a paper with 247 citations. They observe that nanorods and nanowires impregnated with 1% Pt show high sensitivity for 1,000 ppm of bioethanol at or below 150°C, with short recovery and response times.

Na et al. (2011) evaluate the selective detection of bioethanol using a cobalt tetraoxide (Co_3O_4)-decorated ZnO nanowire network (NW) sensor in a paper with 239 citations. They explain the gas selectivity by the catalytic effect of nanocrystalline Co_3O_4 and the extension of the electron depletion layer via the formation of p–n junctions.

Zhu et al. (2018) evaluate the ethanol gas-sensing properties of hierarchical flower-like ZnO nanostructures in a paper with 206 citations. They obtain ZnO nanoparticles, nanoplates, and nanoflowers by a facile hydrothermal route. They note that the nanoplate-assembled nanoflowers exhibited significantly higher gas-sensing performances than the others. This is due to their hierarchical architectures with a large specific area and abundant spaces for gas diffusion. Furthermore, they observe that the concentration of the surfactant CTAB used had an essential effect on the ultimate morphology of the hierarchical nanoflowers.

Kim et al. (2007) evaluate the ethanol-sensing properties of tin dioxide (SnO_2)-ZnO composite thin films in a paper with 202 citations. They prepare these films by alternate deposition of ten droplets of SnO_2 and ZnO sols. They note that this sensor showed a high response to 200 ppm at 300°C. In contrast, S(ethanol) and S(acetone) of pure SnO_2 and ZnO thin films were similar to each other. They assert that the heterostructure between SnO_2 and ZnO was one of the probable reasons for the successful discrimination between bioethanol and acetone.

Hsueh et al. (2007b) evaluate ZnO nanowire bioethanol sensors with Pd adsorption in a paper with 198 citations. They study the growth of high-density single-crystalline ZnO nanowires on patterned ZnO:Ga/SiO_2/Si templates. With Pd adsorption, they observe that measured sensitivities of these sensors increased from 18.5% to 44.5% at 170°C and increased from 36.0% to 61.5% at 230°C.

Rao (2000) evaluates the performance of ZnO ceramic semiconductor gas sensors for ethanol vapor lanthana (La_2O_3) and palladium (Pd) in a paper with 195 citations. He notes that the sensitivity of the sintered elements were promoted by the addition of these elements. Furthermore, the sensor operating temperature was reduced with the noble metal. These elements were highly sensitive and selective for ethanol vapors at 175°C in air atmosphere. He asserts that the promoting effect of the sensitivity of the sensor elements with lanthanum oxide was related to the selectivity to oxidation of ethanol vapors.

Hongsith et al. (2008) evaluate the bioethanol sensors based on ZnO- and Au-doped ZnO nanowires in a paper with 181 citations. They observe the bioethanol sensing properties of ZnO nanowires from the resistance change under ethanol vapor atmosphere. By considering sensitivity and response time, they note that the optimum operating temperature of the ethanol sensor was 240°C. They then observe that the sensitivity of the sensor based on Au-doped ZnO nanowires exhibited higher value than that of the sensor based on undoped ZnO nanowires.

93.3.2 SnO_2-BASED BIOETHANOL FUEL SENSORS

Brief information about seven prolific studies with at least 201 citations each on bioethanol sensing properties of SnO_2 is given in Table 93.2. Furthermore, brief notes on the contents of these studies are also given.

Chen et al. (2006) evaluate the linear ethanol sensing of SnO_2 nanorods with extremely high sensitivity in a paper with 247 citations. They fabricate these nanorods with a diameter down to 3 nm through a hydrothermal route. They observe that the sensitivity is up to 83.8 as the nanorod sensor is exposed to 300 ppm bioethanol vapor in air. Moreover, they observe the linear dependence of the sensitivity on the bioethanol concentration for each of the 20 sensors. They note that compared with the measured results of 80–180 nm SnO_2 particles, such linear dependence is related to the small size effect.

Chen et al. (2005a) evaluate bioethanol sensing characteristics of single-crystalline SnO_2 nanorods in a paper with 228 citations. They fabricate these nanorods with diameters of 4–15 nm and lengths of 100–200 nm using $SnCl_4$ as a precursor. They observe that these sensors exhibited a sensitivity of 31.4 for 300 ppm of bioethanol. Both the response and recovery time are short, around 1 s. Moreover, they observe a linear dependence of the sensitivity on the bioethanol concentration. They attribute these behaviors to the high surface-to-volume ratio of the nanorods.

Hwang et al. (2011) evaluate the effect of Ag nanocluster decoration on bioethanol sensing characteristics on SnO_2 NWs in a paper with 221 citations. They coat the Ag layers with thicknesses of 5–50 nm on the surface of SnO_2 NWs via e-beam evaporation. They observe that the SnO_2 NWs decorated by isolated Ag nano-islands displayed a 3.7-fold enhancement in gas response to 100 ppm bioethanol at 450°C compared to pristine SnO_2 NWs. In contrast, as the Ag decoration layers became continuous, the response to bioethanol decreased significantly. They note that the enhancement and deterioration of the bioethanol sensing characteristics by the introduction of the Ag decoration layer were strongly governed by the morphological configurations of the Ag catalysts on SnO_2 NWs and their sensitization mechanism.

Zhang et al. (2008) evaluate the ethanol sensing properties of electrospun SnO_2 nanofiber-based gas sensors in a paper with 217 citations. They obtain these nanofibers by electrospinning of a poly(vinyl alcohol)/$SnCl_4 \cdot 5H_2O$ solution. They note that the SnO_2 nanofibers with an average diameter of ~100 nm could be directly deposited on a microhotplate by near-field electrospinning. This microgas sensor exhibited large response, low detection limit, fast response/recovery, and good reproducibility. The detection limit was <10 ppb, and the response/recovery time towards 10 ppm ethanol was <14 s.

TABLE 93.2
SnO_2-based Bioethanol Fuel Sensors

No.	Papers	Nanomaterials	Issues	Cits.
1	Chen et al. (2006)	SnO_2 nanorods	SnO_2 nanorod-based gas sensors	247
2	Chen et al. (2005a)	SnO_2 nanorods	SnO_2 nanorod-based gas sensors	228
3	Hwang et al. (2011)	Ag nanoclusters on SnO_2 NWs	Ag nanoclusters on SnO_2 NW-based gas sensors	221
4	Zhang et al. (2008)	SnO_2 nanofibers	SnO_2 nanofiber-based gas sensors	217
5	Jinkawa et al. (2000)	SnO_2	SnO_2-based gas sensors: Effect of oxides	217
6	Kim et al. (2007)	SnO_2-ZnO composite thin films	SnO_2-ZnO composite thin film-based gas sensors	201

NW, nanowire network.

Jinkawa et al. (2000) evaluate the ethanol gas sensitivity and surface catalytic property of SnO_2 sensors modified with acidic or basic oxides in a paper 217 citations. They note that the sensitivity to bioethanol gas at 300°C increased tremendously with an addition of a basic oxide (e.g., La_2O_3), while it hardly changed with that of an acidic oxide (WO_3). The basic metal oxide to SnO_2 brought about enhancement of catalytic activity for the dehydrogenation of ethanol gas to CH_3CHO and for the consecutive oxidation of CH_3CHO to CO_2. On the other hand, the acidic metal oxide enhanced only the dehydration reaction, showing even an adverse effect on consecutive oxidation. They note that the enhancement of the catalytic oxidation activity to an appropriate level could be a reason for the high sensitivity to ethanol gas for the sensors loaded with basic oxides.

93.3.3 Other Bioethanol Fuel Sensors

Brief information about ten prolific studies with at least 186 citations each on the bioethanol sensing properties of sensors other than ZnO and SnO_2 is given in Table 93.3. Furthermore, brief notes on the contents of these studies are also given.

Liu et al. (2005) develop highly selective and stable bioethanol sensors based on single-crystalline divanadium pentoxide (V_2O_5) nanobelts in a paper with 505 citations. They obtain these nanobelts by a simple mild hydrothermal method with high yield. They find that these gas sensors show great potential for the detection of bioethanol molecules at a relatively low temperature. The experiments with variations in relative humidity and tests with other gases indicated no problems of interference with bioethanol.

Shan et al. (2010) carry out the low potential determination of bioethanol based on ionic liquid-functionalized graphene (IL-graphene) in a paper with 285 citations. With alcohol dehydrogenase (ADH) as a model, they fabricate the ADH/IL-graphene/chitosan-modified electrode through a simple casting method. The resulting sensor showed rapid and highly sensitive amperometric response to bioethanol with a low detection limit (5 μm). Moreover, they use this sensor to determine bioethanol in real samples.

TABLE 93.3
Other Bioethanol Fuel Sensors

No.	Papers	Nanomaterials	Issues	Cits.
1	Liu et al. (2005)	V_2O_5 nanobelts	V_2O_5 nanobelt-based gas sensors	505
2	Shan et al. (2010)	Graphene	Graphene-based gas sensors	285
3	Na et al. (2011)	Co_3O_4-decorated ZnO nanowires	Co_3O_4-decorated ZnO nanowire-based gas sensors	237
4	Choudhury (2009)	PANI/Ag nanocomposites	PANI/Ag nanocomposite-based gas sensors	228
5	Hwang et al. (2011)	Ag nanoclusters on SnO_2 NWs	Ag nanoclusters on SnO_2 NW-based gas sensors	220
6	Wang et al. (2016)	Al-doped NiO nanorod-flowers	Al-doped NiO nanorod-flower-based gas sensors	218
7	Wang et al. (2013)	α-Fe_2O_3 nanostructures	α-Fe_2O_3 nanostructure-based gas sensors: Effect of morphology	205
8	Obayashi et al. (1976)	Perovskite-type oxides (Ln, M) BO_3	Perovskite-type oxide-based gas sensors	202
9	Wu et al. (2007)	Soluble CNF	Soluble carbon nanofiber-based gas sensors	193
10	Xu et al. (2008)	PolyHEMA photonic crystals	PolyHEMA photonic crystal-based gas sensors	186

NW, nanowire network; CNF, carbon nanofiber.

Choudhury (2009) evaluate the dielectric properties and ethanol vapor sensitivity of polyaniline (PANI)/Ag nanocomposites in a paper with 228 citations. They prepare these materials by *in situ* oxidative polymerization of aniline monomer in the presence of different concentrations of Ag nanoparticles. They find that the particle size increased with increasing Ag concentration in the composite, owing to the aggregation effect. They observe higher conductivity, dielectric constant, and dielectric loss of PANI/Ag nanocomposites than pure PANI. The conductivity of nanocomposites increased with increasing Ag concentration. Finally, they find that these materials possess superior bioethanol sensing capacity compared to pure PANI, and there is a linear relationship between the responses and bioethanol and/or Ag concentration.

Wang et al. (2016) evaluate a superior bioethanol gas sensor based on Al-doped nickel monoxide (NiO) nanorod-flowers in a paper with 205 citations. They fabricate these nanorod-flowers with uniform sizes and well-defined morphologies by a facile solvothermal reaction. They find that the 2.15 at.% Al-doped NiO nanorod-flowers showed improved bioethanol gas sensing properties compared to those of pure NiO nanorod-flowers. The incorporation of Al ions with NiO nanocrystals adjusts the carrier concentration and induces the change of oxygen deficiency and chemisorbed oxygen of NiO nanorod-flowers.

Wang et al. (2013) evaluate the bioethanol-sensing properties of the three-dimensional hierarchical flower-like α-Fe_2O_3 nanostructures in a paper with 205 citations. They note that the samples are loose and porous with flower-like structure, and the subunits are irregularly shaped nanosheets. The morphology of these structures was tunable as a function of reaction time. They note that this hierarchical sensor exhibited significantly improved sensor performances in comparison with the compact α-Fe_2O_3 structures. They attribute this enhancement of sensing properties to the unique porous and well-aligned nanostructure.

Obayashi et al. (1976) evaluate the perovskite-type oxides as ethanol sensors in a paper with 202 citations. They consider some of perovskite oxides such as (Ln, M) BO_3 (Ln=lanthanoid element, M=alkaline earth metal, and B=transition metal) as oxidation catalysts. When the compounds are maintained at 150°C–400°C and a trace amount of ethanol in air comes in contact with the oxides, the resistivity of the oxides increases in proportion to the gas concentration with a comparatively short response time, and it recovers to the initial value when the gas is removed.

Wu et al. (2007) evaluate the bioethanol sensing performance of soluble carbon nanofiber (CNF) in a paper with 193 citations. With ADH as a model, they fabricate the ADH/CNF-modified electrode by a simple casting process. They note that this biosensor showed rapid and highly sensitive amperometric response to bioethanol with acceptable preparation reproducibility and excellent stability.

Xu et al. (2008) evaluate the polymerized polyhydroxyethyl methacrylate (HEMA) photonic crystals as bioethanol sensor materials in a paper with 186 citations. They coat the surface of monodisperse silica particles with a thin layer of polystyrene. Surface charge groups were attached by grafting polymerization of styrene sulfonate. They then self-assemble these silica particles into crystalline colloidal arrays (CCA) in deionized water. They polymerize HEMA around the CCA to form a HEMA-polymerized CCA (PCCA) to produce a three-dimensional periodic array of voids in HEMA PCCA. Diffraction from the embedded CCA sensitively monitors the concentration of bioethanol in water because HEMA PCCA shows a large volume dependence on bioethanol due to a decreased Flory–Huggins mixing parameter.

93.4 DISCUSSION

93.4.1 Introduction

Crude oil-based gasoline fuels have been widely used in the transportation sector since the 1920s. However, there have been great public concerns over the adverse environmental impact of these fuels. Hence, biomass-based bioethanol fuels have been increasingly used in blending gasoline fuels.

In the meantime, research in nanomaterials and nanotechnology has intensified in recent years to become a major research field in scientific research. In this context, a large number of nanomaterials have been developed nearly for every research field. These materials offer an innovative way to increase the efficiency of the production and utilization of bioethanol fuels as in other scientific fields. Similarly, there has been intense research in the field of sensors in recent years.

However, it is essential to develop efficient incentive structures for primary stakeholders to enhance research in this field. Although there have been a number of review papers on bioethanol fuel sensors, there has been no review of the research of the most-cited 25 articles in this field.

This book chapter presents a review of the most-cited 25 articles with at least 181 citations each in the field of bioethanol fuel sensors. Then, it discusses the key findings of these highly influential papers and comments on future research priorities in this field.

There are three major research fronts for this field: ZnO, SnO_2, and other material-based bioethanol fuel sensors with 12, 6, and 10 papers, respectively.

93.4.2 ZnO-based Bioethanol Fuel Sensors

Brief information about 12 prolific studies with at least 181 citations each on bioethanol sensing properties of ZnO-based sensors is given in Table 93.1. Furthermore, brief notes on the contents of these studies are also given.

Nine of these prolific studies evaluate nanomaterial-based bioethanol sensors (Bie et al., 2007; Hongsith et al., 2008; Hsueh et al., 2007a,b; Na et al., 2011; Rout et al., 2006; Wan et al., 2004; Wang et al., 2012; Zhu et al., 2018). Furthermore, two studies evaluate thin film-based ZnO bioethanol sensors (Kim et al., 2007; Paraguay et al., 2000).

These studies build first on research related to the ensors in general (Brolo, 2012; Turner, 2013) and on research related to ZnO-based bioethanol sensors (Liu and Zeng, 2003; Yang et al., 2002).

On the other hand, the first group of studies related to nanomaterial-based ZnO bioethanol sensors rely on research related to nanomaterials and nanotechnology (Geim, 2009; Geim and Novoselov, 2007). The second group of studies related to thin film-based ZnO bioethanol sensors rely on research related to thin films (Aspnes, 1982; Nix, 1989).

These studies hint that thin film- and nanomaterial-based ZnO bioethanol sensors perform well in sensing bioethanol gas, although there are differences in fabricating them. This is due to their hierarchical architectures with a large specific area and abundant spaces for gas diffusion for nanomaterials.

In the end, these most-cited papers in this field hint that the efficiency of sensing bioethanol gas could be maximized using the structure, processing, and property relationships of nanomaterials and thin films used in the sensing process (Formela et al., 2016; Konur, 2018, 2020b, 2021a–d; Konur and Matthews, 1989).

93.4.3 SnO_2-based Bioethanol Fuel Sensors

Brief information about six prolific studies with at least 201 citations each on bioethanol sensing properties of SnO_2 is given in Table 93.2. Furthermore, brief notes on the contents of these studies are also given.

Four of these prolific studies evaluate nanomaterial-based SnO_2 bioethanol sensors (Chen et al., 2005a, 2006). Additionally, Kim et al. (2007) evaluate thin film-based SNO_2 bioethanol sensors.

These studies build first on research related to sensors in general (Brolo, 2012; Turner, 2013) and on research related to SNO_2-based sensors (Cheng et al., 2004; Park et al., 2007).

On the other hand, the first group of studies related to nanomaterial-based SnO_2 bioethanol sensors rely on research related to nanomaterials and nanotechnology (Geim, 2009; Geim and Novoselov, 2007). The second group of studies related to thin film-based SnO_2 bioethanol sensors rely on research related to thin films (Aspnes, 1982; Nix, 1989).

These studies hint that thin film- and nanomaterial-based SnO_2 bioethanol sensors perform well in sensing bioethanol gas, although there are differences in fabricating them. This is due to their hierarchical architectures with a large specific area and abundant spaces for gas diffusion for nanomaterials.

In the end, these most-cited papers in this field hint that the efficiency of the sensing bioethanol gas could be maximized using the structure, processing, and property relationships of nanomaterials and thin films used in the sensing process (Formela et al., 2016; Konur, 2018, 2020b, 2021c–f; Konur and Matthews, 1989).

93.4.4 Other Bioethanol Fuel Sensors

Brief information about ten prolific studies with at least 186 citations each on the bioethanol sensing properties of sensors other than ZnO and SnO_2 is given in Table 93.3. Furthermore, brief notes on the contents of these studies are also given.

Nine of these prolific studies evaluate nanomaterial-based bioethanol sensors (Choudhury, 2009; Hwang et al., 2011; Liu et al., 2005; Na et al., 2011; Shan et al., 2010; Xu et al., 2008; Wang et al., 2013, 2016; Wu et al., 2007). Additionally, Obayashi et al. (1976) evaluate perovskite-type oxide-based bioethanol sensors.

The studied nanomaterials in these prolific papers are V_2O_5 nanobelts, graphene, Co_3O_4-decorated ZnO nanowires, PANI/Ag nanocomposites, Ag nanoclusters on SnO_2 NWs, Al-doped NiO nanorod-flowers, α-Fe_2O_3 nanostructures, soluble CNFs, and polyHEMA photonic crystals.

These studies build first on research related to sensors in general (Brolo, 2012; Turner, 2013) and on research related to V_2O_5-based sensors (Raible et al., 2005; Yan et al., 2015) or α-Fe_2O_3 sensors (Chen et al., 2005b; Hu et al., 2007) among the materials used for sensors.

On the other hand, the first group of studies related to nanomaterial-based bioethanol sensors rely on research related to nanomaterials and nanotechnology (Geim, 2009; Geim and Novoselov, 2007). The second group of studies related to perovskite-type oxide-based bioethanol sensors rely on research related to perovskite-type oxides (Cherry et al., 1995; Tejuca et al., 1989).

These studies hint that oxide- and nanomaterial-based bioethanol sensors perform well in sensing bioethanol gas, although there are differences in fabricating them. This is due to their hierarchical architectures with a large specific area and abundant spaces for gas diffusion for nanomaterials.

In the end, these most-cited papers in this field hint that the efficiency of the sensing bioethanol gas could be maximized using the structure, processing, and property relationships of nanomaterials and oxides used in the sensing process (Formela et al., 2016; Konur, 2018, 2020b, 2021a–d; Konur and Matthews, 1989).

93.5 CONCLUSION AND FUTURE RESEARCH

Brief information about the key research fronts covered by the 25 most-cited papers with at least 181 citations each in the field of bioethanol fuel sensors is given in Table 93.4.

There are three major research fronts for this field: ZnO, SnO_2, and other material-based bioethanol fuel sensors with 12, 6, and 10 papers, respectively.

It is notable that there is similarity of these research fronts in the sample of reviewed papers and the sample of the most-cited 100 papers (the first column data in Table 93.4) in this field.

The first group of prolific papers focus on ZnO-based nanomaterials for sensing bioethanol gas building on the wider research fronts of both sensors and nanomaterials and nanotechnology. These studies show that all the studied nanomaterials perform well in sensing bioethanol gas. These studies also show that the efficiency of the sensing bioethanol gas could be maximized using the structure, processing, and property relationships of nanomaterials used in the sensing process. It is notable that there is ample room to expand nanotechnology applications in sensing bioethanol gas, especially in using graphene and other two-dimensional nanomaterials.

Similarly, the second group of prolific papers focus on SnO_2-based nanomaterials for sensing bioethanol gas building on the wider research fronts of both sensors and nanomaterials and

TABLE 93.4
Most Prolific Research Fronts in the Bioethanol Fuel Sensors

No.	Research Fronts	Sample Papers (%)	Reviewed Papers (%)
1	ZnO-based bioethanol fuel sensors	40	48
	Nanomaterial-based ZnO	35	36
	Conventional material-based ZnO	5	12
2	SnO_2-based bioethanol fuel sensors	28	24
	Nanomaterial-based SnO_2	24	16
	Conventional material-based SnO_2	4	8
3	Other material-based bioethanol fuel sensors	47	40
3.1.	Nanomaterial-based sensors	39	36
3.2.	Conventional material-based sensors	8	4

Total number of reviewed papers = 25. Sample papers: sample of the most-cited 100 papers.

nanotechnology. These studies show that all the studied nanomaterials perform well in sensing bioethanol gas. These studies also show that the efficiency of the sensing bioethanol gas could be maximized using the structure, processing, and property relationships of nanomaterials used in the sensing process. It is notable that there is ample room to expand nanotechnology applications in sensing bioethanol gas, especially in using graphene and other two-dimensional nanomaterials.

Finally, the third group of prolific papers focus on other material-based nanomaterials for sensing bioethanol gas building on the wider research fronts of both sensors and nanomaterials and nanotechnology. These studies show that all the studied nanomaterials perform well in sensing bioethanol gas. These studies also show that the efficiency of the sensing bioethanol gas could be maximized using the structure, processing, and property relationships of nanomaterials used in the sensing process. It is notable that there is ample room to expand nanotechnology applications in sensing bioethanol gas, especially in using graphene and other two-dimensional nanomaterials.

It is notable that thin films, oxides, and other materials have also been used in bioethanol fuel sensors, besides nanomaterials. However, in light of tremendous advances in nanotechnology in recent years, research in this field shall likely focus on nanomaterial-based bioethanol fuel sensors compared to thin film- and oxide-based bioethanol fuel sensors in future. As there is ample room for the development of this field further.

These findings confirm that the application of nanomaterials and nanotechnology in bioethanol fuel sensors significantly improve the efficiency of these processes through the enhancement of the structure–processing–property relationships. This would make bioethanol fuels as a viable alternative to crude oil-based gasoline and petrodiesel fuels.

It is recommended that such review studies should be performed for other research fronts on both the production and utilization of bioethanol fuels complementing the corresponding scientometric studies.

ACKNOWLEDGMENTS

The contribution of the highly cited researchers in the field of the bioethanol fuel sensors has been gratefully acknowledged.

REFERENCES

Al-Hasan, M. (2003). Effect of ethanol–unleaded gasoline blends on engine performance and exhaust emission. *Energy Conversion and Management*, 44(9): 1547–1561.

Aspnes, D. E. 1982. Optical properties of thin films. *Thin Solid Films* 89:249–262.

Azevedo, A. M., D. M. F. Prazeres, J. M. S. Cabral and L. P. Fonseca. 2005. Ethanol biosensors based on alcohol oxidase. *Biosensors and Bioelectronics* 21:235–247.

Bhardwaj, R. and K. G. Sharma. 2020. Semiconductor metal oxide based ethanol gas sensor using ZnO: A short review. *AIP Conference Proceedings* 2297:020027.

Bie, L. J., X. N. Yan, J. Yin, Y. Q. Duan and Z. H. Yuan. 2007. Nanopillar ZnO gas sensor for hydrogen and ethanol. *Sensors and Actuators, B: Chemical* 126:604–608.

Brolo, A. G. 2012. Plasmonics for future biosensors. *Nature Photonics* 6:709–713.

Burnham, J. F. 2006. Scopus database: A review. *Biomedical Digital Libraries* 3:1–8.

Chaubey, A. and B. Malhotra. 2002. Mediated biosensors. *Biosensors and Bioelectronics* 17:441–456.

Chen, J., L. Xu, W. Li and X. Gou. 2005b. α-Fe_2O_3 nanotubes in gas sensor and lithium-ion battery applications. *Advanced Materials* 17:582–586.

Chen, Y. J., L. Nie, X. Y. Xue, Y. G. Wang and T. H. Wang. 2006. Linear ethanol sensing of SnO_2 nanorods with extremely high sensitivity. *Applied Physics Letters* 88:083105.

Chen, Y. J., X. Y. Xue, Y. G. Wang and T. H. Wang. 2005a. Synthesis and ethanol sensing characteristics of single crystalline SnO_2 nanorods. *Applied Physics Letters* 87:1–3.

Cheng, B., J. M. Russell, W. Shi, L. Zhang and E. T. Samulski. 2004. Large-scale, solution-phase growth of single-crystalline SnO_2 nanorods. *Journal of the American Chemical Society* 126:5972–5973.

Cherry, M., M. S. Islam and C. R. A. Catlow. 1995. Oxygen ion migration in perovskite-type oxides. *Journal of Solid State Chemistry* 118:125–132.

Choudhury, A. 2009. Polyaniline/silver nanocomposites: Dielectric properties and ethanol vapour sensitivity. *Sensors and Actuators, B: Chemical* 138:318–325.

Formela, K., A. Hejna, L. Piszczyk, M. R. Saeb and X. Colom. 2016. Processing and structure-property relationships of natural rubber/wheat bran biocomposites. *Cellulose* 23:3157–3175.

Geim, A. K. 2009. Graphene: Status and prospects. *Science* 324:1530–1534.

Geim, A. K. and K. S. Novoselov. 2007. The rise of graphene. *Nature Materials* 6:183–191.

Hellegouarc'h, F., F. Arefi-Khonsari, R. Planade and J. Amouroux. 2001. PECVD prepared SnO_2 thin films for ethanol sensors. *Sensors and Actuators, B: Chemical* 73:27–34.

Hill, J., E. Nelson, D. Tilman, S. Polasky and D. Tiffany. 2006. Environmental, economic, and energetic costs and benefits of biodiesel and ethanol biofuels. *Proceedings of the National Academy of Sciences of the United States of America* 103:11206–11210.

Hill, J., S. Polasky and E. Nelson, et al. 2009. Climate change and health costs of air emissions from biofuels and gasoline. *Proceedings of the National Academy of Sciences of the United States of America* 106:2077–2082.

Hongsith, N., C. Viriyaworasakul, P. Mangkorntong, N. Mangkorntong and S. Choopun. 2008. Ethanol sensor based on ZnO and Au-doped ZnO nanowires. *Ceramics International* 34:823–826.

Hsieh, W. D., R. H. Chen, T. L. Wu and T. H. Lin. 2002. Engine performance and pollutant emission of an SI engine using ethanol-gasoline blended fuels. *Atmospheric Environment* 36:403–410.

Hsueh, T. J., C. L. Hsu, S. J. Chang and I. C. Chen. 2007a. Laterally grown ZnO nanowire ethanol gas sensors. *Sensors and Actuators, B: Chemical* 126:473–477.

Hsueh, T. J., S. J. Chang, C. L. Hsu, Y. R. Lin and I. C. Chen. 2007b. Highly sensitive ZnO nanowire ethanol sensor with Pd adsorption. *Applied Physics Letters* 91:053111.

Hu, X., J. C. Yu, J. Gong, Q. Li and G. Li. 2007. α-Fe_2O_3 nanorings prepared by a microwave-assisted hydrothermal process and their sensing properties. *Advanced Materials* 19:2324–2329.

Hwang, I. S., J. K. Choi and H. S. Woo, et al. 2011. Facile control of C_2H_5OH sensing characteristics by decorating discrete Ag nanoclusters on SnO_2 nanowire networks. *ACS Applied Materials and Interfaces* 3:3140–3145.

Jinkawa, T., G. Sakai, J. Tamaki, N. Miura and N. Yamazoe. 2000. Relationship between ethanol gas sensitivity and surface catalytic property of tin oxide sensors modified with acidic or basic oxides. *Journal of Molecular Catalysis A: Chemical* 155:193–200.

Kim, K. W., P. S. Cho and S. J. Kim, et al. 2007. The selective detection of C_2H_5OH using SnO_2-ZnO thin film gas sensors prepared by combinatorial solution deposition. *Sensors and Actuators, B: Chemical* 123:318–324.

Konur, O. 2000. Creating enforceable civil rights for disabled students in higher education: An institutional theory perspective. *Disability & Society* 15:1041–1063.

Konur, O. 2002a. Access to nursing education by disabled students: Rights and duties of nursing programs. *Nurse Education Today* 22:364–374.

Konur, O. 2002b. Assessment of disabled students in higher education: Current public policy issues. *Assessment and Evaluation in Higher Education* 27:131–52.

Konur, O. 2002c. Access to employment by disabled people in the UK: Is the Disability Discrimination Act working? *International Journal of Discrimination and the Law* 5:247–279.

Konur, O. 2006a. Participation of children with dyslexia in compulsory education: Current public policy issues. *Dyslexia* 12:51–67.

Konur, O. 2006b. Teaching disabled students in higher education. *Teaching in Higher Education* 11:351–363.

Konur, O. 2007a. A judicial outcome analysis of the Disability Discrimination Act: A windfall for the employers? *Disability & Society* 22:187–204.

Konur, O. 2007b. Computer-assisted teaching and assessment of disabled students in higher education: The interface between academic standards and disability rights. *Journal of Computer Assisted Learning* 23:207–219.

Konur, O. 2012. The evaluation of the research on the bioethanol: A scientometric approach. *Energy Education Science and Technology Part A: Energy Science and Research* 28:1051–1064.

Konur, O. 2015. Current state of research on algal bioethanol. In *Marine Bioenergy: Trends and Developments*, Eds. S. K. Kim and C. G. Lee, pp. 217–244. Boca Raton, FL: CRC Press.

Konur, O. 2016a. Scientometric overview in nanobiodrugs. In *Nanoarchitectonics for Smart Delivery and Drug Targeting*, Eds. A. M. Holban and A. M. Grumezescu, pp. 405–428. Amsterdam: Elsevier.

Konur, O. 2016b. Scientometric overview regarding nanoemulsions used in the food industry. In *Emulsions: Nanotechnology in the Agri-Food Industry*, Ed. A. M. Grumezescu, pp. 689–711. Amsterdam: Elsevier.

Konur, O. 2016c. Scientometric overview regarding the nanobiomaterials in antimicrobial therapy. In *Nanobiomaterials in Antimicrobial Therapy*, Ed. A. M. Grumezescu, pp. 511–535. Amsterdam: Elsevier.

Konur, O. 2016d. Scientometric overview regarding the nanobiomaterials in dentistry. In *Nanobiomaterials in Dentistry*, Ed. A. M. Grumezescu, pp. 425–453. Amsterdam: Elsevier.

Konur, O. 2016e. Scientometric overview regarding the surface chemistry of nanobiomaterials. In *Surface Chemistry of Nanobiomaterials*, Ed. A. M. Grumezescu, pp. 463–486. Amsterdam: Elsevier.

Konur, O. 2016f. The scientometric overview in cancer targeting. In *Nanoarchitectonics for Smart Delivery and Drug Targeting*, Eds. A. M. Holban and A. Grumezescu, pp. 871–895. Amsterdam; Elsevier.

Konur, O. 2017a. Recent citation classics in antimicrobial nanobiomaterials. In *Nanostructures for Antimicrobial Therapy*, Eds. A. Ficai and A. M. Grumezescu, pp. 669–685. Amsterdam: Elsevier.

Konur, O. 2017b. Scientometric overview in nanopesticides. In *New Pesticides and Soil Sensors*, Ed. A. M. Grumezescu, pp. 719–744. Amsterdam: Elsevier.

Konur, O. 2017c. Scientometric overview regarding oral cancer nanomedicine. In *Nanostructures for Oral Medicine*, Eds. E. Andronescu and A. M. Grumezescu, pp. 939–962. Amsterdam: Elsevier.

Konur, O. 2017d. Scientometric overview regarding water nanopurification. In *Water Purification*, Ed. A. M. Grumezescu, pp. 693–716. Amsterdam: Elsevier.

Konur, O. 2017e. Scientometric overview in food nanopreservation. In *Food Preservation*, Ed. A. M. Grumezescu, pp. 703–729. Amsterdam: Elsevier.

Konur, O., Ed. 2018. *Bioenergy and Biofuels*. Boca Raton, FL: CRC Press.

Konur, O. 2019. Nanotechnology applications in food: A scientometric overview. In *Nanoscience for Sustainable Agriculture*, Eds. R. N. Pudake, N. Chauhan and C. Kole, pp. 683–711. Cham: Springer.

Konur, O. 2020a. The scientometric analysis of the research on the bioethanol production from green macroalgae. In *Handbook of Algal Science, Technology and Medicine,* Ed. O. Konur, pp. 385–401. London: Academic Press.

Konur, O., Ed. 2020b. *Handbook of Algal Science, Technology and Medicine.* London: Academic Press.

Konur, O., Ed. 2021a. *Handbook of Biodiesel and Petrodiesel Fuels: Science, Technology, Health, and Environment.* Boca Raton, FL: CRC Press.

Konur, O., Ed. 2021b. *Handbook of Biodiesel and Petrodiesel Fuels: Science, Technology, Health, and Environment. Volume 1. Biodiesel Fuels: Science, Technology, Health, and Environment.* Boca Raton, FL: CRC Press.

Konur, O., Ed. 2021c. *Handbook of Biodiesel and Petrodiesel Fuels: Science, Technology, Health, and Environment. Volume 2. Biodiesel Fuels Based on the Edible and Nonedible Feedstocks, Wastes, and Algae: Science, Technology, Health, and Environment.* Boca Raton, FL: CRC Press.

Konur, O., Ed. 2021d. *Handbook of Biodiesel and Petrodiesel Fuels: Science, Technology, Health, and Environment. Volume 3. Petrodiesel Fuels: Science, Technology, Health, and Environment.* Boca Raton, FL: CRC Press.

Konur, O. 2021e. Nanotechnology applications in diesel fuels and the related research fields: A review of the research. In *Handbook of Biodiesel and Petrodiesel Fuels: Science, Technology, Health, and Environment. Volume 1. Biodiesel Fuels: Science, Technology, Health, and Environment*, Ed. O. Konur, pp. 89–110. Boca Raton, FL: CRC Press.

Konur, O. 2021f. Nanobiosensors in agriculture and foods: A scientometric review. In *Nanobiosensors in Agriculture and Food*, Ed. R. N. Pudake, pp. 365–384. Cham: Springer.

Konur, O. 2023. Bioethanol fuel sensors: Scientometric study. In *Evaluation and Utilization of Bioethanol Fuels. I.: Evaluation of Bioethanol Fuels, Transport Engines, and Bioethanol Sensors. Handbook of Bioethanol Fuels Volume 5*, Ed. O. Konur. Boca Raton, FL: CRC Press.

Konur, O. and F. L. Matthews. 1989. Effect of the properties of the constituents on the fatigue performance of composites: A review. *Composites* 20:317–328.

Liu, B. and H. C. Zeng. 2003. Hydrothermal synthesis of ZnO nanorods in the diameter regime of 50nm. *Journal of the American Chemical Society* 125:4430–4431.

Liu, J., Z. Wang, Q. Peng and Y. Li. 2005. Vanadium pentoxide nanobelts: Highly selective and stable ethanol sensor materials. *Advanced Materials* 17:764–767.

Ma, X., L. Sun and C. Song. 2002. A new approach to deep desulfurization of gasoline, diesel fuel and jet fuel by selective adsorption for ultra-clean fuels and for fuel cell applications. *Catalysis Today* 77:107–116.

Murdoch, M., G. I. N. Waterhouse and M. A. Nadeem, et al. 2011. The effect of gold loading and particle size on photocatalytic hydrogen production from ethanol over Au/TiO_2 nanoparticles. *Nature Chemist*ry 3:489–492.

Na, C. W., H. S. Woo, I. D. Kim and J. H. Lee. 2011. Selective detection of NO_2 and C_2H_5OH using a Co_3O_4-decorated ZnO nanowire network sensor. *Chemical Communications* 47:5148–5150.

Najafi, G., B. Ghobadian and T. Tavakoli, et al. 2009. Performance and exhaust emissions of a gasoline engine with ethanol blended gasoline fuels using artificial neural network. *Applied Energy* 86:630–639.

Newman, P. W. G. and J. R. Kenworthy. 1989. Gasoline consumption and cities: A comparison of U.S. cities with a global survey. *Journal of the American Planning Association* 55:24–37.

Nix, W. D. 1989. Mechanical properties of thin films. *Metallurgical Transactions A* 20:2217.

North, D. C. 1991. Institutions. *Journal of Economic Perspectives* 5:97–112.

Obayashi, H., Y. Sakurai and T. Gejo. 1976. Perovskite-type oxides as ethanol sensors. *Journal of Solid State Chemistry* 17:299–303.

Paraguay, D., F. M. Miki-Yoshida, J. Morales, J. Solis and L. W. Estrada. 2000. Influence of Al, In, Cu, Fe and Sn dopants on the response of thin film ZnO gas sensor to ethanol vapour. *Thin Solid Films* 373:137–140.

Park, M. S., G. X. Wang and Y. M. Kang, et al. 2007. Preparation and electrochemical properties of SnO_2 nanowires for application in lithium-ion batteries. *Angewandte Chemie International Edition* 46:750–753.

Raible, I., M. Burghard, U. Schlecht, A. Yasuda and T. Vossmeyer. 2005. V_2O_5 nanofibres: Novel gas sensors with extremely high sensitivity and selectivity to amines. *Sensors and Actuators B: Chemical* 106:730–735.

Rao, B. B. 2000. Zinc oxide ceramic semi-conductor gas sensor for ethanol vapour. *Materials Chemistry and Physics* 64:62–65.

Rout, C. S., S. H. Krishna, S. R. C. Vivekchand, A. Govindaraj and C. N. R. Rao. 2006. Hydrogen and ethanol sensors based on ZnO nanorods, nanowires and nanotubes. *Chemical Physics Letters* 418:586–590.

Sangeetha, S. V., N. Shilpa and S. S. S. Jain. 2020. A review: Design and fabrication of gas sensor using sol-gel spin coating technique for the detection of ethanol. *Solid State Technology* 63:1438–1445.

Schauer, J. J., M. J. Kleeman, G. R. Cass and B. R. T. Simoneit. 1999. Measurement of emissions from air pollution sources. 2. C_1 through C_{30} organic compounds from medium duty diesel trucks. *Environmental Science & Technology* 33:1578–1587.

Schauer, J. J., M. J. Kleeman, G. R. Cass and B. R. T. Simoneit. 2002. Measurement of emissions from air pollution sources. 5. C_1-C_{32} organic compounds from gasoline-powered motor vehicles. *Environmental Science and Technology* 36:1169–1180.

Shan, C., H. Yang, D. Han, Q. Zhang, A. Ivaska and L. Niu. 2010. Electrochemical determination of NADH and ethanol based on ionic liquid-functionalized graphene. *Biosensors and Bioelectronics* 25:1504–1508.

Tejuca, L. G., J. L. G. Fierro and J. M. Tascon. 1989. Structure and reactivity of perovskite-type oxides. *Advances in Catalysis* 36:237–328.

Turner, A. P. 2013. Biosensors: Sense and sensibility. *Chemical Society Reviews* 42:3184–3196.

Wan, Q., Q. H. Li and Y. J. Chen, et al. 2004. Fabrication and ethanol sensing characteristics of ZnO nanowire gas sensors. *Applied Physics Letters* 84:3654–3656.

Wang, A. L., H. Xu and J. X. Feng, et al. 2013. Design of Pd/PANI/Pd sandwich-structured nanotube array catalysts with special shape effects and synergistic effects for ethanol electrooxidation. *Journal of the American Chemical Society* 135:10703–10709.

Wang, C., X. Cui and J. Liu, et al. 2016. Design of superior ethanol gas sensor based on al-doped NiO nanorod-flowers. *ACS Sensors* 1:131–136.

Wang, L., Y. Kang and X. Liu, et al. 2012. ZnO nanorod gas sensor for ethanol detection. *Sensors and Actuators, B: Chemical* 162:237–243.

Wu, L., X. Zhang and H. Ju. 2007. Detection of NADH and ethanol based on catalytic activity of soluble carbon nanofiber with low overpotential. *Analytical Chemistry* 79:453–458.
Xu, X., A. V. Goponenko and S. A. Asher. 2008. Polymerized polyHEMA photonic crystals: pH and ethanol sensor materials. *Journal of the American Chemical Society* 130:3113–3119.
Yan, W., M. Hu, D. Wang and C. Li. 2015. Room temperature gas sensing properties of porous silicon/V_2O_5 nanorods composite. *Applied Surface Science* 346:216–222.
Yang, P., H. Yan and S. Mao, et al. 2002. Controlled growth of ZnO nanowires and their optical properties. *Advanced Functional Materials* 12:323–331.
Zhang, Y., X. He, J. Li, Z. Miao and F. Huang. 2008. Fabrication and ethanol-sensing properties of micro gas sensor based on electrospun SnO_2 nanofibers. *Sensors and Actuators, B: Chemical* 132(1):67–73.
Zhu, L., Y. Li and W. Zeng. 2018. Hydrothermal synthesis of hierarchical flower-like ZnO nanostructure and its enhanced ethanol gas-sensing properties. *Applied Surface Science* 427:281–287.

Index

FSC
www.fsc.org
MIX
Papier | Fördert gute Waldnutzung
FSC® C083411

Kolibri 360 GmbH
Ferdinand-Jühlke-Straße 7
99095 Erfurt, Deutschland
produktsicherheit@kolibri360.de